[redacted]

Computer-Assisted Lead Finding and Optimization

Current Tools for Medicinal Chemistry

Editors:
Han van de Waterbeemd
Bernard Testa
Gerd Folkers

Verlag Helvetica Chimica Acta · Basel

WILEY-VCH

Weinheim · New York · Chichester
Brisbane · Singapore · Toronto

Dr. Han van de Waterbeemd
F. Hoffmann-La Roche Ltd.
Dept. PRPC 65/314
CH-4070 Basel

Prof. Dr. Bernard Testa
Ecole de Pharmacie
Université de Lausanne
BEP
CH-1015 Lausanne-Dorigny

Prof. Dr. Gerd Folkers
Abteilung Pharmazie
ETH-Zürich
Winterthurerstrasse 190
CH-8057 Zürich

Published jointly by
VHCA, Verlag Helvetica Chimica Acta, Basel (Switzerland)
WILEY-VCH Weinheim (Federal Republic of Germany)

(Cover illustration: HIV-1 protease inhibitor saquinavir bound to the enzyme.)

Library of Congress Card No. applied for.

A CIP catalogue record for this book is available from the British Library.

Die Deutsche Bibliothek – CIP-Einheitsaufnahme
Computer assisted lead finding and optimization : current tools for medicinal chemistry / ed.: Han van de Waterbeemd ... – Basel : Verl. Helvetica Chimica Acta ; Weinheim ; New York ; Chichester ; Brisbane ; Singapore ; Toronto : Wiley-VCH, 1997
ISBN 3-906390-16-0

Printed on acid-free paper.

Printing: Freiburger Graphische Betriebe, D-79015 Freiburg
Printed in Germany

PREFACE

Computer-Assisted Lead Finding and Optimization

The 11th European Symposium on *Quantitative Structure-Activity Relationships* was held September 1-6, 1996, on the beautiful campus of the University of Lausanne, Switzerland. The first meeting in this series was held in 1973 in Prague, the last symposia taking place in Interlaken (1988), Sorrento (1990), Strasbourg (1992) and Barcelona (1994).

The objective of the Symposium, whose timely subtitle was *Computer-Assisted Lead Finding and Optimization,* was to bring together scientists creative and active in the field of molecular design using computer and information technologies of capital interest in drug discovery and optimization. The methods and advances discussed included the design of biomaterials, and the understanding and prediction of physicochemical and biological properties such as solubility, lipophilicity, biotransformation, receptor affinity and toxicity.

About 325 participants from all over the world, representing most of the major pharmaceutical companies and leading academic groups, met in Lausanne for intense scientific exchanges. The Symposium featured 22 invited lectures, 14 oral communications and over 150 posters.

The Proceedings you now hold contain the many new ideas, concepts and findings that were presented and evaluated during the Symposium and which should render more efficient the design of promising bioactive lead compounds and their subsequent optimization to useful products such human and veterinary drugs, fragrances and agrochemicals. In the present book an account is found of the current strategies used in computer-assisted drug design. Important topics include progress in chemometrics, molecular modeling and three-dimensional QSAR approaches. Relatively new mathematical methods such as genetic algorithms, artificial neural networks and fuzzy logic all have found their application in rational molecular design.

As amply illustrated by recent developments in these disciplines, significant progress has been made in lead finding strategies. This is of great importance to the pharmaceutical industry. Modern methods of molecular similarity and diversity and the use of appropriate two-dimensional and three-dimensional metrics form a cornerstone in the rational design of combinatorial libraries and rational selection of compounds for high throughput screening. The combination of database searching and molecular modeling has resulted in three-dimensional database screening techniques, a further exciting approach for lead finding.

Since high affinity alone does not make a drug, part of the book is devoted to computational approaches to drug delivery. A deeper understanding of metabolism, toxicity, pharmacokinetics and the role of physicochemical properties in absorption processes should bring medicinal chemists closer to rational delivery design. Some of the methods reported are reaching a status of matureness, while others are just emerging and need further refinement and practical validation. This is the promise of the next Symposium, to be held in 1998.

Han van de Waterbeemd,
Bernard Testa,
and Gerd Folkers

CONTENTS

Introduction

Medicinal Chemistry. The Challenges for the Future

by **Paul M. Vanhoutte**

I.R.I.S., 6 place des Pléiades, F-92415 Courbevoie Cedex, France

Medicinal Chemistry. The Challenges for the Future

Paul M. Vanhoutte

I.R.I.S., 6 place des Pléiades, F-92415 Courbevoie Cedex, France

To researchers of the academic world, drug discovery may perhaps seem an easy process. Indeed, once a viable target has been defined, the extraordinary power of modern medicinal chemistry should make identification of the best possible drug to modulate the function of the selected target straight forward. However, in reality from the selection of the target onwards, the discovery process of a new chemical entity is lengthy, costly and cumbersome. This short paper outlines some of my views on what constitue the important factors in the design of new chemical entities and on what means are necessary to accelerate their discovery.

Therapeutic Targets. - One of the great dangers in pharmaceutical research is to fall in love with science. The scientific journals are full of exuberant accounts of exciting break-throughs, many of which lead nowhere, or to extremely small therapeutic niches. Considering the effort, time and, unavoidably the amount of money involved in bringing a new chemical entity to therapeutic fruition, it is obviously crucial at the onset to select therapeutic goals, and hence pharmacological targets, that correspond to a real medical need. It is currently agreed, that for the beginning of the 21st century, the medical demand for more effective therapies will remain very high in the areas (in descending rank order of importance) of diabetes, osteoporosis, AIDS, osteoarthritis, Alzheimer disease, cardiovascular disease and most forms of cancer. Hence, these areas constitute reasonably safe pharmaceutical research domains, although they are clearly open to fierce competition between major drug discovery research teams.

For any director of drug discovery a key question remains : how many projects can the company afford at any one time ? The answer is complex. Probably, the popular wisdom of not putting all of one's eggs into the same basket prevails. However, the overall strategy of the company will play a dominant role (e.g. a willingness to address antiviral therapy), as will its industrial capabilities (e.g. access to in house aerosol technology). I believe that a major responsibility of any drug discovery team is to ensure the succession for the company's compounds that are successful on the market, but which will unavoidably face generic competition. Obviously, every research director hopes to open a major new therapeutic application for his or her marketing colleagues.

Ultimately, the number of selected therapeutic goals, and hence research projects, must reflect the critical mass of people (medicinal chemists and pharmacologists) available. Indeed, if insufficient human and material resources are dedicated to each project, the outcome is diffusion and delay. This implies that the number of effective projects is determined by the research investment of the company, in proportion to its global turnover. Does that mean that only "the megagiants" formed in recent years will survive in terms of discovery research ? I believe that the answer is most certainly no, and for several reasons. First, it takes only one brilliant mind to design a discovery or to recognize serendipity, and in drug discovery quantity has never made up for quality. Second, smaller companies, through industrial networking and resources sharing, have powerful means to compete with the giants, with the advantage of greater efficiency, flexibility, and speed in the decision making process.

Tissue Targets. - The ideal target for a drug discovery program is a well defined protein (e.g. receptors transporter or enzyme) that abounds in only one type of diseased tissue. Unfortunately, this is but a dream. Since all eukaryotic cells stem from the same genetic pool, they are bound to be able to express a diversity of receptors and enzyme systems. Hence, any drug interfering with a event in one cell type, will ultimately affect the others. The more ubiquitous the targeted systems, the more side effects can be expected. The resulting undesirable reactions may undermine the therapeutic potential of the new entity. Hence, the pharmaceutical researchers together with colleagues in development, have to be pragmatic in estimating the safety margin of every new drug and evaluating its cost-benefit ratio-while, bearing in mind the old saying that if it is not a poison, it is not a drug (after all, in Ancient Greek "φ α ρ μ α κ ο η" means poison, and not remedy !). This makes tissue targeting, or the determined effort to deliver a therapeutic agent specifically to a diseased organ or tissue, a priority in drug discovery. The most compelling demonstration of the advantage of organ targeting is provided by the aerosol administration of drugs affecting the bronchial tree.

Cellular Targets. - Whereas for a number of decades several drug companies have thrived on "fast following" (me-tooism), the global changes in healthcare policy now make this approach increasingly less profitable. There is thus a real need for innovation and the urge to be the first to explore new cellular targets. The advent of molecular biology as a powerful tool in drug discovery has accelerated the identification of new cellular components that are amenable to drug design. It is somewhat humbling to realize that ultimately, the importance, if any, of a newly defined and cloned protein to the function of the human body can be best determined by the design of a simple, selective antagonist or inhibitor. The estimation of the true medicinal interest of a novel protein is rendered difficult by the paucity of relevant animal models of many diseases, and by the total lack of appropriate models of the most important complicating factor in human pathology, aging.

Chemistry, - In drug discovery, everything begins with a laboratory code number, and hopefully ends with a new chemical entity on the market. The close interaction between the medicinal chemist and the pharmacologist is critical in achieving such success. However, the starting point of the chemical optimizing process is progressively changing. Indeed, as molecular biology produces a seemingly never ending source of new molecular targets, with as yet undefined physiological or pathological roles, chemical selectivity has to be indicated by either combinatorial procedures or by high-throughput screening of vast chemical libraries. Of course, these procedures will not discover new drugs. They merely provide the lead to initiate the conventional pharmaco-chemical modulation. The latter is helped considerably by molecular modelling, which permits acceleration of the optimization process. Together, these three tools (combinatorial chemistry, high throughput screening and molecular design) can considerably accelerate the discovery of new chemical entities. I strongly believe that they will never replace the power of invention of a well trained medicinal chemist partnered by an imaginative pharmacologist.

Conclusion: What About The Future? - The greatest challenge for the years to come is to achieve tissue selectivity either by identifying a critical target involved exclusively in a given disease process, or by tightly controlling the organ delivery of a molecule acting more broadly. We will in both cases have to better understand how the ageing process, which is so complex for humans, affects the interaction of drugs with their cellular targets, and their disposition by the body. Although technologies derived from molecular genetics are helpful to predict future therapeutical applications and to validate targets, I personally do not believe that gene therapy will constitute a major industrial break through, because so few diseases are caused by a single genetic defect. However, properly tissue-targeted gene delivery or antisense therapy may become helpful, provided these manoeuvres meet the criteria required from conventional therapies. We need to accelerate the discovery and development processes, to guarantee a sufficient lifespan under industrial protection for new chemical entities, otherwise the enormity of the financial investment required to bring a molecule from the bench to the patient will soon become unbearable. To achieve such acceleration we have to maintain the association between the medicinal chemist and the pharmacologist as actively as possible. But we also have to equip both with the modern tools to optimize their mission: molecular biology for the pharmacologist, molecular modeling and high throughput screening (including combinatorial approaches) for the chemist. It is my conviction that much can still be gained from an effort relying on conventional heterocyclic chemistry.

PART I. CHEMOMETRICS

Chapter 1

A General View on Similarity and QSAR Studies

by **Hugo Kubinyi**

Drug Design, ZHB/W- A30, BASF AG,
D-67056 Ludwigshafen

A General View on Similarity and QSAR Studies

by **Hugo Kubinyi**

Drug Design, ZHB/W- A30, BASF AG, D-67056 Ludwigshafen

Quantitative Structure-Activity Relationships (QSAR) describe biological activities in terms of hydrophobic, electronic, polar and steric properties of the molecules within a congeneric data set. Compounds with similar physicochemical properties are expected to have also similar biological activities. On the other hand, dissimilar compounds should differ in their biological activities. This inherent assumption is confirmed by a systematic investigation of the correlation of certain columns of 2D and 3D similarity matrices with their activity values. Such columns describe the similarities of all molecules of a data set to certain molecules within this data set.

In linear QSAR's, the correlation of one or very few columns of the similarity matrices gives an equally good or even better result than classical Hansch analysis. Distance matrices **D** as well as covariance matrices **C** can be used as similarity measures. Also in nonlinear relationships similarity matrices can be correlated with biological activities, without the need to select a certain nonlinear mathematical model. In this case, the distance matrices **D** are clearly superior to the covariance matrices **C**.

3D QSAR methods, *e.g.* CoMFA, describe molecular properties in terms of 3D fields. Distance and covariance matrices can also be applied in such studies. The calculation of molecular similarity index matrices from classical CoMFA fields and from SEAL (similarity) fields allows a significant compression of the data, without a major loss of the original information. The quality of the resulting correlations is comparable to the results of CoMFA and CoMSIA studies. The main advantage of 3D similarity matrices seems to be that, instead of a common alignment of all molecules, the 3D superpositions may be performed in a more or less unique manner for each pair of compounds. The terms **QSiAR** (**Q**uantitative **Si**milarity-**A**ctivity **R**elationships) and **3D QSiAR** are proposed for such relationships.

The last part of this paper discusses some difficulties in the definition of the "similarity" of seemingly related molecules. Sometimes chemical similarity is not reflected by the resulting biological activities. There are even examples where different molecules are "similar" with respect to certain target proteins, while they are "dissimilar" with respect to others.

Introduction. - The patterns of steric, electronic, and hydrophobic properties of a drug determine its interactions with the binding site of a receptor. In addition, the overall hydrophobic properties as well as ionization and dissociation are responsible for its transport and distribution within the biological system. QSAR models describe biological properties by correlating them with different physicochemical properties of the compounds or of individual groups, or by using indicator variables for certain structural features [1-9].

The basic concept of all QSAR models is the similarity of biological properties of compounds with similar structures and similar physicochemical properties, in a qualitative manner and on a quantitative scale. Linear Hansch analysis correlates the similarity or dissimilarity of one or several properties of two compounds with a corresponding similarity or dissimilarity in their activities. The situation is different in nonlinear structure-activity relationships: Compounds with significantly different properties may have similar activities. However, as shown below, also nonlinear structure-activity relationships can be easily described by molecular similarity matrices.

3D QSAR methods [10, 11], *e.g.* Comparative Molecular Field Analysis (CoMFA), define molecular properties in terms of 3D fields. In CoMFA studies three most difficult problems have to be solved for a proper application of these powerful new tools, *i.e.*

- the correct alignment of flexible ligands,
- the definition and characteristics of the 3D fields, and
- the selection of relevant variables in the PLS analyses.

How to Define the Similarity of Molecules for QSAR Studies? - Most organic chemists regard the similarity of molecules in terms of 2D structural formulas. Whenever the molecular frames are similar and the attached substituents are not too different, two molecules are considered to be structurally related. The situation changes when, *e.g.*, an electrophilic substitution at an aromatic ring shall be achieved. In such a case the chemist will take a closer look, whether electron donor substituents ease such a reaction and in which order different substituents shall be introduced into the system.

Medicinal chemists have, from their experience, a different point of view. They are trained to think in terms of pharmacophores, *i.e.* certain three-dimensional arrangements of hydrogen bond donor and acceptor groups, other polar groups and more lipophilic parts of a molecule. If the 3D structure of the biological target, *e.g.* an enzyme, is known, the degree of similarity of the ligands can be defined in absolute terms: It is the structural complementarity, in size, shape and electronic properties, to the binding site. The situation is much more difficult if such a structure is not known. Then sets of structurally diverse molecules must be investigated to derive a pharmacophore hypothesis and an alignment rule, which helps to explain the activity or lack of activity of the different compounds.

Several approaches, not discussed here in detail, have been used to define and describe the similarity of molecules [12-15]. In 1991, Rum and Herndon already used N x N similarity index matrixes, derived from 2D topological descriptors, and stepwise regression analysis to correlate several columns of these similarity matrices with biological activities [16].

In 1993, Good and Richards [17-19] performed more systematic investigations on different approaches to correlate the 3D similarities of molecules with their biological activities. First, 3D structures of all molecules were generated and aligned in space, as in the CoMFA approach. Then similarity indices between all pairs of molecules were calculated, using either the Carbo or the Hodgkin electrostatic potential similarity index (see below). In the last step, the resulting N x N similarity matrices were correlated with the biological activities of the molecules, using either neural networks [17] or PLS analysis [17, 18].

The Carbo index R_{AB} was originally developed as a method to compare molecules in terms of their electron density ρ [20]. In a more general version this similarity index can be applied to compare any structural properties P_A and P_B of two molecules A and B (eq. 1).

$$R_{AB} = \frac{\int P_A P_B dn}{(\int P_A{}^2 dn)^{1/2} (\int P_B{}^2 dn)^{1/2}} \tag{1}$$

Alternatively, a Gaussian approximation of the Carbo index [18, 19] or the Hodgkin index H_{AB} (eq. 2) [21] can be used. While the Carbo index is sensitive to the shape of the property distribution, the Hodgkin index reflects more its magnitude.

$$H_{AB} = \frac{2\int P_A P_B dn}{\int P_A{}^2 dn + \int P_B{}^2 dn} \tag{2}$$

PLS analysis [22, 23] was used to correlate the similarity matrices of several data sets with their biological activities. Better results were obtained after the elimination of noise variables (*i.e.* variables not contributing to prediction) with the variable selection program GOLPE [24].

This concept was adapted, in a very simple manner, to the lipophilicity similarity of molecules, which can be easily defined as the negative values of the absolute differences between their partition coefficients, $a_{ij} = - | \log P_i - \log P_j |$ ($\log P_i$, $\log P_j$ = logarithms of the partition coefficients of the molecules I and J). For various examples of nonlinear lipophilicity-activity relationships, PLS analyses of the lipophilicity similarity matrices, in combination with a genetic algorithm selection of variables, gave as good results as the application of the parabolic or the bilinear model [25]. Several other applications of similarity index analyses in QSAR studies, most often combined with the application of PLS analysis, have been published in the last few years [26-35].

An interesting approach for the property-based alignment of molecules, developed by Kearsley and Smith, is the public domain computer program SEAL [36] which calculates, according to the relative orientations of two molecules in 3D space, the steric and electronic similarity between all pairs of atoms in a distance-dependent manner. Other properties, like hydrophobic character, hydrogen bond donor and acceptor properties, *etc.*, may be implemented in this approach [37]. The better the properties overlap in the two molecules, the higher will be their calculated similarity. Thus, orientation of the two molecules in space is guided by this parameter, until the best mutual alignment is achieved. As an additional option, the alignment can be improved by allowing some internal flexibility, *i.e.* combining a torsional angle force field with the property-based superposition of the molecules (program TORSEAL [37]).

Comparative Molecular Similarity Indices Analysis (CoMSIA) uses the program SEAL to calculate "similarity fields" of the molecules to different standard probes in a regular grid [38]. As compared to CoMFA, the big advantage of the Gaussian-type fields used in CoMSIA studies is that no cut-off conditions are needed close to the surface of the molecules and within the molecules (see below, section 5). As a consequence, smoother and less fragmented contour maps are obtained in CoMSIA studies. Whereas CoMFA contour maps are always located outside the molecules, CoMSIA contour maps are also located around and "inside" functional groups and are therefore much easier to interpret.

Classical QSAR's Correlate the Similarity of Properties and Activities. - QSAR correlates one or several physicochemical properties with biological activities. In the case of linear one-variable or even linear multiple regression analyses, a trivial consequence of these relationships is that molecules which are similar in their respective physicochemical properties also have similar biological activities.

Correspondingly, the similarity coefficients of molecules can be correlated with biological activities. Two different functions have been tested as similarity measures: Distance matrices **D**, *e.g.* the Euclidian distances d_{ij} between a molecule I and a molecule J (eq. 3), and covariance matrices **C**, made up of covariance indices c_{ij} (eq. 4). If **X** is the original N x M matrix of explanatory data (N rows, M columns), then all x_{ik} values must be normalized (*i.e.*, mean-value-centered and standardized) before, column by column. Afterwards, every column has a mean value of zero and a standard deviation of one. Next, the distance matrix **D** is calculated from the transformed x values by eq. 3, where x_{ik} and x_{jk} are the kth properties of the ith and the jth molecules.

$$d_{ij} = \sqrt{\sum_{k=1}^{k=m} (x_{ik} - x_{jk})^2} \qquad (3)$$

Alternatively, covariances c_{ij} (eq. 4) between molecules I and J can be used to describe the similarity of molecules. Also in this case, all x_{ik} values must be mean-value-centered and standardized before the c_{ij} values are calculated.

$$c_{ij} = \sum_{k=1}^{k=m} x_{ik} \cdot x_{jk} \qquad (4)$$

It is reasonable to expect that the "similarity" of one or several molecules to all other molecules correlates with their biological activities. As an example, the antiadrenergic activities of *N,N*-dimethyl-α-bromo-phenethylamines were investigated. The biological activities C of this data set have been explained [39, 40] by π and σ^+ (eq. 5; Q^2 and s_{PRESS} are the results of a leave-one-out cross-validation).

$$\log 1/C = 1.151\ (\pm 0.19)\ \pi - 1.464\ (\pm 0.38)\ \sigma^+ + 7.817\ (\pm 0.19) \qquad (5)$$

$(n = 22;\ r = 0.945;\ s = 0.196;\ F = 78.63;\ Q^2 = 0.841;\ s_{PRESS} = 0.238)$

If the distance matrix **D** is calculated, starting from the two mean-value-centered and standardized physicochemical properties, an N x N similarity matrix is obtained. Using an evolutionary algorithm for variable selection [41-43], the best correlation with biological activities is obtained by a combination of the distance vectors to the molecules 1 and 6 (eq. 6), *i.e.* the unsubstituted parent compound and the *p*-methyl analog (for the numbering of the compounds, see [40]).

$$\log 1/C = 0.801\ (\pm 0.15)\ d_{1,j} - 0.583\ (\pm 0.15)\ d_{6,j} + 8.231\ (\pm 0.24) \qquad (6)$$
$$(n = 22;\ r = 0.929;\ s = 0.221;\ F = 59.70;\ Q^2 = 0.820;\ s_{PRESS} = 0.253)$$

The different signs of the $d_{1,j}$ and $d_{6,j}$ regression coefficients indicate that the most active molecules should - in their properties - be distant (*i.e.*, dissimilar) to molecule #1 and close (*i.e.*, similar) to molecule #6. This is indeed the case: Most active compounds are disubstituted compounds and most of them bear a *p*-methyl group. Only very few highly active compounds are disubstituted ones without a *p*-methyl group (then, *e.g.*, bearing a *m*-methyl group) or a monosubstituted analog. In addition, such an easy-to-understand interpretation of a regression model is exactly what the organic chemists want to hear from the theoreticians.

Some other combinations of two columns of the distance matrix give slightly worse results. If the covariance matrix **C** is calculated instead, the combination of any two columns of this matrix, *e.g.* $c_{1,j}$ and $c_{2,j}$ (eq. 7), gives the same quality of fit (of course, with different regression coefficients), which is also identical with the result of the two-variable Hansch analysis (eq. 5).

$$\log 1/C = -1.812\ (\pm 0.35)\ c_{1,j} + 1.744\ (\pm 0.36)\ c_{2,j} + 8.696\ (\pm 0.09) \qquad (7)$$
$$(n = 22;\ r = 0.945;\ s = 0.196;\ F = 78.63;\ Q^2 = 0.841;\ s_{PRESS} = 0.238)$$

In principle, not the complete similarity matrix needs to be calculated. To check certain hypotheses, only one or several similarity vectors may be used. Even a molecule outside the data set could be selected to calculate a similarity vector. This corresponds to an analogous strategy in Free-Wilson analysis [44-46]. In such a case, the molecule must belong to the same series of compounds, *i.e.* it must have the same parent structure with any combination of structural features already included in the series.

Another example, the correlation of ET_B receptor affinities of endothelin analogs [43, 47], including some more variables, is chosen to prove the generality of this approach. For the endothelin analogs, the best three-variable combination, out of seven different physicochemical parameters, is given by eq. 8.

$$\log 1/C_{ET\text{-}B} = 1.194\ (\pm 0.27)\ \pi - 0.053\ (\pm 0.033)\ MR - 0.175\ (\pm 0.17)\ W_d + 8.858\ (\pm 0.84) \qquad (8)$$
$$(n = 16;\ r = 0.946;\ s = 0.434;\ F = 34.30;\ Q^2 = 0.826;\ s_{PRESS} = 0.560)$$

Calculation of the distance and covariance matrices, **D** and **C**, including all original x variables, and selection of the "best" combinations of columns, leads to eqs. 9 and 10, respectively.

$$\log 1/C_{ET\text{-}B} = -0.663\ (\pm 0.20)\ d_{6,j} + 0.265\ (\pm 0.22)\ d_{10,j} + 0.897\ (\pm 0.19)\ d_{13,j} + 6.618\ (\pm 0.70) \qquad (9)$$
$$(n = 16;\ r = 0.964;\ s = 0.359;\ F = 52.15;\ Q^2 = 0.836;\ s_{PRESS} = 0.544)$$

$$\log 1/C_{ET\text{-}B} = 1.017\ (\pm 0.58)\ c_{6,j} - 0.527\ (\pm 0.49)\ c_{7,j} - 0.507\ (\pm 0.13)\ c_{13,j} + 9.073\ (\pm 0.24) \qquad (10)$$

$(n = 16;\ r = 0.945;\ s = 0.439;\ F = 33.55;\ Q^2 = 0.819;\ s_{PRESS} = 0.571)$

What is the difference, where are the advantages of similarity and covariance matrices, as compared to regression and PLS analyses? First of all, regression analysis selects variables, with the risk of generating a chance correlation. PLS analysis, as well as distance and covariance matrices, use all the available information. In the derivation of the SAMPLS algorithm, Bush and Nachbar [48] have proven that PLS analysis, starting from the covariance matrix **C** of all properties x_{jk}, gives exactly the same result as PLS analysis of the complete **X** block.

PLS analysis, starting from the original 7 x 16 **X** block, gives four significant components, with $r = 0.946$; $s = 0.454$; $F = 23.56$; $Q^2 = 0.728$; $s_{PRESS} = 0.732$, *i.e.*, a slightly worse internal predictivity, as compared to eqs. 9 and 10. It must be admitted that mainly the selection of certain vectors, out of all vectors of the similarity matrix, gives such a better result. The vectors are similarity measures to selected molecules. Obviously, the similarity or dissimilarity to these molecules is responsible for the biological activity.

For the Selwood data set [49], including 31 compounds and 53 variables, it has been shown that, in regression as well as in PLS analyses, only proper variable selection produces satisfactory results [41, 42, 50]. Whereas the consideration of all variables gives a PLS result with low internal predictivity ($r = 0.929$; $s = 0.335$; $F = 31.58$; $Q^2 = 0.279$; $s_{PRESS} = 0.768$; Q^2 criterion, five significant components) [41], selection of only 11 explanatory variables (from the 10 best three-variable regressions) leads to a much better PLS result ($r = 0.909$; $s = 0.376$; $F = 23.91$; $Q^2 = 0.671$; $s_{PRESS} = 0.519$; Q^2 criterion, five significant components) [42]. This is a further indication for the dependence of PLS results on the presence of too many non-explanatory variables [51].

The situation is slightly different with distance and covariance matrices. Although a certain influence of the variables, which were selected for the calculation of the **D** and **C** matrices, is observed, it is much less pronounced than in the case of PLS analysis. In addition, the number of rows of the **D** and **C** matrices, included by the variable selection procedure, is smaller than the number of vectors in the PLS analysis:

Distance matrix, all x variables (rows 3, 4, 30)

$r = 0.797$; $s = 0.526$; $F = 15.68$; $Q^2 = 0.576$; $s_{PRESS} = 0.567$

Distance matrix, variables of the 10 best three-variable regressions (rows 4, 5, 30)

$r = 0.845$; $s = 0.466$; $F = 22.44$; $Q^2 = 0.622$; $s_{PRESS} = 0.535$

Covariance matrix, all x variables (rows 9, 24)

$r = 0.773$; $s = 0.542$; $F = 20.82$; $Q^2 = 0.536$; $s_{PRESS} = 0.583$

(rows 4, 27, 28)

$r = 0.813$; $s = 0.508$; $F = 17.52$; $Q^2 = 0.574$; $s_{PRESS} = 0.559$

Covariance matrix, variables of the 10 best three-variable regressions (rows 3, 9, 10)

$r = 0.897$; $s = 0.385$; $F = 37.18$; $Q^2 = 0.722$; $s_{PRESS} = 0.460$

Other combinations of certain rows of the **D** and **C** columns give comparable results, some of them with even slightly better internal predictivity. Overall, and from the calculation of several other examples it can be concluded that the use of distance or covariance matrices has its position somewhere between regression analysis and PLS analysis. It is characterized by two main features:

- There is no arbitrary selection of certain x variables, like in regression analysis. All original variables or a certain selection may be included, like in PLS analysis.
- The results can easily be interpreted as the similarity and/or dissimilarity to certain analogs of the series.

Hansch Analysis - The Nonlinear Case. - So far, only linear and linear multiple relationships have been considered in this investigation. For nonlinear relationships between log 1/C and log P it has been shown that similarity matrices, derived from lipophilicity distances $a_{ij} = -|\log P_i - \log P_j|$, can be correlated with biological activities, using PLS analysis, with or without variable selection [25]. This has recently been confirmed by Martin *et al.* [31], who demonstrated that for their data distance-based PLS analysis performed better than either (a) ordinary regression analysis using a squared term, (b) PLS analysis based on the covariance matrix, or (c) PLS analysis based on an electrostatic field similarity matrix. No variable selection was applied by Martin *et al.* [31]. It should be noted that for such univariate cases the original formulation of the lipophilicity similarity index a_{ij} corresponds to d_{ij} (eq. 3), with the exception that the x_{ij} values are not normalized before the a_{ij} values are calculated.

A simple example of a nonlinear lipophilicity-activity relationship is presented in *Figure 1*, which describes the neurotoxic activities of homologous primary alcohols in the rat [52]. Correlation of the biological data with lipophilicity, using either a parabolic or a bilinear model, gives eqs. 11 and 12 [53].

$$\log 1/C = -0.269\ (\pm 0.038)\ (\log P)^2 + 1.030\ (\pm 0.14) \log P + 1.674\ (\pm 0.11) \qquad (11)$$

optimum log P = 1.92 (1.82 / 2.02)

$(n = 10;\ r = 0.989;\ s = 0.101;\ F = 154.9;\ Q^2 = 0.951;\ s_{PRESS} = 0.149)$

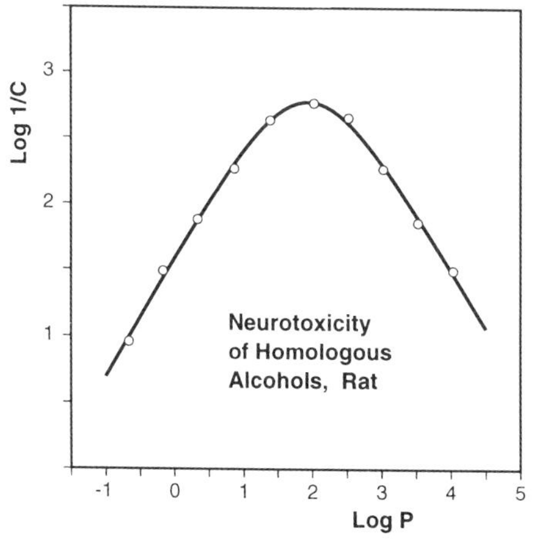

Figure 1. *Nonlinear dependence of neurotoxic activities (rat, i.p.) of homologous primary alcohols (from the left to the right: methanol to decanol) on their lipophilicity [52]. Within "short" distances, similar activities result from similar properties.*

$$\log 1/C = 0.892\ (\pm 0.050) \log P - 1.766\ (\pm 0.10) \log(\beta P + 1) + 1.586\ (\pm 0.049) \quad (12)$$

$\log \beta = -1.933$ optimum $\log P = 1.94$

$(n = 10;\ r = 0.998;\ s = 0.041;\ F = 637.6)$

Calculation of the lipophilicity distance matrix and variable selection leads to eqs. 13 and 14 as the best one- and two-variable models.

$$\log 1/C = 0.702\ (\pm 0.10)\ a_{6,j} + 2.968\ (\pm 0.16) \quad (13)$$

$(n = 10;\ r = 0.984;\ s = 0.112;\ F = 246.3;\ Q^2 = 0.935;\ s_{PRESS} = 0.161)$

$$\log 1/C = 0.408\ (\pm 0.055)\ a_{5,j} + 0.402\ (\pm 0.044)\ a_{7,j} + 3.153\ (\pm 0.092) \quad (14)$$

$(n = 10;\ r = 0.997;\ s = 0.055;\ F = 531.1;\ Q^2 = 0.985;\ s_{PRESS} = 0.083)$

Whereas the one-variable model (eq. 13) is slightly inferior to the two-variable parabolic model (eq. 11), the two-variable model (eq. 14) is much better and even comparable to the three-variable bilinear model (eq. 12). In its definition, the model expressed by eq. 13 corresponds to the so-called "cut-off model" (eq. 15), formulated several years ago [54]. In eq 15, $\log (P > P_x)$ is zero for all log P values below a critical $\log P_x$ value, whereas it is $(\log P - \log P_x)$ for all larger log P values.

$$\log 1/C = a \log P - b \log (P > P_x) + c \quad (15)$$

For several data sets, which were correlated by PLS analysis of the lipophilicity distance matrices [25], the results of regression analyses using certain columns of these matrices are given in *Table 1*. For the monoamine oxidase inhibitors and the antihypertensive clonidine analogs (eqs. 16 and 17), additional variables of the original data set were treated separately, *i.e.* not considered in the calculation of the lipophilicity distance matrix. However, no significantly different results are obtained, if the distance matrix is calculated from all variables, as can be seen, *e.g.*, from a comparison of eqs. 18-20 (C_{25} = CNS-mediated antihypertensive activity, in %; C_{60} = peripheral hypertensive activity, in mm Hg; pC values are the negative logarithms of these C values).

Clonidine analogs, parabolic Hansch analysis

$$pC_{25} = -0.387\ (\pm 0.16)\ (\log P)^2 + 0.895\ (\pm 0.33) \log P + 0.887\ (\pm 0.21)\ pC_{60} - 0.098\ (\pm 0.33) \quad (16)$$

optimum $\log P = 1.16\ (0.95 / 1.48)$

$(n = 21;\ r = 0.945;\ s = 0.376;\ F = 47.02;\ Q^2 = 0.847;\ s_{PRESS} = 0.449)$

Clonidine analogs, bilinear model

$$pC_{25} = 0.784\ (\pm 0.26) \log P - 3.685\ (\pm 1.39) \log (\beta P + 1) + 0.830\ (\pm 0.20)\ pC_{60} - 0.189\ (\pm 0.30) \quad (17)$$

$\log \beta = -2.078$ optimum $\log P = 1.51$

$(n = 21;\ r = 0.954;\ s = 0.354;\ F = 40.52)$

Table 1. *Nonlinear Lipophilicity-Activity Relationships. Results from Regression Analyses, Using Lipophilicity Similarity Matrices* $a_{ij} = -|\log P_i - \log P_j|$ *and Variable Selection.*

Data set	Variables[a)]	Fit r	Fit s	Internal Prediction Q^2	Internal Prediction s_{PRESS}	Ref.
Amines and alcohols,	11, I[d)]	0.982	0.436	0.950	0.520	
MAO inhibition, *in vitro*	1, 9, I	0.997	0.172	0.992	0.219	25
Barbiturates, *in vitro*	8, 20	0.992	0.077	0.979	0.091	
absorption	11, 20	0.992	0.078	0.978	0.093	25
Acid and base, different	6	0.979	0.128	0.945	0.148	
pH's, buccal absorption	5, 7	0.986	0.109	0.957	0.138	25
Carbamates, gastric	6	0.972	0.024	0.905	0.032	
absorption	6, 8[e)]	0.977	0.024	0.734	0.058	25
Carbamates, intestinal	3	0.962	0.047	0.845	0.067	
absorption	4, 8	0.997	0.015	0.986	0.022	25
Different drugs, blood-	5, 11	0.951	0.100	0.875	0.115	
placenta transfer rates	6, 8	0.951	0.100	0.876	0.115	25
Homologous primary	6	0.984	0.112	0.935	0.161	
alcohols, neurotoxicity[b)]	5, 7	0.997	0.055	0.985	0.083	53
Clonidine analogs, blood	5, pC_{60}[f)]	0.943	0.372	0.854	0.425	
brain barrier penetration[c)]	1, 17, pC_{60}	0.955	0.339	0.883	0.392	25
Amines, inhibition of	10	0.990	0.189	0.972	0.223	
Rhinocladium beurmanni	8, 11	0.997	0.107	0.990	0.141	25
cis-9-Alken-1-yl acetates	4	0.972	0.280	0.897	0.382	
as butterfly pheromones	2, 4	0.994	0.143	0.973	0.215	
(*Polia pisi*)	3, 4	0.993	0.155	0.967	0.235	54
Mandelic acid esters, spas-	8	0.986	0.165	0.957	0.202	
molytic activity, *in vitro*	8, 9	0.993	0.122	0.970	0.179	25
Phorbol diesters, inflam-	1, 4	0.998	0.098	0.981	0.211	
matory activity, mice	4, 6	0.998	0.098	0.981	0.211	25

a) numbers indicate the a_{ij} columns of the corresponding lipophilicity similarity matrix (identical with the numbers of the compounds in the cited references)
b) eqs. 13 and 14
c) CNS-mediated antihypertensive activity, after i.v. application; eqs. 18 and 19
d) indicator variable for alcohols (I = 1) and amines (I = 0)
e) not significant at the 95% level
f) peripheral hypertensive activity, after i.v. application

Clonidine analogs, lipophilicity similarity matrix and pC_{60} values

$$pC_{25} = 0.714\ (\pm 0.26)\ a_{5,j} + 0.889\ (\pm 0.20)\ pC_{60} + 0.612\ (\pm 0.40) \quad (18)$$

$(n = 21;\ r = 0.943;\ s = 0.372;\ F = 72.00;\ Q^2 = 0.854;\ s_{PRESS} = 0.425)$

$$\log 1/C = 1.921\ (\pm 0.71)\ a_{1,j} - 1.345\ (\pm 0.60)\ a_{17,j} + 0.808\ (\pm 0.20)\ pC_{60} + 0.287\ (\pm 0.35)$$

$(n = 21;\ r = 0.955;\ s = 0.339;\ F = 59.13;\ Q^2 = 0.883;\ s_{PRESS} = 0.392)$ (19)

Clonidine analogs, distance matrix (including all original x variables)

$$pC_{25} = -1.324\ (\pm 0.26)\ d_{1,j} - 0.714\ (\pm 0.45)\ d_{14,j} + 0.722\ (\pm 0.35)\ d_{17,j} + 3.521\ (\pm 0.82)$$

$(n = 21;\ r = 0.950;\ s = 0.357;\ F = 52.91;\ Q^2 = 0.869;\ s_{PRESS} = 0.416)$ (20)

The main advantage of such distance matrices seems to be that no initial assumptions on the linearity or nonlinearity of the lipophilicity-activity relationships and no complex mathematical models are required. In this respect, they offer the same advantage as neural networks. In contrast to the linear case, covariance matrices give much worse results in nonlinear cases, a fact that was already observed by Martin *et al.* [31].

The "Hidden Secret" of the Topological Indices. - Molecular connectivity parameters and other topological indices [55, 56] have been extensively used to describe quantitative structure-activity relationships. In the past, the scientific community was split into two groups, the one being in favor of these values, the other, larger one, being more or less sceptical about their real value and meaning [57]. In principle, every organic molecule can be described in a certain (unique?) manner by a set of such parameters. Obviously, topological indices are extremely useful to analyze the 2D similarity of molecules. Although not yet achieved, the inverse problem, to derive a molecular structure (or several structures) from a set of topological indices, may be solved one day.

Why can topological parameters successfully be applied in at least some quantitative structure-activity relationships? First of all, most of these relationships just describe unspecific structure-activity relationships, like narcotic, bactericidal, fungicidal, and neurotoxic properties, which can equally well be described by linear or nonlinear lipophilicity-activity relationships. It is well known (*e.g.* [58, 59]) that within simple series of organic compounds certain connectivity indices perfectly reflect all properties which are, as a first approximation, related to the size of the molecules (number of atoms, partition coefficient, molar refractivity). The second explanation can directly be derived from the results of this investigation: Within congeneric series, they just describe the similarity of the molecules. For this purpose they are perfectly suited, especially if the polarizability and the electronic properties do not vary too much within a group of molecules.

Where are the limitations? Like other QSAR parameters, they do not describe the differences between different configurations of molecules, stereoisomers as well as enantiomers. Such differences are only considered in 3D QSAR studies. Much more important is the other limitation: The topology and topography of a molecule may be compared to the skeleton of a human being. How boring would life be, if we recognize each other only by our skeletons and not by shape, behavior and spirit! The situation with molecules may be not as complex as in our case - but so far, with our limited knowledge, it is. We only have a rough understanding of the interactions between a ligand and its binding site, including a large number of different enthalpic and entropic energy changes during the binding process. Topological parameters are not as well suited to describe local hydrophobic properties, electron donor and acceptor properties, or molar polarizability of certain functional groups, as the classical QSAR parameters are. But these and no other properties are responsible for the interactions between a ligand and its binding site.

3D QSAR Studies: Hard *vs.* Soft Molecular Fields. - In addition to the proper alignment of all molecules within a series (section 1), one of the major problems in CoMFA studies is the fact that, close to the surface of the molecules, the potentials for the calculation of the molecular fields, *e.g.* the Lennard-Jones potential for the steric field and the Coulomb potential for the electrostatic field, are very steep. At the atom positions they approach infinite values. Within a distance of 2 Å, the default grid distance in CoMFA studies, the Lennard-Jones potential may change its value from about zero to extremely large numbers (*Figure 2*) [60]. Cut-off values have to be defined to avoid such unreasonably large positive or negative values, which might cause serious problems in PLS analysis [22, 23] and in a GOLPE variable selection [24].

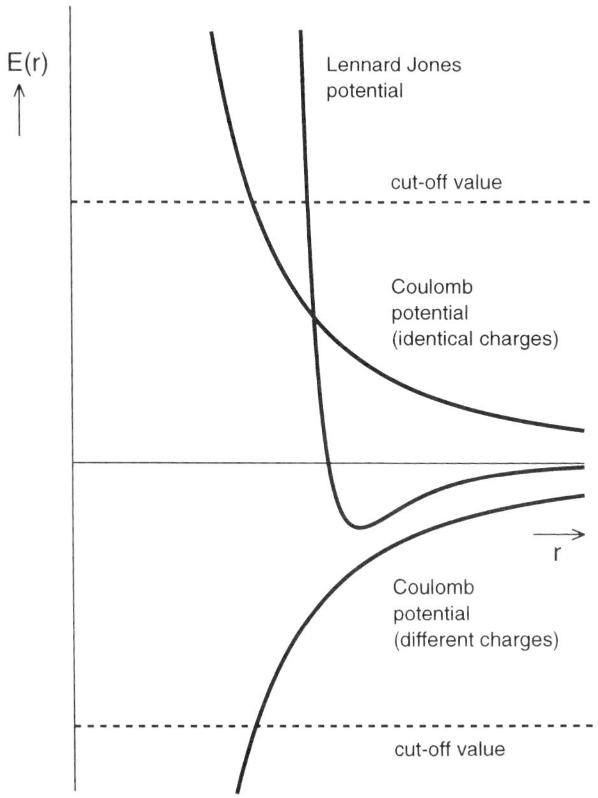

Figure 2. *The Lennard-Jones potential describes the dependence of the polarizability interaction between two atoms on their distance. Close to the surface of the atoms, the attraction (energy minimum) changes very rapidly to a steric repulsion. The sign of the Coulomb potential depends on the sign of the charges of both atoms. Although less step, it approaches infinite values at the atom positions, too. Thus, arbitrary cut-off values have to be defined to avoid unreasonably large positive or negative values of these "hard" potentials.*

In addition, small shifts in the alignment of the molecules, reorientations of the box or minor changes of the grid distance may produce very different results. Only two observations of such dramatic effects shall be mentioned here: Norden described a better CoMFA result after random orientations of his molecules, as compared to a rational alignment [61]. Cho and Tropsha obtained a broad range of very different Q^2 values as a consequence of the rotation of the box around different sets of molecules, without changing their relative alignment [62].

Several different modifications of the CoMFA procedure have been proposed to avoid such problems:

a) a cross-validated Q^2-guided region selection to achieve consistent results [62],
b) reorientations of the molecules for which too low activities are calculated, to achieve an "improved" (but more subjective) alignment [63],
c) single and domain mode variable selection in the PLS analysis [64], and
d) region selection by the GOLPE program [65].

Comparative Molecular Similarity Indices Analysis (CoMSIA) is another CoMFA modification, developed by Klebe, Abraham and Mietzner [38]. The CoMSIA fields are derived from SEAL-type Gaussian functions, describing in each grid point the similarity to a certain probe atom or group. These potentials differ significantly from the CoMFA field potentials, especially close to the surface of the molecules (*Figure 3*). On the other hand, the definition of cut-off values in CoMFA potentials can be regarded as a rough approximation to Gaussian functions, which have the advantage that no inifinite values result and no arbitrary cut-off value selections are needed. As a more or less unexpected consequence, CoMSIA PLS analyses can be performed without variable selection. Not only the potentials are smooth functions, also the resulting contour maps are much smoother and less fragmented than the corresponding CoMFA maps (especially in comparison to the maps resulting after variable selection) and therefore easier to interpret.

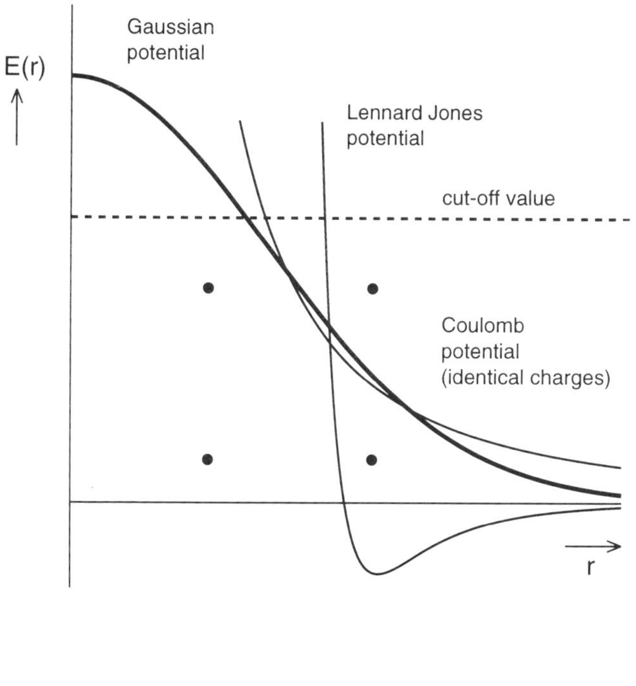

Figure 3. *Gaussian functions are much smoother than Lennard Jones and Coulomb potentials. The distances of the black circles indicate the 2Å default value of a CoMFA grid. Whereas shifts of tenths of an Å dramatically change the values of the CoMFA potentials, Gaussian functions are much less sensitive.*

The most important question is: How to explain this superiority of the CoMSIA fields? If one considers the situation of a ligand in a binding site, with its conformational flexibility and at least some flexibility of parts of the backbone and of side chains of the protein, a smooth potential should better describe the real situation than the extremely hard Lennard-Jones and Coulomb potentials. Starting from a rigid alignment, the Gaussian function "apologizes" smaller uncertainties in the orientation of the molecules to a much larger extent than the Lennard-Jones and Coulomb potentials. Although not yet systematically tested, it is reasonable to assume that in CoMSIA studies minor changes of the box orientation around the molecules or of the grid distance will have a much smaller influence on the results than in classical CoMFA studies.

CoMFA Analyses - Problems of the Steroid Data Set. - The binding affinities of 31 steroids to CBG (corticosteroid binding globulin; training set #1-21, test set #22-31) have been selected by Cramer *et al.* in their first description of the CoMFA approach [10]. Since then, several groups have re-analyzed this data set, using different CoMFA modifications. In comparing the results, many problems arise. First and most important, some papers contain wrong steroid structures, *e.g.* of estriol (16α-OH *vs.* 16β-OH) and testosterone (3-OH *vs.* 3 >C=O). Second, the predictive $r^2 = 0.65$ [10] for the test set cannot be reproduced. Using the values given in the Tables 3-5 of ref. 10 and the formula $r^2_{pred} = 1 - \Sigma(y_i - y_{pred})^2/\Sigma(y_i - y_{mean})^2$ [66], an $r^2_{pred} = 0.309$ is observed instead (compare [38]; compound #7 is misplaced in Figure 7 of ref. 10); it should be noted that the y_{mean} value in the calculation of the r^2_{pred} refers to the training set [66]. Klebe *et al.* obtained in their CoMSIA analyses r^2_{pred} values in the range of 0.36-0.40 [38], using the definition given in ref. 65.

If a one-parameter Free-Wilson equation is derived for the training set (eq. 21; the term 4-C=C- indicates the presence or absence of a cycloaliphatic 4,5-double bond in ring A of the steroids), the internal predictivity of this model is comparable to the results of many different CoMFA and other 3D QSAR studies, using the same data set.

$$\log 1/CBG = 2.022 (\pm 0.52)\ 4,5\text{-C=C-} + 5.186 (\pm 0.36) \quad (21)$$
$$(n = 21;\ r = 0.882;\ s = 0.568;\ F = 66.41;\ Q^2 = 0.726;\ s_{PRESS} = 0.630)$$

The test set predictivity of eq. 21 is at least as good as in the CoMFA and CoMSIA analyses: $r^2_{pred} = 0.477$; $s_{PRESS} = 0.733$ (n = 10). It has been pointed out that compound #31, a 9-fluoro-substituted steroid, is outside the structural space of the training set. The same argument applies to compound #23, with its 21-acetoxy group. In addition, the variance of the y-values in the test set is smaller than in the training set. If, *e.g.*, compounds #1-12 and #23-31 are used as the training set (n = 21), eq. 22 is obtained.

$$\log 1/CBG = 1.667 (\pm 0.75)\ 4,5\text{-C=C-} + 5.306 (\pm 0.65) \quad (22)$$
$$(n = 21;\ r = 0.731;\ s = 0.697;\ F = 21.82;\ Q^2 = 0.454;\ s_{PRESS} = 0.754)$$

Despite the slightly inferior fit, as compared to eq. 21, the reliability of this model is proven by its excellent test set (compounds #13-22) predictivity: $r^2_{pred} = 0.909$; $s_{PRESS} =$

0.406; n = 10). The differences between both models, especially in their test set predictivity, are a further striking evidence for the influence of the training and test set selections on the obtained results. Thus, a careful selection of the training set molecules is of utmost importance. A broad variety of structural features should be included in these molecules, in order to allow reliable predictions for the test set compounds.

3D-Structure-Activity Relationships Based on 3D Similarity Matrices. - The similarity concept can also be applied in CoMFA- and CoMSIA-type 3D QSAR studies. As already mentioned above, the SAMPLS algorithm [48] converts the original **X** block (*i.e.*, the matrix of all x_{ij} values) into an N x N covariance matrix **C**, to derive from this matrix exactly the same numerical result as "classical" PLS analysis does from the original **X** block. Accordingly, the similarity matrix analysis approach should also be applicable to derive 3D QSAR models. This can be performed, *e.g.*, by calculating the covariance matrices **C** from different 3D molecular fields (no scaling of the original x_{ij} values). As a first application of this new approach, covariance matrices were generated for the steroid data set from different SEAL fields [38]. After variable selection, the "best" two-parameter models (fit criterion) were used to derive the statistical parameters for the internal predictivity, for the whole set and the training set (compounds #1-21), as well as for the prediction of the test set compounds (#22-31) by the training set models.

These very first and preliminary studies showed that equally good and even better internal predictions are obtained, with Q^2 values up to 0.7. However, the external predictivities of the models (r^2_{pred} values) varied to a large extent. This confirms the often observed fact that the simpler a model is, the better may be its external predictivity. Complex models which give a better fit most often fail in their external predictivity (*e.g.* [64]). Further systematic investigations with different molecular fields are necessary to evaluate the full potential of this new approach. In addition, better training set selections (compare eqs. 21 and 22) seem to be necessary to obtain stable test set predictions.

Further progress should result from an individual alignment of all compounds, in such a way that every molecule is superimposed to every other one in a unique manner, in order to achieve the best mutual fit, even according to different physicochemical properties. Work in this direction is in progress. A certain disadvantage is the fact that (so far) no contour maps can be derived. However, a direct interpretation of the equations is possible, because the regression coefficients of the selected covariance matrix columns are a measure for the similarities to the corresponding compounds.

Similarity Through the Eyes of a Medicinal Chemist. - The definition and quantitative description of chemical similarity is a critical issue, not only in QSAR but also in combinatorial chemistry. Many different groups are involved in research programs devoted to the development of similarity and dissimilarity measures. However, we should not forget that sometimes minor chemical differences may have a dramatic effect on biological activities, in a qualitative as well as in a quantitative manner.

A well-known example are the analogs norepinephrine (**1**, R = H), epinephrine (**1**, R = methyl), and isoproterenol (**1**, R = *i*-propyl), with decreasing α-adrenergic and increasing β-adrenergic activities. The exchange of the phenolic hydroxy groups of isoproterenol against chlorine atoms led to the first β-adrenergic blocker dichloro-isoproterenol **2**, *i.e.*

to the conversion of an agonist into an antagonist. The close structural relationships between male and female sex hormones and some anabolic steroids, or the gluco- and corticosteroids are other striking examples for the qualitatively very different actions of compounds with only minor variations in their chemical structures.

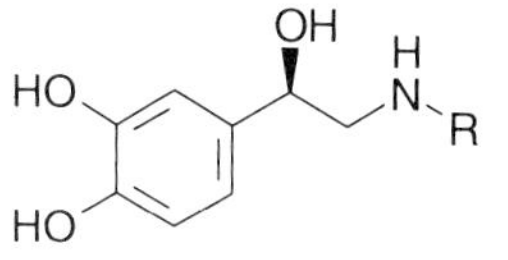

1

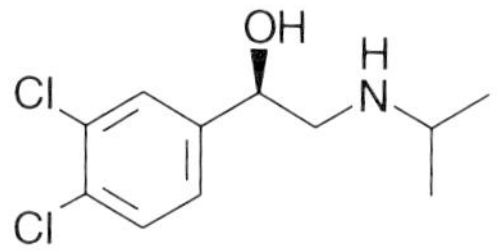

2

A further example of different kinds of biological activities of compounds with very similar structures are the tricyclic compounds promethazine **3** (an H_1-antihistaminic), chlorpromazine **4** (a dopamine-antagonistic neuroleptic) and imipramine **5** (an anti-depressant, which is acts as a norepinephrine- and serotonine-uptake inhibitor).

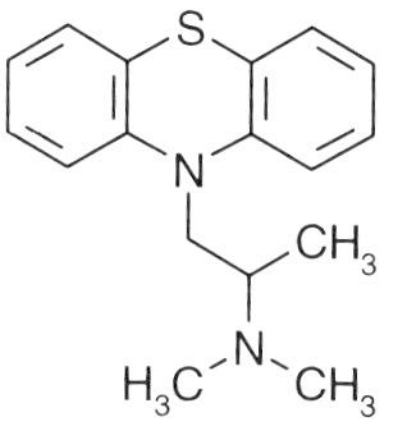

3 Promethazine
H_1-Antagonist

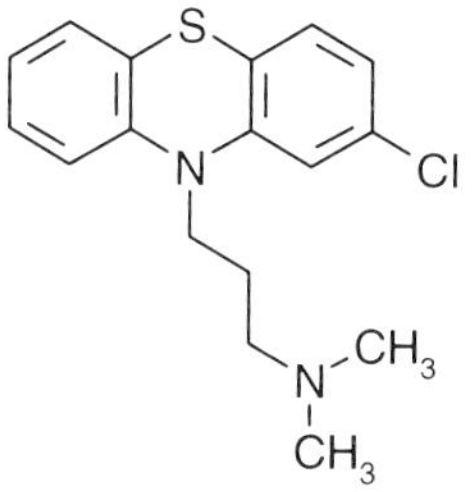

4 Chlorpromazine
Neuroleptic

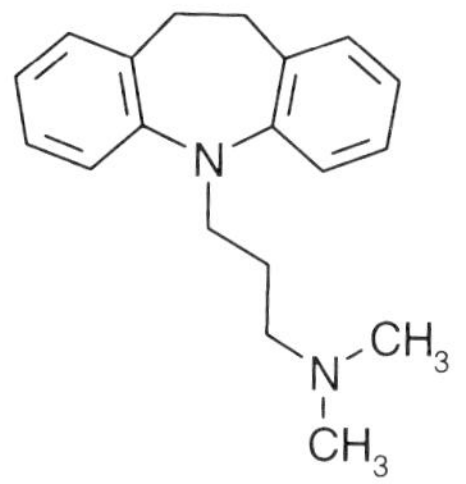

5 Imipramine
Antidepressant

Despite their chemical similarity, closely related analogs may bind in a different orientation into the binding pockets of their proteins. Whereas the different binding behavior of dihydrofolate and methotrexate can be easily explained by their different patterns of hydrogen bond donor and acceptor groups [67, 68], the different orientations of the pair *N*-carbobenzoxy-L-phenylalanine (**6**, X = O) and *N*-β-phenylpropionyl-L-phenylalanine (**6**, X = CH_2) [69], of some rhinovirus coat protein ligands, *e.g.* **7** [70], and of three isomeric phenylimidazoles **8** (R = Phenyl, H, H) [71] in their binding pockets are much more difficult to understand.

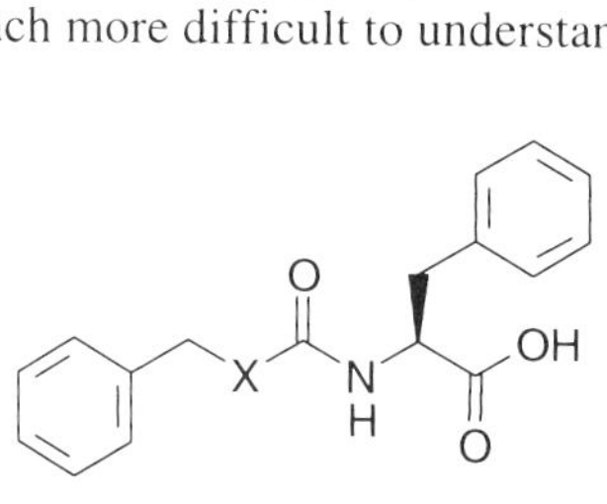

6

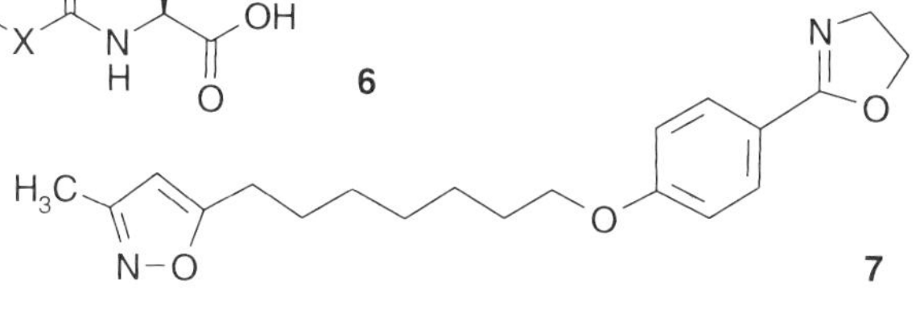

7

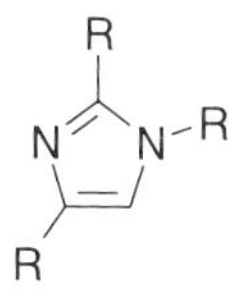

8

The pair thiorphan **9** and *retro*-thiorphan **10** [72] differs only by the direction of the amide group which connects the benzyl-substituted thiol with the acidic part of the molecule. Can these compounds be considered to be similar or not?

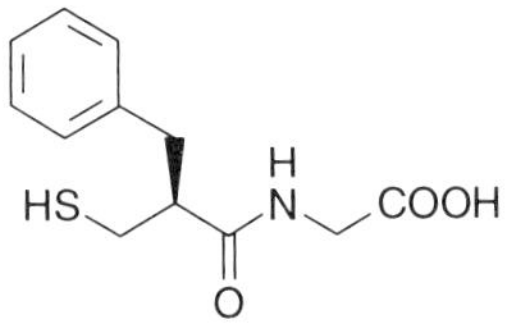

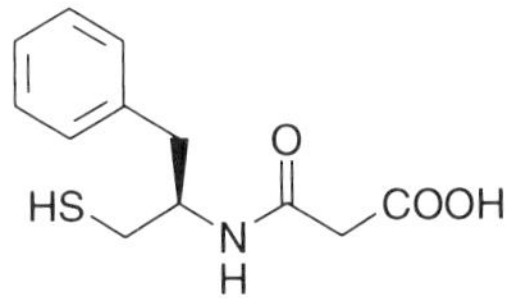

Enzyme	K_i, in µM	
Thermolysin	1.8	
NEP 24.11	0.0019	**9** Thiorphan
ACE	0.14	
Thermolysin	2.3	
NEP 24.11	0.0023	**10** *retro*-Thiorphan
ACE	>10	

Difficult to answer, especially if the biological target is unknown. Fortunately the 3D structures of the complexes of both compounds with the bacterial zinc protease thermolysin have been determined. Both compounds show identical binding modes, which is reflected by their identical thermolysin inhibitor activities: The compounds may be considered to be "similar". They also inhibit neutral endopeptidase, NEP 24.11, with even higher affinity constants. However, against angiotensin converting enzyme (ACE), a zinc protease related to thermolysin and NEP 24.11, they show very different inhibitory activities [73]: Here they are dissimilar! The structural reasons for this different behavior will remain unknown, as long as we do not know the exact 3D structure of ACE.

Minor structural changes of an active analog may even destroy its biological activity. An example are the thermolysin inhibitors **11** [74], where the exchange of X = NH or CH_2 against X = O reduces inhibitory activities by three to four orders of magnitude. A similar but much less pronounced effect is observed for the thrombin inhibitors **12** [75].

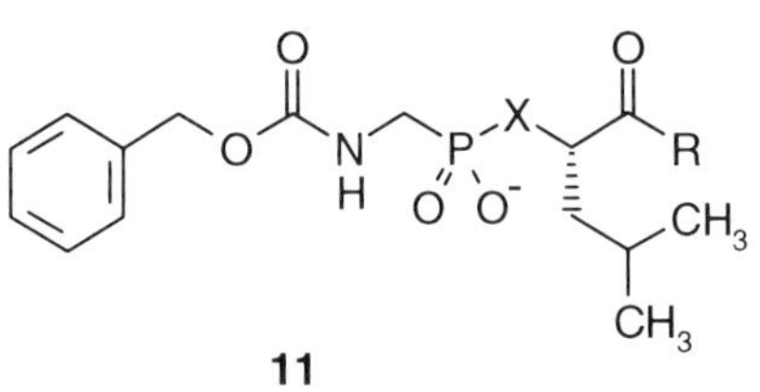

11

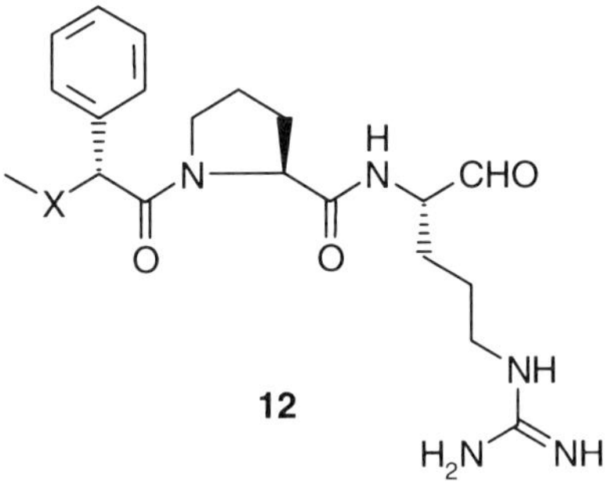

12

So far the biggest effect of a small structural change is observed for some transition state inhibitors. The introduction of a single hydroxy group into the weakly active cytosine desaminase inhibitor **13** leads to analog **14**, which is several orders of magnitude more active as an inibitor than compound **13** [76].

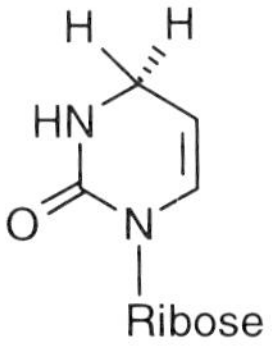

13 K_i = 30 µM

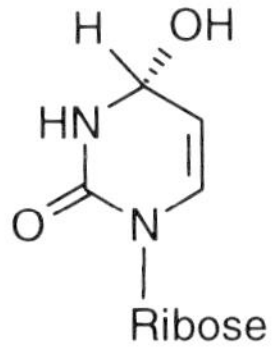

14 K_i = 1.2 pM

Again, the explanation for this striking difference can be derived from the X-ray structures of the protein-inhibitor complexes. Compound **14** corresponds to the transition state of the enzymatic reaction. Its hydroxy group is optimally suited to form several hydrogen bonds and to replace the water molecule which is coordinated to the Zn^{2+} ion and added to cytosine during the desamination. On the other hand, compound **13** does not replace this water molecule and obviously fits less well into the binding site.

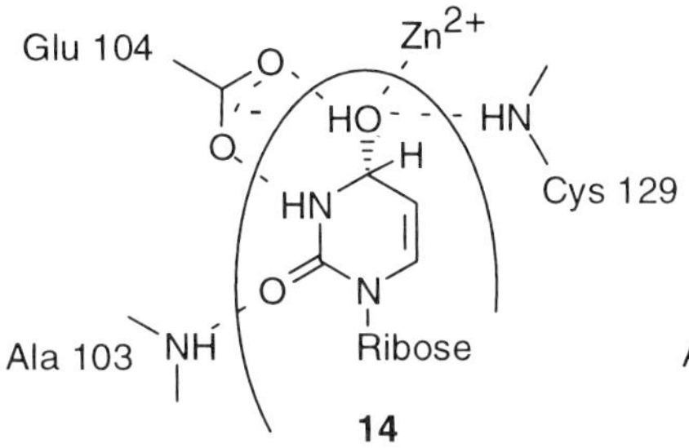

14

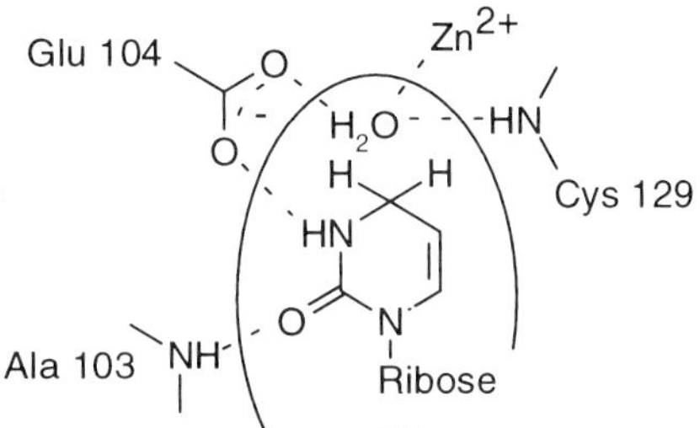

13

The surprising effect that nebularine **15**, without a hydroxy group, is a picomolar inhibitor of adenosine desaminase (K_i = 0.3 pM) is explained by the hypothetical intermediate **16** [77]. In comparison with the affinity of the natural substrate, adenosine, the binding affinity increases by 8 orders of magnitude!

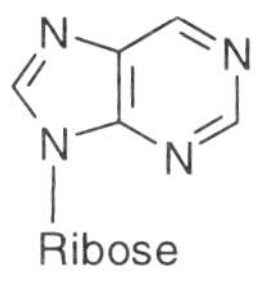

15 Nebularine

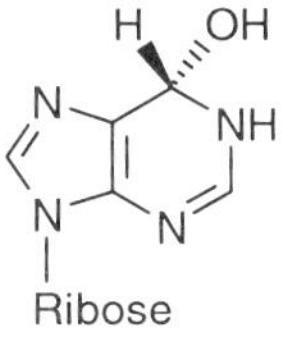

16 hypothetical active form of **16**

Discussion and Conclusions. - As shown in the last section, the definition of the similarity and dissimilarity of drugs is a rather difficult issue and can only be defined with respect to a given biological target. It should be considered only in chemically closely related series that interact with an identical protein in the same manner. In such series, or when only transport and distribution of the drugs are responsible for the structure-activity relationship, distance matrices are suited to describe any kind of linear, linear multiple and nonlinear structure-activity relationships as good as other structure-activity models. The term **QSiAR** (**Q**uantitative **Si**milarity **A**ctivity **R**elationships) is proposed for all approaches, where a distance or covariance matrix is derived from the original variables or a subset of these variables to describe the pairwise similarities of compounds, and to correlate certain columns of this similarity matrix with the biological activity values. While there are several examples of different binding modes of chemically related ligands, caused by the specific structural features of the binding sites of the enzymes and receptors, there are indeed many quantitative relationships which prove the simple concept of similarity analyses.

The name **3D QSiAR** is proposed for approaches which correlate the biological activities of molecules with similarity index matrices, calculated from 3D molecular fields. The chance to perform individual alignments of the compounds, even according to different molecular properties, broadens the applicability of this method and should improve the predicitive ability of the resulting models. However, further investigations are necessary to confirm this optimistic expectation.

An extremely difficult problem in the design of combinatorial libraries is the proper definition of the similarity of molecules with respect to biological activities. How similar two molecules are, depends on the underlying question: Shall the similarity be defined in terms of global properties, of size and shape, or of certain structural features (*e.g.*, a pharmacophore)? As long as we do not know the structure of the binding site, it appears impossible to define an objective similarity measure to predict biological activities. If this fact is taken into account, many sophisticated similarity comparisons become rather meaningless. Whether we like this situation or not: We can never be sure that a certain similarity measure really applies to a particular ligand-protein interaction.

Acknowledgements: Many stimulating discussions with my colleagues Gerhard Klebe, Ute Abraham and Thomas Mietzner have contributed to the development of the ideas and concepts discussed in this paper and are thus gratefully acknowledged.

REFERENCES

[1] C. A. Ramsden, Ed., Quantitative Drug Design, Volume 4 of: Comprehensive Medicinal Chemistry, C. Hansch, P. G. Sammes, J. B. Taylor, Eds., Pergamon Press, Oxford, 1990

[2] H. Kubinyi, QSAR: Hansch Analysis and Related Approaches, VCH, Weinheim, 1993

[3] C. G. Wermuth, Ed., Trends in QSAR and Molecular Modelling 92 (Proceedings of the 9th European Symposium on Structure-Activity Relationships: QSAR and Molecular Modelling, Strasbourg, 1992), ESCOM Science Publishers B. V., Leiden, 1993

[4] C. Hansch, A. Leo, Exploring QSAR. Fundamentals and Applications in Chemistry and Biology, 2 Volumes, American Chemical Society, Washington, 1995

[5] H. van de Waterbeemd, Ed., Chemometric Methods in Molecular Design, VCH, Weinheim, 1995

[6] H. van de Waterbeemd, Ed., Advanced Computer-Assisted Techniques in Drug Discovery, VCH, Weinheim, 1995

[7] T. Fujita, Ed., QSAR and Drug Design: New Developments and Applications, *Pharmacochem. Libr.* **1995**, *23*

[8] F. Sanz, J. Giraldo, F. Manaut, Eds., QSAR and Molecular Modelling: Concepts, Computational Tools and Biological Applications (Proceedings of the 10th European Symposium on Structure-Activity Relationships: QSAR and Molecular Modelling, Barcelona, 1994), Prous Science Publishers, Barcelona, 1995

[9] H. van de Waterbeemd, Structure-Property Correlations in Drug Research, Academic Press, 1996
[10] R. D. Cramer III, D. E. Patterson, J. D. Bunce, *J. Am. Chem. Soc.* **1988**, *110*, 5959-5967
[11] H. Kubinyi, Ed., 3D QSAR in Drug Design: Theory, Methods and Applications, ESCOM Science Publishers B. V., Leiden, 1993
[12] M. A. Johnson, G. M. Maggiora, Eds., Concepts and Applications of Molecular Similarity, Wiley, New York, 1990
[13] C. Pepperrell, Three-Dimensional Chemical Similarity Searching, Wiley, Chichester, 1994
[14] P. M. Dean, Ed., Molecular Similarity in Drug Design, Chapman & Hall, New York, 1995
[15] K. Sen, Ed., Molecular Similarity I and II, Topics in Current Chemistry **1995**, *173* and *174*
[16] G. Rum, W. C. Herndon, *J. Am. Chem. Soc.* **1991**, *113*, 9055-9060
[17] A. C. Good, S.-S. So, W. G. Richards, *J. Med. Chem.* **1993**, *36*, 433-438
[18] A. C. Good, S. J. Peterson, W. G. Richards, *J. Med. Chem.* **1993**, *36*, 2929-2937
[19] A. C. Good, ref. 14, pp. 24-56
[20] R. Carbo, L. Leyda, M. Arnau, *Int. J. Quantum Chem.* **1980**, *17*, 1185-1189
[21] E. E. Hodgkin, W. G. Richards, *Int. J. Quantum Chem., Quantum Biol. Symp.* **1987**, *14*, 105-110
[22] S. Wold, A. Ruhe, H. Wold, W. J. Dunn III, *SIAM J. Sci. Stat. Comput.* **1984**, *5*, 735-743
[23] S. Wold, E. Johansson, M. Cocchi, in: 3D QSAR in Drug Design. Theory, Methods and Applications, H. Kubinyi, Ed., ESCOM Science Publishers B. V., Leiden, 1993, pp. 523-550
[24] G. Cruciani, S. Clementi, M. Baroni, in: 3D QSAR in Drug Design. Theory, Methods and Applications, H. Kubinyi, Ed., ESCOM Science Publishers B. V., Leiden, 1993, pp. 551-564
[25] ref. 2, pp. 175-177
[26] R. Benigni, A. Giuliani, *Quant. Struct.-Act. Relat.* **1993**, *12*, 397-401
[27] R. G. A. Bone, H. O. Villar, *J. Mol. Graphics* **1995**, *13*, 201-208
[28] I. T. Cousins, M. T. D. Cronin, J. C. Dearden, C. D. Watts, *SAR and QSAR in Environ. Res.* **1995**, *3*, 279-292
[29] R. Benigni, M. Cotta-Ramusino, F. Giorgi, G. Gallo, *J. Med. Chem.* **1995**, *38*, 629-635
[30] M. Hahn, D. Rogers, *J. Med. Chem.* **1995**, *38*, 2091-2102
[31] Y. C. Martin, C. T. Lin, C. Hetti, J. DeLazzer, *J. Med. Chem.* **1995**, *38*, 3009-3015
[32] D. C. Horwell, W. Howson, M. Higginbottom, D. Naylor, G. S. Ratcliffe, S. Williams, *J. Med. Chem.* **1995**, *38*, 4454-4462
[33] R. Benigni, A. Giuliani, ref. 8, pp. 55-56
[34] W. C. Herndon, G. Rum, ref. 8, pp. 380-384
[35] C. A. Montanari, M. S. Tute, A. E. Beezer, J. C. Mitchell, *J. Comput.-Aided Mol. Design* **1996**, *10*, 67-73
[36] S. K. Kearsley, G. M. Smith, *Tetrahedron Comput. Methodol.* **1990,** *3*, 615-633
[37] G. Klebe, T. Mietzner, F. Weber, *J. Comput.-Aided Mol. Design* **1994**, *8*, 751-778
[38] G. Klebe, U. Abraham, T. Mietzner, *J. Med. Chem.* **1994**, *37*, 4130-4146
[39] S. H. Unger, C. Hansch, *J. Med. Chem.* **1973**, *16*, 745-749
[40] ref. 2, p. 60
[41] H. Kubinyi, *Quant. Struct.-Act. Relat.* **1994**, *13*, 285-294
[42] H. Kubinyi, *Quant. Struct.-Act. Relat.* **1994**, *13*, 393-401
[43] H. Kubinyi, *J. Chemometrics* **1996**, *10*, 119-133
[44] S. M. Free Jr., J. W. Wilson, *J. Med. Chem.* **1964**, *7*, 395-399
[45] H. Kubinyi, *Quant. Struct.-Act. Relat.* **1988**, *7*, 121-133
[46] H. Kubinyi, ref. 1, pp. 589-643
[47] K. Hasegawa, M. Hirata, T. Koshi, T. Ohshima, Y. Miyashita, S. Sasaki, T. Okabe, *Bioorg. Med. Chem. Letters* **1994**, *4*, 1157-1160
[48] B. L. Bush, R. B. Nachbar Jr., *J. Comput.-Aided Mol. Design* **1993**, *7*, 587-619
[49] D. L. Selwood, D. J. Livingstone, J. C. W. Comley, A. B. O'Dowd, A. T. Hudson, P. Jackson, K. S. Jandu, V. S. Rose, J. N. Stables, *J. Med. Chem.* **1990**, *33*, 136-142
[50] D. Rogers, A. J. Hopfinger, *J. Chem. Inf. Comput. Sci.* **1994**, *34*, 854-866
[51] H. Kubinyi, U. Abraham, ref. 11, pp. 717-728

[52] M. J. McCreery, W. A. Hunt, *Neuropharmacology* **1978**, *17*, 451-461
[53] H. Kubinyi, *Arzneim.-Forsch. (Drug Res.)* **1979**, *29*, 1067-1080
[54] H. Kubinyi, in: QSAR in Design of Bioactive Compounds (Proceedings of the 1st Telesymposium on Medicinal Chemistry, 1984), M. Kuchar, Ed., Prous Science Publishers, Barcelona, 1984, pp. 321-346
[55] L. B. Kier, L. H. Hall, Molecular Connectivity in Chemistry and Drug Research, Academic Press, New York, 1976
[56] L. B. Kier, L. H. Hall, Molecular Connectivity in Structure-Activity Analysis, Research Studies Press, Letchworth, 1986
[57] ref. 1, p. 251; ref. 2, p. 53
[58] A. K. Saxena, *Quant. Struct.-Act. Relat.* **1995**, *14*, 142-148
[59] H. Kubinyi, *Quant. Struct.-Act. Relat.* **1995**, *14*, 149-150
[60] G. Folkers, A. Merz, D. Rognan, in: 3D QSAR in Drug Design. Theory, Methods and Applications, H. Kubinyi, Ed., ESCOM Science Publishers B. V., Leiden, 1993, pp. 583-618
[61] B. Norden, 10th European Symposium on Structure-Activity Relationships: QSAR and Molecular Modelling, Barcelona, 1994, personal communication
[62] S. J. Cho, A. Tropsha, *J. Med. Chem.* **1995**, *38*, 1060-1066
[63] R. T. Kroemer, P. Hecht, *J. Comput.-Aided Mol. Design* **1995**, *9*, 396-406
[64] U. Norinder, *J. Chemometrics* **1996**, *10*, 95-105
[65] G. Cruciani, M. Pastor, S. Clementi, in: H. van de Waterbeemd, B. Testa, G. Folkers, Computer-Assisted Lead Finding and Optimization (Proceedings of the 11th European Symposium on Quantitative Structure-Activity Relationships, Lausanne, 1996), in print
[66] S. A. DePriest, D. Mayer, C. B. Naylor, G. R. Marshall, *J. Am. Chem. Soc.* **1993**, *115*, 5372-5384
[67] C. Bystroff, S. J. Oatley, J. Kraut, *Biochemistry* **1990**, *29*, 3263-3277
[68] H.-J. Böhm, G. Klebe, *Angew. Chem.* **1996**, *108*, in press
[69] W. R. Kester, B. W. Matthews, *Biochemistry* **1977**, *16*, 2506-2516
[70] G. D. Diana, A. M. Treasurywala, T. R. Bailey, R. C. Oglesby, D. C. Pevear, F. J. Dutko, *J. Med. Chem.* **1990**, *33*, 1306-1311
[71] T. L. Poulos, A. J. Howard, *Biochemistry* **26**, 8165-8174 (1987)
[72] S. L. Roderick, M. C. Fournie-Zaluski, B. P. Roques, B. W. Matthews, *Biochemistry* **1989**, *28*, 1493-1497
[73] B. P. Roques, F. Nobel, V. Daugé, M. Fournié-Zaluski, A. Beaumont, Pharmacol. Rev. **1993**, *45*, 87-146
[74] B. P. Morgan, J. M. Scholtz, M. D. Ballinger, I. D. Zipkin, P. A. Bartlett, *J. Am. Chem. Soc.* **1991**, *113*, 297-307
[75] R. T. Shuman *et al.*, in: J. A. Smith, J. E. Rivier, Eds., Peptides. Chemistry and Biology (Proceedings of the 12th American Peptide Symposium, Cambridge, MA, USA, 1991), ESCOM Science Publishers B. V., Leiden, 1992, pp. 801-802
[76] S. Xiang, S. A. Short, R. Wolfenden, C. W. Carter, *Biochemistry* **1995**, *34*, 4516-4523
[77] W. M. Kati, S. A. Acheson, R. Wolfenden, *Biochemistry* **1992**, *31*, 7356-7366

PART I. **CHEMOMETRICS**

Chapter 2

Prediction of Acute Mammalian Toxicity from Molecular Structure for a Diverse Set of Substituted Anilines Using Regression Analysis and Computational Neural Networks

by **Stephen Roger Johnson** and **Peter C. Jurs***

Department of Chemistry, 152 Davey Laboratory,
Pennsylvania State University, University Park, PA 16802,
USA

erful as a descriptor for mammalian toxicity [2,3]. It has been suggested that this is due to nonspecific effects involved in aquatic toxicity. The mechanism of the toxicity would seem to be significantly more important in the study of mammalian toxicity. Still, several mammalian toxicity QSAR studies have been performed relating log P to the toxicity [2,3,13,14]. While this may, to some extent, resolve some of the difficulties in determining mammalian toxicity, it does not address the difficulty of determining the toxicity for a large number of compounds. In the present study, a large number of numerical descriptors were calculated directly from structure. Using several different methods of feature selection, small, information rich, subsets of descriptors were chosen, which were then mapped to the log(LD_{50}) values. Therefore, molecular structure became the sole basis for predicting the toxicity of new compounds.

Although the current study concerns the prediction of acute mammalian toxicity, a biological activity, the approach employed here has also been used in the development of many different quantitative structure-property relationships (QSPR). These include aqueous solubility [15], reduced ion mobility constants [16], boiling points [17], and carbon-13 nuclear magnetic resonance spectra [18]. Indeed, the generality of the method to both activity and physical property prediction enhances its utility in pharmaceutical and other chemical design endeavors.

Experimental. - *Data Set.* The mouse oral LD_{50} acute toxicity data of substituted anilines were taken from the RTECS database [5]. The RTECS database is a compilation of toxicity data taken from many sources. The data set was collected from the RTECS data base with the goal of obtaining a highly diverse set of substituted anilines, containing many different functional groups. *Table 1* shows the diversity of the data set chosen. The data set consisted of 115 substituted anilines with known LD_{50} values, and is presented in *Table 2*. Due to the number of original sources from which the data were derived, an accurate estimation of the error was not possible. However, for the reasons mentioned above, it is not unreasonable to assume that a fair amount of error is present in the data.

Table 1. *Diversity of the Data Set*

Functional Group	Number of Cmpds. Containing at least one
Ether	54
Bromine	1
Chlorine	22
Nitro	14
Thio	7
Fluorine	3
N-Substituted	25
Ring (other than anilino ring)	44

Note: Some compounds contain more than one type of functional group

For the purposes of developing and validating multiple linear regression models, the data set was split randomly into a 103-member training set and a 12-member external

Table 2. *Observed and Calculated LD_{50} Values for the Substituted Anilines Used in This Study*

No.	Compound Name	LD50[a] (mg/kg)	LD50[b] (mg/kg)	LD50[c] (mg/kg)
1	Aniline[(d)]	464	315	511
2	4-(3',5',5'-trimethylhexyloxy)-m-anisidine[(f)]	800	1225	560
3	N-Benzylidene-p-butoxy aniline[(d)]	500	494	750
4	4,4'-(3'-Aminopropylidene)di-aniline[(d)]	2300	654	1213
5	N,N-bis(2'-chloroethyl) aniline[(d)]	123	141	169
6	p-Bromo aniline[(d)]	289	440	396
7	p-Butoxy aniline[(d)]	500	655	610
8	p-Butoxy-N-methyl aniline[(e)]	500	713	547
9	m-Chloro aniline[(d)]	334	438	430
10	o-Chloro aniline[(d)]	256	434	491
11	p-Chloro aniline[(f)]	100	423	427
12	2-Chloro-4-nitro aniline[(d)]	1250	844	947
13	2-Chloro-5-nitro aniline[(e)]	1600	870	930
14	4-Chloro-2-nitro aniline[(d)]	800	942	580
15	4-Chloro-3-nitro aniline[(d)]	800	861	722
16	3-Chloro-4-octyloxy aniline[(d)]	1500	1017	1296
17	p-Cyano aniline[(d,g)]	650	35	487
18	p-(2'-Cyclohexenyl) aniline[(d,g)]	30	744	119
19	p-Decyloxy aniline[(d)]	1500	2046	1386
20	2,4-Dichloro aniline[(d)]	400	536	573
21	2,5-Dichloro aniline[(e)]	1600	548	566
22	3,4-Dichloro aniline[(d)]	740	525	1163
23	2,6-Dichloro-4-nitro aniline[(d)]	1500	981	3596
24	2,6-Dichloro-4-thiocyanato aniline[(d)]	250	288	147
25	2,4-Dimethoxy aniline[(d)]	1000	641	532
26	2,5-Dimethoxy aniline[(d)]	120	512	565
27	3,4-Dimethoxy aniline[(f)]	681	460	530
28	3,4-Dimethyl-2,6-dinitro-N-(1'-ethylpropyl)aniline[(d)]	1340	1366	1616
29	N,N-Dimethyl-p-phenylazo aniline[(d)]	300	304	318
30	N,N-Dimethyl-p-thiocyanato aniline[(d)]	150	141	132
31	2,4-Dinitro aniline[(d)]	370	619	950
32	p-(5'-(p-Ethoxyphenoxy)pentyloxy) aniline[(d)]	3000	2572	2350
33	4-Ethyl aniline[(d)]	500	634	423
34	p-Hexyl aniline[(d)]	500	708	385
35	p-Hexyloxy aniline[(d)]	500	625	264
36	3-Methoxy-4-methoxymethoxy aniline[(d)]	1800	455	693
37	p-(5'-(p'-methoxyphenoxy)pentyloxy) aniline[(d)]	1500	2149	1757
38	4,4'-Methylenedianiline[(d)]	745	654	955
39	N-Methyl-p-nitroso aniline[(d)]	46	40	43
40	N-(3-Methyl-1-pentyn-3-yl) aniline[(d)]	500	852	282
41	p-(5'-(2'-napthyloxy)pentyloxy) aniline[(d)]	4000	3240	2221
42	m-Nitro aniline[(d)]	308	700	490
43	o-Nitro aniline[(f)]	1070	783	530
44	p-Nitro aniline[(d)]	810	690	489
45	4,4'-oxydianiline[(d)]	685	578	967
46	p-(5'-phenoxypentyloxy) aniline[(d)]	750	1390	822
47	p-(5-(p-phenoxyphenoxy)pentyloxy) aniline[(d)]	1000	1020	3059

No.	Compound Name	LD50[a] (mg/kg)	LD50[b] (mg/kg)	LD50[c] (mg/kg)
48	p-(3'-Phenylpropoxy) aniline[(d,g)]	80	649	160
49	4,4'-methylene bis(2-chlorobenzenamine)[(d)]	880	631	964
50	2-Methoxy-4-propylthiobenzenamine[(d)]	680	582	672
51	4,4'-Thiodianiline[(d)]	620	565	927
52	2,4,6-Trichloroaniline[(d,g)]	5800	631	3752
53	2,4,6-Trimethyl aniline[(e)]	590	755	521
54	4,4'-(Trimethylenedioxy)Bis(N,N-dimethylaniline)[(f)]	4000	1364	2335
55	4-(p-Anilinobenzyl) morpholine[(f)]	1000	614	928
56	N-Phenyl-2-naphthylamine[(d)]	1450	748	588
57	m-Toluidine[(d)]	740	486	432
58	o-Toluidine[(f)]	520	509	481
59	p-Toluidine[(d)]	330	484	432
60	3-Chloro-p-toluidine[(d)]	316	577	661
61	N-Propyl-N-(2'-chloroethyl)-2,6-dinitro -4-trifluoromethylaniline[(d)]	730	1523	921
62	3,5-Dinitro-o-toluidine[(d)]	1522	936	1445
63	3,5-Dinitro-p-toluidine[(d)]	1318	865	1426
64	3-Trifluoromethyl aniline[(d)]	220	380	583
65	N,N-Dipropyl-4-trifluoromethyl-2,6-dinitro aniline[(d)]	3197	1538	1027
66	o-Anisidine[(d)]	1400	463	469
67	4-Benyloxy-m-anisidine[(d)]	700	547	244
68	4-(2'-(Benzyloxy)ethoxy)-m-anisidine[(d)]	700	808	784
69	4-Butoxy-m-anisidine[(e)]	400	557	247
70	3-Chloro-p-anisidine[(f)]	650	494	541
71	4-(2'-Cyclohexylethoxy)-m-anisidine[(d,g)]	80	655	195
72	4-(5'-(p-chlorophenylthio)pentyoxy)-m-anisidine[(d)]	2000	2585	2293
73	4-(5'-cyclohexylpentyloxy)-m-anisidine[(e)]	3500	2752	1488
74	4-(cyclopentyloxy)-m-anisidine[(d)]	400	656	413
75	4-Decyloxy-m-anisidine[(d)]	3000	2800	2531
76	4-(3'-Heptenyloxy)-m-anisidine[(d)]	750	696	586
77	4-Heptyloxy-m-anisidine[(d)]	800	738	641
78	4-Hexyloxy-m-anisidine[(d)]	150	584	269
79	4-(3'-(p-methoxyphenoxy)propoxy)-m-anisidine[(e)]	3000	1196	1684
80	4-(5'-(p-methoxyphenyl)pentyloxy)-m-anisidine[(d)]	4000	2558	2319
81	N-Methyl-4-(5'-phenylpentyloxy)-m-anisidine[(d)]	1500	1855	1880
82	N-(6-(4,6-dimethyl-2-pyrimidinylthio)hexyl)benzenamine[(d)]	2000	2211	1948
83	N,N-dimethyl-4-(2'-(4-pyrimidinyl)ethenyl) benzenamine[(d)]	1090	551	960
84	4-(5'-(methylthio)pentyloxy)-m-anisidine[(d)]	1200	638	746
85	4-(1-Napthylmethoxy)-m-anisidine[(d)]	1200	771	1234
86	5-nitro-o-anisidine[(f)]	1060	874	740
87	4-Nonyloxy-m-anisidine[(d)]	2500	1885	1909
88	4-Octyloxy-m-anisidine[(d)]	1400	1157	1239
89	4-Pentyloxy-m-anisidine[(d)]	100	548	175
90	4-(7'-phenoxyheptyloxy)-m-anisidine[(d)]	1200	2161	3057
91	4-(5'-phenoxypentyloxy)-m-anisidine[(d)]	3000	2083	2024
92	4-(3'-phenoxypropoxy)-m-anisidine[(d)]	750	780	731
93	4-(7'-phenylheptyloxy)-m-anisidine[(f)]	3000	3186	2626

No.	Compound Name	LD50[a] (mg/kg)	LD50[b] (mg/kg)	LD50[c] (mg/kg)
94	4-(6'-phenylhexyloxy)-m-anisidine[(f)]	2500	3047	2038
95	4-(8'-phenyloctyloxy)-m-anisidine[(d)]	4000	2409	3099
96	4-(5'-phenylpentyloxy)-m-anisidine[(d)]	1500	1855	1440
97	4-(3'-phenylpropoxy)-m-anisidine[(d)]	400	744	396
98	2-(p-aminophenyl)-N-ethyl-2-phenyl propionamide[(d)]	1750	1155	2468
99	4-((5'-(phenylthio)pentyl)oxy)-m-anisidine[(e)]	1500	2382	2081
100	4-(3'-(phenylthio)propoxy)-m-anisidine[(d)]	750	1014	928
101	4-propoxy-m-anisidine[(d)]	900	565	515
102	4-((5'-o-tolylpentyl)oxy)-m-anisidine[(e)]	2000	2743	2683
103	2,3-Xylidene[(d)]	836	613	383
104	2,4-Xylidene[(d)]	250	648	485
105	2,5-Xylidene[(f)]	841	649	485
106	2,6-Xylidene[(e)]	707	669	628
107	3,4-Xylidene[(d)]	707	607	501
108	3,5-Xylidene[(d)]	421	625	389
109	p-Anilinophenol[(e)]	2120	948	684
110	2-Anilino-4-chloro-6-(4-methylpiperazino) -5-methylthiopyrimidine[(d)]	600	704	1260
111	1-o-anisidino-2-propanone[(d)]	434	477	532
112	1-Anilino-3-(2-propynyloxy)-2-propanol[(d)]	900	922	1055
113	2-Anilino-N,N-diethylpropionamide[(d)]	1360	732	1248
114	3-Anilino-N,N-diethylpropionamide[(d)]	1100	844	1376
115	2-(p-aminophenol)-2-phenyl propionamide[(d)]	260	746	238

[a] Observed toxicity, from RTECS database [5].
[b] Toxicity, Calculated from Model II, neural network
[c] Toxicity, Calculated from Model III
[d] Member of the Training Set
[e] Member of the Cross Validation Set for CNN, Training Set for Linear Model
[f] Member of the Prediction Set
[g] Assigned weight of zero by least median squares regression

prediction set. It is important to note that the training set was used to develop regression models, and that the external prediction set was used only to validate the models. A low rms error for external prediction indicates that the model would be useful for predicting the toxicity of unknowns of similar structure. A cross validation set of 11 compounds was randomly chosen from the training set for use in preventing the overtraining of the computational neural networks. This will be discussed in more detail below.

Structure Entry and Modeling. The 115 compounds in the data set were sketched and entered into HYPERCHEM [19] where they were stored as connection tables. After entry, the semi-empirical molecular orbital program, MOPAC [20], was used to optimize
the geometry of each compound using the PM3 Hamiltonian [21]. As some of the descriptors calculated are geometry-dependent, accurate three dimensional structures were needed.

Descriptor Generation. Once the geometry of each structure was optimized, it was possible to calculate the descriptors. The descriptors are meant to numerically encode the important features of the compounds, and they can be classified as topological, electronic and geometric, or hybrids of these three categories. Topological descriptors range from simple features, such as atom counts and molecular weights, to somewhat more complicated features like molecular connectivity [22] and kappa indices [23], which encode information about size and shape. Electronic properties of the data set are encoded in the electronic descriptors. These include features such as HOMO and LUMO energies and dipole moments. The geometric descriptors calculated included the principal moments of inertia, molecular volume and surface area. In addition, several descriptors were calculated that are combinations of the classes listed above. These descriptors encoded features such as hydrogen bonding and the potential for other intermolecular interactions. In total, 166 descriptors were calculated for each compound in the data set.

Descriptor Reduction. In order to reduce the number of features considered during model building, the descriptors were analyzed using several objective tests. Those descriptors that were found to contain 90% or more identical values were removed from the descriptor pool as they contained little useful information. Following this, descriptors were tested for pairwise correlations. Descriptors with pairwise correlation exceeding 0.90 were investigated, and one of the two descriptors was removed from the pool. The descriptor pool was reduced to 73 descriptors using these methods. Topliss and Edwards [24] suggest that the number of independent variables be no greater than 60% of the observations to avoid chance correlations. Since 73 descriptors (71% of the observations in the training set) remained, the pool was further reduced using vector space descriptor analysis [25].

Vector space descriptor analysis is a technique meant to reduce the size of the descriptor pool by maximizing the mutual orthogonality of the descriptor vectors. If there is no information overlap among descriptors, the vectors will be orthogonal. Therefore, choosing a reduced set of descriptors which possesses the maximum amount of mutual orthogonality will lead to a smaller pool that still contains most of the variance in the data. This is accomplished by choosing an initial basis vector (descriptor) and adding the most orthogonal vector to the subspace using a Gram-Schmidt orthogonalization technique. Next, the vector most orthogonal to the new subspace defined by the vectors in the current basis is added. This is continued iteratively until the stopping criteria is reached, in this case when the projection angle between the current descriptor vector and the current subspace fell below one degree. Following this procedure, 58 descriptors remained in the reduced pool, which is within the guidelines of Topliss and Edwards.

Model Building. Once descriptor reduction was complete, three types of models were developed. Type I models, or linear models, were developed using feature selection routines which use simulated annealing [26] and genetic algorithms [27] to find a small, information rich subset of descriptors via a directed search of the reduced feature space. Both routines use the rms error from an ordinary least sum of squares multiple linear regression to evaluate the fitness of each subset of descriptors. For reasons discussed below, the descriptor pool was later screened using a feature selection routine based on simulated annealing and the rms error from the least median of squares [28] multiple linear regression.

Once a suitable Type I model had been found, a Type II model was developed using the same subset of descriptors. The subset of descriptors found in the best Type I model were submitted to computational neural networks in an attempt to improve the prediction over that of the linear model. The computational neural networks were trained using a quasi-Newton update function, and a cross validation set was used to guard against network overtraining. External prediction was performed for compounds not included in training set. A favorable rms error for the external prediction set is an indication of the utility of the model for use in predicting future unknowns similar in character to the training data.

In addition to the linear feature selection routines, the reduced descriptor pool was also submitted to a feature selection routine using the genetic algorithm which uses a CNN to evaluate the fitness of a subset of descriptors. It is believed that such a routine may find subsets of descriptors that are more suited to take advantage of the non-linear nature of computational neural networks. The models developed using features chosen using this criteria are known as Type III models. The routine attempts to find a small, information rich subset of descriptors using a directed search based on the rms error from a computational neural network training. The starting point for each CNN training was determined using an optimization routine based on generalized simulated annealing [26]. Additionally, a hash code library was used to avoid training identical subsets multiple times, thus speeding the algorithm. As with the Type I and Type II models, external prediction was performed with the trained Type III model.

Descriptor generation, model development, and all statistical analysis was performed using the Automated Data Analysis and Pattern Recognition Toolkit (ADAPT) developed in our laboratory [29,30].

Results and Discussion. - *Type I (Linear) Model Construction.* Many potential linear models were investigated using the evolutionary feature selection routines based on simulated annealing (SA) and the genetic algorithm (GA). The regression model containing the smallest subset of descriptors with the most favorable F value, multiple correlation coefficient, R, and rms error was chosen for further investigation.

The model (labeled model 1) chosen using these routines is shown in *Table 3.* As the statistics demonstrate, the overall quality of this model is low. *Figure 1* shows a plot of the calculated $\log(LD_{50})$ vs. observed $\log(LD_{50})$ for the training and prediction sets

Table 3. *Regression Synopsis for Features Selected Using Ordinary Least Squares Criteria (Model 1)*

Coeff.	S.E. Coeff.	Label	Description
6.673	1.116	CONS	Y-intercept
-2.552	0.7039	ALLP-1	Number of self-avoiding connected paths[31]
7.522×10^{-2}	1.684×10^{-2}	NAB	Number of aromatic bonds
0.8095	0.1627	QPOS	Charge on the most positive atom[32]
-4.319×10^{-2}	8.974×10^{-3}	RNCS	Relative negative charged surface area[33]
0.4096	0.1226	V5PC	5th Order Path-Cluster Valence Molecular Connectivity Index[22]

F = 10.590 R = 0.60 N = 98 RMS = 0.310

using this model. There are two important observations that should be made when interpreting this plot. First, there is wide scatter in the plot, with few points actually lying on the 1:1 correlation line. Second, the general trend in the data is somewhat horizontalacross the plot rather than the expected positive slope. Although the plot represents a 1:1 correlation line rather than a regression line, this horizontal structure clearly represents a poor fit.

The model was subjected to several statistical tests designed to detect outliers: residual, standardized residual, studentized residuals, leverage, DFFITS, and Cook's Distance [34]. From these tests, several compounds were considered outliers. When these compounds were removed, and a new regression model was calculated, the rms error improved. However, the overall statistical quality of the model did not significantly improve. Furthermore, it was noted that several compounds not flagged as outliers also had high leverage values. Removal of these compounds from the model led to model coefficients which did not differ significantly from zero. In other words, descriptors that were found to be important when the high leverage data were in the model were unnecessary when these compounds were removed.

As is generally well-known, least sum of squares regression (i.e., ordinary least squares) is very sensitive to points with high leverage. Both the SA and GA feature selection routines used the rms error from the ordinary least sum of squares regression to evaluate the fitness of a subset of descriptors. For this reason, it was theorized that the presence of data with high leverage may have detrimentally influenced the feature selection. The method known as least median of squares, developed by Rousseeuw and Leroy [28], attempts to determine the best regression equation while de-emphasizing data points with high leverage.

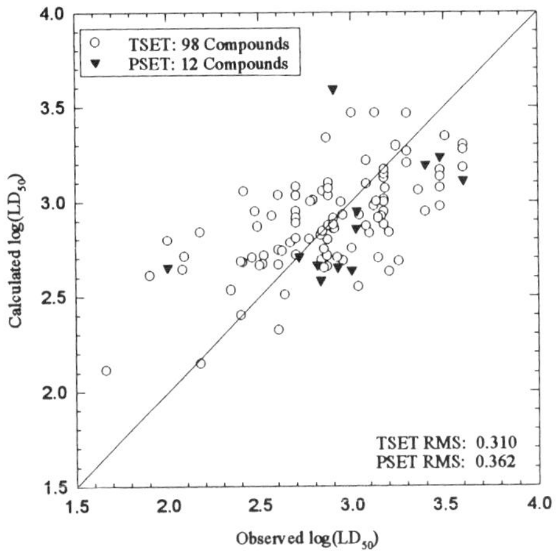

Figure 1. *Calculated log(LD$_{50}$) versus Observed log(LD$_{50}$) for Model 1 Chosen Using Ordinary Least Squares (Model 1)*

In an attempt to perform feature selection that was not overly influenced by compounds with high leverage, a feature selection routine was created which uses least median of squares (LMS) to evaluate the fitness of a subset of descriptors found using simulated annealing. The LMS algorithm works by minimizing the median squared residual through an iterative procedure. For a model containing k independent variables (descriptors), k+1 random observations are chosen from the data. A regular least sum of squares regression is performed using only these k+1 observations, however the squared residuals are calculated for *all* n observations. The median squared residual is maintained as the cost function and a new subset of k+1 observations are chosen randomly. This procedure is repeated many times. For the subset of observations with the lowest median squared residual, all observations with a standardized residual greater than 2.5 are given a weight of zero. All other observations are assigned a weight of one. A weighted least squares regression is performed and the rms error is calculated.

Due to the iterative nature of this algorithm, regression using the LMS method is done at high computational expense. In an effort to improve the computational speed of the algorithm, the k+1 observations are selected using a directed search (simulated annealing). A similar approach has been published recently using the genetic algorithm [35,36]. This least median squares algorithm was then used to evaluate the fitness of subsets of descriptors in a feature selection routine. In the feature selection routine, the rms error from the weighted least squares regression is used as the cost function for the SA routine [26], and the LMS regression is repeated for each iteration of the SA algorithm.

Using this feature selection routine (LMS-SA), many descriptor subsets were investigated. Each subset of descriptors resulted in a different set and number of compounds being given a weight of zero. Depending on the features chosen, the number of compounds weighted zero ranged from 5 to 21. Again, the model with the most favor-

Table 4. *Regression Synopsis for Features Selected Using Least Median Squares Criteria (Model 2)*

Coeff	S.E. Coeff	Label	Description
1.491	0.2290	CONS	Y-intercept
0.8280	0.1587	QPOS	Charge on the most positive atom
-2.187×10^{-2}	8.422×10^{-3}	RNCS	Relative Negative Charged Surface Area
1.159×10^{-3}	1.787×10^{-4}	VOL	Solvent Accessible Volume
-7.864×10^{-2}	3.048×10^{-2}	CTAA	Count of H-bond Acceptors
3.899	0.9233	CHDH-2	Sum of Charges on Donatable Hydrogens/ Number of Donatable Hydrogens

F = 18.719 R = 0.710 N = 98 RMS = 0.269

able F-value, R, and rms error, as well as a consideration of the number of compounds assigned a weight of zero, was chosen for further investigation. The best overall set of features chosen using the LMS-SA selection routine are shown in *Table 4*. The LMS method weighted 5 points as zero for the weighted least squares: p-cyanoaniline (Compound **17**, *Table 2*), p-(2'-cyclohexenyl) aniline (**18**), p-(3'-phenylpropoxy) aniline (**48**), 2,4,6-trichloroaniline (**52**), and 4-(2'-cyclohexylethoxy)-m-anisidine (**71**). These five compounds were among the compounds noted as having high leverage in Model 1.

Removal of the zero weighted compounds from the model (model 2) resulted in a significant improvement in the statistics of the model, which is shown in *Table 4*. As a further test, the variance inflation factor (VIF) was calculated for each descriptor in the model to test for the presence of multicollinearities [37]. The VIF is $1/(1-R^2)$, when each independent variable is regressed onto the others. None of the descriptors had a VIF that exceeded the recommended cutoff value [37] of 10.

It is interesting to note that two of the descriptors chosen by the LMS-SA routine were also chosen by the feature selection routines based on ordinary least squares regression. In addition, a third descriptor, the solvent accessible volume, encodes some information regarding the size of the molecule, as do the connectivity descriptors found in Model 1. These observations do not seem unreasonable, as it is likely that the leverage points affected the selection of some descriptors in an effort to bring the regression line closer to these data points.

Figure 2 is a plot of the calculated $\log(LD_{50})$ vs. observed $\log(LD_{50})$ for the training and prediction sets. Although the plot still shows a fair amount of deviation from the 1:1 correlation line, the general characteristics of the plot are significantly improved over that found for Model 1. Furthermore, the removal of any of the observations that were identified as having large leverage values did not affect the statistical quality of the model. Finally, the model was used for external prediction, resulting in an rms error of 0.274 log units for the 12-member external prediction set.

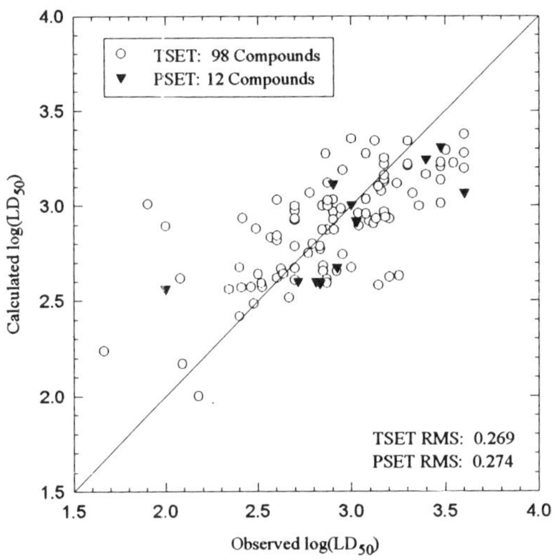

Figure 2. *Calculated $\log(LD_{50})$ versus Observed $\log(LD_{50})$ for Model 2 Selected Using Least Median Squares Criteria*

Type II Model Construction. These five descriptors were then submitted to computational neural networks, in an attempt to further improve the model. When used in a

QSAR study, a computational neural network can be thought of as a non-linear modeling technique. The training set used for development of the linear model was split randomly into an 85 member training set (TSET) and an 11 member cross validation set (CVSET). The same external prediction set (PSET) used in the development of the linear model was used for validation of the neural network. Many different network architectures were investigated, and the network architecture with the lowest rms error and the fewest adjustable parameters (weights and biases) was used. The network used was a fully-connected, three-layer, feed-forward network with a 5-3-1 architecture (22 adjustable parameters). To avoid chance effects, it is recommended [38] that the ratio of observations to adjustable parameters be over 2.0, and this network architecture fulfills that requirement.

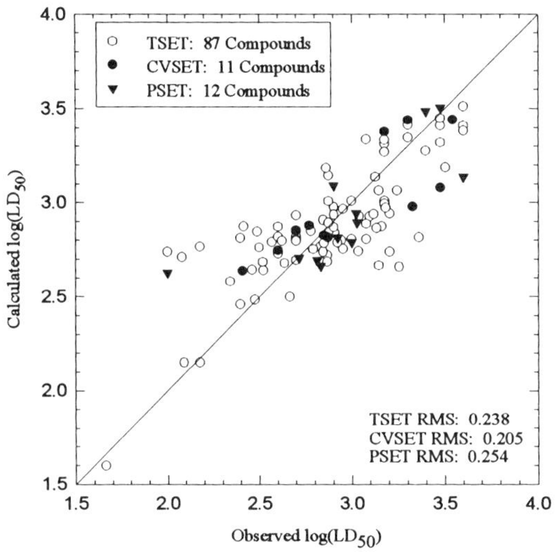

Figure 3. *Calculated log(LD_{50}) versus Observed log(LD_{50}) for Model 2 Chosen Using LMS, Computational Neural Network Training*

The computational neural networks used in this study were trained using a quasi-Newton method (BFGS) [39] to update the weights and biases. To prevent overtraining of the network, the CVSET error was monitored throughout training. Training was halted when the rms error of the CVSET no longer decreased but began to rise. At this point it was assumed that the network was learning only idiosyncrasies of the training set members. Since the final result is highly dependent upon the starting weights and biases of the network, training was performed several times using different starting weights and biases. The best set of weights and biases was used for prediction of the external prediction set. *Figure 3* shows a plot of the calculated log(LD_{50}) versus observed log(LD_{50}) for the 5-3-1 computational neural network. The training set rms error was 0.238 (mean absolute error 0.178), an improvement of 12% over the linear model. The prediction set rms was 0.254 (mean absolute error 0.186), a 7.3% improvement over the linear model.

The improvement of the computational neural network model is due to two factors: the increased number of adjustable parameters, and the ability of the network to exploit non-linearities. The calculated log(LD_{50}) values resulting from this model are presented in *Table 2*.

Although the descriptors found using the feature selection routines do not represent causal relationships, they can be helpful in furthering the understanding of the toxic mechanism of the compounds. The final model found using linear methods of feature selection in this study contained five descriptors (see *Table 4*): one electronic, one geometric, and three charged-partial surface area (CPSA) descriptors [33]. The electronic descriptor, the charge on the most positive atom (QPOS), could be interpreted as indicating the presence of various functional groups. The solvent-accessible volume [40] was calculated using water as the solvent of interest. This descriptor (VOL) is encoding information regarding the apparent size of the molecule, which may be an important factor in such processes as transport across membranes in the body, or certain receptor events.

The CPSA descriptors combine electronic and geometric information in an attempt to encode the features of molecules that may be responsible for intermolecular interactions. Three such descriptors were found to be important in this study. The first CPSA descriptor, the relative negative charged surface area (RNCS), also helps to differentiate among the various functional groups present, as well as encoding information regarding the ability of those functional groups to participate in intermolecular interactions. This may help encode information regarding the molecules affinity for binding sites, or other interactions, which may be important in the toxic mechanism. The remaining two CPSA descriptors encode information regarding hydrogen bonding [33,41]. The descriptor CTAA is a simple count of atoms in the molecule that are capable of accepting hydrogen bonds. The second (CHDH-2) is the sum of the charges on all donatable hydrogens divided by the number of donatable hydrogens.

Type III (CNN) Model Construction. In an attempt to further exploit the non-linear behavior of computational neural networks, a feature selection routine (GENDES) has been developed which uses CNN's to evaluate the fitness of subsets of descriptors chosen by the genetic algorithm [42]. The genetic algorithm performs a directed search of the descriptor space using a Darwinian inspired evolutionary method. A random population of subsets is chosen, and the fitness of each subset is evaluated, in this case by a CNN. The best population members are chosen for mating and crossover, and the fitness of the new subsets are evaluated. Using this procedure, many subsets of descriptors are investigated. In addition to the mating procedure, random mutations can also occur in the subset strings. These mutations are used to keep the algorithm from becoming trapped in local minima on the response surface. The mutations occur by randomly removing one descriptor from a subset, and replacing it with a new randomly-chosen descriptor.

The fitness of each subset chosen by the genetic algorithm is evaluated by training a three-layer, fully-connected, feed-forward computational neural network using the BFGS algorithm. As the results of any training of a CNN is dependent upon the starting point on the response surface, and multiple trainings of a given subset would be prohibitively expensive, a generalized simulated annealing routine is used to find a reasonable starting point from which to perform the training [26]. A hash code library is also used, to avoid training a subset of descriptors a second time, as this would be overly time consuming.

Table 5. *Features Selected Using Computational Neural Networks as the Fitness Criteria (Model 3)*

Label	Description
V5C	5th Order valence cluster molecular connectivity
FLEX	Flexibility Index
SEMN	Minimum electrophilic superdelocalizability
FPSA-3	Atomic charge weighted fractional positive surface area
CHDH-3	Sum of the charges on donatable hydrogens/total molecular surface area

Many subsets of descriptors were investigated using this method, and the subsets which had acceptable rms errors for both the training set and the cross validation set were retained for further investigation. The subsets chosen were then trained using the BFGS algorithm, as explained above. Graphical analysis of the calculated versus observed plots for the training set was used to select the final model. The subset of descriptors with the best behaved calculated versus observed plot was then used for external prediction. The descriptors chosen by the genetic algorithm feature selection routine

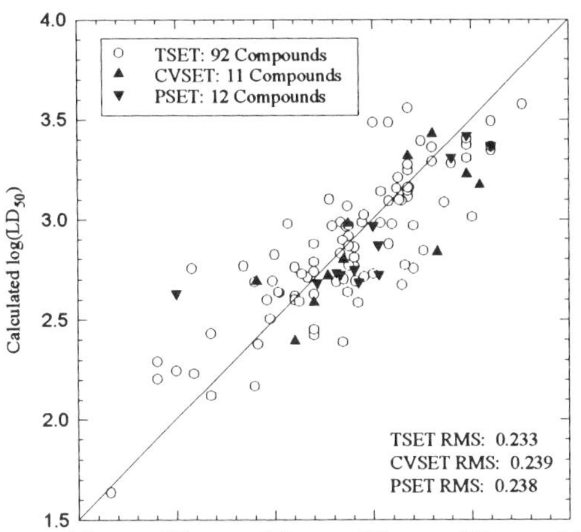

Figure 4. *Calculated log(LD_{50}) versus Observed log(LD_{50}) for Model 3 5-3-1 Neural Network*

are shown in *Table 5*. *Figure 4* shows the calculated log(LD_{50}) versus observed log(LD_{50})for this model. Although the rms error for the training set (0.233 log units, mean absolute error 0.176) was not significantly improved over the rms error found for the training set using the LMS-SA routine, it is important to realize that, unlike the LMS-SA model, no compounds were removed from the training set. In addition, the prediction set rms (0.238 log units, mean absolute error 0.177) was improved over the LMS-SA model. Again, the calculated values for this model are presented in *Table 2*.

As for the linear model, the descriptors used in this model do not represent a causal relationship between structure and the acute mammalian toxicity. The descriptors chosen by GENDES encode information that is somewhat similar to that encoded by the descriptors chosen by the LMS-SA routine. The model contained two CPSA descriptors (FPSA-3 and CHDH-3), one electronic descriptor (SEMN), and two topological descriptors. Although none of the descriptors are pure geometric descriptors, an accurate three-dimensional geometry is important for the calculation of the CPSA descriptors.

Although the complex nature of CNNs makes the process of descriptor interpretation somewhat more difficult, it is still possible to gain some understanding of the toxic mechanism from them. The atomic charge weighted fractional positive surface area (FPSA-3) encodes information regarding the presence of functional groups in the molecule. For example, the presence of an electron withdrawing group creates a region of positive charge on the atom to which it is connected. The addition of atomic charge weighting increases the information content of the descriptor as different functional groups create regions of positive charge which differ in magnitude. Quantification of fractional accessible surface area of weighted positive charge may also encode information regarding the ability of the molecule to participate in intermolecular interactions [33]. In many respects the information encoded by FPSA-3 is similar to that contained in the RNCS descriptor in the linear model (model 2).

As in the linear model, a descriptor which encodes information regarding hydrogen bonding (CHDH-3) is also present in the Type III model. This descriptor encodes information regarding the ability of a compound to participate in solute-solvent interactions, when the solvent is capable of hydrogen bonding (for example, water) [33,41]. To some degree, this descriptor may be encoding the presence of N-substitution of the aniline molecule. The amino hydrogens are, for many of the compounds in the data set, the only donatable hydrogens present. In any case, the presence of hydrogen bonding could have significant implications in any biological event.

Two topological descriptors are present in the model. The first is the valence path-5 cluster molecular connectivity descriptor [22]. This descriptor encodes information regarding the size and degree of branching present in the molecule, while also containing information regarding bond types present in the molecule due to the valence correction. The second topological descriptor is derived from Kier's index of flexibility descriptor [43]. The flexibility index used here is defined as ${}^1\kappa_\sigma{}^2\kappa_\alpha\,(MW)^{-1/2}$, where ${}^1\kappa_\sigma$ and ${}^2\kappa_\alpha$ are the first and second atom corrected kappa indices, and MW is the molecular weight. Essentially, this descriptor encodes information regarding cyclicity, spatial density of the molecule, and molecular size. Smaller values of FLEX indicate a lower flexibility, or a more rigid molecule. A cursory glance at the data set used in this study shows many compounds with long side chains, and/or a high degree of branching. It is possible that this descriptor is encoding information regarding receptor events in which a compound may or may not be able to participate.

The only electronic descriptor present in the model is the minimum electrophilic superdelocalizability (SEMN) [44]. Electrophilic superdelocalizability is a measure of the probability of a particular atom in a compound being subjected to an electrophilic attack. Derived from MO theory and frontier electron theory [45], electrophilic superdelocalizability is a measure of the concentration of occupied orbitals at each atom. This allows comparison between corresponding atoms in different molecules, probing potential reactive sites [46]. The descriptor stored here is the value for the atom with the lowest electrophilic superdelocalizability. Essentially, this descriptor is a measure of the reactivity of the compounds which may have important consequences in the toxic mechanism.

Investigation of Functional Dependence. The improved results yielded by computational neural networks are typically ascribed to two factors. Computational neural networks have more adjustable parameters than are present in multiple linear regression for the same number of descriptors. In addition, computational neural networks utilize non-linear relationships between descriptors and the activities being predicted. This ability to utilize non-linearity may be demonstrated by a functional dependence plot for a fully trained neural network [42,47,48]. A functional dependence plot is generated by varying one input while the others are held constant. The scaled output of the CNN is then plotted against the scaled input.

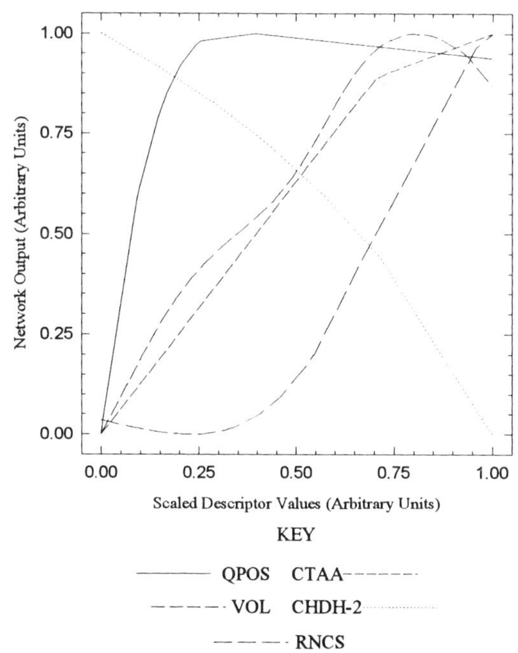

Figure 5. *Functional Dependence Plot of Descriptors chosen using linear criteria (model 2)*

The functional dependence plot for model 2 is shown in *Figure 5*. These descriptors were chosen using a linear selection criteria. As can be seen in *Figure 5*, three descriptors (VOL, CTAA, CHDH-2) behave in a nearly linear fashion. The remaining two descriptors do indeed show some non-linear functional dependence. A similar plot is shown in *Figure 6* for the descriptors in model 3, which were chosen using the rms error from a computational neural network as the selection criteria. The behavior shown in this plot is decidedly more non-linear than that shown for the descriptors chosen using linear methods. The non-linear feature selection routine did select descriptors that have inherent non-linear relationships to the acute toxicity.

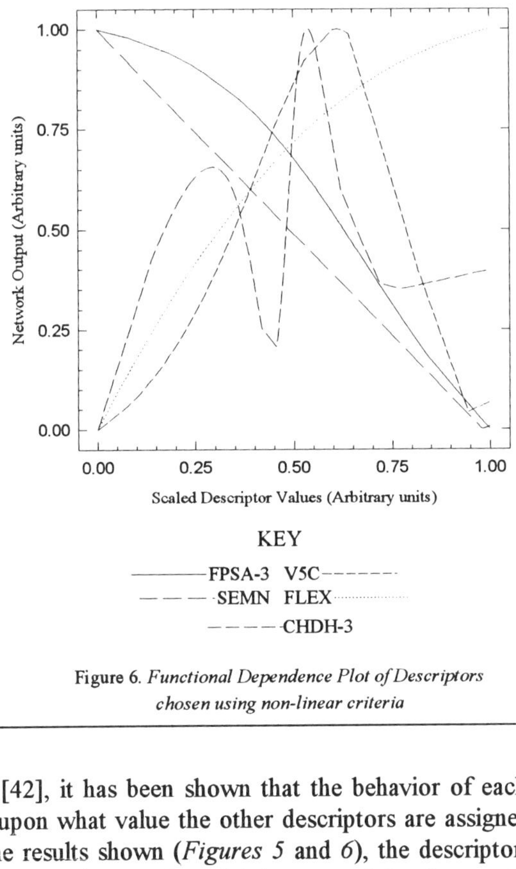

Figure 6. *Functional Dependence Plot of Descriptors chosen using non-linear criteria*

Recently [42], it has been shown that the behavior of each descriptor is somewhat dependent upon what value the other descriptors are assigned while they are held constant. For the results shown (*Figures 5* and *6*), the descriptors held constant were assigned a value 0.5. When a value of 0.1 was used for the constant, the particular response did, in fact, change (not shown). The degree of non-linearity, however, was not significantly altered from the results presented here.

Conclusion. - A five-descriptor model was developed which accurately predicts acute mammalian oral toxicity for an extremely diverse set of substituted anilines. In contrast to other QSAR toxicity studies, no experimentally derived descriptors were used in model development. The ability to estimate acute mammalian toxicity would have many benefits, including increased speed, and decreased cost. Computational neural networks improved the linear model, resulting in an rms error of 0.238 log units for the training set, and 0.254 log units for the external prediction set. It is believed that this improvement stems from the non-linear nature of the CNN, as well as the increased number of adjustable parameters. A computational neural network trained using features chosen to exploit non-linearities resulted in a training set rms error of 0.233 log units and an rms error of 0.238 log units for the external prediction set. The low rms error for the external prediction set indicates a high future utility for predicting the toxicity of compounds similar to those found in the training and prediction data. Feature selection was performed using a variety of methods to evaluate the fitness of subsets of descriptors chosen by evolutionary optimization routines. The LMS algorithm attempts to de-emphasize data with high leverage, although due to its iterative nature it does so at high computational cost. In the future, it is possible that the LMS algorithm could be used to help identify compounds that are too dissimilar from the bulk of the training set to be used in model construction. It is believed that this would be very useful when investigating highly diverse, or highly erroneous, data sets, such as the one used in this study. Additionally, a feature selection routine which uses computational neural networks to evaluate the fitness of subsets of descriptors chosen using the genetic algorithm was employed. It is believed that subsets chosen using this algorithm may be able to better take advantage of the non-linear relationships due to the nature of computational neural networks. For complex problems, such as mammalian toxicity, this may result in significantly improved prediction.

REFERENCES

[1] K. Sun, G. F. Krause, F. L. Mayer Jr., M. R. Ellersieck, A. P. Basu, *Environ. Toxicol. Chem.* **1995**, *14*, 1745.

[2] H. Jäckel, W. Klein, *Quant. Struct.-Act. Relat.* **1991**, *10*, 198.

[3] M. T. D. Cronin, J. C. Dearden, *Quant. Struct.-Act. Relat.* **1995**, *14*, 117.

[4] K. Enslein, T. R. Lander, M. E. Tomb, P. N. Craig, P. N. 'A Predictive Model for Estimating Rat Oral LD50 Values', Princeton Scientific Publishers, Princeton, 1983.

[5] Registry of Toxic Effects of Chemical Substances (RTECS). Compiled by the National Institute of Occupational Safety and Health, USA, 1993.

[6] T. W. Schultz, D. T. Lin, L. M. Arnold, *Sci. Total Environ.* **1991**, *109/110*, 569.

[7] A. Harada, M. Hanzawa, J. Saito, K. Hashimoto, *Eniron. Toxicol. Chem.* **1992**, *11*, 973.

[8] M. Nendza, C. L. Russom, *Xenobiotica* **1991**, 21, 147.

[9] T. W. Schultz, M. Cajina-Quezada, S. K. Wesley, *Bull. Environ. Contam. Toxicol.* **1989**, *43*, 564.

[10] V. J. Furay, S. Smith, *Bull. Environ. Contam. Toxicol.* **1995**, *54*, 36.

[11] L. D. Newsome, D. E. Johnson, R. L. Lipnick, S. J. Broderius, C. L. Russom, *Sci. Total Environ.* **1991**, *109/110*, 537.

[12] L. M. Arnold, D. T. Lin, T. W. Schultz, *Chemosphere* **1990**, *21*, 183.

[13] M. Nendza, *Chemosphere* **1991**, *22*, 613.

[14] H. Tanii, *Toxicology Letters* **1995**, *76*, 27-30.

[15] J. M. Sutter, P. C. Jurs, *J. Chem. Inf. Comp. Sci.*, **1996**, *1*, 100.

[16] M. D. Wessel, P. C. Jurs, *Anal. Chem.*, **1994**, *66*, 2480.

[17] M. D. Wessel, P. C. Jurs, *J. Chem. Inf. Comp. Sci.*, **1995**, *5*, 841.

[18] B. E. Mitchell, P. C. Jurs, *J. Chem. Inf. Comp. Sci.*, **1996**, *1*, 58.

[19] Hypercube, Inc., Waterloo, ON.

[20] Stewart, J. P. P. Mopac 6.0, Quantum Chemistry Program Exchange; Indiana University, Bloomington, IN, Program 455.

[21] J. P. P. Stewart, *J. Comput.-Aided Mol. Des.* **1990**, *4*, 1.
[22] L. B. Kier, L. H. Hall, in 'Molecular Connectivity in Structure-Activity Analysis', Research Studies Press Ltd., John Wiley & Sons, **1986**, 18-20.
[23] L. B. Kier, *Quant. Struct.-Act. Relat. Pharmacol. Chem. Biol.*, **1985**, *4*, 109.
[24] J. G. Topliss, R. P. Edwards, *J. Med. Chem.* **1979**, *22*, 1238.
[25] C. J. Russell, S. L. Dixon, P. C. Jurs, *Anal. Chem.* **1992**, *64*, 1350.
[26] J. M. Sutter, P. C. Jurs, in 'Data Handling in Science and Technology (Vol 15). Adaption of Simulated Annealing to Chemical Optimization Problems', Ed. J. H. Kalivas, Elsevier: Amsterdan, 1995, Chapter 5.
[27] B. T. Luke, *J. Chem. Inf. Comput. Sci.* **1994**, *34*, 1279.
[28] P. J. Rousseeuw, A. M. Leroy, 'Robust Regression and Outlier Detection' John Wiley & Sons: New York, 1987.
[29] A. J. Stuper, W. E. Brugger, P. C. Jurs, 'Computer-Assisted Studies of Chemical Structure and Biological Function' Wiley Interscience: New York, 1979.
[30] P. C. Jurs, J. T. Chou, M. Yuan, in 'Computer-Assisted Drug Design', Eds. E. C. Olsen and R. E. Christoffersen, The American Chemical Society, Washington, DC, 1979; pp 103-129.
[31] M. Randic, *Comput. Chem.*, **1979**, *3*, 5.
[32] S. L. Dixon, Ph.D. Thesis, The Pennsylvania State University, Aug. 1994, Chapter 4.
[33] D. T. Stanton, P. C. Jurs, *Anal. Chem.* **1990**, *62*, 2323.
[34] D. A. Belsey, E. Kuh, R. E. Welsch, 'Regression Diagnostics', Wiley, New York, 1980.
[35] D. L. Massart, *Engineering Applications of Aritificial Intelligence*, **1995**, *8*, 177.
[36] C. L. Karr, B. Weck, *Fluid/Particle Separation Journal*, **1995**, *8*, 117.
[37] D. T. Stanton, P. C. Jurs, *J. Chem. Inf. Comp. Sci.* **1991**, *31*, 301.
[38] D. J. Livingstone, P. T. Manallack, *J. Med. Chem.* **1993**, *36*, 1295.
[39] L. Xu, J. W. Ball, S. L. Dixon, P. C. Jurs, *Environ. Toxic. Chem.*, **1994**, *13*, 841.
[40] R. S. Pearlman, in 'Physical Chemical Properties of Drugs', Eds. S. H. Yalkowsky, A. A. Sinkula, and S. C. Valvani, Marcel Dekker, Inc., New York, 1980, Chapter 10.
[41] S. N. Vinogradov, R. H. Linnell, 'Hydrogen Bonding', Van Nostrand Reinhold, New York, 1971.
[42] M. D. Wessel, P. C. Jurs, in 'Applications of Neural Networks in Chemistry', Ed. J. Gasteiger, American Chemical Society, Washington D.C., In Press.
[43] L. B. Kier, in 'Progress in Clinical and Biological Research (Vol. 291). QSAR: Quantitative Structure-Activity Relationships in Drug Design', Ed. J. L. Fauchère, Alan R. Liss, Inc., New York, 1989, pp. 105-109.
[44] K. Fukui, T. Yonezawa, C. Nagata, *Bull. Chem. Soc. Japan,* **1954**, *27*, 423.
[45] K. Fukui, T. Yonezawa, H. Shingu, *J. Chem. Phys.*, **1952**, *20*, 722.
[46] O. Kikuchi, *Quant. Struct.-Act. Relat.*, **1987**, *6*, 179.
[47] S.-S. So, W. G. Richards, *J. Med. Chem.*, **1996**, *35*, 3201.
[48] T. Suzuki, M. Ishida, *Pharm. Sci.*, **1995**, *1*, 297.

PART I. **CHEMOMETRICS**

Chapter 3

Modeling without Boundary Conditions: An Issue in QSAR Validation

by **Alessandro Giuliani***[a]) and **Romualdo Benigni**[b])

[a])Institute for Research on Senescence Sigma-Tau, Via Pontina Km 30.400, I-00040 Pomezia (Roma)
[b])Instituto Superiore di Sanità, Viale Regina Elena 299, I-00161 Roma

Modeling without Boundary Conditions: An Issue in QSAR Validation

by **Alessandro Giuliani***

Institute for Research on Senescence Sigma-Tau, Via Pontina Km 30.400, I-00040 Pomezia (Roma)

and **Romualdo Benigni**

Instituto Superiore di Sanità, Viale Regina Elena 299, I-00161 Roma

Some methodological issues related to QSAR are investigated. Particular emphasis is put on the scaling behavior and the context dependence of QSAR models. One of the main problems with QSAR is the passage from the general statistical properties of the models to the local predictions at the single compound level: the paper highlights some traps hidden in this crucial step and sketches some theoretical solutions.

Introduction. - QSAR models are mathematical models, i.e. representations of a particular phenomenon using quantitative relations between its component parts [1].

The character of Modeling heavily depends on the relative "hard" or "soft" character of the domain of application [2]. The location of QSAR at the interface between "hard" and "soft" sciences [3], while making this field very interesting and stimulating, creates several methodological problems. These problems have important consequences both for the construction of reliable predictions of biological activity starting from physicochemical information, and, more generally, for the modeling of complex systems like biological systems [2], [4], [5], [6].

In this paper we will try to sketch a general theoretical framework relative to some methodological problems posed by the QSAR procedures.

Hard and Soft. - Traditionally, sciences may be grouped into two categories: "hard" sciences, where repeatable events can be predicted from a mathematical formalism expressing laws of nature; and "soft" sciences of complex systems, where in many occasions only a narrative account of distinguishable events is possible [7].

While mathematical modeling of hard sciences, like physics, is generally based upon differential equations, this style of modeling can hardly be applied to soft sciences like biology [8], [2].

Modeling through functional laws, typical of physics, is based on a number of requirements that are seldom satisfied in biology. Particularly important requirements are 1) previous identification of the objects that play a role in the phenomenon under study and 2) identification of the boundary conditions and of the applicability domain of the model. The fulfilling of the above criteria makes it possible to correctly evaluate the validity domain of an empirical finding [1], [9].

In order to relax the above constraints and then make the "soft" sciences amenable to a quantitative treatment, different strategies can be devised. The most commonly used approach consists in trying to build probabilistic models in which the analysis of the single atomisms of the system is replaced by the study of time and/or ensemble averages over a huge number of events [10], [11], [12], [6].

The probabilistic approach allows us to build models without having a thourough theoretical knowledge of the system at hand: we need only to fulfill the two above described requirements at the distributional level [6], [13].

Obviously, the information we can derive from such kinds of modeling only holds at the distributional level: we can not describe the system at the level of the single atomisms.

Examples of such kinds of modeling in biological sciences are: the analysis of time dependent signals like ECG [14], [15] or EEG [16], population genetics [17], neural transmission [18], [19], quantitative ecology [20], [21].

In all these cases, the quantitative approach works very well because it is possible to identify a global phenomenological level, which is largely autonomous from the particular microscopic level realizations, and is easily described by statistical formalisms [12], [22]. In order to predict a pathological status, we only need to know the degree of variability of heartbeat over thousands of beats and not the causes underlying the duration of each single beat [14]. The problem is that the QSAR procedures do not permit such kind of modeling: we need predictions at the level of the individual compound and there is no distributional level to be analyzed. The probabilistic character inherent in the QSAR models derives from our imperfect knowledge of the studied phenomena, and does not imply any meaningful "average" observable [23], [24], [25]. In fact, while it is absolutely correct to investigate the statistical distribution of kinetic energies over an ensemble of molecules, it is devoid of any interest to define an average carcinogenic potency over a lot of different chemicals. Thus we must follow a different approach in order to approximate the conditions necessary to build a reliable quantitative model of our observations.

QSAR: Mapping Between Two Very Different Worlds. - In general mathematical terms, a QSAR procedure is a mapping: we look for some rule (either an explicit function, or an expert system, or a black-box neural model) which, given a description of an object in language A, returns us the description of the same object in language B [26], [27]. Language A is physicochemical in nature, language B is biological. Even if the two languages can have a common syntactic structure, since they are expressed in vectorial form by ordered strings representing physicochemical properties and the outcomes of biological experiments respectively, they are fundamentally different.

The physicochemical side of the mapping can be formalized at different definition levels, going from ab initio quantum-chemical features to purely topological indices of the structural formula. This goes from a maximum information density carried by the physicochemical description, hence maximum richness of mechanistic consequences derivable from the validation of a QSAR model, to a purely formal description of the molecule as that provided by an oriented graph, i.e. as a mathematical object not necessarily linked to any physical reality.

In all these cases, the link between the object (the particular molecule) and its description is totally transparent and follows the same construction rules for any molecule.

The information scientists call these descriptions "finite grammars": classical examples of finite grammars are programming languages like FORTRAN or PASCAL. Natural languages, like Italian or English, are not finite grammars: they have an open structure and there are no general rules to form meaningful words [27], [28].

The consequences of using a finite grammar are [26]:

1) The physical meaning of the descriptors is independent from the data set: there is the possibility of a context free characterization of the molecules.

2) We can transform a particular description into another one only relying on physical theory. This property can be very useful in order to compare different QSAR's making use of different physicochemical descriptions.

3) We can directly proof the generalization power of a model by designing other molecules with exactly the desired physicochemical properties.

The biological side of the mapping is completely different: given the complexity of the biological material, it appears to us more like a natural language than a formal one [7] [3], so:

1) There is no possibility of deriving context-free descriptions: the meaning of the biological descriptors is heavily dependent on unknown experimental facts and on the mechanism of action of the molecules we study [7], [29].

2) There is no general rule to go from a particular description to another: the relations between different biological properties are purely empirical and context-dependent [30].

3) There does not exist any stability domain within which we can vary the particular molecule, while being sure of not altering the biological activity.

Analogously to what happens in system analysis, where the general rate of a process is the rate of the slower step, the QSAR procedures inherit the features of the biological side of the mapping. Going ahead in this analogy, this means that, with the only remarkable exception of the efforts to improve the quantification of the general "shape" of the molecules [31], [32], there is little room for improving the potency of QSAR models working on the refinement of chemical descriptions (fast step), while a lot of things can be done on the biological side (slow step) and, perhaps, on general statistical methodology (environmental conditions).

Taming the Wild Side. - Classically, the strategies to optimize the QSAR procedures by taming the "wild" biological side of the mapping follow two approaches [23] :

1) Trying to artificially fix the context of the studied phenomenon, by limiting the QSARs to the analysis of homogeneous series of compounds.

2) Trying to reduce the amount of "biology" in the structure-activity relationship, by choosing very simple biological end-points like in vitro receptor binding studies.

Obviously the two approaches are not mutually alternative and in the great majority of cases they are simultaneously applied in QSAR science.

It is worth noting that the two approaches are indirect, so their efficacy is only statistical in character: hence it could be useful to explore the limits of these two milestones of good QSAR practice. It is important to stress the fact that we fully recognize the heuristic value of the above prescriptions, and that, exploring the limits of a principle is, in our opinion, an unescapable step to maximize its power.

Point 1) is based on the possibility to define a series in chemical terms.

There are a lot of reasonable and scientifically sound ways by which chemists build series of homogeneous compounds. All these methods are based on the definition of a common skeleton carrying the general shape and fundamental functionalities information of the series. The series elements are then generated by relatively small variations on this common theme. The goal is to keep invariant the "semantic" part of the biological message conveyed by the molecules allowing only for "syntactic" variations [23], [33], [24]. The semantic is identified by the "pharmacophore" which is considered the chemical message endowed with a biological meaning, i.e. an identifiable activity [34].

This approach restricts the QSARs to model the relatively small and continuous variations of activity in a context of presence of activity (at different degrees) for all the members of the series [35]. In a well designed series, no singular points are supposed to exist and all the members of the series must be biologically active at varying degrees. As a consequence, inactivity simply becomes the lowest limit of activity. There can only exist "asymptotic inactive molecules" but not "true inactives".

This way of reasoning is reminescent of classical information science: the biological activity of a chemical is equated to the transfer of information in a noisy channel, and QSAR is aimed at finding the best way of maximizing the fidelity of the message (measured in terms of biological response) [36]. The elements of the series correspond to slightly different copies of the same message; each copy is a sub-optimal (noisy) version of the message, the QSAR equation is the quantitative expression of the noise distribution in a certain physicochemical space.

This way of reasoning implies that a series must be such from a biological point of view: a series is a collection of chemicals all sharing the same biological message, i.e. the same mechanism of action. The boundary condition is the commonality of action mechanism and thence the presence of a biological activity (even if at different degrees) [2].

We can only infer that chemically similar compounds constitute a series from a biological point of view [29]. The biological space is not globally isomorphic with the chemical space, this results in sudden falls of the validity of a QSAR model for negligible variations in the chemical space. We have no reliable way to understand what the homology criteria (i.e. the metrics) adopted by the biological system are. The research in molecular modeling based on the simulation of receptor sites is now trying to fill this gap, but we are still very far from the solution [37], [3]. *Figure 1* gives a graphical picture of this behavior.

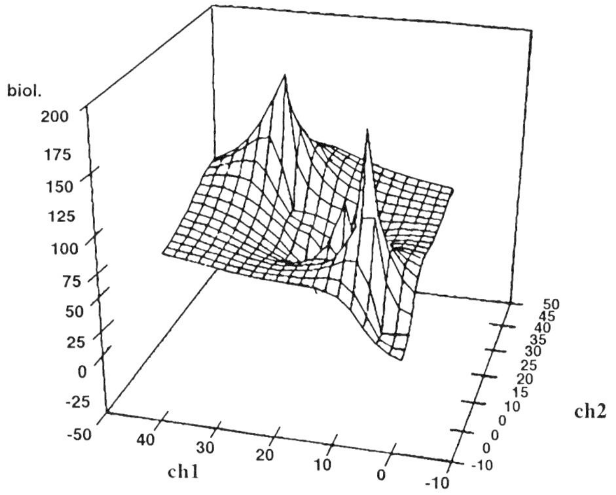

Fig. 1. *The lack of a global continuity between chemical and biological spaces gives rise to abrupt discontinuities in the structure activity space. These discontinuities make chemically similar molecules differ a lot in terms of biological activity. The biological activity surface over the chemical space appears as a rugged landscape: the chemical series is not a biological series.*
In the figure the z-axis (biol.) represents biological activity, x and y axes (che1 and che2) span a generical chemical space.

Beside the discontinuity between biological and chemical similarity, the lack of isomorphism between chemical and biological spaces can derive from the non hasymptotic character of the inactive compounds. When the constraint of the limitation of QSAR models to the modulation of activity without any singularity linked to the presence of "true inactive compounds" is violated, the model is suddenly pushed out of its validity domain.

An example coming from toxicity QSAR modeling can help to better appreciate this point [38]: Debnath et al. [39] studied the modulation of the mutagenic potency of the aromatic and heteroaromatic nitrocompounds in the TA98 and TA100 Salmonella strains. They highlighted very good and reliable QSAR models indicating a major role for log P and a secondary role for LUMO energy [39]. These models were made more cogent by the lateral validation constituted by a very similar equation obtained with a different bacterial system [40].

We checked the behavior of this well established QSAR by applying it to other compounds, pertaining to the same chemical series, but lacking any detectable mutagenic activity [38]. If the lack of activity could be considered the asymptotic limit of potency, the model had to behave like *Fig. 2a* with the lack of activity correspondent to the lower limit of potency value. In this case the chemical concept of a series had been sufficient for defining reliable boundary conditions for the QSAR equation. Otherwise, the singular, non-asymptotic character of the lack of activity, would destroy the validity of the QSAR with a consequent spread of inactive compounds over the entire domain of the relation *(Fig.2b).*

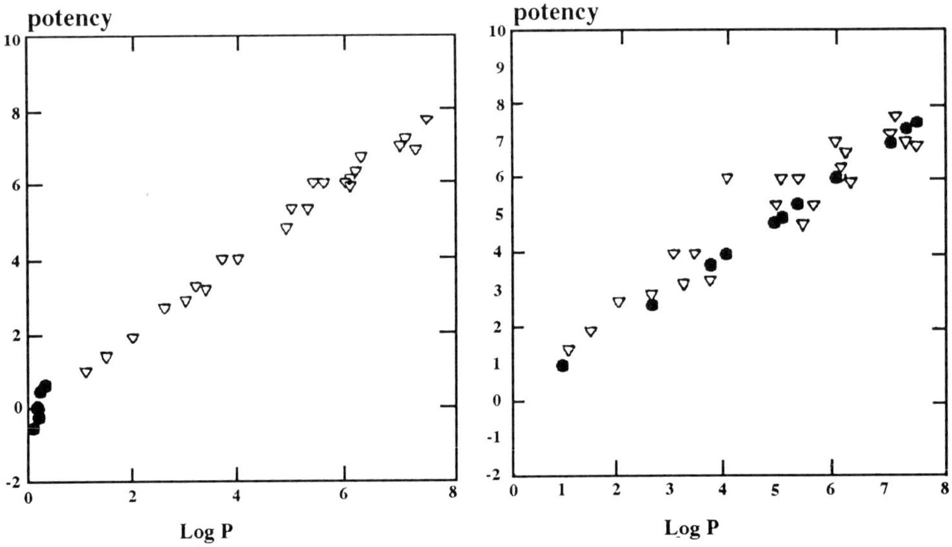

Fig. 2. *The circles represent the predictions of the model relative to the inactive compounds, the triangles are the observed values of compounds with a detectable activity. In a) the asymptotic character of lack of activity make the predictions fit well with the preexistent model. In b) the "true inactive" character of the compounds destroys the validity of the model. The qualitative aspect of activity necessitates a separate dimension with respect to the potency.*

This (*Fig. 2b*) was actually the case, demonstrating the inadequacy of a purely chemical definition of a homogeneous series and hence of the boundary conditions of a QSAR equation [38].

This is explained by recalling that the activity is exerted via one or more specific interactions with highly structured molecular material (like receptors or enzymes). Thus the biological structure imposes a metrics to the chemical space [3], [41], [42].

We can explain this behavior using a metaphor: two particles floating on the stream of a river which are nearby at time zero, will continue to stay close to each other as long as the river flows without any serious obstacle *(Fig. 3a)*, this behavior is analogous to that of a continuous relation. When the river reaches some emerging rocks, the flow is shaped by the rocks which impose a metrics to the flow, and the two particles can be set very distant apart *(Fig. 3b)*.

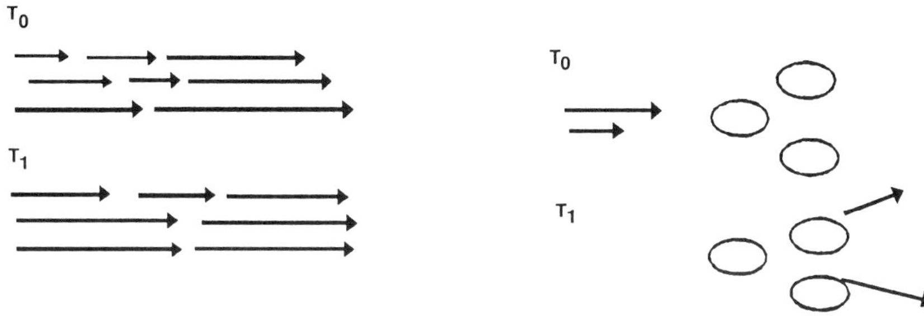

Fig. 3. *In figure a) a laminar flow is depicted. The distance between the two particles (asterisks) is invariant in time: the space is continuous.*
In b) the presence of structured material (a pattern of emerging rocks indicated as circles) imposes a metrics to the flow. The distances between the two particles heavily depends on their interaction pattern. The ordered phase (biology) dominates the relation.

This behavior is analogous to the singularities imposed on the chemical space by the interactions of the compounds with complex biological material.

In all those cases where the activity is not strictly dependendent on highly anisotropic interactions, like in ecotoxicology, the QSARs are much more well behaved, and we can relax the constraints of strictly congeneric series, being pretty sure that we will not encounter sudden interruptions in continuity. In the case of ecotoxicology, the weight of "biology" on the mapping is rather limited, and the measured activity can be approximated to a chemical reaction: the only important thing is, in many cases, the absorption characteristics of the molecule [43].

In our opinion, even very "simple" biological tests like "in vitro" binding studies have a degree of complexity sufficient for a discontinuous behavior to be manifested. Any interaction involving biological macromolecules, especially in the case of receptor molecules where the biological activity is mediated not only by a stereospecific interaction but by a consequent dynamical modification of the macromolecule induced by the drug, involves highly non-linear effects [37].

Relaxing the Mathematics. - Another way of adapting quantitative models to the complexity of biological systems is to sacrifice the transparency and the wealth of information of classical mathematical models, so as to achieve more freedom in terms of boundary conditions.

Since a large spectrum of methodological tools for generating a QSAR mapping is available, we can imagine this spectrum distributed over a space spanned by the methods transparency and robustness axes *(Fig.4)*.

The transparency axis quantifies the amount of mechanistic information obtainable from the methodology; it goes from automatic self-adapting methodologies which are, by definition, black boxes systems [44], [45], to explicit Hansch-like QSAR equations, which are fully transparent in terms of interpretation [33].

The robustness axis parametrizes the range of applicability of the method, and hence the relative freedom from boundary conditions: it goes from structural equations to unsupervised learning procedures. *Figure 4* must be interpreted in terms of general properties of the methods, independent from the particular physicochemical variables adopted for the analysis. Moreover, all the methods are imagined as applied in a cross-validation context with separate test and training sets.

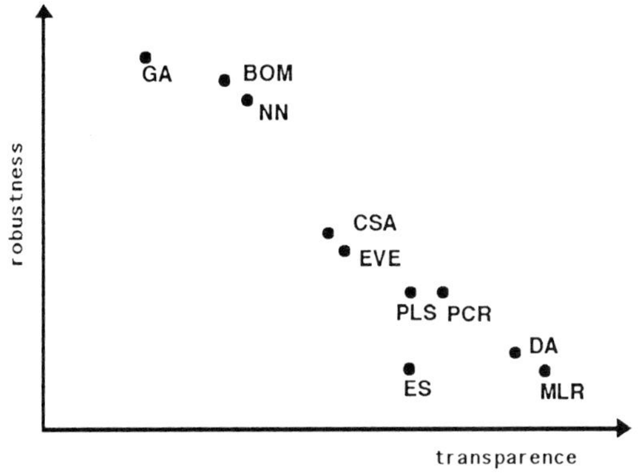

Fig. 4. *A temptative subjective classification of some statistical methodologies commonly used in QSAR is sketched. The axes represent the relative transparency (easy of intepretation) and robustness (lack of hard constraints of applicability) of the methods. The labels are to be interpreted as follows:*

CSA: Cluster Significance Analysis, DA : Discriminant Analysis, ES : Expert Systems

EVE: Evaluation of discriminating Variables in the Euclidean distance space,

PCR: Principal Components Regression, PLS: Partial Least Squares, SOM: Kohonen Self-Organizing-Maps.

It is worth noting that more than a single method can be applied in a QSAR procedure. Going back to *Fig.4*, we can imagine the Hansch approach through explicit equations as the baseline method: it is obviously an empirical approach, but it is totally transparent in terms of mechanistic interpretation. The equation directly expresses a fully quantifiable mapping between a vectorial representation using physicochemical variables, and a vectorial representation of the same objects in the biological space.

The wealth of information derivable from such an approach is balanced by the low robustness of the method: the constraints in terms of validity domain of the method are very strict. Moreover, the particular choice of physicochemical regressors deeply influences the results. It is worth noting that MLR is not a synonymous of the Hansch approach. Here, with MLR, we indicate a parametric model, defined in an absolute "external" metrics, having explicit chemical information as regressors.

A first step towards relaxing the constraints of the classical Hansch approach is the use of multivariate data analysis methods like principal components regression (PCR) or partial least squares (PLS) methods [46], [47]. Transforming the original variables into factorial indices lowers the transparency of the results but allows one to take into account the correlative structure of the data set, so obtaining a more stable result. PCR and PLS imply a different modeling strategy: while the variables deriving from a PCR procedure reflect the "natural" structure of the data, the variables of PLS are driven by the discrimination task [48]. This is the reason why, in our opinion, PCR is more "transparent" than PLS *(Fig. 4)*. The regressors of PCR are the principal components of the physicochemical description, so they reflect only the correlative structure of the data set and give an unsupervised description of the physicochemical information at hand. The regressors of PLS are driven by the task of optimizing the efficiency of QSAR model in terms of prediction of biological activity, so they do not reflect the "natural" structure of the data but constitute a supervised structuring of chemical information.

Thus with the increased power of PLS there is a resultant decrease in mechanistic information derivable from QSAR, and a less naturalistic model.

A further step away from the classical mathematical modeling style is represented by all those methods which eliminate the necessity to formalize the mapping from chemistry to biology by using an equation. All the methods based on distance spaces like Cluster Significance Analysis (CSA) [49] or Evaluation of discriminating Variables in the Euclidean distance space (EVE) [50] relax the constraint to derive a functional mapping between chemical and biological descriptions. At the basis of these methods is the partition of the chemical space into homogeneous areas (in an unsupervised way) and the search for the relative congruence of this partition with biological information.

The idea is: The higher the chemical similarity, the higher the biological similarity, whatever the functional form of the mapping.

The ultimate step toward the progressive relaxing of the constraints is constituted by the use of self-organizing dynamical algorithms like neural networks (NN), genetic algorithms (GA) and Kohonen maps (SOM) [51], [52], [53], [54].

These methods eliminate any explicit analogy between the shape of the mathematical model and the reality of interest. The problem becomes a pure pattern recognition task and the solution of the mapping relies on the trajectory of a dynamical system (the NN, GA or SOM) toward its attractor [44]. It is important to keep in mind that, even if the explicit links between the model and the phenomenon are lost, the implicit abstract links are kept alive. In fact, the formal basis of these algorithms is the same as that of classical multivariate methods, so the SOMs can be equated to cluster analyses procedures and the hidden units of back-propagation networks give a picture of the space very similar to PLS or PCR [55], [56]. But these links are not transparent to the researcher.

The expert systems (ES) are in a certain sense eccentric with respect to the other methods: in this case the mapping between chemistry and biology is more logical than mathematical in nature. There is no explicit metrics defined on biological and chemical spaces: the different chemical pieces of information are joined in a Bayesian context, generating a probability statement about a biological property [27].

The attempts to improve statistical methods in order to make biology more amenable to a quantitative treatment cannot automatically solve the problems we showed in the preceding section but may help to make them less stringent.

It is important to stress the fact that a data analysis tool is something different from a measurement device like a spectrophotometer or a microscope. Endowed in a particular technique, there are a lot of theoretical assumptions which deeply influence the results and the interpretation of a QSAR. Too often in QSAR literature there appears a plethora of new "smart" data analysis strategies which combine in fantastic ways different methodologies, with the goal towards obtaining the ultimate, "most powerful", data analysis method. The relative merits of these new approaches are judged by their performances on classical data sets in a way similar to that of an athletic tournament. Like in athletics, the differences between the methods are very small: it is possible to read that method A is better than method B because it gave a correlation coefficient of 0.92 compared to a coefficient of 0.87 on a certain data set. [25], [57].

This is not very sensible, because differences of this order of magnitude are totally out of scale with respect to the biological variability and have no informative power at all!

This is reminescent of a spectrophotometer-like approach to statistics, where the only goal is to get the most powerful instrument: the researcher applies statistics in an "external" way, instead of using it for the study of the inherent structure of the data. Statistics is a language by which we know the world by relying on the isomorphism between statistical concepts and physical reality. So a statistically sound method can not be better or worse than another one, it can only give another view to the same data set, and the relative merit depends on the nature of the problem at hand [25].

What we need is a "constitutive" use of statistics as opposed to an "external" one. If an investigator performs a principal components analysis he must know that PCA is not only a way to reduce the number of descriptors in order to avoid the risk of chance correlations, but that the obtained components are the constituting modes of the correlation structure of its data set, with a well defined chemical or biological interpretation [56], [11].

Conclusions. - QSAR science deals with approximate and purely phenomenological models: this implies that the relation between the physicochemical and the biological descriptions is indirect and based on unknown laws and principles. This also means that we can obtain many different QSARs for the same phenomenon, each of them pointing to different mechanistic explanations. There is no contradiction in this behavior, and in principle we cannot speak of a QSAR as being more realistic than another. QSAR procedures model the noise disturbing a given biological "signal", and there are plenty of possible noise generating sources. The important point is the consistency of the forecasts generated by the different models in a given domain of application.

Neverthless, QSAR is actually a scientific enterprise and not a mere pattern recognition task, so we must be able to derive from a QSAR model some hints about the explanation of the system at hand. In other terms, we must generate some usable knowledge about the modeled systems.

In our opinion, QSARs models could be very useful tools with which to delve into the knowledge of biological systems. A good QSAR is a quantitative description, in a physicochemical language, of the biological system it models. Different QSARs relative to the same biological system are different descriptions of the system. If we suceed in finding the way to compare these QSARs, so as to highlight the commonality among them we could have a physicochemical model of the biological system itself.

Acknowledgments. - Ms. Eve Sylvester and Ms. Anna Maria Mancini are gratefully acknowledged for the patient editing of the material.

REFERENCES

[1] A. Morton, *Br. J. Phil. Sci.* **1993**, *44*, 659.
[2] R. Benigni, A. Giuliani, *Am. J. Physiol.* **1994**, *266*, R1697.
[3] H. Frauenfelder, P.G. Wolynes, *Physics Today* **1994**, *58*, 1.
[4] W.T. Williams, G.N. Lance, *Statistician* **1968**, *18*, 31.
[5] K.G. Wilson, *Sci. Am.* **1979**, *8*, 140.
[6] B.J. West, *Physica* **1995**, *86D*, 12.
[7] P. Bak, M. Paczuski, *Proc. Natl. Acad. Sci. U.S.A.* **1995**, *92*, 6689.
[8] E.F. Yates, *Canad. J. Physiol. Pharmacol.* **1982**, *60*, 217.
[9] N. Oreskes, K. Shrader-Frechette, K. Belitz, *Science* **1994**, *263*, 641.
[10] S. Banerjee, P. Sibbald, J. Maze, *J. Theor. Biol.* **1990**, *143*, 91.
[11] D.S. Broomhead, G.P. King, *Physica* **1986**, *20D*, 217.
[12] E. Basar, *Am. J. Physiol.* **1983**, *245*, R510.
[13] B. Bassingthwaighte, R.B. King, S.A. Roger, *Circulation Res.* **1989**, *65*, 578.
[14] A.L. Goldberger, V. Bhargava, B.J. West, A.J. Mandell, *Biophys. J.* **1985**, *48*, 525.
[15] W.T. Miller, D.B. Geselowitz, *Circ. Res.* **1987**, *43*, 310.
[16] C.A. Skarda, W.J. Freeman, *Behav. Brain Sci.* **1987**, *10*, 161.
[17] J.C. Allen, W.M. Schaffer, D. Rosko, *Nature* **1993**, *364*, 229.
[18] G. LeMasson, E. Marder, L.F. Abbott, *Science* **1993**, *259*, 1915.

[19] A.H. Delcour, D. Lipscombe, R.W. Tsien, *J. Neurosci.* **1993**, *13*, 181.
[20] M.P. Hassell, N.H. Comins, *Nature* **1991**, *353*, 255.
[21] D.A. Rand, H.B. Wilson, *Proc. R. Soc. Lond.* **1995**, *B 259*, 111.
[22] E. Jablonka, *J. Theor. Biol.* **1994**, *170*, 301.
[23] R. Franke, Theoretical Drug Design Methods. Elsevier, Amsterdam, **1984**
[24] Y.C. Martin, *J. Med. Chem.* **1978**, *24*, 229.
[25] R. Benigni, A. Giuliani, *Quant. Struct. Act. Relat.* **1991**, *10*, 99.
[26] K.S. Fu, Applications of Pattern Recognition. CRC Press, Boca Raton FL., **1983**.
[27] A. Barr, E.A. Feigenbaum, "The Handbook of Artificial Intelligence", Los Altos CA, Morgan Kauffmann, **1981.**
[28] R. De Mori, *Signal. Process.* **1979**, *1*, 1.
[29] M.D. Barratt, J.V. Castell, M. Chamberlain, R.D. Combes, J.C. Dearden, J.H. Fentem, I. Gerner, A. Giuliani, T.B.J. Gray, D.J. Livingstone, W.M. Provan, A.J.L. Fons, J.M. Verhaar, P. Zbinden, *ATLA* **1995**, *23*, 410.
[30] A. Giuliani, O. Ghirardi, A. Caprioli, S. Di Serio, M.T. Ramacci, L. Angelucci, *Behav. Neur. Biol.* **1994**, *61*, 110.
[31] G. Gilat, Y. Gordon, *J. Math. Chem.* **1994**, *16*, 37.
[32] S. Leicester, J.Finney, R. Bywater, *J. Math. Chem.* **1994**, *16*, 315.
[33] C. Hansch, T. Fujita, *J. Am. Chem. Soc.* **1964**, *86*, 1616.
[34] E.J. Ariens, *Drug Metab. Rev.* **1984**, *15*, 425.
[35] C. Hansch, *J. Med. Chem.* **1976**, *19*, 1.
[36] C.E. Shannon, N. Weaver, "The Mathematical Theory of Communication". University of Illinois Press, Urbana IL, **1949**.
[37] H. Frauenfelder, S.G. Sligar, P.G. Wolynes, *Science* **1991**, *254*, 1598.
[38] R. Benigni, C. Andreoli, A. Giuliani, *Environ. Mol. Mutagen.* **1994**, *24*, 208.
[39] A.K. Debnath, R.L. Lopez De Compadre, A.J. Shustermann, C. Hansch, *J. Med. Chem.* **1991**, *34*, 786.
[40] A.K. Debnath, C. Hansch, *Environ. Mol. Mutagen.* **1992**, *20*, 140.
[41] D.I. Sidelnikov, O.V. Gritsenko, A.P. Simonova, N.G. Rambidi, A.A. Polezhaev, D.S. Chernavskii, *Biol. Cybern.* **1992**, *68*, 53.
[42] I. Procaccia, *Nature* **1988**, *333*, 618.
[43] J. Devillers, R.L. Lipnick, In "Practical Applications of QSAR in Environmental Chemistry and Toxicology", Eds. W. Karcher and J. Devillers, Dordrecht: Kluwer Academic Publishers: **1990**, 129.
[44] J.J. Hopfield, D.W. Tank, *Science* **1986**, 625.
[45] T. Aoyama, Y. Suzuki, H. Ichikawa, *J. Med. Chem.* **1990**, *33*, 905.
[46] P.J. Brown, *Anal. Proceed.* **1990**, *27*, 303.
[47] L.A. Lebart, A. Morineau and K.M. Warwick, "Multivariate Descriptive Statistical Analysis". Wiley, **1984.**
[48] E.S. Soofi, *J. Am. Stat. Ass.* **1994**, *89*, 1243.
[49] J.W. McFarland, D.J. Gans, *J. Med. Chem.* **1987**, *30*, 46.
[50] R. Benigni, *Quant. Struct. Act. Relat.* **1994**, *13*, 406.
[51] D.E. Rumelhart, J.L. McClelland, "Parallel Distributed Processing. 1. Foundations". MIT Press, **1986**.

[52] C.B. Lucasius, G. Kateman, *Chemom. Intell. Lab. Syst.* **1993**, *19*, 1.
[53] T. Kohonen, Self-Organization and Associative Memory. Springer-Verlag, **1984.**
[54] D.G. Bounds, *Nature* **1987**, *329*, 215.
[55] E. Oja, *Neural Networks* **1992**, *5*, 92
[56] J.A. Anderson, S.W. Silverstein, S.A. Ritz, R.S. Jones, *Psychol. Rev.* **1977**, *84*, 413.
[57] R. Benigni, A. Giuliani, *Mut. Res.* **1994b**, *306*, 181.

PART I. **CHEMOMETRICS**

Chapter 4

INLR (Implicit Non-linear Latent variable Regression). II. Blockscaling of Expanded Terms with QSAR Examples

by **Anders Berglund** and **Swante Wold**

Research Group for Chemometrics, Department of Organic Chemistry, Umeå University, S-901 87, Sweden

INLR (Implicit Non-linear Latent variable Regression). II. Blockscaling of Expanded Terms with QSAR Examples

by **Anders Berglund** and **Svante Wold**

Research Group for Chemometrics, Department of Organic Chemistry,
Umeå University, S-901 87, Sweden

We here extend the INLR [1] method with an adjustable scaling of the blocks. This approach has many attractive features. It is robust; all PLS statistics and plots still apply; the solution converges to a linear model if the relationship is linear; and it is simple to use. The X-block is expanded so that the squares and/or cubes and cross-terms of the latent variables are implicitly included in the expanded X-block. This is done by simply adding the squares and/or the cubes of the original x-variables, with cross-terms excluded. The main assumption is that the X-block has a latent structure, the rank of X being much smaller than the number of x-variables. After this expansion the extended blocks are weighted so that prediction errors are minimised.

The performance of INLR is demonstrated on three QSAR examples. INLR improves the predictions for the external test set in all three cases.

Introduction. - In QSAR there is sometimes a curvature in the relationship between the predictors (**X**) and the responses (**Y**), i.e., Y is a non-linear function of X. For example in a study where the side chain of a molecule is varied, an optimum of size on the side chain will lead to a non-linear relationship. With large structural variation at least a quadratic relationship between X and Y may be expected. Multiple responses, which commonly occur in QSAR, also have a tendency to generate non-linear relationships. This may sometimes be handled by making separate models for each response. The drawback is that more models have to be calculated and a simple model for all responses is not available.

There are several approaches to modelling non-linearities between the X and Y variables. The best approach depends on structures of the data. In chemical applications, including QSAR, there are often more variables than there are objects, and the x-variables may be strongly collinear. The X-matrix contains not only relevant information but also noise and irrelevant structure. There may also be missing values in the X-matrix. The Y-matrix can include one or several responses and may also have missing values if it has multiple responses.

When to use a non-linear method, instead of a linear, is often a subjective decision. Usually, the result from a linear model give indications that a non-linear model will improve the results, but the indications are not always clear and can easily be mistaken for noise.

Non-Linear Methods. - There are certain requirements for any non-linear regression method. It should give good predictions and be stable. It must also be able to differentiate between non-linearity and noise. The method should give interpretable results and be easy to use. It should preferably converge to a linear model if there is a linear relationship.

Some possible approaches are given below with their advantages and drawbacks.

a. Transformations [2]. Sometimes the transformation of responses (y) and/or factors (x) make a non-linear relationship linear or almost so. If a variable spans a large interval, a transformation, i.e., log, square root, or fourth root can often improve the results. This approach is useful, particularly if there is some prior knowledge of what type of non-linearity is present. The drawback with transformations is that they also may transform the error distribution in an undesirable way.

b. Expansion of the X-matrix [3]. The predictor block can be expanded with the square-terms, cross-terms and cubic-terms of each variable. The drawback is that the cross-terms will dominate the X-block if there are many variables. To work well, this requires that the X-block is derived from an appropriate experimental design.

c. Neural Networks (NN) [4]. NN is a classical non-linear regression method. NN do not require any knowledge about the type of non-linearity that is present and it is flexible. It requires that there are many more objects (N) than there are variables (K) (N>>K), since there is large numbers of free coefficients. The results of a NN modelling are also hard to interpret, NN is still a black box method. NN is also time consuming and often too flexible causing serious overfit.

d. Non-linear PLS (Spline-PLS [5], NN-PLS [6] etc.). Non-linear PLS combines the features of PLS with a non-linear model of the X and Y scores relationship. The results are also hard to interpret since there are two models that make the relationship. Most of these methods rely on cross-validation since more parameters have to be estimated, than only number of components as for ordinary linear PLS. As with NN it is easy to overfit the data with these methods. The results are also unstable since the models are so flexible, and the chances for spurious models are large.

e. Implicit Non-linear Latent variable Regression (INLR) [1]. This is a simple approach where the squares and the cross-terms of the **latent variables** are included in the extended X-matrix. The X-matrix is expanded with square-terms and/or cubic-terms of the original X-variables, but no cross-terms are included. All PLS-plots and PLS-statistics still apply. Here we shall review this approach and show some interesting and simple scaling approaches making it more flexible but still robust. The approach is illustrated with three QSAR examples.

Theory of INLR. - In linear PLS [7-9] the main assumption is that Y, the predictor variables, can be described by a linear model in the latent variables, T, derived as a linear combination of the X-block.

$$x_{ik} = \sum_{a=1}^{A} t_{ia} * p_{ak} + e_{ik} \qquad (1)$$

Here A is number of components, $\mathbf{p_a}$ is the a:th loading vector of X and **e** is the residuals.

The latent variables (t_a) are derived so that they maximise the covariance between X and Y. In the inner relationship in PLS the X-block is related to the Y-block via the latent variables,

$$u_a = t_a + h_a \qquad (2)$$

Here $\mathbf{u_a}$ is the latent variable of Y and $\mathbf{h_a}$ is a vector of residuals.

We give only a short review of the theory here, the complete theory is in *reference 1*. If there is a non-linear relationship between X and Y, *Equation 2* is no longer valid. If the X matrix is still well described by the latent variables, however, and thus the rank of X is much smaller than the number of variables (K) (A<<K), the cross-terms of the X-variables do not give any unique information. This can be seen from the following which follows from *equation 1*:

$$x_{ik}^2 = \sum_{a=1}^{A} \sum_{b=1}^{A} t_{ia} * t_{ib} * p_{ak} * p_{bk} + e_{ik} \qquad (3)$$

This shows that both the squares and the cross-terms of the latent variables (t_a) are implicitly included in the squared X-block, X_k^2. The results are analogous for the cubic X-block, X_k^3. Thus, in INLR the squared X-block and the cubic X-block are appended to the original X-block, forming the new expanded X-block, *Figure 1*. This will triple the number of variables. After the expansion ordinary linear PLS is used to find the relationship between the extended X-matrix and the Y-matrix. This makes it possible to model relationships when Y is a quadratic or cubic function of the latent variables.

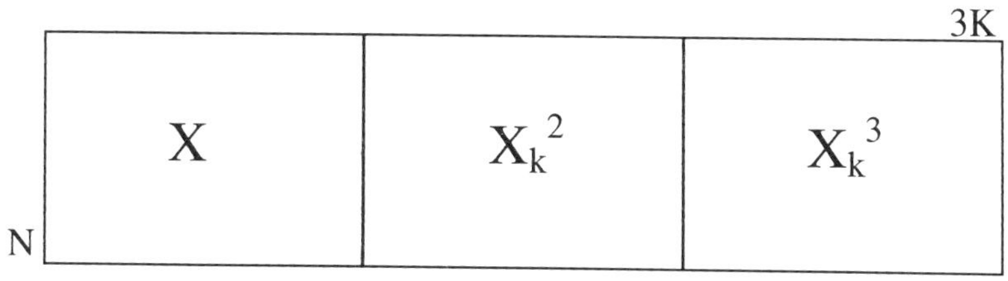

Fig. 1. *The extended X-matrix.*

The calculations of the squares and the cubes are performed after the centring of the original variables. Calculating the square terms on the centred variables will make

them almost orthogonal against the original variables. The differences between calculations before and after centring are described, in terms of latent variables, in the following example.
Assume that X is a matrix with rank two (A=2)

$$X = X_m + t_1*p_1' + t_2*p_2' + E \tag{4}$$

Let us now consider the first element in X, x_{11} and with E=0
In the case where x_{11} is centred

$$xc_{11} = (x_{11}-x_m) = t_{11}*p_{11} + t_{12}*p_{12} \tag{5}$$

Here xc_{11} is the centred value, x_{11} is the non-centred and x_m is the mean value for variable x_1.
The square of xc_{11} then becomes

$$(xc_{11})^2 = (t_{11}*p_{11} + t_{12}*p_{12})^2 = (t_{11}*p_{11})^2 + (t_{12}*p_{12})^2 + \\ + 2(t_{11}*p_{11})(t_{12}*p_{12}) \tag{6}$$

This equals the result from *Equation 3*, both the squares and the cross-terms of the latent variables are included in the squared x-variable.
Let us now consider if the x-variable is not centred.

$$x_{11} = x_m + t_{11}*p_{11} + t_{12}*p_{12} \tag{7}$$

For x_{11} the square becomes

$$(x_{11})^2 = (x_m + t_{11}*p_{11} + t_{12}*p_{12})^2 = x_m^2 + (t_{11}*p_{11})^2 + (t_{12}*p_{12})^2 + \\ + 2(t_{11}*p_{11})(t_{12}*p_{12}) + 2*x_m(t_{11}*p_{11} + t_{12}*p_{12}) \tag{8}$$

Comparing *Equations 6* and *8* shows that the differences are the terms x_m^2 and $2*x_m(t_{11}*p_{11} + t_{12}*p_{12})$. x_m^2 is only a constant and will not make any differences to the results since also the squares are centred after they are calculated. The second term shows that if centring is not done before expansion, the squared variable will be more collinear with respect to the original variable. This is due to the fact that the squared variable also contains information about the linear latent variables t_1 and t_2. It is also notable that this effect will depend on the mean value, x_m. If x_m is large the effect also becomes larger. The same can be shown for the cubic terms.

Block Scaling of Extended Terms. - One possible drawback of this approach is that the expanded terms may add noise to the X-matrix. The positive effect of adding more variables are then cancelled out by the negative effect of more noise. This drawback is relatively more pronounced if only a small non-linearity is present. Also the amount of noise and irrelevant information in the X-block will increase this negative effect. Our new approach to minimise this effect is that the square and/or the cubic blocks are weighted according to their importance. This will also make INLR converge to the linear PLS solution if there are no non-linearity present, that is, the relationship is linear.

This weighting is done in the following way: After centring and scaling, if so is warranted, the matrix of squares is calculated. Before appending the square block to X it is weighted with a scalar, β and then the linear PLS model is calculated with the expanded X-matrix. This is done for different β values starting with e.g., $\beta=0$ to give the linear PLS-solution and stepping up to some β value, e.g., $\beta=1$. Cross-validation [10] is used for deciding the number of components and the best β-value. This results in a matrix of cross-validation values where each column corresponds to a β value and each row to a certain number of components.

If also the cubes are included, those are weighted by a parameter γ. This gives then a matrix of different cross-validation values for each component, *Figure 2*. The points in *Figure 2* corresponds to one model with a specific β and γ value for A components. The points can be placed in different ways, *Figure 2* shows a CCF design (Central Composite Face). A response surface can be calculated, with Q^2 and R^2 as y and β and γ as factors and an optimum can be found. The spacing between the points, number of points, and type of design can be varied according to the data set.

The benefit of setting the β and γ before the calculations is that cross-validation still can be used for estimating the number of components. This is not the case if some parameter has to be estimated using cross validation for each component. This makes also INLR robust since only two parameters have to be estimated (number of components and the β value) for a quadratic model and for a cubic model there are three parameters.

This new approach does not make any differences in the interpretation of the results compared to INLR without weighting. This is also due to the fact that the weighting is done before calculations of the PLS model. A β or γ value near zero will indicate that only a small non-linearity is present or that the amount of noise in X and/or Y makes it difficult to distinguish between noise and a non-linear relationship.

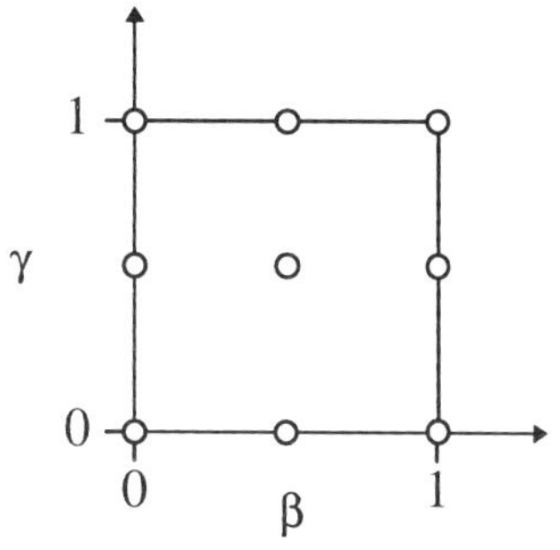

Fig. 2. *The design for investigating the β and γ values.*

Scaling. - The weighting of β and γ values make the choice of scaling of the original variables before calculating the squares less important. An example is if no scaling is used and the expanded X-blocks consist of large values. This will make the expanded X-blocks of even larger values. The reverse happens if the X-block consists of small values. The β and γ adjust for this difference in total variance between the original X-block and the extended-block. Another way of scaling is to scale each variable so that they have the same variance in all blocks, i.e., original X-block, squared X-block and in the cubic X-block.

Collinearities of Cubic Terms. - There are problems with the cubic terms since they are much more collinear to the original x-variables than the square terms are. In other words, the cubic block contains information that already is in the original X-block. Orthogonalising the cubic terms against the original x-variables and squares variables minimises this effect. This matter will be further investigated in future work.

Examples. - Three QSAR data sets are used, all having an independent a test set. This ensures that the improvements are adequately verified.

Example 1. The first example is a QSAR example published by *Hellberg et al.* [11] 58 di-peptides inhibiting the angiotensin converting enzyme (ACE), are described by 58 physico-chemical variables, 29 variables for each amino acid. A training set of 29 compounds was selected, the z-scales [12] derived from principal component analysis and the response was used. The training set was selected using D-optimal design. The same transformation of **y** was used as in *reference* [11], i.e., log[1/(I-50)], where I is the inhibition. The X-matrix and Y-vector were centred and scaled to unit variance before the analysis.

Example 2. The second example concerns the same di-peptides, but now described with GRID[13,14] data[9]. The response and the training and test set were the same as previous example. The same pre-treatment was used as in *reference [2]*. The X-matrix and Y-vector was centred before the analysis.

Example 3. The third data set contains 30 technical non-ionic surfactants characterised by 19 descriptors [15]. The compounds were divided into a training set with 21 objects and a test set with 9 objects. Their toxicity is tested on two bioassays, *Thamnocephalus platurus*, y1, and *Brachionus calyciflorus*, y2. The responses were transformed with MDS-scaling [16]. The X-matrix and Y-matrix were scaled to unit variance and centred before analysis.

Results. - For all the examples only the training set was used to develop the model. After this the test set was used to validate the model.
The calculated statistics are R^2Y which is the explained variance in Y, Q^2_{cum} which is the cross-validated R^2Y, and the root mean square error of prediction (RMSEP) for the test set.

Example 1. The linear PLS analysis results in a three component model with Q^2_{cum}=0.72, R^2Y=0.88 and a RMSEP for the test set of 0.45. The observed versus predicted/calculated plot, *Figure 3*, shows a non-linear tendency.

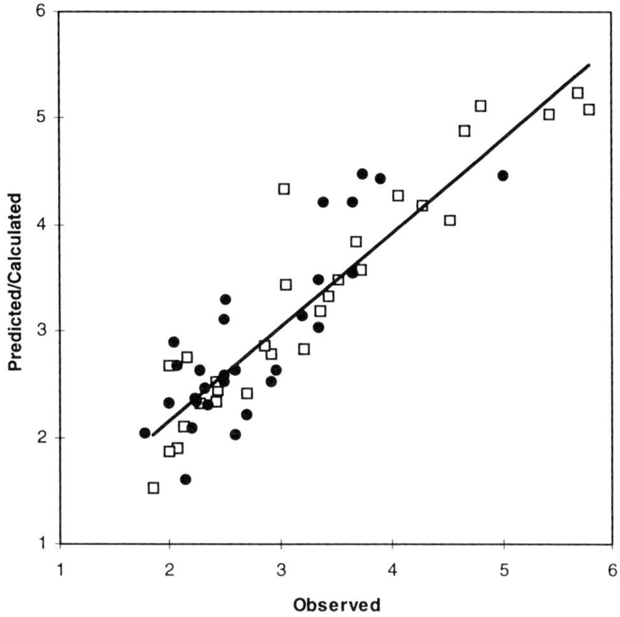

Fig. 3. *Predicted/Calculated vs. Observed for linear PLS on Ex. 1.* • - *Test set,* □ - *Training set*

With INLR, the X-block was expanded with square terms since *Figure 3* indicated a quadratic behaviour. The β-value was varied from β=0 to β=1 with a step value of 0.1. The number of components was set to four. This gives 11 PLS models each with four components. *Figure 4* describes the Q^2_{cum} value as a function of the β-value and number of PLS-components. It is seen that the optimum Q^2_{cum} value is after two components and with a β=0.7. The smoothness of the curve indicates that the changes in Q^2_{cum} are not only random variation. The optimum in the plot indicates that this method minimises the effect of adding more noise to the model.

The statistics for the corresponding INLR model are: A=2, Q^2_{cum} =0.77 and R^2Y=0.92. The INLR gives a RMSEP=0.34 for the test set. This is a 22% improvement compared to linear PLS with only two components instead of three for linear PLS. The observed vs. Predicted/calculated plot, *Figure 5*, shows that the predictions have improved. The relation is now more linear, compared to linear PLS, *Figure 3*.

The loading plot, *Figure 6*, shows, for example, that variable 46 is negatively correlated to the response, V59, and has a minimum since V46 is negatively correlated and the square S46*46 is positively correlated to the response. Variable 46 is the R_f value for salt chromatography. More information can be extracted even though the large number of variables makes the plot not so easy interpret.

In order to see if a cubic model could improve the predictions such a model was also calculated. The β and γ values were varied from 0 to 1 with a step value of 0.1. The cubes were orthogonalised against the original and squared variables. *Figure 7* shows that the cubes according to cross-validation does not improve the model. This result indicates that this approach converges to the simpler model.

The example shows how easy it is to improve the predictions with INLR, if there is a non-linear relationship between X and Y, and still retain all features of PLS.

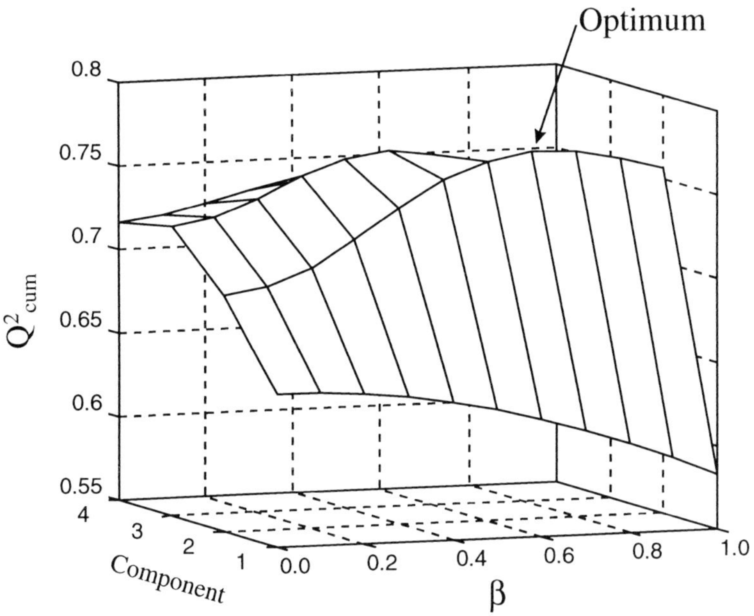

Fig. 4. *Q^2_{cum} as a function of the β-value and number of PLS components.*

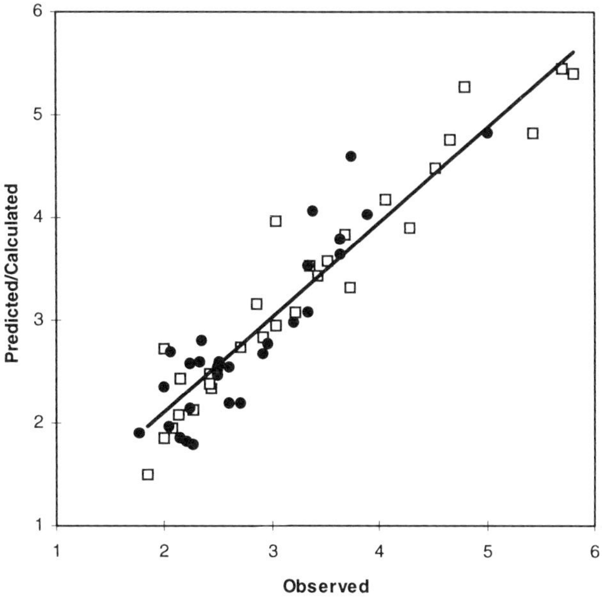

Fig. 5. *Predicted/Calculated vs. Observed for INLR on Ex. 1* • *- Test set,* □ *- Training set*

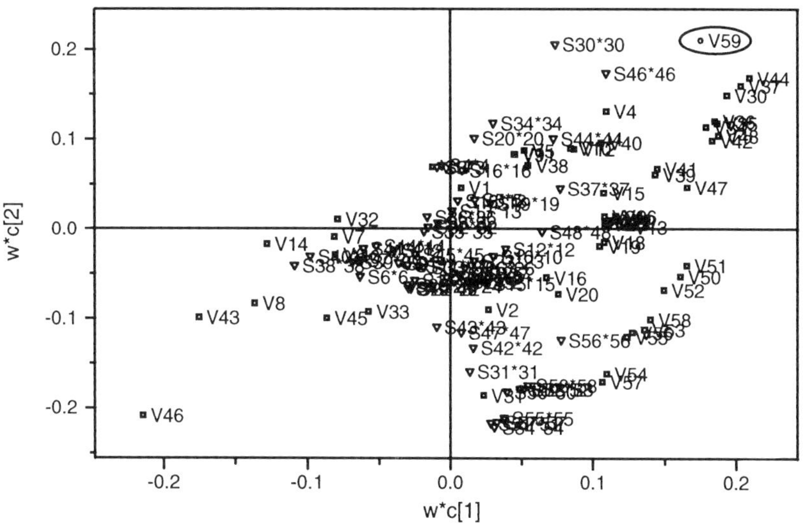

Fig. 6. *Weight plot for the two first components. V59 is the response.*

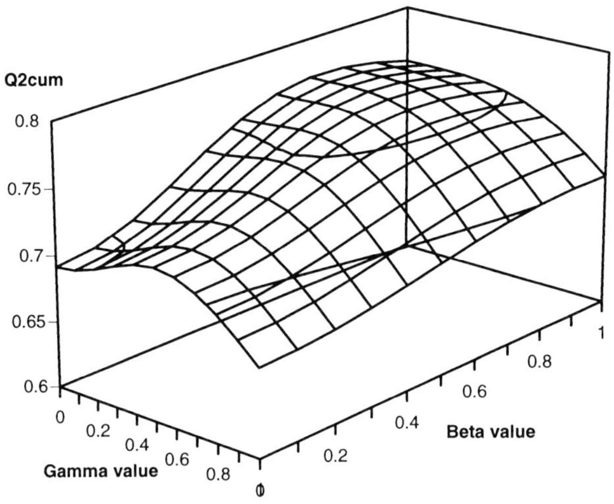

Fig. 7. *Q^2_{cum} as a function of the β and γ values for the second component.*

Example 2. Here, linear PLS gives a two component model with a Q^2_{cum} = 0.76, R^2Y = 0.87 and a RMSEP = 0.45. This is the same RMSEP as for the linear PLS model in the previous example. The differences are that the observed vs. predicted/calculated plot is more linear and based on only two components. The GRID data gives a more linear model than the data in the previous example.

In order to see if it was possible to improve the model still further, the same procedure was used as described above. The INLR model with β = 0.9 gives a two component model with Q^2_{cum} = 0.79, R^2Y = 0.89, which is only a small improvement compared to linear PLS. The same tendency may be seen in the predictions for the test set, with RMSEP=0.41 and a 9% improvement. This improvement is much smaller than the previous example, because the linear PLS model with GRID data was more linear from the outset.

Example 3. In the third example, multiple responses PLS was used, since the two responses are correlated with each other. The best linear PLS model gave two significant components with a Q^2_{cum} =0.85 and a R^2Y=0.91. The RMSEP for the test set where 0.85 for y1 and 0.68 for y2. Both y1 and y2 showed curvature, *figure 8*. A third component for linear PLS did not improve the model. A single linear PLS model for each response did not improve the model.

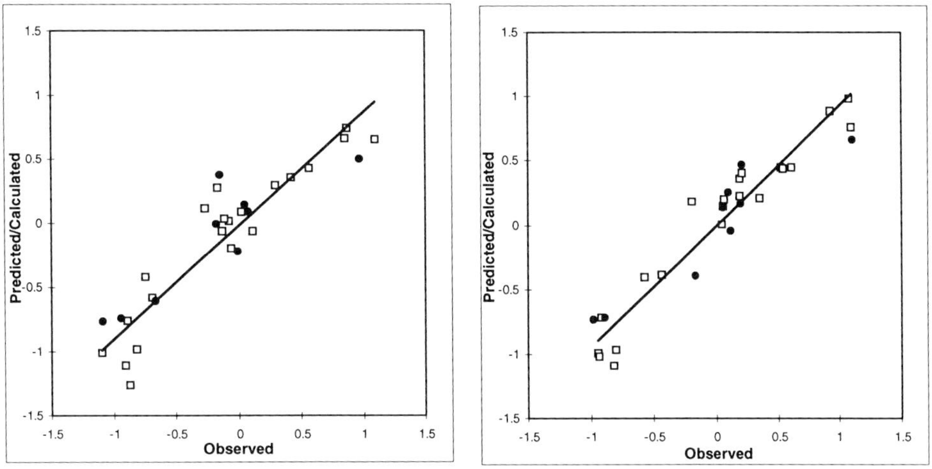

Fig. 8. *Predicted/Calculated vs. Observed for linear PLS on Ex. 3. Y1 is to the left and Y2 to the right.* • - *Test set,* □ - *Training set*

The same procedure was used as in the two previous examples. *Figure 9* shows that also in this surface there is a clear optimum with β=0.55 and with three PLS components. This model gives a Q^2_{cum} =0.86 and a R^2Y=0.95. The RMSEP for the different responses are RMSEP=0.78 for y1 and RMSEP=0.63 for y2. The improvements in percent are 8% for Y1 and 7% for Y2. The observed vs. Predicted/calculated plot, *Figure 10*, shows that the predictions have improved. The relation is now more linear, compared to linear PLS, *Figure 8,* especially for Y1.

This example shows that INLR also can improve multiple responses models, where non-linearity is more frequent. For this example INLR needs one more component which makes the solution more complex. This may be due to the fact that the x-variables are more noisy than in the previous examples. The explained variance in X for the INLR model is 63 %.

An interesting fact about the surface plot, *figure 9*, is that for the first component is the Q^2_{cum} value decreasing with an increasing β-value. This was also seen in the surface plot from example one, *figure 4*. For the third example is it clear that it is only in the third PLS-component that there is an improvement in the Q^2_{cum} value. An obvious conclusion from this is that separate β-values for each PLS-components might improve the predictions, since as seen in *figure 4 and 9*, a β-value of 0.55 gives worse predictions for the first component. This would not impair on the final interpretation for the PLS-model. The drawback is that there are more free parameters which has to be estimated making the cross-validation even more crucial. This aspect of improving the prediction will be investigated further.

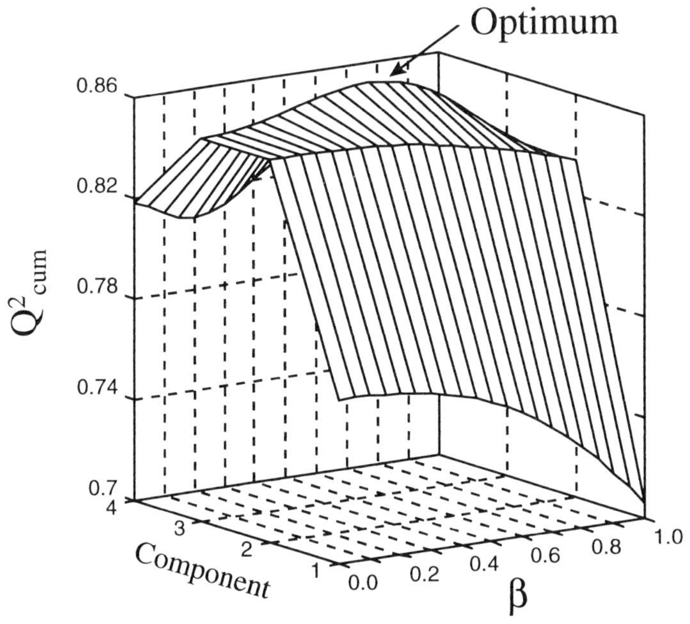

Fig. 9. *Q^2_{cum} as a function of the β-value and number of PLS components*

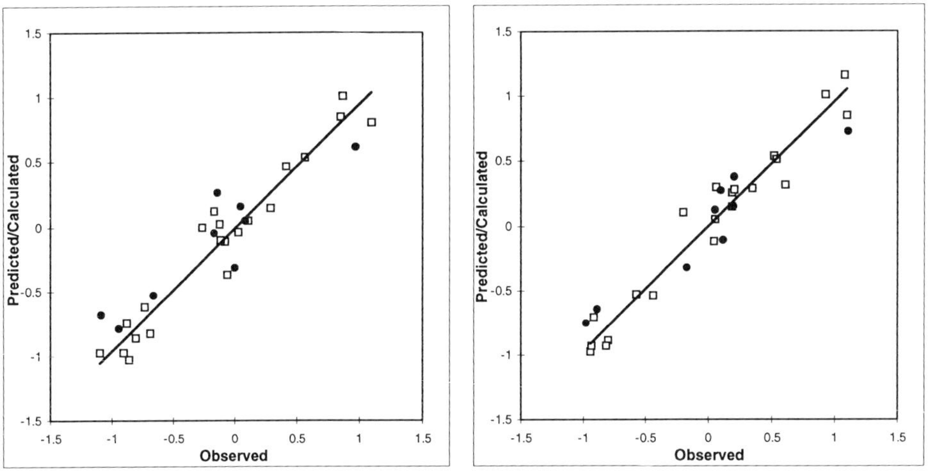

Fig. 10. *Predicted/Calculated vs. Observed for INLR on Ex. 3. Y1 is to the left and Y2 to the right.* • - *Test set,* □ - *Training set*

Conclusions. - We have shown a new non-linear regression method, INLR, with some QSAR examples. INLR has many good features, the results are simple and interpretable, it is easy to use, the model is robust and it gives good predictions.

It is important to remember that this approach is only possible when the number of components, A, is much smaller than K (the number of x-variables), but this is often the case in chemistry.

The new approach with weighting the squared variables and/or cubic variables makes INLR converge to a linear PLS solution if the relationship is linear. It also improves the predictions compared to our first approach without weighting the expanded block. The results show that there is a balance between the positive effect of adding more information to the X-matrix, and the negative effect of more noise. *Figure 4 and 9* shows this clearly.

The robustness is due to that INLR is using a single regression method, linear PLS, to develop the relationship between the expanded X-matrix and Y. The non-linearity being calculated in terms of the latent variables also makes the solution more stable.

The use of PLS allows of all of the plots and statistics of linear PLS, and the calculation of the regression coefficients as in linear PLS. The performance of INLR was shown on three QSAR data sets. INLR improved the predictions in all three examples compared to linear PLS.

REFERENCES

[1] A. Berglund, S. Wold, *J. Chemometrics,* in press.

[2] G. E. P. Box, N. R. Draper, *Empirical Model-Builldinq and Response Surfaces*, Wiley, Chichester **1987**

[3] O. M. Kvalheim. *Chemometrics Intell. Lab. Syst.* **1990**, *8*, 59-67.

[4] J. Zupan, J. Gasteiger, *Anal. Chim. Acta.*, **1991**, *248*, 1-30.

[5] S. Wold.. *Chemometrics Intell. Lab. Syst.* **1992**, *14*, 71-84.

[6] T. R. Holcomb, M. Morari. *Computers chem. Engng.* **1992**, *16*, 379-391.

[7] A. Höskuldsson, *Predictions Methods in Science and Technology Vol. 1. Basic Theory.* Thor Publishing, Arnegaards Alee 7, Denmark, **1996**

[8] S. Wold. In 'QSAR: Chemometric Methods in Molecular Design', Ed. H. van de Waterbeemd, Methods and Principals in Medicinal Chemistry, Vol 2., Verlag Chemie, Weinheim, Germany, **1995**, pp. 195-218.

[9] S. Wold, E. Johansson, and M. Cocchi, In '3D QSAR in Drug Design, Theory, Methods, and Applications' H. Kubinyi (Ed.),. ESCOM Science Publishers, Leiden, 1993, pp. 523-550.

[10] S. Wold, *Technometrics*, **1978**, *20*, 397-405.

[11] L. Hellberg, J. Ericksson, F. Jonsson, Lindgren, M. Sjöström, B. Skagerberg, S. Wold, P. Andrews. *Int. J. Peptide Protein Res.* **1991**, *37*, 414-424.

[12] S. Hellberg, M. Sjöström, B. Skagerberg, S. Wold. *J. Med. Chem.* **1987**, *30*, 1126-1135.

[13] P. J. Goodford. *J. Med Chem* **1985**, *28*, 849-857.

[14] P. J. Goodford. *J. Chemometrics,* **1996**, *10*, 107-117.

[15] L-L. Uppgård, Å. Lindgren, M. Sjöström, S. Wold, In preparation.

[16] Å. Lindgren, M. Sjöström, S. Wold, *Quant. Struct. Act. Relat.* Submitted.

PART I. **CHEMOMETRICS**

Chapter 5

MOLSURF - a Generator of Chemical Descriptors for QSAR

by **Per Sjöberg**

Qemist AB, Hertig Carls alle 29, S-691 41 Karlskoga, Sweden

MOLSURF - a Generator of Chemical Descriptors for QSAR

by **Per Sjöberg**

Qemist AB
Hertig Carls alle 29, S-691 41 Karlskoga, Sweden

MolSurf calculates properties widely used by chemists for characterizing molecules chemical behavior. The calcualated parameters enable a simple and straight forward interpretation of the derived QSAR model for medicinal chemists which is of considerable importance during the preclinical drug development phase.

History and Definitions. - I developed the first version of MolSurf [1] as graduate student in Peter Politzer's research group. Unlike many other quantum groups we focused on the chemical applications. Since the pioneering work by Peter Politzer and his collaborators the calculated electrostatic potential has become a well-established guide to molecular reactive behavior and also a widely used tool in drug design [1,2] Writing the MolSurf code was truly my best effort to follow the groups tradition. The first version of MolSurf was completed in 1989 [2]. The purpose is to predict a broad spectrum of wet chemistry from quantum mechanically calculated electronic energies in the molecules valens region. This region is represented with a surface of constant calculated electronic density encompassing the molecule. The method seemed too have potential as a tool for the commercial search for better drugs. A patent was therefore applied for in 1989 [3]. Tore Brinck improved the MolSurf code resulting in algorithms for predicting more aspects of wet chemistry. He also wrote the first graphical interface for MolSurf. In 1991 Astra AB bought the first license. Encouraged I raised risk capital and founded the company Qemist AB in 1993. Since then MolSurf has been further developed in collaboration between Qemist AB, Astra AB, and Medivir AB. This has resulted in the mature drug design program MolSurf is today.

The software market for computational chemistry is very competitive. A new program has to offer special advantages to be well received. MolSurf has three.

- *Communication.* MolSurf calculates descriptors chemists relates to. A chemist does not describe the world of molecules with HOMO, LUMO, electrostatic potential, force fields, or other terms common in the quantum vocabulary. Instead the chemist uses the terms hydrogen bonds, nucleophiles, electrophiles, bases, acids, hydrophobicity, polarizability, π-interactions, and so on. MolSurf produces descriptors for each and one of these. A model you understand will trigger your creativity in your search for the next and better molecule, a string of math will not.
- *Correlations to wet chemistry.* The quality in MolSurf stands for god correlation between the ***descriptors*** that are calculated and experimental data. Some correlations has been published by Peter Politzer, Tore Brinck, and myself. Some will be presented here.

- *Focusing on substituents.* MolSurf allows you to define substituents in a molecules for which descriptors will be calculated. Assume a project with molecules having a common "mother-molecule" and two points of substitution. You can then use for example the calculated logP for the two substituents instead of log P for the molecule. Focusing on substituents will render a more flexible model and a more precise description on the variations between the molecules. This will likely increase the cross validity for the model and hence the predictability.

A molecules' chemical behavior depends on the distribution of electrons and energy in the valence region. This region is in MolSurf modeled with a molecular surface defined as an envelope encompassing the molecule with constant calculated electronic density of 0.02 au. On this surface are the electrostatic potential, V(r), and the local ionization energy, I(r), calculated. The calculation of V(r) arising from the nuclei and the electrons is well described in the literature [2]. The electrostatic potential reflects the *electrostatic forces* between interacting molecules. I(r) was used for the first time in MolSurf [2]. I(r) is defined within the framework of SCF-Molecular Orbital theory by

$$I(r) = \sum_{i=1}^{N} \frac{\rho_i(\bar{r}) \cdot |\varepsilon_i|}{\rho(\bar{r})} \qquad (1)$$

The summation is made over all occupied molecular orbitals. $r_i(\mathbf{r})$ is the electronic density of the *i*th molecular orbital at the point **r**, e_i is the orbital energy of the ith molecular orbital, and r(**r**) is the total electronic density function. A quantum expert can see that the equation is inspired by Koopmans theorem [2]. I(r) reflects the *charge transfer* component in intermolecular interactions. This can be understood from the approach that a valence regions with low values for I(r) indicates sites with loose electrons and vice versa.

MolSurf also uses a calculated quantity reflecting the *steric* situation around the reactive functionalities in the molecule.

The descriptors calculated in MolSurf are all defined as functionals of the three properties introduced above. The functions themselves will not be described since they are the hart of the code in MolSurf. The idea with MolSurf is to use the descriptors as parameters for QSAR.

See MolSurf as a descriptor-machine! Your input is a wave function, from any of the most common quantum programs, of the molecule. MolSurf calculates descriptors of all its important chemical properties. The descriptors come out of MolSurf in a format adopted to any of the widely used multi-variable programs. The descriptors describing properties of sites such as base strength, charge-transfer, hydrogen bonds, and steric factors are calculated for the atoms in the molecules. The hydrophobicity, polarity, polarizability, and size are calculated for the molecule and for any of its substituent defined by the user.

Currently Implemented Descriptors. - *Hydrogen Bonding, Wet Chemistry and MolSurf.* - Probably a majority of biological processes involves the breaking or forming of hydrogen bonds. The energy varies between 5 - 30 Kcal/mole. A strong to medium strong hydrogen bond is usually formed between a hydrogen bonded to an electronegative atom, A, and another electronegative atom, B, which can offer at least one pair of electrons.

Table 1. *Hydrogen Bonds*

Donator	Exp.	MolSurf
$F_3CCH(OH)CF_3$	1.96	2.3
Water	1.00	1.0
Acetic acid	0.97	1.1
Isopropanol	0.70	0.79
Nitromethane	0.22	0.16
Acetone	0.07	0.01

$-\delta \quad +\delta \quad -\delta$

A—H- - -B

The bonding obviously has an electrostatic nature. The short distance between the hydrogen and B indicates a charge transfer. Here we refer to *A-H* as the donor and *B* as the acceptor. The activity for the donor, α, and for the acceptor, β, refers to the experimentally-based 'solvatochromic parameters'. They quantifies a solvent's ability to donate or accept a proton in a solute-solvent hydrogen bond [9]. *Table 1* lists experimental and calculated a for donors. The numbers refers to the most active hydrogen (MolSurf calculates a for all hydrogens in a molecule).

Charge Transfer - Chemical Sense and MolSurf. - Covalent interaction involves transfer of electronic charge from a nucleophile to an electrophile. A typical example is π-interaction between benzene and an electropositive molecule or ion. As a trained and experienced organic wizard you can from the drawing of the molecules see where the 'philes' are hiding. You know that a nitrated aromatic will not donate its π electrons due to the electron withdrawing effect from the nitro-group. The chemist who do not recognize the carbon in chloroacetic acid as a strong electrophile should seek a new career. However, you would need numbers if you intend to include charge transfer in your QSAR project. A problem is that there exist no general experimental definition of charge transfer to correlate a model against as there are for hydrophobicity, pKa, polarizability and a few more properties. A site in a molecule with loose valence electrons will react as a nucleophile. Logically I defined a descriptor for charge transfer, CT, as a functional of I(r) at the molecular surface. The descriptor is scaled to give the value 0.0 for methane. Positive values are more nucleophilic and negative more electrophilic than methane. Since methane itself is inert only large absolute numbers indicate reactive sites. In the case of an aromatic system, or other π-system, CT is obtained by averaging over the carbons contributing with π-electrons. Listed in *Table 2 and 3* are some examples on CT and a phrase of 'sense' to compare with. *Table 2* is a list of molecules with well known reactivity at the carbon. *Table 3* lists some π-systems. Do your sense agree with CT ?

Table 2. *Charge Transfer*

Molecule	Chemical sense	CT
Methyl anion	a strong nucleophile	1.66
Methyl	not reactive	0.00
Acetic amide	not reactive carbon	–0.08
Acetone	weak electrophile	–0.12
Acetic aldehyde	electrophile	–0.16
Acide fluoride	reactive electrophile	–0.38

Table 3. *Charge Transfer for Unsaturated Systems*

Molecule	Chemical sense	CT
Toluol	stronger than benzene due to the donating methyl group	0.23
Benzene	nucleophilic	0.18
Ethene	nucleophilic	0.15
Ethyl	stronger nucleophile than ethane	0.19
1,3,5-Trinitrobenzene	a strong electrophile forming Meisenheimer complexes	–0.36

Base Strength, pKa - Wet chemistry and MolSurf. Bases are often molecules carrying atoms with lone pairs. The most important base in biological system are molecules with nitrogen atoms. A QSAR-study involving interaction with nitrogen bases will not rend a useful model without a good estimate of pKa. The lone pairs at nitrogen forms a covalent bond with the proton. The pKa value for different nitrogen can vary from 11 in ammonia to much below zero in the nitro-group. The differences are explained by inductive effects from the chemical environment within the molecule. The oxygens in the nitro-group will engage the lone-pairs in covalent resonance structures. The lone pair in ammonia are more accessible to form a bond with a proton. In the quantum world the simple rational is that the lone-pair at nitrogen in the nitro-group has a lower energy in the and thereby more inert for protonation. The ionization energy at the surface of the nitrogen is the obvious choice to construct the descriptor from. Also a minor contribution from the electrostatic terms is mixed in. *Figure 1* shows the calculated pKa plotted against the experimental for fifty bases spanning from strong bases to isoaxazole with pKa = -3. A model similar to the model used in MolSurf was used for predicting pKa for heterocyclic bases [9].

Polarity - Chemical Sense and MolSurf. A molecule with a separation of positive and negative charge is polar. Water is an example of a very polar molecule. A free atom is completely non-polar. The distribution of charge in an atom is completely symmetric. The center of gravity for the distribution of the electrons is positioned at the nuclei. The electrostatic potential is generated by the protons in the nuclei and the electrons distributed in their orbitals. The sign of the potential in any particular region depends upon whetear the effects of the nuclei or the electrons are dominant there. In a free atom is the electrostatic potential constant on a surface of constant radius. That is not the case for water. The potential at the molecular surface of water is negative at the oxygen and positive at the hydrogens. The negative potential associated with the oxygen atom is attributed to its lone pairs. The potential at the surface of the hydrogens are dominated by the nuclei. The internal charge separation within a molecule as well as the descriptor for polarity has been defined as the average deviation of V(r) on the surface:

$$\prod = \frac{1}{A}\int_0^A |V(r) - V_m| da \qquad (2)$$

V(r) is the potential at the coordinate **r** = (x,y,z) on the surface and Vm is the average surface-potential for the molecule. It has been shown by Brinck *et al* that for a group of twenty solvents P correlates well with Kamlet's experimentally-based polarity [10] and with the dielectric constant [11]. P ranges from 0.0 for a neutral atom and 21.6 Kcal/mol for water. Skilled chemists have no problem to judge from a drawing of the molecule if it is polar or not. Polarity is more used as an intellectual tool than experimentally. The experimental scales mentioned above are not widely used. In MolSurf P is scaled by dividing the numbers with the value for water. The polarity for water then becomes 1.00. *Table 4* lists molecules with calculated polarity. Does it make sense?

A challenge for the reader: can you find a molecule that has P larger than water and that is not a zwitterion and where all non-carbon atoms are bonded to carbons? The last constraint will prohibit you to suggest mineral acids or molecules bordering the concept of zwitterions. E-post your answer and I will revert the polarity!

Fig. 1. *Following molecules are represented (numbers are experimental pKa values): isoaxazole /* ***-2.93*** */ 1,3,5-triazine /* ***-2.31*** */ 4-nitropyrazole /* ***-1.96*** */ 2-nitroimidazole /* ***-0.81*** */ 4-nitroimidazole /* ***-0.16*** */ pyrazine /* ***0.40*** */ oxazole /* ***0.80*** */* $NCNH_2$ */* ***1.10*** */ pyrimidine /* ***1.10*** */ 3-aza-pyrazole /* ***1.17*** */ 3-nitropyridine /* ***1.18*** */ 4-nitropyridine /* ***1.39*** */ 5-amino-3,4 diazapyrazole /* ***1.76*** *pyridazine /* ***2.10*** */ 4-aza-pyrazole /* ***2.27*** */ 2-amino-3-nitropyridine /* ***2.33*** */pyrridazole/* ***2.45*** */ pyrazole /* ***2.52*** */ thiazole /* ***2.52/*** *2-t-Butylthiazole /* ***3.15*** */ 2-aminopyrimidine /* ***3.66*** */ 2,4 diamino-1,3,5-triazine /* ***3.91*** */ 7-aza-indole /* ***4.57*** */ 3,4,5-trimethylpyrazole /* ***4.63*** */ 2-hydroxy-4-aminopyrimidine/****4.67*** *pyridine/* ***5.21/*** *2-aminothiazole /* ***5.32*** */ 2-aza-indolizine /* ***5.54*** */ 2-phenylimidazole /* ***5.83*** */ 4,5 -diaminopyrimidine /* ***6.04*** */ benzimidazole /* ***6.11*** */ 1-aza-indolizine /* ***6.79*** */ 2,4,5,6-tetraaminopyrimidine /* ***6.82*** */ 2-aminopyridine /* ***6.86*** */ 4-aza-indole /* ***6.92*** *imidazole /* ***6.95*** */ imidazole /* ***6.99*** */ 2,4 diaminopyrimidine /* ***7.40*** */* CH_3NHNH_2 */* ***7.87*** */ 6-aza-indole /* ***7.93*** */ 5-aza-indole /* ***8.24*** */ morpholine /* ***8.49*** */ 4-aminopyridine /* ***9.12*** */ amonia /* ***9.24*** */ piperazine /* ***9.73*** */ 2-phenyl-2-imidazoline /* ***9.90*** */ 1-aminobutane /* ***10.61*** */ methylamine /* ***10.62.***

Table 4. *Polarity*

Molecule	P
Free atom	0.0
H_2	0.12
CCl_4	0.22
Benzene	0.23
Diethyl ether	0.31
tert-Butanol	0.36
Pyridine	0.40
2-Propanol	0.42
Acetone	0.47
Ethanol	0.51
Dimethylacetamide	0.52
Dimethylformamide	0.57
CO	0.59
Methanol	0.75
Formamide	0.83
Water	1.0

Lipophilicity or log Po/w - Wet Chemistry and MolSurf. - Octanol/water partition coefficients are among the most frequently used parameters in QSAR. The approach for hydrophobicity in MolSurf is to provide a model for calculating theoretical values for log Po/w. The model implemented in MolSurf expressed in formalized math looks like this:

$$\log P_{o/w} \propto F\left[\Pi, \alpha, \beta, S\right] \qquad (3)$$

Π is the descriptor for polarity, α and β the descriptors for hydrogen bond donors and acceptors. All three defined above. S is the molecular size. A similar model was used on various organic molecules and nitroaromatics in two published studies with very promising results [12,13]. *Figure 2* shows the correlation between the descriptor and experimental partition coefficients (as log P values) for 50 molecules. The molecules span from the very hydrophilic ethyldiol to saturated hydrocarbons.

Polarizability - Wet Chemistry and MolSurf. - Polarization can play a role in a biological process involving the interaction between a positively charge electrophile and a nucleophile. You might therefore want to include a descriptor for this effect in your QSAR-study. The electron cloud in a molecule or atom will be polarized when submitted to an external electric field. The degree of polarization is roughly linearly dependent on the molecular volume [14,15]. Tore Brinck and Peter Politzer have shown the tightness of electron binding also affect the polarizability. Take for example C_6F_6. It's volume is about 30 % larger than the volume for benzene. But the experimental polarizability is larger for benzene (10.32 compared to 9.58 Å3) [16]. A chemist knows that florines will tie the electrons in the π-system very tight. Fluorinated molecules will therefore be less polarized by an external field. Brinck et al have shown that one effective approach is to include a measure of the tightness of binding of the electrons on the molecular surface. The measure used was the I(r) on the surface. The correlation coefficient for 25 molecules was 0.992 compared to 0.911 when only the volume was used in the model [17]. A similar model is used in MolSurf for the descriptor for the polarizability.

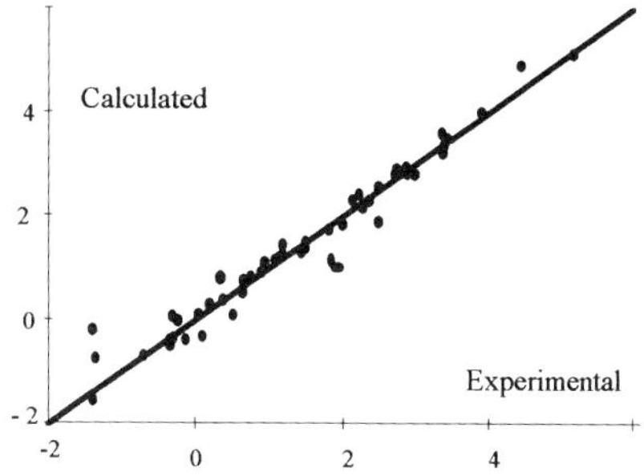

Fig. 2. *Prediction of log P. Molecules represented (numbers are experimental log P values):* $HOCH_2CH_2OH$/**1.39**/$(CH_3)_2O$/**0.10**/C_2H_4/**1.14**/C_6H_5F/**2.277**H_2O/**1.38**/CH_2F_2/**0.20**/CF_4/**1.18**/C_3H_8/**2.36**/$(CH_3)_2SO$/**1.35**/C_2H_2/**0.37**/$C_6H_3(NO_2)_3$/**1.18**/1,3C_6H_4ClOH/**2.49**/CH_3OH/**0.70**/CF_3CH_2OH/**0.37**/CH_3CH_2Cl/**1.43**/C_6H_5Cl/**2.49**/CH_3NO_2/**0.34**/CH_3F/**0.51**/1,3$C_6H_4(NO_2)_2$/**1.49**/CCl_4/**2.72**/CH_3CN/**0.34**/CHF_3/**0.64**/$C_6H_3(NO_2)_3$/**1.49**/$C_6H_5CH_3$/**2.74**/CH_3CH_2OH/**0.31**/pyridine/**0.65**/C_2H_6/**1.81**/2napthol/**2.87**/cCH_2OCH_2/**0.30**/CH_3CHF_2/**0.75**/$C_6H_5NO_2$/**1.85**/$CH_3(CH_2)_2CH_3$/**2.89**/CH_3COOH/**0.24**/$(CH_3CH_2)_2O$/**0.89**/C_6H_5COOH/**1.95**/C_6H_5Br/**2.99**/CH_3COCH_3/**0.24**/$C_6H_5NH_2$/**0.93**/C_2F_6/2.00/Napthalene/**3.36**/$CH_3CH_2NH_2$/**0.13**/$HCCCH_3$/**0.94**/C_6H_6/**2.13**/$C_6H_4Cl_2$/**3.38**/$CH_3CHOHCH_3$/**0.05**/CH_4/**1.09**/C_6F_6/**2.00**/n-pentan/**3.39**/c-C_6H12/**3.44**/n-hexan/**3.90**/anthracene/**4.45**/$CH_3(CH_2)_6CH_3$/**5.18** .

The Steric Factor. - A site in a molecule might be inhibited to take part in chemical reactions due to a crowded surrounding hindering the counter molecule to get close enough. This is refereed too as steric hindering. MolSurf will quantifies this factor for you. If the value is smaller than 1.0 it means this atom is somewhat hindered in the actual molecule than in the reference molecule. Hindered bases are examples on how steric hindering can be used as an advantage. A nitrogen can react both as a nucleophile and a base. In some reactions you need a nitrogen base to pick up the protons but at the same time there are electrophilic molecules in the reaction you do not want to interfere with. Then you use a hindered base. Listed in *Table 5* are the calculated values from MolSurf quantifying the effect of hindering a nitrogen with crowding substituents.

Table 5. *Steric Factors*

Base	SF
Ammonia	1.0
Methylamine	0.71
Di-*sec*-propylamine	0.10

Using MolSurf for QSAR. - MolSurf has been used in several drug design projects. Two of them are published and will be discussed below. The projects has so far involved using the descriptors for establishing a model for the relation between structure and biological activity. Chemist have a basic knowledge how the properties calculated in MolSurf depend on the molecular structure. Consequently MolSurf has the potential to generate a model from which chemists can predict better molecules using their basics. As an example a model predicts high activity if

- a substituent has an acidic hydrogen.
- an unsaturated functionality in the mother molecule is nucleophilic, probably in order to π-bind.
- when a number of constraint on the geometry for the molecule are met.

Starting from the structure of the molecules and the statements above the chemists in the project will suggest new molecules for testing.

MolSurf also has a graphical interface where the molecules are color coded with respect to the descriptors. This feature can be a useful tool for lead finding. This aspect of MolSurf will not be described in this article.

The geometries and the wave functions for all the molecules in the projects are computed with any of the widely used programs MOPAC, GAUSSIAN, or SPARTAN. MolSurf reads the outputfiles with the wave function from any of these programs as they come out. The output-files carrying the descriptors comes out from MolSurf in a format that is easily converted to matrices to be used by the multi variable programs. Up till this point in the projects the computations are straight forward and no difficult decisions are needed to be made. In the MolSurf projects I participated in we employed the method of Partial Least Square (PLS) in order to study the relationship between structure and activity. The multi variable programs do not have complete algorithms for finding a model. Much is left to your own judgment. My advice is to have a skilled multi-variable-expert in the project. Here are rules we have practiced:

- The quality of the model is defined as a high cross validity coefficient.
- Few parameters enable a simple and straight forward interpretation of derived model(s) which is of considerable importance for you during the preclinical drug development phase. You should at least get down to less than 10, including the parameters describing the geometry. The chemical interaction terms should be very few. Usually there are not more than two or at the most three principal bonds in a biological interaction each described only by one descriptor. As a start you can use the VIP[18] method to exclude parameters. To exclude more you might need to excercise some of the tips listed below. Sometimes you might consider to sacrify cross validity in order to get fewer parameters. Look for the descriptors reflecting specific bonds and that has a large impact on the cross validity. They might be the key attraction forces in your projects.
- Leave out 'overlapping' descriptors from the model parameters. Polarizability often 'overlaps' with molecular size. When polarity, log P and size are used at the same time for the molecule or the same substituent you can usually exclude at least one of

them. If this causes the cross validity term to decrease significantly they did not 'overlap'.

- Often it is more efficient to use the log P and polarity descriptor for the substituents than for the molecule. Say for example that you have a hydrogen bond in the interaction. This might require an acid proton. The molecule will therefore need to be polar at this site. The rest of the molecule might work best if it is non-polar or have a large log P. If you now use the polarity and log P for the whole molecule you have a dual situation that might rend a model difficult to understand. If you instead use the log P and the polarity for all the substituents except the substituent that should accommodate the acid proton the model will be straight forward with respect to log P and polarity and the cross validity will increase.
- Do not pronounce molecules as outliers only to slightly improve the validity coefficient if you do not have a good reason. Exclude only molecules that drastically decrease the validity!
- List the max. and min. for all descriptors in an Excel file. Identify descriptors that are not varied and exclude them. For example if all the hydrogen bond donor parameters are smaller than 0.2 for all molecules in the project (1.0 for water) you can never make any conclusions on hydrogen bonds.
- Assess the predictivity of the derived model by dividing the set of investigated molecules into a training set and a test set. This part of the QSAR is in detailed described in the two posters that are referred to in the examples below.

Quantitative Structural Activity Study on PETT-Compounds using Calculated Descriptors from Ab Initio Wave Functions [22]. - It is well known that derivatives from PhenylEthylThiazole-Thiorea, PETT (**1**) inhibit the replication of HIV viruses by binding to the viruses' RT enzyme. The problem with this therapy is the rapid development of resistance due to mutations on the RT enzymes. Therefore the cell tests where performed on the most important clone, CL22. The descriptors where calculated with MolSurf for all 61 molecules that had been tested. SIMCA was used for the PLS analyses. Two outliers where taken out from the set leaving 59 molecules.
The number of descriptors where reduced from 125 till only 10(!) Six of them described the influence from the geometry. The bonding sites where :

- a Lewis base or a hydrogen bond acceptor at R1
- a nucleophilic aromatic ring donating in p-interaction
- a hydrogen bond donor and/or acceptor at any one of the meta and ortho substituents.

The cross validity coefficient using the 'leave-one-out' technique for this model was 0.813.

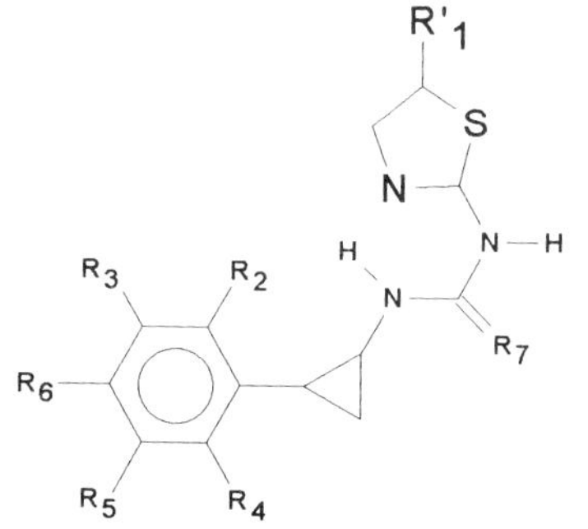

1

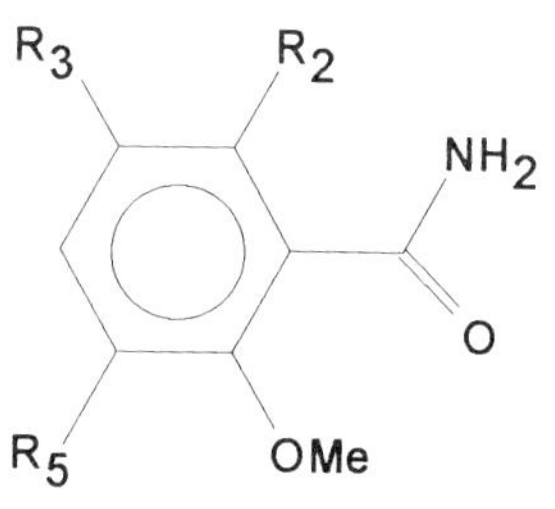

2

QSAR on Substituted Salicyamides Using MolSurf Descriptors [23]. - The binding affinity to [^{3}H]spiperone for 60 substituted salicylamides where modeled. The idea was to compare a study with MolSurf with an earlier work using physico-chemical variables of traditional type [24]. The calculations where made on model compounds (**2**) where the pyrrolidone side chain where omitted since this part is constant in the investigated series. The number of descriptors where reduced from 71 till 6. One is the size of the molecules two where log P for substituents, one the polarizability for the molecule, and three descriptors indicated specific bonding. The bonds where nucleophilic π-interaction from the aromatic, hydrogen bonding to R5 and the methoxy group (R5 and methoxy = acceptors). Note that all three specific attractions are of the same character; binding to an electrophile or an acid hydrogen at the receptor. It is likely they are all interacting with the same electrophile. Note also that we picked up the bonding to the methoxy despite this part is constant in the set of investigated molecules. The reason is inductive effects from the varied substituents will change the charge distribution in the methoxy groups. This effect would not shown up in any method where you use physico-chemical parameters. The cross validation coefficient using 'leave-one-out' was 0.83.

REFERENCES

[1] MolSurf is marketed by Qemist AB, Sweden. E-mail for inquires to: *par.sjoberg@mbox309.swipnet.se*

[2] P. Politzer, D.G. Truhlar. Eds., 'Chemical Applications of Atomic and Molecular Electrostatic Potentials', Plenum Press, New York, 1981.

[3] P. Politzer, J.S. Murray, in 'Reviews of Computational Chemistry', VCH Publishers, New York, 1991.

[4] P. Sjöbergs Ph.D. Dissertation 'Molecular Surface Properties - A Guide for Chemical Reactivity', Department of Chemistry, University of New Orleans, La., Sept. 1989.

[5] G06F 15/46, G01N 33/00. The patent claims the use of calculated molecular properties at the molecular surface as a tool for designing molecules with biological activity.

[6] P. Politzer, D.G. Truhlar, Eds., 'Chemical Applications of Atomic and Molecular Electrostatic Potentials', Plenum Press, New York, 1981.

[7] P. Sjöberg, J.S. Murray, T. Brinck, P. Politzer, *Can. J. Chem.* **1990**, *68*, 1440.

[8] T.A. Koopmans, *Physica* **1933**, *1*, 104.

[9] T. Brinck, J.S. Murray, P. Politzer, R.E. Carter, *J. Org. Chem.* **1991**, *56*, 2934-2936.

[10] Kamlet, M.J., Abboud, J.-L. M., Abraham, M.H., Taft, R. W., *J. Phys. Chem.* **1988**, *92*, 5244.

[11] Brinck, T.; Murray, J. S.; Politzer, P. *Mol. Phys.* **1992**, *76*, 609.

[12] T. Brinck, J.S. Murray, P. Politzer, *J. Org. Chem.* **1993**, *58*, 7070-7073.

[13] J.S. Murray, T. Brinck, P. Politzer, *J. Phys. Chem.* **1993**, *97*, 13807-13809.

[14] K.M. Gough, *J. Chem. Phys.* **1989**, *91*, 2424.

[15] K. E. Laidig, R.F.W. Bader, *J. Chem. Phys.* **1990**, *93*, 7213.

[16] K.J. Miller, *J. Am. Chem. Soc.* **1990**, *112*, 8533, 8543.

[17] T. Brinck, J. S. Murray, P. Politzer, *J. Chem. Phys.* **1993**, *98*.

[18] Variable of importance used in SIMCA.

[19] This project was a collaboration between Qemist AB, P. Lind, Medivir AB, and U Norinder, Astra Pain Control AB.

[20] This project was a collaboration between Qemist AB, U Norinder, Astra Pain Control AB, and S. Rännar, Dep. Org. Chem., University of Umeå.

[21] U. Norinder, T. Högberg, *Acta Parm. Nordica* **1992**, *4*, 73-78.

PART I. CHEMOMETRICS

Chapter 6

Kohonen Neural Network: A Novel Approach to Search for Bioisosteric Groups

by **Soheila Anzali**[a]), **Gerhard Barnickel**[a]), **Michael Krug**[a]), **Markus Wagener**[b]) and **Johann Gasteiger**[b])

[a])E. Merck, Department of Medicinal Chemistry/Drug Design, D-64271 Darmstadt, Germany
[b])Computer-Chemie-Centrum, University Erlangen-Nürnberg, D-91052 Erlangen, Germany

Kohonen Neural Network: A Novel Approach to Search for Bioisosteric Groups

Soheila Anzali, Gerhard Barnickel, Michael Krug
E. Merck, Department of Medicinal Chemistry/ Drug Design,
D-64271 Darmstadt, Germany

Markus Wagener, Johann Gasteiger
Computer-Chemie-Centrum, University Erlangen-Nürnberg,
D-91052 Erlangen, Germany

It will be shown how a self-organizing neural network such as the one introduced by Kohonen can be used to analyse features of molecular surfaces such as shape and the molecular electrostatic potential. On the one hand, two-dimensional maps of molecular surface properties can be generated and used for the comparison of a set of molecules. On the other hand, the surface geometry of one molecule can be stored in a network and this network can be used as a template for the analysis of the shape of various other molecules. The bioisosteric-database by Istvan Ujvary (BIOISOSTER version 1.3), a database for analogue design, including 1515 bioisosteric groups has been probed. A fragment library of small "fragments and functional groups" together with their corresponding Kohonen maps (Kohonen-Map-Library, KML) has been generated. The comparison of the electrostatic potential patterns of the maps of such a library can be used to cluster the bioisoteric groups and consequently improve the efficiency of the 3D-design of bioactive molecules. This will be shown using an example of 5HT1a-agonists.

Introduction.- The shape of a molecule and electronic properties such as the electrostatic potential strongly influence many chemical and biological properties of a molecule. The intuitively appealing picture of a key fitting into a lock for describing substrate-receptor interactions emphasises the role of shape in biological activity. However, not only the geometry of a molecule must be appropriate to fit it into a receptor, but also certain properties on the molecular surface such as hydrophobicity, the electrostatic or the hydrogen bonding potential must fit to ensure an effective binding. Several approaches based on shape analysis have been developed to study structure-activity relationships within a series of individual compounds. In the last decade, the three-dimensional comparison of molecular electrostatic fields and volumes was successfully applied in many cases to reveal effective structure-activity relationships [1]. The human brain is highly efficient in performing an analysis of three-dimensional objects and of properties on the surface of these objects such as shape, color, and texture. This is achieved by the generation of sensory maps of the environment in the visual, auditory, or somatosensory cortex. Artificial neural networks are models for the information processing in the human brain. The application of neural networks in chemistry has increased dramatically in recent years [2-4]. The neural network method developed by Kohonen [5-8] is rather efficient in modeling the generation of sensory maps in the brain. In the Kohonen network the artificial neurons self-organize in an unsupervised learning process and thus can be used to generate topological feature maps. This property of a Kohonen neural network has already been used for the mapping of molecular surface properties such as the molecular electrostatic potentials (MEP) into two dimensions [9-11]. The use of a Kohonen network in SAR and drug design is also given [12]. We will show here that a Kohonen neural network can also be used to analyse the shape and the surface properties of molecules simultanously to give a quantitative definition of similarity for the molecular surfaces. In addition, the mapping of molecular surface properties of several hunderd small fragments and functional groups by Kohonen networks (a Kohonen-Map-Library, KML) can be used to find bioisosteric groups for a given fragment from a highly active compound.

Methods.- The structures were built from connection tables using the program CORINA [13]. Charge distributions were calculated by the PM3 (MOPAC version 5.0). The MEP is calculated for the van der Waals surface by Coulomb's law. A Kohonen net is trained by random-sampling of points on a van der Waals surface as input using the KMAP simulator (Li et al., unpublished results, 1993). The trained neurons were then colored according to the MEP values at these points.
The paramters of the network were : size = 50*50, step (Δ) = 400, learning rate (h_0) = 0.3, span (σ_0) = 15, reduction speed for learning rate (λ_η) = 0.95, reduction speed for span (λ_σ) = 1, training steps = 20000. The topology of the network is a torus. The main advantage of the toroidal network (compared to a rectangular network) is that there is no boundary on a torus and therefore all neurons in the net are treated equally and the result is a map without boundaries, which may be circurlarly shifted along the vertical or horizontal direction.
Also depending on the random initialization of the Kohonen network and the way the datapoints are presented to the network, as well as to the sites chosen for making the cuts into the torus, the Kohonen maps can be oriented in variety of ways. Through a toroidal network, the maps can be shifted, mirrored and rotated against each other to achieve a similar position of their patterns.
η_0, σ_0, λ_η, λ_σ and step are parameters which have to be chosen according to the individual tasks. The η_0 determines the initial maximal modification rate, the σ_0, called the learning span, determines the initial size of the neighbourhood around the winner neuron in which the connection weights will be modified. The parameters step, λ_η, λ_σ determine together the speed of the convergence of the training process. The choice of proper training parameters depends generally on the target object, the net configuration and the individual requirements. It demands some experience. To determine the proper values for these parameters, several molecules were examined. For each molecule 50 Kohonen maps were trained. The experiment was repeated with different set of parameters and it was discovered that the 50 obtained Kohonen maps for every molecule with the selected parameters above show solid constant patterns. But nevertheless, the results obtained with other proper parameters show very little variations to the above parameters.

Kohonen Network.- The Kohonen neural network is a self-organizing network, which can be used to generate a non-linear projection of objects from a high-dimensional space into a lower-dimensional space, usually a two-dimensional plane. This method enables the decrease of the dimensions of the information space while conserving the topology of the information as well as possible. Learning in a Kohonen network is an unsupervised competitive process so that an object is mapped into a neuron. Each neuron has as many weights w_{ij} as there are input variables x_{si} for an object, s. The winning neuron, c_s, will be the one that has weights that come closest to the input variables (*Eq.1*).

$$c_s \Leftarrow \min\left[\sum_{i=1}^{n}(x_{si} - w_{ji})^2\right] \qquad (1)$$

The learning process adjusts the weights so that they become even closer to the input variables, but it does so with a factor that decreases with distance from the winning neuron.

We have shown that a Kohonen network can be used for the generation of maps of molecular surface properties by projecting these surfaces of three-dimensional objects into two dimensions [9-11]. For example, maps of the molecular electrostatic potential (MEP) can thus be obtained. Input to the Kohonen network consists of the Cartesian coordinates of points on the molecular surface. With three input variables each neuron also has three weights. The neurons of such a Kohonen network are arranged in two-dimensions. In the projection of the molecular surfaces of the molecules studied here, an arrangement of 50*50 neurons was used. The surface of a molecule is continuous, without a beginning and end. Thus, it seemed desirable to also have an arrangement of neurons without beginning and without end. This can be achieved by arranging the neurons on the surface of a torus. For visualization, this torus is cut along two perpendicular lines and then the torus surface is spread into a plane. Thus the neurons placed at the outskirts of the square shown in *Figure 1* are made the neighbors of the ones located on the other side. Since these cuts can be made at arbitrary lines, the maps can be shifted into any direction.

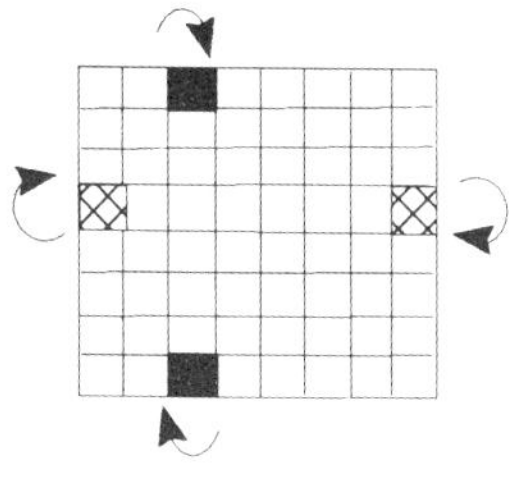

Fig.1. *Plane obtained by making two cuts into a torus, which can be shifted into any direction. Thus, the two neurons marked by a cross are direct neighbors as are the two neurons shown as black squares.*

The method for obtaining two-dimensional maps of molecules consists of training the Kohonen network with the three Cartesian coordinates of points of the molecular surface randomly spotted on the Van der Waals surface of the molecule. With a Kohonen network of 50*50 neurons, there are 3*50*50 = 7500 weights. The architecture of the Kohonen network is shown in *Figure 2*.

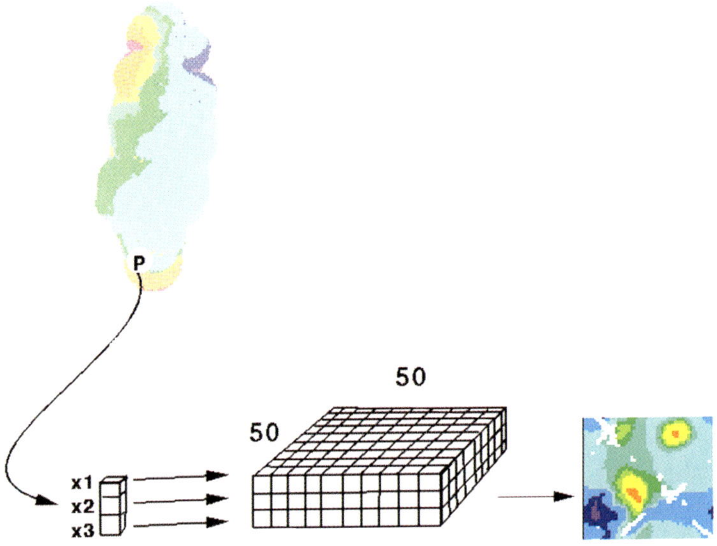

Fig.2. *Mapping of a molecular surface into a Kohonen network leading to a map of the molecular electrostatic potential.*

As the size of the network is much smaller than the size of the dataset (number of points from the molecular surface), a neuron is normally excited by several points of the surface. In this case, the neuron obtains the average value of the information of those points that excite it. The information on the molecule surface - the properties on the surface, in our case, the individual electrostatic values of each of the coordinates of the van der Waals surface - is not used in the learning process. Thus, during processing by unsupervised learning, the points with similar coordinates are put closely together by the network into the same or adjacent neurons.

Clearly, the topology of a molecular surface which is more like a sphere or an ellipsoid is different from the topology of a torus. This necessarily must lead to topological distortions during the projection of the surface of a molecule onto the surface of a torus. During learning in a Kohonen network the torus adjusts itself as well as possible to the molecular surface. Nevertheless, the topological distortions manifest themselves in neurons that are not used during projection [14]. These empty neurons show up as white areas in the resulting Kohonen maps as can be seen in the following figures.

From a 3D-Space to a 2D-Map.- In order to give an idea of the correspondence between a 3D-space and its 2D-map, we show in *Figure 3* an example of a projected Kohonen map of the Van der Waals surface of corticosterone. This representation permits one to realize the correspondence between the spatial arrangement of the functional groups of the steroid and the colors in the 2D Kohonen map. Corticosterone has two sites with a large negative value of the MEP, the carbonyl group at position 3 (3) and the side chain COCH2OH at position 17 (1). Consistent with this, the Kohonen map shows two spaces with a red-yellow color for these sites. The spatial distance of these groups is reflected by two different shapes of the projection of the MEP into the Kohonen network. The third site with a negative value of the MEP stems from the hydroxyl group at position 11 (4). For this group a space with a yellow color is reserved in this map. Furthermore, the large positive MEP area of corticosterone is below the D-ring and the sidechain at position 17 (2). The projection of the MEP into the Kohonen map indicates the location of this space (violet color) close to the space of the negative MEP area of COCH2OH at position 17.

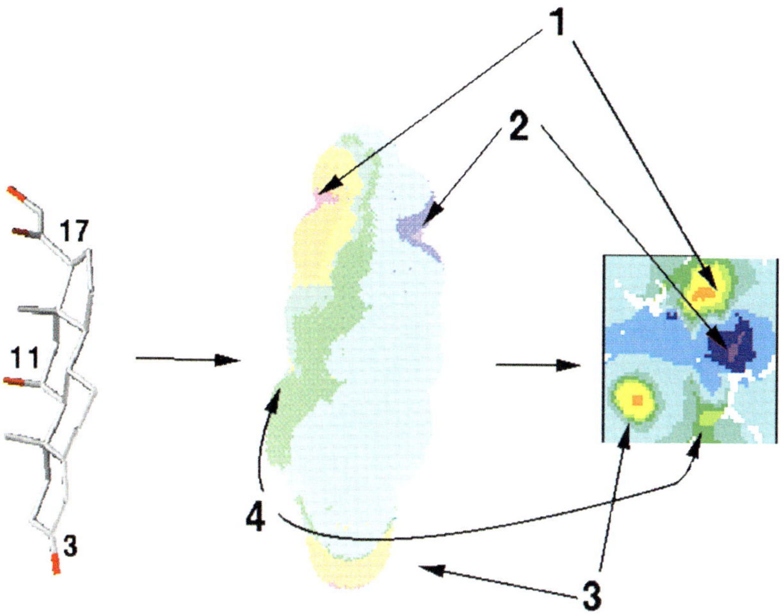

Fig.3. *3D-model of corticosterone, parallel projection of the electrostatic potential onto the Van der Waals surface of corticosterone, and the corresponding Kohonen map.*

Bioisosteric Effect can be Deduced by a Comparison of the ESP-Maps. - The bioisosteric-database by Istvan Ujvary (BIOISOSTER version 1.3), a database for analogue design, including 1515 bioisosteric groups was analysed. The question was if some coherency between the physicochemical properties and the bioisosteric effect can be deduced by looking at the calculated Kohonen maps of such groups. *Figure 4* shows an example of such a structure-pair in this database. The squares show the part of these structures, which are defined as bioisosteric groups.

Several hunderd pairs from this database were selected. The selection of these compounds was based on diverse structural fragment-pairs as far as possible. From the selected database the bioisosteric groups are then cut out. The 3D-structures of the selected

BIOISOSTERIE-DB **ACI039**

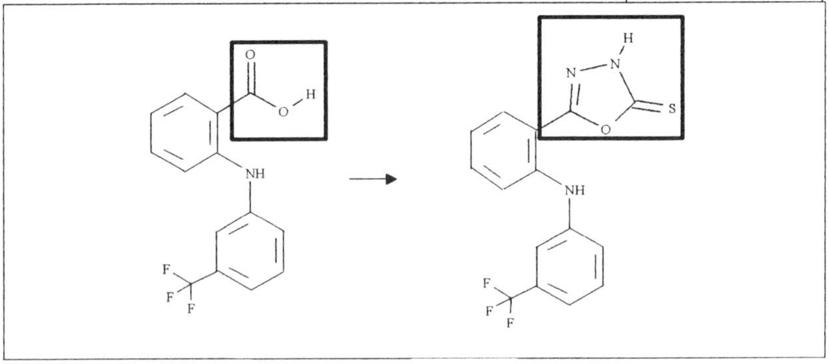

Transform:

carboxylic acid - 1,3,4-oxadiazole-2- thione	Bio_Activity: **antiinflammatory cyclooxygenase inhibitor lipoxygenase inhibitor**

Reference:

Boschelli, D.H. et al.: Bioorg.Med.Chem.Lett. 1992, 2, 69-72; Boschelli, D.H. et al.: J.Med.Chem. 1993, 36, 1802-1810; see also: Boschelli, D.H. et al.: Bioorg.Med.Chem.Lett. 1992, 2, 705-708.

Fig.4. *A structure-pair of the Bioisostere-Database. The squares show the bioisosteric-groups.*

fragments are calculated using the program CORINA. The MEP were calculated on the Van der Waals surface. Thus we calculated the Kohonen maps for each bioisosteric-groups with a unique color plate.
More than 50 per cent of the calculated map-pairs show similarity on the electrostatic potential patterns. *Figure 5* shows some examples of the calculated fragments. As shown here the fragment-pairs are structurally very different, but their maps show a high similarity of the electrostatic potential patterns.

This is an interesting result because of the possibility to select fragment-pairs of this database, which can have a general validity : "*true bioisosteric-groups*". The result also could be seen as a proof for the use of the Kohonen maps for such a comparison of any fragment-pair.

Therefore we constructed a 3D-database of several hunderd fragments and functional groups including their corresponding Kohonen maps. The comparison of the electrostatic potential patterns of the maps of such a library can be used to cluster the

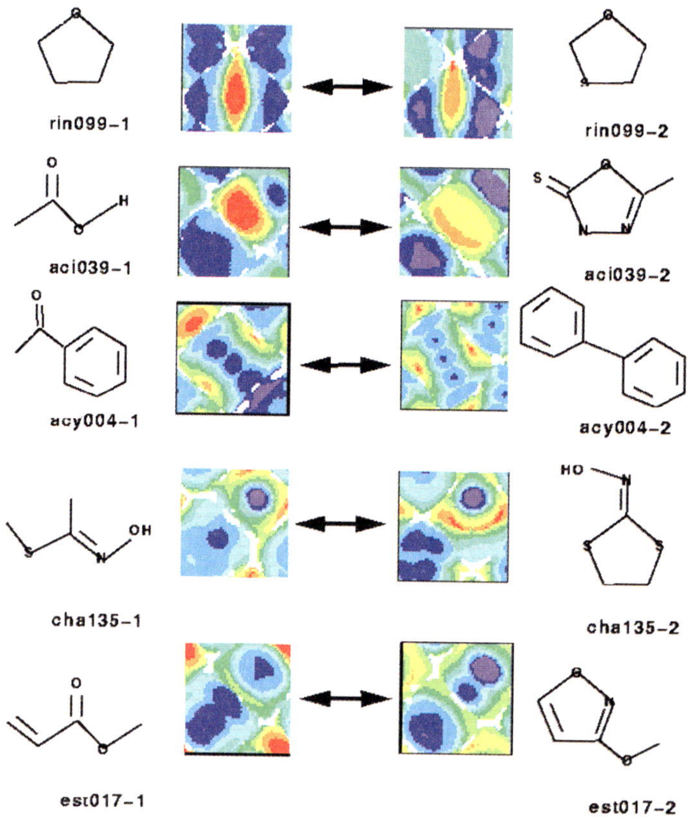

Fig.5. *Some examples of the bioisosteric-groups and their calculated Kohonen maps.*

bioisosteric groups and consequently improve the efficiency of the 3D-design of bioactive molecules. As an example in *Figure 6*, the most active compound of 3-(1,2,5,6-tetrahydropyridin-4-yl)indole derivative [15], a 5HT1a-ligand, is shown.

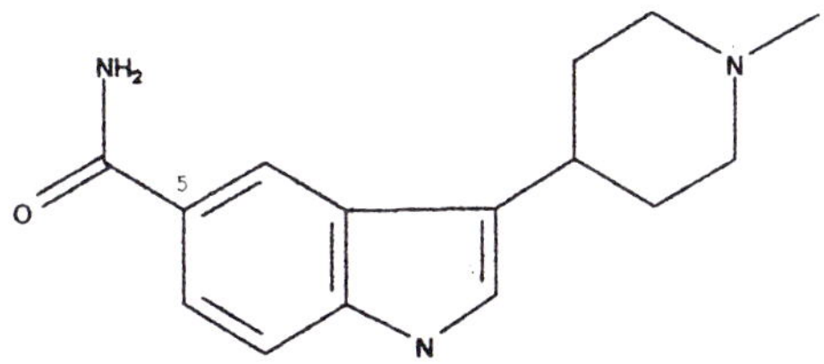

Fig.6. *The most active compound of 3-(1,2,5,6-tetrahydropyridin-4-yl)indole derivative.*

It is known that the amide group at position 5 is an essential part of the interaction of 5HT1a-ligands with the receptor. So, we cut out the amide group and calculated it's Kohonen map. All of the maps of the Kohonen-map-library were then probed for any similar electrostatic potential patterns to the map of the amide group. *Figure 7* shows 4 fragments with their corresponding Kohonen maps, which are selected as bioisosteric-candidates for the amide group with similar map-patterns. Such a selection makes it possible, that the optimization of compounds could be done faster and the results are more effective for further investigation at the 3-dimensional level.

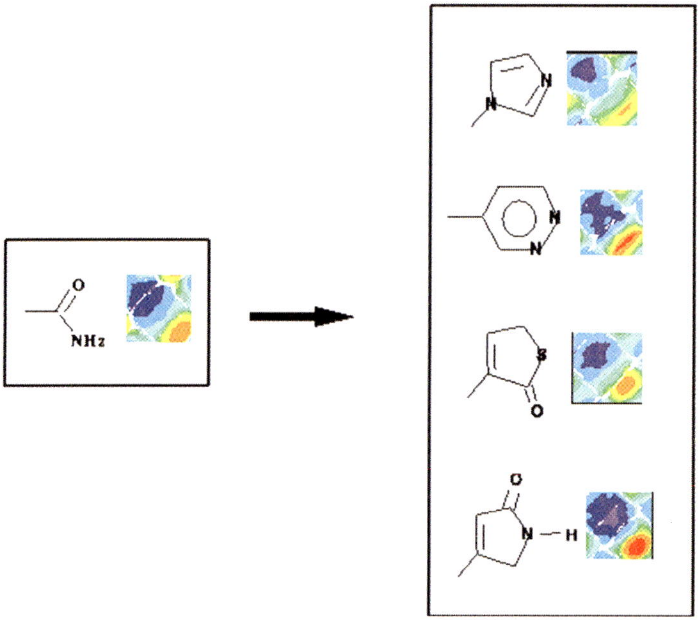

Fig.7. *Four selected bioisoteric-candidates for the amide-group by the comparison of their calculated Kohonen maps.*

Template Approach. - The binding of a ligand to a receptor is a combination of steric and electrostatic factors. Using the template approach both factors can be considered. The basis of the template approach is an analysis of the shapes of molecules and the quantification of a shape-similarity or-dissimilarity within a series of compounds using a reference molecule. A reference network is trained with coordinates of points from the van der Waals surface of the most active and rigid compound. Then the analogs are filtered through this network and thus produce comparative maps. The dissimilarity of the comparative maps to the reference map can then be deduced by the size of the blank areas (empty neurons) in these maps. The larger the difference in shape between the reference molecule and the compared molecule, the more empty neurons (blank areas) are obtained in the comparative map. A preceding superposition of the molecules is required for this approach. A model of the template approach is given in *Figure 8*.

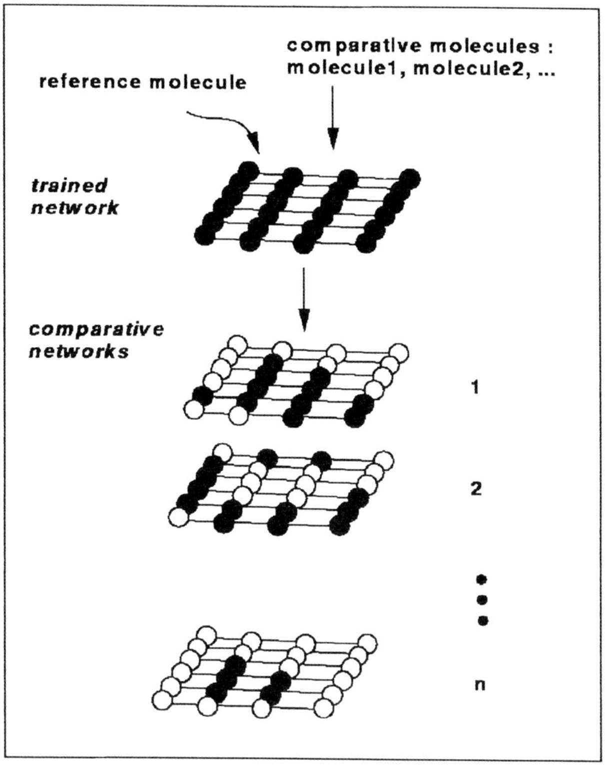

Fig.8. *A model of the template approach. The reference molecule provides a template network, molecules that are to be compared are sent through this network leading to comparative networks.*

In Search of the Best Superposition. - For comparison of the information included in the Kohonen maps we used the following two measure :

I_{AB} : Scaled number of neurons, a_{ij}, from the reference map A and neurons, b_{ij}, from the compared map that have quite similar electrostatic potential values as expressed by the same color range ($c(a_{ij})$, $c(b_{ij})$).

D_{AB} :Scaled mean difference of neurons, a_{ij}, from the reference map A and neurons, b_{ij}, from the compared map that have different electrostatic potential values.

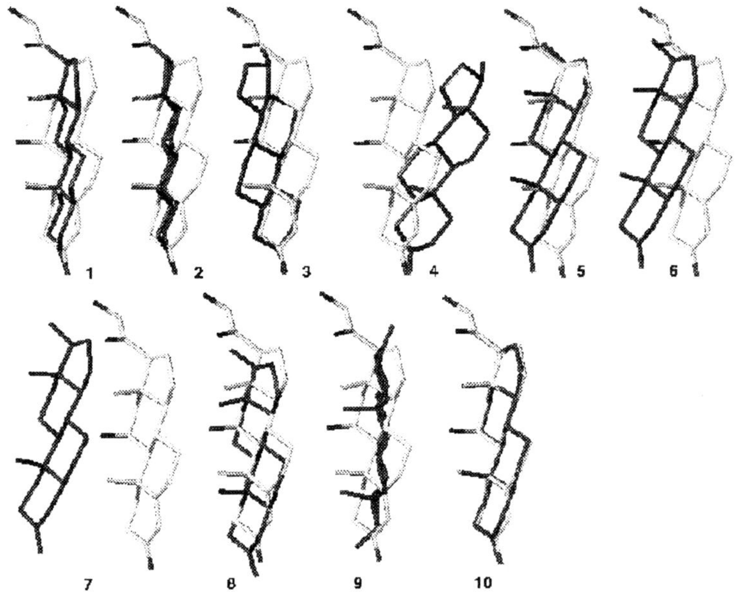

Fig.9. *Ten various spatial positions of 5-androstendiol (gray color) with regard to corticosterone (black color)*

These are given by *Eq. 2* and *3* :

$$I_{AB}[\%] = \frac{100}{n \cdot m} \cdot i_{AB} \quad \text{with} \quad i_{AB} = \sum_{i}^{n} \sum_{j}^{n} \delta_{ij} \quad \text{and} \quad \delta_{ij} = \begin{cases} 1 & \text{for } c(a_{ij}) = c(b_{ij}) \\ 0 & \text{for } c(a_{ij}) \neq c(b_{ij}) \end{cases} \qquad (2)$$

$$D_{AB}[\%] = \frac{100}{n \cdot m - i_{AB}} \cdot \frac{1}{NC} \sum_{i}^{n} \sum_{j}^{n} |c(a_{ij}) - c(b_{ij})| \qquad (3)$$

with NC = number of colors
n . m = total number of neurons

Thus a similarity index between two Kohonen maps can be calculated by :

$$S_{AB}[\%] = \frac{I_{AB}}{D_{AB}} \cdot 100 \qquad (4)$$

Through the simlarity index, the MEP and the molecular shape are simultaneously taken into account.

The next example shows the use of a template approach for finding the best alignment of two structures. Here, we compare two steroid molecules from a series of compound, that bind to the corticosteroid binding globulin (CBG) [16]. The reference molecule is corticosterone, a higly active compound. The molecule to be compared is 5-androstendiol, a low active compound. 5-androstendiol is taken in ten different spatial positions generated manually by a random tranformation of each of the three coordinate axis (see *Figure 9*).

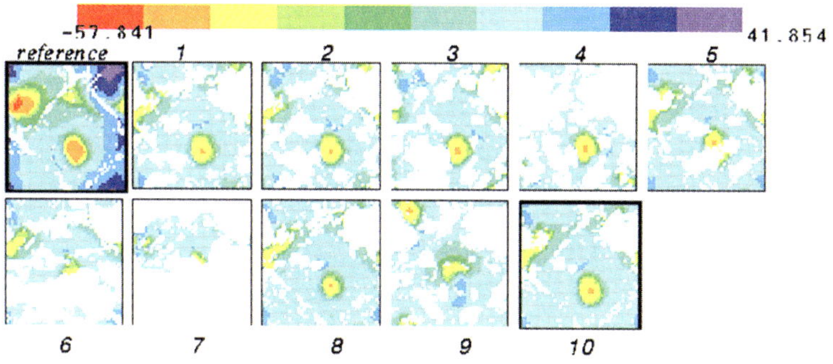

comparative map	N_e	S_{AB}
1	918	71.6
2	1112	50.9
3	1195	39.5
4	1663	40.0
5	826	88.8
6	1484	34.6
7	2011	15.1
8	739	93.4
9	1089	52.5
10	**561**	**134.7**

Fig.10. *The maps obtained for ten various spatial positions of 5-androstendiol (comparative maps) using the Van der Waals surface of corticosterone as the reference molecule.* (N_e) *= the number of empty neurons,* (S_{AB}) *= the similarity index.*

The maps obtained are shown in *Figure 10* together with their count of empty neurons. The best superposition (compare *Figure 9 and 10*), is deduced from the similarity index and by comparison of the number of empty neurons of the comparative maps. Thus, the best alignment between the compared and the reference molecule can be determined from the number of white neurons of the comparative map : the smaller the number of white neurons, the better the molecular surfaces are matched.

Conclusions. - Using a Kohonen Neural Network the transformation of the calculated information on the molecular surface can be done onto a 2-dimensional plane. For the comparison of the calculated Kohonen maps an alignment of 3D-structures is not required and that is the advantage of this method.
We constructed a 3D-database of various small fragments and functional groups including their Kohonen maps. The use of such a Kohonen Map Library (KML) is shown in an example of prediction of new 5HT1a-ligands.
The template approach is a useful method for comparison of shape and properties of molecules simultanously.
Also, a similarity index used in this approach could supply the best alignment possible between two molecules with respect to their molecular electrostatic potential and shape. The results obtained are encouraging to develop an automatic procedure and open promising perspectives for the fast and accurate comparison of large datasets of molecules by Kohonen neural networks.

Acknowledgement. - The authors acknowledge the Bundesminister für Bildung, Wissenschaft, Forschung und Technologie (BMBF), FRG, (ÄBAV project) grant # 01IB305D for financial support.

REFERENCES

[1] "3D QSAR in Drug Design", Ed. H. Kubinyi, Escom, Leiden, 1993.
[2] J. Zupan, J. Gasteiger, *Anal. Chim. Acta* **1991**, *248*, 1-30
[3] J. Gasteiger, J. Zupan, *Angew. Chem.* **1993**, *105*, 510-536; Angw. Chem. Int. Ed. Engl. 1993, 32, 502-527.
[4] J. Zupan, J. Gasteiger, Neural Networks for Chemists: An Introduction; VCH-Publishers, Weinheim, FRG, 1993.
[5] T. Kohonen, *Biol. Cybern.* **1982**, *43*, 59-69.
[6] T. Kohonen, Self-Organisation and Associative Memory; third Edition, Springer, Berlin, 1989.
[7] T. Kohonen, *Proc. IEEE* **1990**, *78*, 1460-1480.
[8] T. Kohonen, Self-Organizing Maps, Eds. T. S. Huang, T. Kohonen, M. R. Schröder, Springer, Berlin, 1995.
[9] J. Gasteiger, X. Li, C. Rudolph, J. Sadowski, J. Zupan, *J. Am. Chem. Soc.* **1994**, *116*, 4608-4620.
[10] J. Gasteiger, X. Li, A. Uschold, *J. Mol. Graph.* **1994a**,*12*, 90-97.
[11] J. Gasteiger, X. Li, *Angew. Chem. Int. Ed. Engl.* , **1994**, *33*, 643-646.
[12] S. Anzali, G. Barnickel, M. Krug, J. Sadowski, M. Wagener, J. Gasteiger; In 'Neural Networks in QSAR and Drug Design' (Ed. J. Devillers), Academic Press, London, 1996, pp. 209-222.
[13] J. Sadowski, C. Rudolph, J. Gasteiger, *Anal. Chim. Acta*, **1992**, *265*, 233-241.
[14] X. Li, J. Gasteiger, J. Zupan, *Cybern.*, **1993**, *70*, 189-198.
[15] A. Agarwal, Ph. P. Pearson, E. W. Taylor, H. B. Li, T. Dahlgren, M. Herslöf, Y. Yang, G. Lambert, D. L. Nelson, J. W. Regan, A. R. Martin, *J Med. Chem.* **1993**, *36*, 4006-4014
[16] A.C. Good, S.S. So, W.G. Richards, *J. Med Chem*, **1993**, *36*, 433-438.

PART I. **CHEMOMETRICS**

Chapter 7

Genetic Selection Test Series

by **James Devillers*** and **Daniel Domine**

CTIS, 21 rue de la Bannière, F-69003 Lyon (France)

Genetic Selection of Test Series

by **James Devillers*** and **Daniel Domine**

CTIS, 21 rue de la Bannière, F-69003 Lyon (France)

A genetic algorithm (GA) was used for the optimal selection of a set of aliphatic substituents highly representative of a physicochemical parameter space. The parameters considered were the hydrophobic constant for aliphatic substituents Fr, H-bonding acceptor (HBA) and donor (HBD) abilities, the molar refractivity (MR), and the inductive parameter F, respectively. The GA results were evaluated by means of graphical displays derived from a nonlinear mapping of the data set. They agreed with those obtained with hierarchical cluster analysis, Kohonen network, N2M, and ART-2-A network.

Introduction. - In drug design it is often required to search through a huge number of possibilities for solutions. Problems to be solved need computer programs to be adaptive to continue to work well in a changing environment or, to be innovative to construct something truly new and original. Under these conditions, biological evolution appears as an appealing source of inspiration for addressing numerous problems in computer-aided molecular design [1]. Indeed, the natural evolution can be viewed as a process allowing the search for "solutions" among an enormous number of possibilities. The solutions are represented by organisms occupying different trophic levels in the environment and able to survive and reproduce while the set of possibilities deals with the genetic sequences of these organisms. The natural evolution can also be seen as a process for designing innovative solutions to complex problems [2, 3]. The fitness of an organism depends of many physiological, ecological, and ethological factors. The fitness criteria continually change as organisms evolve. Thus, natural evolution is searching a constantly changing set of possibilities. Searching for optimal solutions in the face of changing conditions by using principles of the Darwinian evolution is precisely what it is performed by genetic algorithms (GAs) [1, 4-8]. Invented by John Holland in the 1960s, GAs have proved their usefulness in QSAR and drug design [1]. Under these conditions, the aim of this study was to underline their interest for designing optimal test series from a data matrix of 103 aliphatic substituents (*Table 1*) described by the hydrophobic constant for aliphatic substituents Fr, H-bonding acceptor (HBA) and donor (HBD) abilities, the molar refractivity (MR), and the inductive parameter F, respectively [9].

GA Implementation. - In the spirit of analogy with biology, GAs work on populations of chromosomes and involve three types of operators: selection, crossover, and mutation (*Fig. 1*). Each chromosome typically takes the form of a bit string. It is divided into genes encoding a trait and located at a particular locus (position) on this chromosome. GAs require a fitness function that assigns a score to each chromosome in the population. The fitter the chromosome, the more times it is likely to be selected to reproduce. Basically, crossover consists of exchanging subsequences of genetic material between two single-chromosome

Table 1. *Aliphatic Substituents.*

No.	Substituent	No.	Substituent
1	Br	2	Cl
3	F	4	I
5	NO_2	6	H
7	OH	8	SH
9	NH_2	10	CBr_3
11	CCl_3	12	CF_3
13	CN	14	SCN
15	CO_2^-	16	CO_2H
17	CH_2Br	18	CH_2Cl
19	CH_2I	20	$CONH_2$
21	CH=NOH	22	CH_3
23	$NHCONH_2$	24	OCH_3
25	CH_2OH	26	$SOCH_3$
27	OSO_2CH_3	28	SCH_3
29	$NHCH_3$	30	CF_2CF_3
31	C≡CH	32	CH_2CN
33	$CH{=}CHNO_2$-(trans)	34	$CH{=}CH_2$
35	$COCH_3$	36	$OCOCH_3$
37	CO_2CH_3	38	$NHCOCH_3$
39	$C{=}O(NHCH_3)$	40	CH_2CH_3
41	OCH_2CH_3	42	CH_2OCH_3
43	SOC_2H_5	44	SC_2H_5
45	$CH_2Si(CH_3)_3$	46	NHC_2H_5
47	$N(CH_3)_2$	48	CH=CHCN
49	Cyclopropyl	50	COC_2H_5
51	$CO_2C_2H_5$	52	$OCOC_2H_5$
53	$EtCO_2H$	54	$NHCO_2C_2H_5$
55	$CONHC_2H_5$	56	$NHCOC_2H_5$
57	$CH(CH_3)_2$	58	C_3H_7
59	$OCH(CH_3)_2$	60	OC_3H_7
61	$CH_2OC_2H_5$	62	SOC_3H_7
63	SC_3H_7	64	NHC_3H_7
65	$Si(CH_3)_3$	66	2-Thienyl
67	3-Thienyl	68	$CH{=}CHCOCH_3$
69	$CH{=}CHCO_2CH_3$	70	COC_3H_7
71	$OCOC_3H_7$	72	$CO_2C_3H_7$
73	$(CH_2)_3CO_2H$	74	$NHCOC_3H_7$
75	$CONHC_3H_7$	76	C_4H_9
77	$C(CH_3)_3$	78	OC_4H_9

Table 1. *(Continued)*.

No.	Substituent	No.	Substituent
79	$CH_2OC_3H_7$	80	NHC_4H_9
81	$N(C_2H_5)_2$	82	$CH{=}CHCOC_2H_5$
83	$CH{=}CHCO_2C_2H_5$	84	C_5H_{11}
85	$CH_2OC_4H_9$	86	C_6H_5
87	OC_6H_5	88	$SO_2C_6H_5$
89	NHC_6H_5	90	2-Benzthiazolyl
91	$CH{=}CHCOC_3H_7$	92	$CH{=}CHCO_2C_3H_7$
93	COC_6H_5	94	$CO_2C_6H_5$
95	$OCOC_6H_5$	96	$NHCOC_6H_5$
97	$CH_2C_6H_5$	98	$CH_2OC_6H_5$
99	$CH_2Si(C_2H_5)_3$	100	$CH{=}CHC_6H_5$-(trans)
101	$CH{=}CHCOC_6H_5$	102	Ferrocenyl
103	$N(C_6H_5)_2$		

haploid parents. Mutation randomly flips some of the bits in a chromosome. A classical GA works as follows [3-8]:

1. Start with a randomly generated population of *n* chromosomes representing candidate solutions to a clearly defined problem.
2. Calculate the fitness $f(x)$ of each chromosome x in the population.
3. Generate offspring from parent chromosomes selected in the current population. Use crossovers and mutations for optimizing the process.
4. Replace the current population with the new population.
5. Go to step 2.

Each iteration of the above process is called a generation. A GA is typically iterated from 50 to 1000 generations. The entire set of generations is called a run. At the end of a run, there are often one or more highly fit chromosomes in the population.

For our GA analyses, data were centered (i.e., zero mean) and reduced (i.e., unit variance). Different fitness functions were tested in order to compare their performances for selecting sets of substituents presenting a maximum information. They were derived from the mean Euclidean coefficient [10]. This coefficient was calculated as shown in Eq. (1).

$$d_{12} = \frac{\sqrt{\sum_{i=1}^{m} \left(z_{i1} - z_{i2}\right)^2}}{m} \qquad \textbf{(1)}$$

where z_{i1} and z_{i2} represent the values of the ith variable for the genes 1 and 2 in a chromosome and m represents the total number of variables.

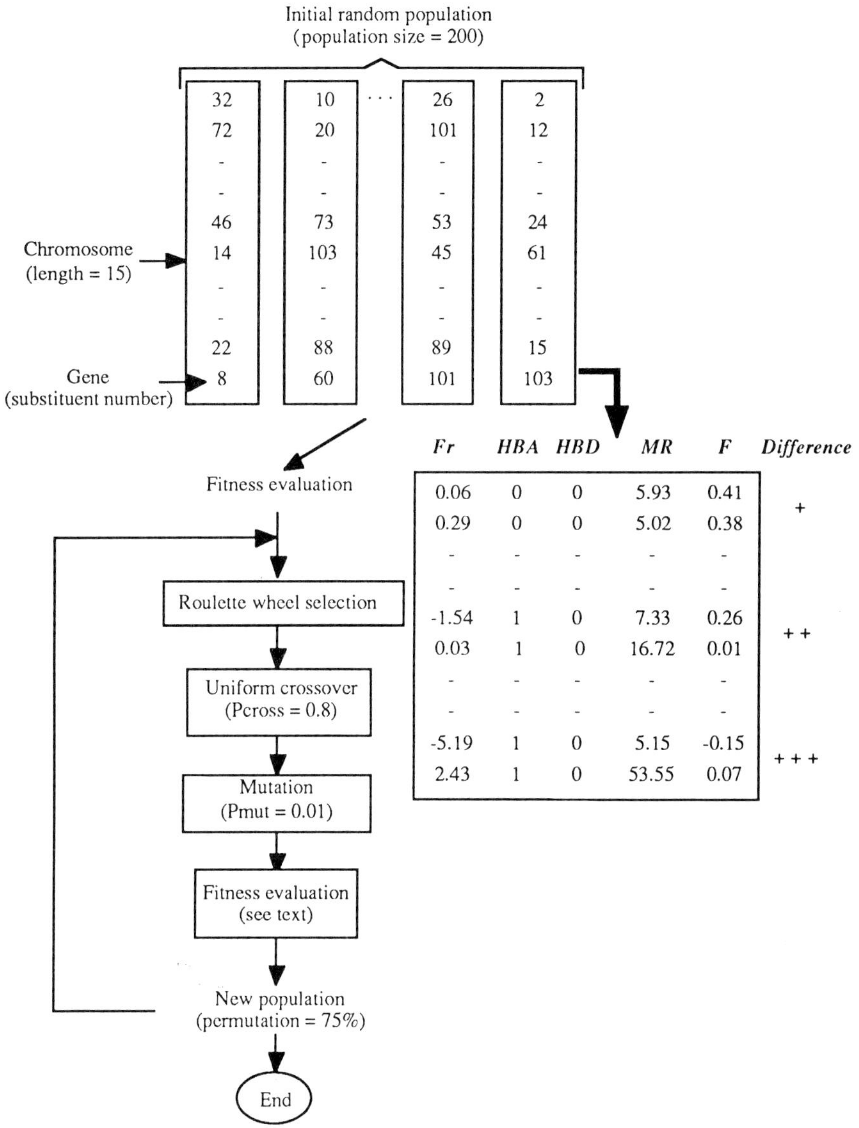

Figure 1. *GA set-up. The parameters used to run the GA were optimized by means of a trial and error procedure from a huge number of runs. The GA was stopped after 1000 generations or if no improvement of the best chromosome was observed during 300 generations.*

This coefficient was calculated for each possible couple of genes in the chromosome. The number of combinations (N) for n genes is given by Eq. (2).

$$N = C_n^2 = \frac{n!}{2!(n-2)!} = \frac{n(n-1)}{2} \quad (2)$$

Four different fitness functions were designed from Eq. (1). In all cases, the aim was to obtain a maximal d_{ij} value. The first one was the mean of the N calculated distances (Eq. (3)).

$$Fitness = \frac{\sum_{ij} d_{ij}}{N} \quad (3)$$

The second fitness function was derived by adding a constraint (Eq. (4)) implying that at least one of the terms of the distance equation had to be superior to a given value.

$$C_i = x \times \sqrt{(\max_i - \min_i)^2} \quad (4)$$

In Eq. (4), x is a coefficient whose value is determined from a trial and error procedure. "$\max_i$" and "$\min_i$" represent the maximum and minimum values for Fr, HBA, HBD, MR, or F. If for all couples of genes in a chromosome, the constraint (Eq. (4)) is satisfied, the chromosome is maintained in the population otherwise it is eliminated. This strategy allowed to remove chromosomes having couples of substituents which were relatively distant but presented only average differences for all the parameters.
The last two fitness functions used the minimum value of the N calculated mean Euclidean distance (Eq. (5)). In both cases, the aim was to maximize this value but in one case, the constraint previously described (Eq. (4)) was added.

$$Fitness = \min(d_{ij}) \quad (5)$$

The GA analyses were performed with the STATQSAR Package [11]. Due to the valuable results obtained in a previous study [12], the performances of the different GA configurations were estimated from the graphical display of the best chromosomes on the nonlinear map [13] of the 103 aliphatic substituents described by means of Fr, HBA, HBD, MR, and F.

Results and Discussion. - The correct use of a GA requires to search for an optimal configuration with several parameters which have to be tuned by a trial and error procedure. *Fig. 1* depicts those selected in our study.

The nonlinear map of the 103 aliphatic substituents described by means of the five substituent constants (Fr, HBA, HBD, MR, and F) is shown in *Fig. 2.1*. Note that, this 2-dimensional map is an optimal representation of the original data matrix after scaling since the total mapping error (E) is low (i.e., 5.1 10^{-2}) and inspection of the individual mapping errors [13] reveals that no points carry a large part of the total mapping error (results not shown).

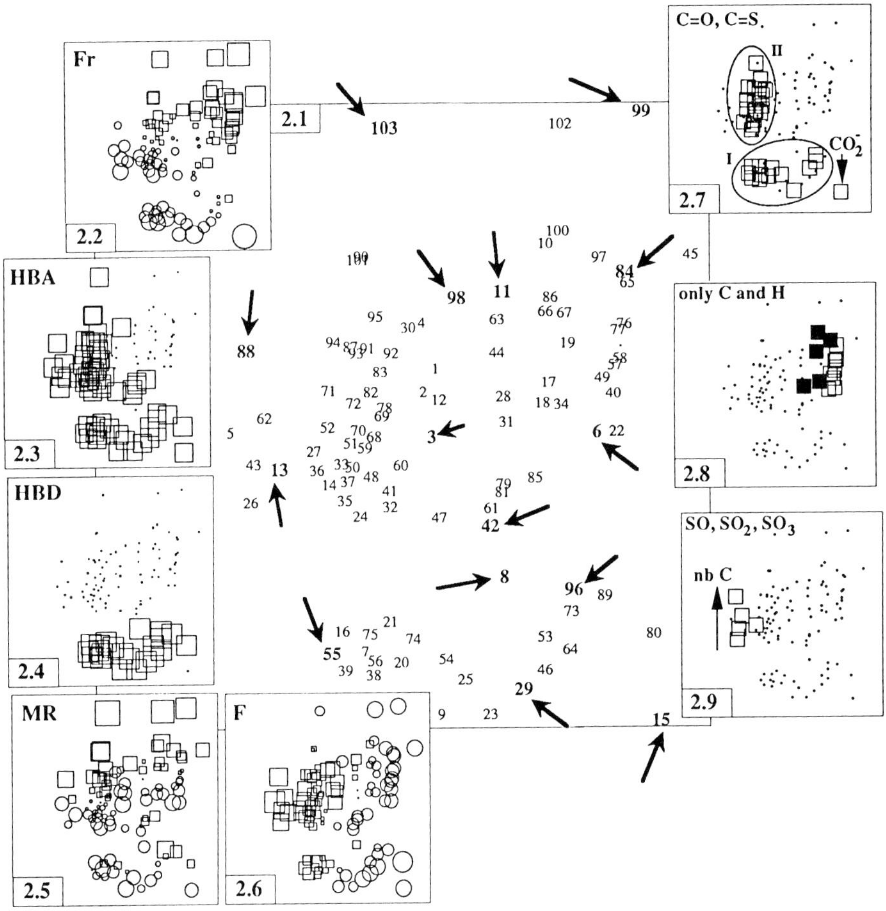

Figure 2. 2.1. *Nonlinear map of the 103 aliphatic substituents described by Fr, HBA, HBD, MR, and F. The arrows show the selection made by the GA. See Table 1 for correspondence between the numbers and the substituents.*
2.2 to 2.6. *Plots of the scaled values of the five parameters on each substituent of the nonlinear map. Squares (positive values) and circles (negative values) are proportional in size to the magnitude of the parameters. In Fig. 2.3 and Fig. 2.4, the dots indicate the substituents which have not the ability to accept and donate H-bonds, respectively.*
2.7 to 2.9. *Plots of the presence of some atoms or functional groups in the aliphatic substituents. The absencé of an element is represented by a dot.*

To confirm that the nonlinear map obtained (*Fig. 2.1*) is representative of the physicochemical space studied and is chemically coherent, collections of maps were drawn by representing the physicochemical data (*Figs. 2.2-2.6*) used to run the NLM and some structural information (*Figs. 2.7-2.9*) on the nonlinear map. *Figs. 2.2-2.6* show that the substituents are distributed and clustered on the nonlinear map according to the values of the parameters. Thus, for example, Fr values of the substituents increase from the bottom to the top of the map (*Fig. 2.2*). In the same way, there are obvious clusterings of HBA and HBD substituents in *Fig. 2.3* and *Fig. 2.4*. *Fig. 2.5* reveals that MR values of substituents generally increase from the bottom to the top of the map. Last, *Fig. 2.6* shows an increasing gradient of the F values running from the right-hand side to the left-hand side of the map. As regards the structural information, substituents containing C=O or C=S groups form two clusters (*Fig. 2.7*). The substituents of the bottom cluster (cluster I) are characterized by HBA and HBD abilities (*Figs. 2.3* and *2.4*) while those of the top cluster (cluster II) can only accept H-bonds. Only substituent no. 15 (CO_2^-) transgresses this rule. *Fig. 2.8* shows that substituents containing only C and H atoms cluster in the right-hand side of the map. These substituents are characterized by low F values (*Fig. 2.6*). It can be noted that the unsaturated substituents (no. 31, 34, 86, 97, and 100 indicated in black) are located on the left side of the cloud of squares while H (no. 6), cyclopropyl (no. 49), and the alkanes (no. 22, 40, 57, 58, 76, 77, and 84) are located on the right-hand side. Inspection of *Fig. 2.6* reveals that this results from slightly higher F values for the unsaturated substituents except for substituent no. 97. *Fig. 2.9* shows that substituents containing SO, SO_2, or SO_3 groups (no. 26, 27, 43, 62, and 88) are preferentially located in the left hand-side of the cloud of points indicating that their Fr values are generally low (*Fig. 2.2*) and that they have the ability to accept H-bonds (*Fig. 2.3*). A closer inspection of this figure allows to stress a gradient linked to the number of carbon atoms of the alkyl group (nb C in *Fig. 2.9*). Therefore, analysis of *Fig. 2* shows that the nonlinear map summarizes the main part of the information contained in the data table of 103 aliphatic substituents and it is coherent in terms of chemical information. As a result, we assume that this map can be used for estimating the quality of the selections performed by the GA.

By plotting on *Fig. 2.1* the chromosomes obtained from the different fitness functions, we showed that Eq. (5) provided test series with the maximum information content. Although the difference was slight, it appeared that when using this equation, it was preferable to include a constraint (C = 0.3) in our study in order to optimize the results. It is noteworthy that these results are broadly in accordance with those previously reported for a set of aromatic substituents [12].

An interesting best chromosome obtained with the selected configuration (i.e., Eq. (5) and C = 0.3) is depicted on *Fig. 2.1* by means of arrows. The repartition of the substituents selected by the GA is wide so that we can consider that the space of physicochemical properties is efficiently represented. Confirmation of this optimal selection can be obtained by plotting on the nonlinear map (*Fig. 3.1*) the physicochemical properties of the 15 selected substituents (*Figs. 3.2-3.6*). These figures show squares and circles of different size for each property and for different locations. This reveals that the substituents of the chromosome are very different and therefore this best chromosome has a high information content. It is important to note that other valuable chromosomes have been obtained. This underlines the heuristic potency of GAs for selecting test series.

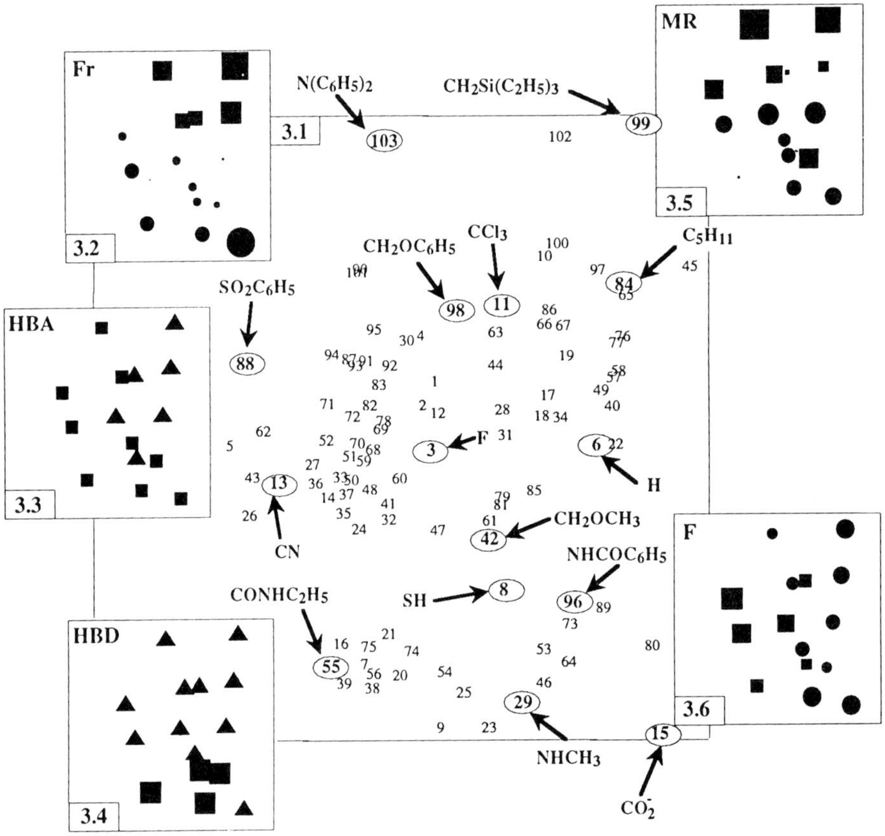

Figure 3. 3.1. *Nonlinear map of the 103 aliphatic substituents described by Fr, HBA, HBD, MR, and F. The selection made by the GA is represented by circles and arrows. See Table 1 for correspondence between the numbers and the substituents.*

3.2 to 3.6. *Plots of the scaled values of the five parameters on each of the 15 substituents selected by the GA. Squares (positive values) and circles (negative values) are proportional in size to the magnitude of the parameters. In Fig. 3.3 and Fig. 3.4, the triangles indicate the substituents which have not the ability to accept and donate H-bonds, respectively.*

The data set under study has been analyzed by means of hierarchical cluster analysis (HCA) [9], Kohonen network [14], N2M [14], and ART-2-A network [15]. Therefore, it was interesting to compare the GA results with those obtained from the above different methodological approaches. *Table 2* lists the ten clusters designed by Hansch and Leo [9] using HCA. In this table, the substituents included in the best chromosome are represented in bold. The presence of at least one substituent selected by the GA in each cluster shows that both methods are in accordance. Similar conclusions can be drawn when the GA results are compared with those obtained with the Kohonen network (*Fig. 4*).

Table 2. *Aliphatic Substituent Constants-Ten Cluster Sets [9]. The Substituents Constituting the Chromosome Selected by the GA are Indicated in Bold.*

Cluster	n*	Substituents
1	6	Br, Cl, CF_3, **F**, I, CF_2CF_3
2	27	NO_2, **CN**, $SOCH_3$, SOC_2H_5, SOC_3H_7, SCN, CO_2CH_3, $OCOCH_3$, $CH{=}CHNO_2$-(trans), COC_2H_5, CH=CHCN, OSO_2CH_3, OCH_3, $COCH_3$, CH_2CN, OCH_2CH_3, $CO_2C_2H_5$, $OCH(CH_3)_2$, $CH{=}CHCOCH_3$, COC_3H_7, OC_3H_7, $OCOC_2H_5$, $OCOC_3H_7$, $CH{=}CHCO_2CH_3$, OC_4H_9, $CO_2C_3H_7$, $CH{=}CHCOC_2H_5$
3	20	**H**, CH_3, CH_2Br, CH_2Cl, $CH{=}CH_2$, SCH_3, CH_2CH_3, Cyclopropyl, $CH(CH_3)_2$, C_3H_7, C_4H_9, $C(CH_3)_3$, CBr_3, **CCl_3**, SC_2H_5, SC_3H_7, CH_2I, 2-Thienyl, C_6H_5, 3-Thienyl
4	11	OH, CO_2H, CH=NOH, $CONH_2$, $NHCOCH_3$, $C{=}O(NHCH_3)$, $NHCO_2C_2H_5$, **$CONHC_2H_5$**, $NHCOC_2H_5$, $NHCOC_3H_7$, $CONHC_3H_7$
5	2	**SH**, C≡CH
6	11	NH_2, CH_2OH, **$NHCH_3$**, NHC_2H_5, $NHCONH_2$, $EtCO_2H$, NHC_3H_7, $(CH_2)_3CO_2H$, NHC_4H_9, NHC_6H_5, **$NHCOC_6H_5$**
7	7	**CH_2OCH_3**, $CH_2OC_2H_5$, $N(CH_3)_2$, $CH_2OC_3H_7$, $N(C_2H_5)_2$, $CH_2OC_4H_9$, **$CH_2OC_6H_5$**
8	7	$CH_2Si(CH_3)_3$, $Si(CH_3)_3$, **C_5H_{11}**, $CH_2C_6H_5$, $CH{=}CHC_6H_5$-(trans), **$CH_2Si(C_2H_5)_3$**, Ferrocenyl
9	11	$CH{=}CHCO_2C_2H_5$, $CH{=}CHCO_2C_3H_7$, $OCOC_6H_5$, OC_6H_5, $CH{=}CHCOC_3H_7$, COC_6H_5, $CO_2C_6H_5$, **$SO_2C_6H_5$**, 2-Benzthiazolyl, $CH{=}CHCOC_6H_5$, **$N(C_6H_5)_2$**
10	1	**CO_2^-**

* number of substituents in the considered cluster.

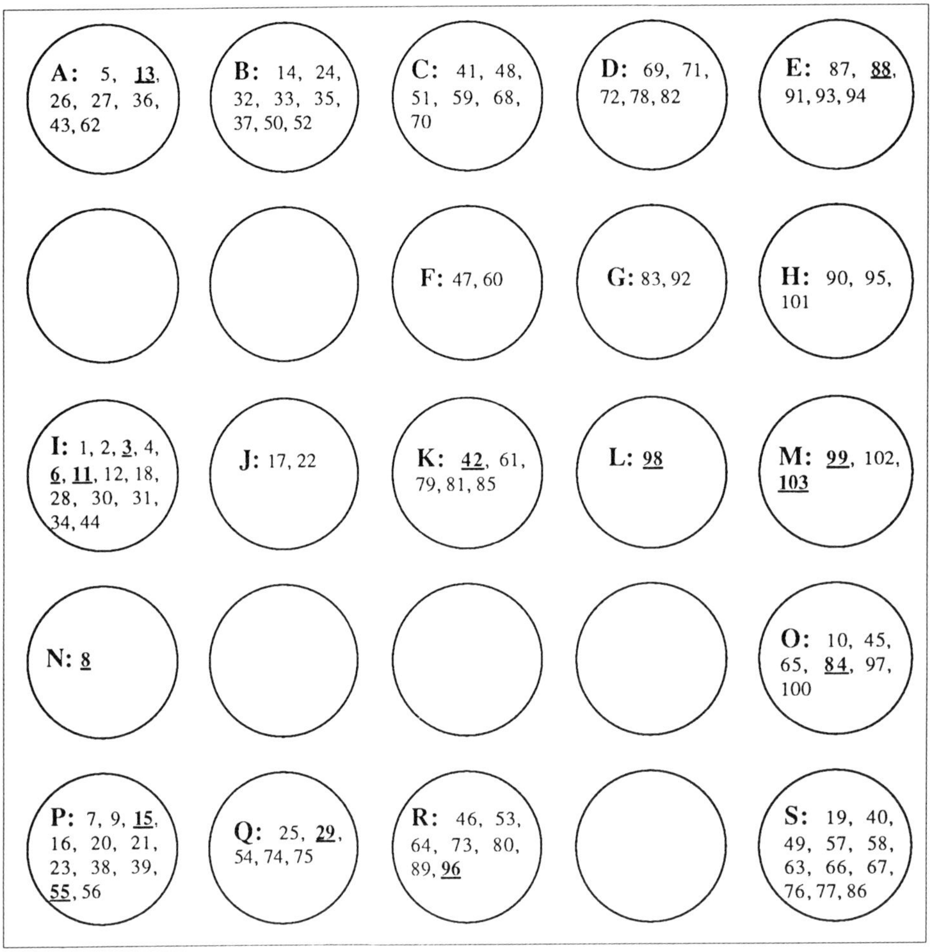

Figure 4. *Kohonen map of the aliphatic substituents. Underlined bold numbers represent the substituents selected by the GA. See Table 1 for correspondence between the numbers and the substituents.*

Fig. 4 shows a 5 x 5 Kohonen map run for 10,000 cycles and with a final radius of 0. It is worth noting that by adjusting the size of the network, it was possible to obtain more or less clusters (results not shown). In the present study, the 5 x 5 array of neurons provided 19 clusters (*Fig. 4*). From *Fig. 4*, it would be possible to select optimal test series by taking one or several substituents in each of the clusters formed. For comparison purposes, we have represented on this figure, the GA selection (underlined bold numbers). It is noteworthy that the results obtained are in accordance with those previously obtained since there are only one, two, or in one case three substituents selected per neuron.

N2M is a combination of Kohonen network, minimum spanning tree (MST), and NLM [16]. The N2M display is a nonlinear mapping of "loaded"

Kohonen neurons (i.e., neurons containing substituents) on which an MST has also been projected. The main advantage of this new graphical technique is that MST and NLM allow to clearly show the distances separating the different clusters of the Kohonen network. *Fig. 5* is the N2M display of the 103/5 data matrix under study. On this figure, the substituents included in the best chromosome have been represented by arrows. Their location in very different clusters on the map clearly shows the coherency of the results.

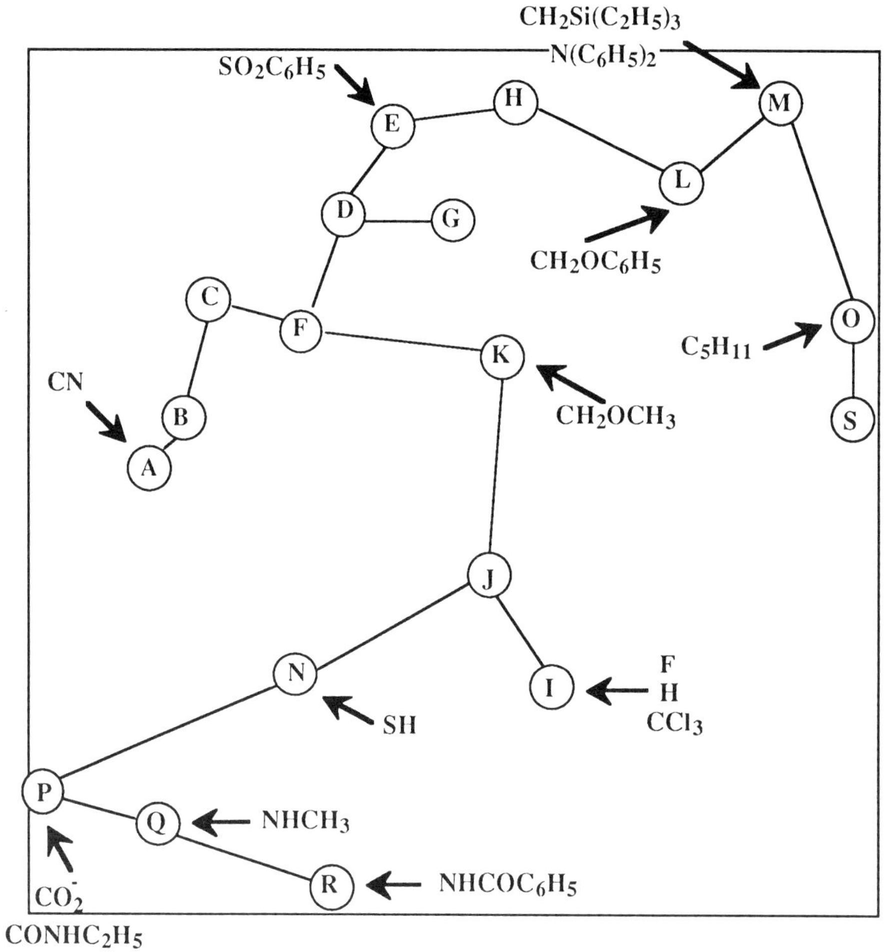

Figure 5. *N2M display of the aliphatic substituents. The arrows show the location of the substituents selected by the GA. See Fig. 4 for correspondence between the letters and the clusters.*

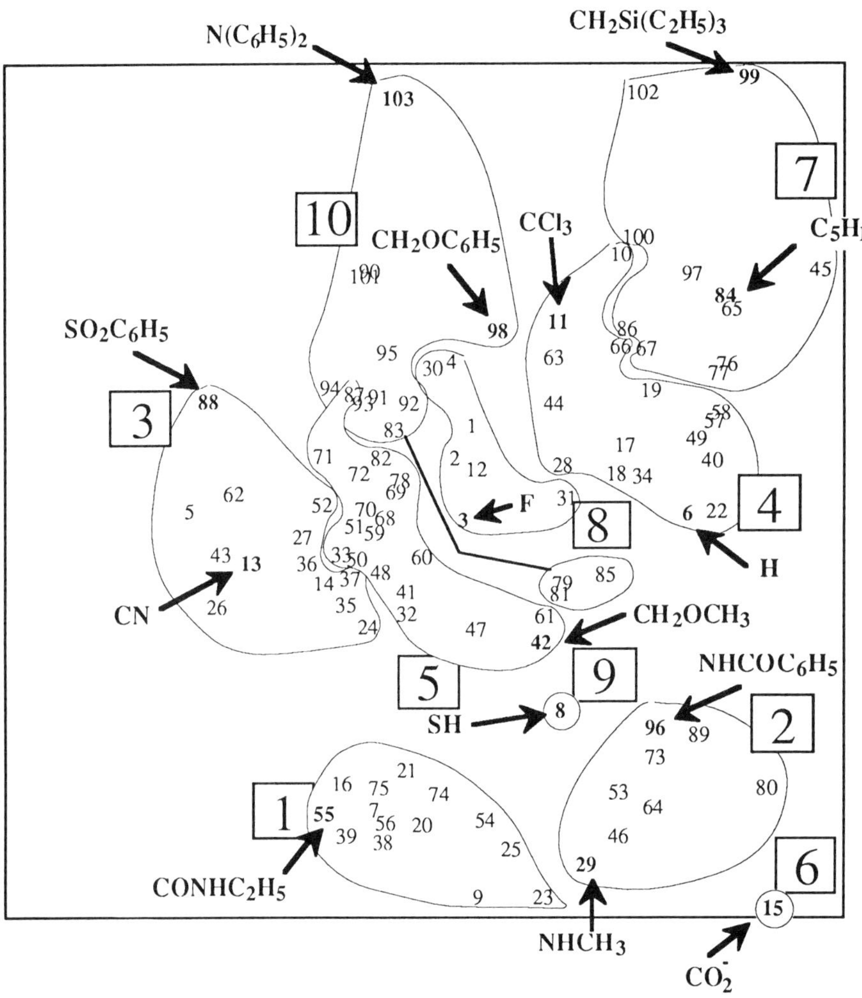

Figure 6. *Representation of the ART 2-A clusters on the nonlinear map of the 103 substituents (Fig. 2.1). See Table 1 for correspondence between the "small" numbers and the substituents. The large numbers inside a square correspond to the cluster indices of ART 2-A [15]. Cluster 10 is split into two groups linked by a line. The arrows indicate the substituents selected by the GA.*

Fig. 6 represents, on the nonlinear map of the 103 aliphatic substituents (*Fig. 2.1*), ten clusters derived by the neural network ART 2-A [15]. The unsupervised clustering performed by ART 2-A can be briefly summarized as follows. ART 2-A is divided into two subsystems: an attentional subsystem, which processes familiar stimuli, and an orienting subsystem that detects unfamiliar input patterns and resets the attentional subsystem when it detects such a pattern. The algorithm compares input patterns with the existing network structure to check whether they fit into ("resonate with") it. In case of resonance, an "adaptation" of the existing weights is performed, otherwise the structure is adapted by creation of new weight vector(s) (i.e., cluster(s)). Hence, the network learns by modification or creation of a set of coefficients (weights). ART 2-A offers the possibility of interpretation of the clusters derived in terms of the original variables by means of their weights which provide the characteristic profiles of the individuals constituting the clusters (results not shown). Numerical results allowing to stress the similarities between the clusters and the individuals and the distances separating the clusters are also computed (results not shown). In addition, it is possible to gradually inspect the structure of the data sets at hand. This is also made possible by the fact that convergence is rapidly obtained. In this study, the network parameters were fixed in such a way that ten clusters were obtained [15]. On *Fig. 6*, the GA selection has been also represented by means of arrows. *Fig. 6* shows that the results obtained from the GA are in accordance with those obtained from ART 2-A since there are only one or two substituents selected per cluster. It is also worth noting that the substituents isolated in a cluster (i.e., substituents no. 8 and 15 in clusters no. 9 and 6, respectively) are also selected by the GA.

From the above comparisons, the complementarity between GAs and other linear and nonlinear methods can be also noted. Our GA allows to guide the selection of substituents among the clusters or on the maps derived. Deviations from an optimal selection can be envisioned by exploiting the proximities of the individuals in the clusters or on the maps. Last, it must be noted that one advantage of GAs is that they do not produce one unique solution but populations of optimal test series which give complementary or supplementary information and allow to take into account other constraints such as ease for synthesis or the necessity to design safer chemicals.

Conclusion. - Our results confirm those recently obtained with the same methodology on a data matrix of 166 aromatic substituents [12]. They show that GAs can be used for the rational selection of test series with high information content. It is obvious that the correct use of GAs requires a lot of parameters to tune and therefore a classical method such as statistical experimental design [17] can appear more appealing. However, we have to mention that, even if in this study, we have only represented the best chromosome for the selected fitness function, the advantage of GAs, is that they provide a population of valuable chromosomes. All of them can be useful to optimize a selection taking into account other constraints such as synthetic feasibility, cost, toxicity, and so on. These constraints cannot be neglected in the rational design of new molecules.

REFERENCES

[1] J. Devillers, *Genetic Algorithms in Molecular Modeling*, Academic Press, London, 1996.

[2] C.G. Langton, *Artificial Life. An Overview*, The MIT Press, Cambridge, 1995.

[3] M. Mitchell, *An Introduction to Genetic Algorithms*, The MIT Press, Cambridge, 1996.

[4] D.E. Goldberg, *Genetic Algorithms in Search, Optimization & Machine Learning*, Addison-Wesley Publishing Company, Inc., Reading, 1989.
[5] L. Davis, *Handbook of Genetic Algorithms*, Van Nostrand Reinhold, New York, 1991.
[6] Z. Michalewicz, *Genetic Algorithms + Data Structures = Evolution Programs, 2nd (extended) edition*, Springer-Verlag, New York, 1994.
[7] J.J. Grefenstette, J.E. Baker, in '*Proceedings of the Third International Conference on Genetic Algorithms*', Ed. J.D. Schaffer, Morgan Kaufmann Publishers, San Mateo, 1989, pp. 20-27.
[8] J. Holland, *Pour la Science* **1992,** *179,* 44.
[9] C. Hansch, A. Leo, *Substituent Constants for Correlation Analysis in Chemistry and Biology*, John Wiley & Sons, New York, 1979.
[10] D.B. Turner, P. Willett, A.M. Ferguson, T.W. Heritage, *SAR QSAR Environ. Res.* **1995,** *3,* 101.
[11] STATQSAR, CTIS, Lyon, France, 1994.
[12] C. Putavy, J. Devillers, D. Domine, in '*Genetic Algorithms in Molecular Modeling*', Ed. J. Devillers, Academic Press, London, 1996, pp. 243-269.
[13] D. Domine, J. Devillers, M. Chastrette, W. Karcher, *J. Chemometrics* **1993,** *7,* 227.
[14] D. Domine, J. Devillers, D. Wienke, L. Buydens, *Quant. Struct. Act. Relat.* **1996,** *15*, 395-402.
[15] D. Domine, J. Devillers, D. Wienke, L. Buydens, *J. Chem. Inf. Comput. Sci.* **1996,** (in press).
[16] D. Domine, D. Wienke, J. Devillers, L. Buydens, in '*Neural Networks in QSAR and Drug Design*', Ed. J. Devillers, Academic Press, London, 1996, pp. 223-253.
[17] M. Sjöström, L. Eriksson, in '*Chemometric Methods in Molecular Design*', Ed. H. van de Waterbeemd, VCH, Weinheim, 1995, pp. 63-90.

PART II. **LEAD FINDING STRATEGIES**

Chapter 8

Combinatorial Docking

by **Hans-Joachim Böhm**

F. Hoffmann-La Roche Ltd, Pharmaceuticals Division,
Computational and Structural Chemistry, CH-4070 Basel,
Switzerland

Combinatorial Docking

by **Hans-Joachim Böhm**

F. Hoffmann-La Roche Ltd, Pharmaceuticals Division, Computational and Structural Chemistry, CH-4070 Basel, Switzerland
(hans-joachim.boehm@roche.com)

The prediction of synthetic accessibility represents a major challenge in the further development of computational methods for *de novo* ligand design. One way to tackle the problem is to restrict the molecules to be build to a certain subset that can be obtained from a small number of chemical reactions. Combinatorial Docking is a computational procedure that builds ligands in the protein binding site by combining building blocks from a well-behaved chemical reaction. For example, amides can be constructed from amino acids as building blocks. The extension of this approach to other simple chemical reactions is straightforward. Possible strategies for combinatorial docking are discussed and some examples are presented.

Introduction. - Computational *de novo* ligand design is a new exciting research area in medicinal chemistry [1-4]. The goal is to exploit the knowledge of the 3D-structure of the target protein in an efficient way to design novel protein ligands without resorting to prior knowledge about other existing ligands. Several years ago, a number of computational approaches have been described in the literature for de novo ligand design [5-23]. The goal of the first pioneering approaches was to design new ligands that are complementary to the protein binding site. The initial focus was on structure generation or retrieval. Our current experience indicates that this first milestone has been met. The existing computer programs for *de novo* ligand design have already proven to be useful as idea generators. Several successful applications have been reported. For example, the program LUDI was used successfully to design novel ligands for the protein FKBP [24] and for the enzyme phospholipase A2 [25]. Other reported successful applications include the design of renin inhibitors using the program GROW [8] and the design of HIV-protease inhibitors using the program MCDNLG [7,26]. Some of them are now widely used in pharmaceutical industry.

An important aspect of the further development of computational tools for *de novo* ligand design is to take into account the synthetic accessibility of the designed compounds. There are two possible approaches to this problem. First, one can design complete molecules and then find out afterwards whether this compound can be synthesized or not. However, current experience especially with the atom-based approaches indicates that a large percentage of the designed compounds will have to be rejected because no easy synthetic route can be found by the computational methods for synthesis prediction. While offering a very large diversity of the generated structures, this approach therefore appears to be inefficient. The alternative approach is to build knowledge about chemical synthesis into the design step. This basicly means doing 'combinatorial chemistry on the computer'. The advantage of this approach is that a very high percentage of the generated compounds can actually be synthesized because only those structures will be accepted that result from well-known synthetic procedures. The idea is to focus on a limited set of chemical reactions that are well understood and also amenable to automated parallel synthesis and to build molecules that are accessible by these reactions. The advantage of this approach is its efficiency and that the designed compounds can be synthesized easily. A possible disadvantage is the limited structural diversity of the constructed molecules. Not all known biologically active compounds are accessible by the current methods of combinatorial chemistry.

Technical Aspects of Combinatorial Docking. - Combinatorial docking is a computational procedure that builds protein ligands in the protein binding site by combining building blocks from a well-behaved chemical reaction. It may be regarded as a special variant of fragment-based *de novo* design. Examples of such simple chemical reactions are given in *Figure 1*.

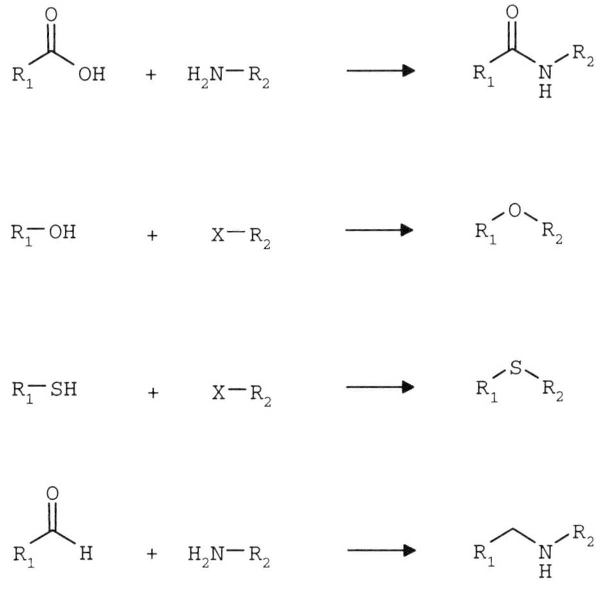

Fig. 1. *Examples of simple chemical reactions suitable for combinatorial docking.*

Amides can be formed from carboxylates and amines. Ethers can be obtained from the reaction of an alcohol with a halogen-compound. Databases of commercially available compounds such as the Available Chemicals Directory (ACD) [27] contain large number of compounds which are suitable as building blocks for these reactions. For example, more than 10,000 carboxylates are contained in the ACD. The number of available amines is in the same range. A significant fraction of these compounds may not be suitable as building block for various reasons (molecular weight too high, additional reactive side chains interfere with desired chemical reaction, etc.). Nevertheless, the number of amides that can be obtained from the combination of a carboxylate with an amine, is on the order of 5-50 million - at least in theory. The problem of the large number of potential structures get even more serious when other chemical reactions are considered with more than two educts involved. In the Ugi four-component reaction, an amine, a carboxylate, an aldehyde and an isonitrile are used as educts. The ACD contains a large number of amines, carboxylates and aldehyde that are suitable for use in the Ugi-reaction. Obviously, a very large number of different chemical compounds is available from the Ugi reaction [28,29].

There are two possible approaches to combinatorial docking. The first option is to generate the complete structure in advance and then dock it into the protein binding site. This has the advantage that synergistic effects determining the bound conformation can be taken into account. One may also hope that the construction of a 3D-model for the complete ligand will produce more realistic conformations. However, this possible advantage can only be exploited if ligand flexibility is accounted for by generating multiple conformations. The generation of more than 10^4-10^5 3D-structures with subsequent docking is impracticable at present. The second approach is to dock the building blocks individually into the binding site and connect them in a synthetically reasonable way. The big advantage of this approach is that very large virtual libraries can be handled.

A technical problem relates to the placement of fragments into the binding site. The basic idea of *de novo* design is to construct high affinity protein ligands from small pieces, either atoms or fragments. One underlying assumption is that the position of a small fragment in the binding site is the same as when it is part of a larger molecule. In several cases, this has been shown to be correct. For example, in the system thrombin-NAPAP, the position of the bound benzamidine remains unchanged even if the rest of the NAPAP inhibitor structure is removed [30]. We have investigated this assumption for the complex streptavidin-biotin. First, the position of biotin in the binding site was calculated by molecular mechanics energy minimization [31] with the protein kept fixed. The rms deviation of the force-field-optimized position of the ligand from the X-ray structure is only 0.25 Å. Then the same calculation was performed for the monocyclic urea part only. The rms deviation between force-field-optimized position from the X-ray structure is now 0.92 Å. This result indicates that the monocyclic ureido group moves significantly away from the position observed in the X-ray structure of the streptavidin-biotin complex. Therefore, it may not always be advisable to start from very small pieces because the favorable positions of these very small groups may not coincide with their position when embedded into a larger ligand. The problem is not so severe with larger fragments. The use of larger building blocks is therefore advantageous. Alternatively, one can allow atoms or very small fragments to move in the course of the fragment build-up procedure. However, this procedure will add computational complexity to the combinatorial docking and will it make more time-consuming.

Examples. - In the following, two examples of combinatorial docking are described. In both cases, the computer program LUDI [22,23] was used. This program builds protein ligands by fragment-joining. The fragments are simple small organic molecules with typically up to 50 atoms. The program can construct a large diversity of possible structures if a sufficiently large fragment library is used. Initially, a standard library consisting of roughly 1000 fragments was used. Recently, also larger fragment libraries have been employed such as for example a 30000 compound subset of the fine chemicals directory (FCD) [32]. In principle, LUDI can be used to construct specific classes of ligands by using the appropriate fragment libraries. For example, peptides can be constructed by using amino acids as fragments.

Two new libraries were constructed for this purpose [33]. First, amino acid fragments $HCONHCH(R)CONH_2$ were extracted from high resolution protein structures taken from the PDB. The conformational flexibility of the amino acids is taken into account by using several conformers for each amino acid. On average, 100 conformers are stored per amino acid. Second, a library of commercially available non-natural amino acids and peptidomimetics was constructed. In this case, the conformers were generated using the computer program MIMUMBA [34].

The design of a ligand by combinatorial docking can be done as follows. The first step is the selection of a seed group. The seed group can be placed using LUDI. In the second example, LUDI was used to place the seed group benzamidine. Alternatively, one may start from a known protein-ligand complex and just keep a small part of the known ligand and use this as seed in the design of novel ligands. This approach was followed in the first example. A third option is to use interactive computer graphics following a molecular mechanics optimization of the position in the binding site for the initial placement of the seed group.

The basic computational methodology is the same as described previously [22,23]. First, a new fragment is appended. Then, the position of this new fragment is checked for steric overlap with the protein and with other parts of the ligand, for electrostatic repulsion and for the presence of buried unpaired polar groups. Optionally, one can also specify some further requirements. For example, it can be requested that a minimum percentage of the fragment surface has to be in contact with the protein or that an interaction with a specified atom of the protein is made (e.g. interaction with the zink atom of a metalloprotease).

All hits satisfying the criteria mentioned above are scored using an empirical scoring function described previously [35]. The full fragment library is searched and only the top scoring hits are kept. The construction of a ligand proceeds similar to the program GROW described by Moon and Howe [8]. The result of a LUDI run is a list of putative ligands with a predicted 3D-structure and a predicted binding constant.

Design of Peptides as Inhibitors of Elastase [33]. - Elastase inhibitors may be useful in the treatment of emphysema and other diseases. We used the 3D-structure of human leukocyte elastase complexed with the peptidic chlormethylketone inhibitor Suc-Ala-Ala-Pro-Ala-CH_2Cl (**1**, PDB-refcode 1HNE [36]) in our calculations. The C-terminal amino acid Ala with the covalent link to the protein (**2**) was kept without modifications of the heavy atom positions.

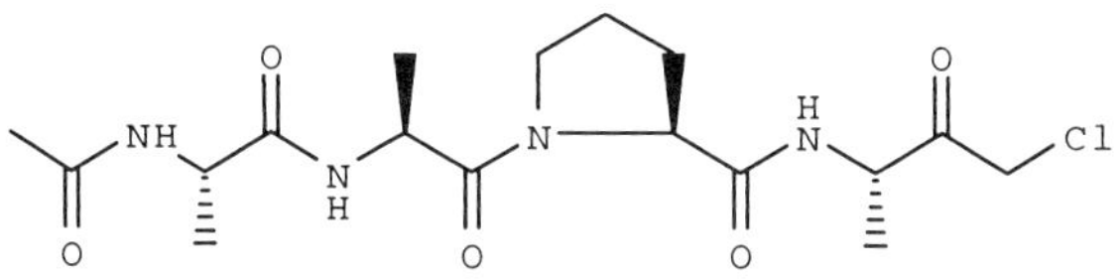

1

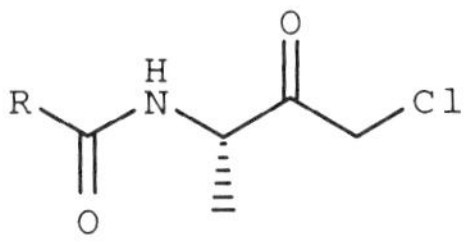

2

LUDI was then used to search the library of natural amino acids for suitable groups for P_2 - P_4. In the first step the position P_2 was considered. LUDI finds the amino acids Pro, Ala and Gly. The highest score is found for proline, in very good agreement with published binding data on peptides [37]. The docked proline was then used to append a further amino acid at position P_3. The LUDI-hits for P_3 are Leu, Gly and Ala. This leucine was then again used to search for P_4-substituents. However, no hit was found in this calculation. It should be noted, the one requirement for LUDI to append another amino acid is that at least one hydrogen bond is formed by the peptide backbone. LUDI therefore failed to find an amino acid backbone conformation that could both be attached to the P_3-Leu and hydrogen bonds with elastase. Therefore, the best peptide for P_3-P_2 obtained from LUDI is Leu-Pro. The docked conformation of this dipeptide is compared with the experimentally observed conformation of Suc-Ala-Ala-Pro-Ala-CH_2Cl in *Fig. 2.* The rms deviation of the backbone atomes for the P_3-P_2 section is 0.82 Å.

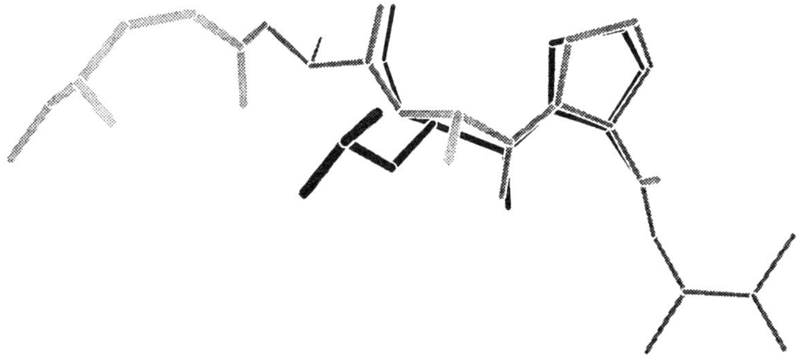

Fig. 2. *Comparison of the fitted P_2-P_3 dipeptide Leu-Pro (dark bonds) with the protein-bound conformation of Suc-Ala-Ala-Pro-Ala-CH_2Cl.*

In a further calculation, the library of non-natural amino acids and peptidomimetics was searched for suitable groups at position P_3. In particular, we were interested in the ability of LUDI to dock the 3-amino-pyridone into the binding site of elastase. This group was recently disclosed as a useful peptidomimetic [38] leading to potent orally active inhibitors of human leukocyte elastase [39]. Using Pro or Ala at position P_2, the pyridone is rejected by LUDI due to intramolecular steric overlap between the pyridone and the side chain at position P_2. Using Gly at postion P_2, 3-amino-pyridone is docked successfully into the binding site of elastase. The docked conformation is compared with Suc-Ala-Ala-Pro-Ala-CH_2Cl in Fig. 3. Each run took about 2 CPU minutes on a Silicon Graphics workstation with a R4000 CPU.

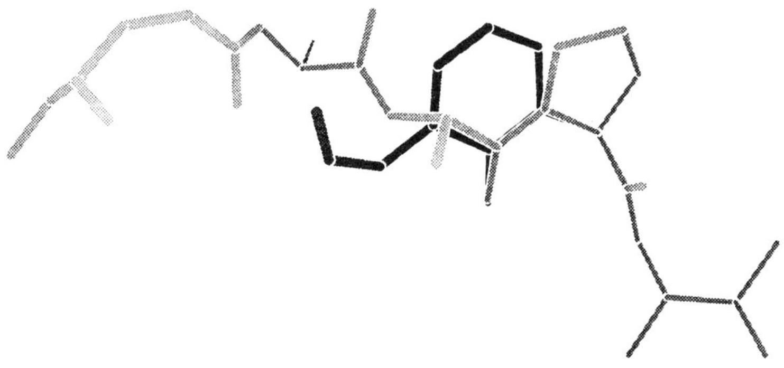

Fig. 3. *Comparison of the fitted 3-amino-pyridin-2-one derivative (dark bonds) with the protein-bound conformation of Suc-Ala-Ala-Pro-Ala-CH_2Cl.*

Automatic Design of a Nonpeptidic Thrombin Inhibitor. - At BASF in Ludwigshafen, the program LUDI was used to design several different classes of novel nonpeptidic thrombin inhibitors. One example, that originates from this work is the design of ethers as thrombin inhibitors. Starting from benzamidine **3** as a known inhibitor which occupies the P1-pocket, the idea was to design ether derivatives which can be easily synthesized. Compound **4** was designed with LUDI. The lipophilic substituent of Compound **4** should occupy the pockets P2 and P3. **4** was synthesized by Thomas Pfeiffer [40] and tested in a thrombin binding assay. It is a potent thrombin inhibitor with a hundred-fold improvement in binding affinity over the unsubstituted benzamidine **3.** Subsequently, the predicted binding mode of **4** was confirmed by X-ray structure determination [40].

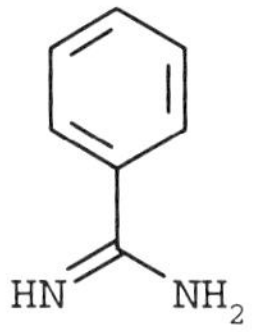

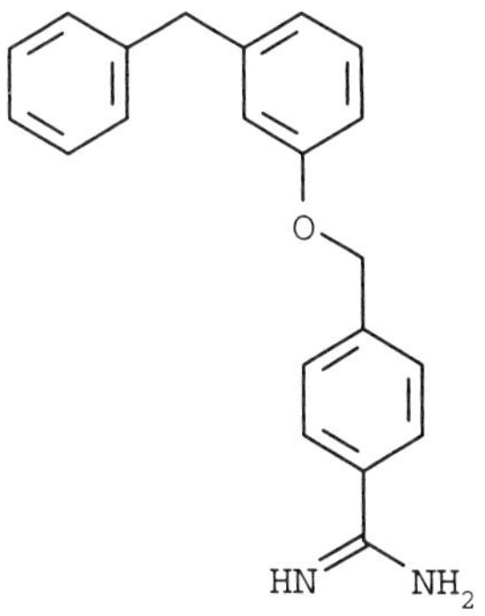

3 **4**

Summary. - Combinatorial docking builds new ligands in the binding site by docking building blocks from a known chemical reaction and linking them in the binding site. Some possible strategies for combinatorial docking are discussed and two examples have been presented.

The present approach should prove to be particularly useful to guide combinatorial chemistry approaches. The abundance of potential building blocks is one major challenge in the construction of combinatorial libraries which calls for computational schemes for the selection of building blocks. If the 3D-structure of the target protein is known, the present approach can be used for this purpose. Due to the build-up procedure, the present algorithm can easily be extended to a very large number of diverse monomers and is therefore useful for monomer selection. If a compound from a combinatorial library shows detectable biological activity, its docked conformation suggested by LUDI immediately offers a starting point for further optimizations. On the other hand, some classes of enzyme inhibitors may not be easily amenable to combinatorial chemistry approaches. In such a situation, the present approach can be used to guide the design and synthesis of protein ligands using classical organic chemistry.

Of course, as with any of the previously published algorithms, the present work faces a number of limitations. First, the protein is kept rigid. Therefore, if ligand binding is accompagnied by a conformational change being significantly different from that of known ligands, the present approach may run into problems. Furthermore, we cannot rule out the possiblity that we miss important conformations of the amino acids. Finally, our scoring function is of limited accuracy.

In summary, a new method for the automatic design of synthetically accessible protein ligands is described. In the present communication, the design of peptides and ethers is presented. Currently, work is in progress to extend the approach to incorporate other simple chemical reactions. The ultimate goal is to develop a robust and reliable method that can automatically design protein ligands that are synthetically easily accessible.

Acknowledgements. - I would like to thank Gerhard Klebe for many fruitful discussions. Most of the work described in the present paper was carried out in the main laboratory of BASF AG in Ludwigshafen.

REFERENCES

[1] Y.C. Martin, *J.Med.Chem.* **1992**, *35*, 2145-2154.
[2] R.A.Lewis, A.R.Leach, *J.Comput.Aided Molec.Des.* **1994**, *8*, 467-475.
[3] H.J.Böhm, in 3D-QSAR in Drug Design, ed. H. Kubinyi, Escom Science Publishers, Leiden, **1993**, pp.386-405
[4] C.L.M.J. Verlinde, W.G.J. Hol, *Structure* **1994**, *2*, 577-587
[5] Y.Nishibata, A.Itai, *J.Med.Chem.* **1993**, *36*, 2921-2928
[6] R.S.Bohacek, C.McMartin, *J.Am.Chem.Soc.* **1994**, *116*, 5560-5571
[7] D.K.Gehlhaar, K.E.Moerder, D.Zichi, C.J.Sherman, R.C.Ogden, S.T.Freer, *J.Med.Chem.* **1995**, *38*, 466-472
[8] J.B.Moon, W.J.Howe, *Proteins* **1991**, *11*, 314-328
[9] V.Tschinke, N.C.Cohen, *J.Med.Chem.* **1993**, *36*, 3863-3870
[10] S.H.Rotstein, M.A.Murcko, *J.Med.Chem.* **1993**, *36*, 1700-1710
[11] M.B.Eisen, D.C.Wiley, M.Karplus, R.E.Hubbard, *Proteins* **1994**, *19*, 199-221
[12] A.Caflish, A.Miranker, M.Karplus, *J.Med.Chem.* **1993**, *36*, 2142-2167
[13] R.A.Lewis, D.C.Roe, C.Huang, T.E.Ferrin, R.Langridge, I.D.Kuntz, *J.Mol.Graphics* **1992**, *10*,66-78
[14] V.Gillet, P.Johnson, P.Mata, S.Sike, P.Williams, *J.Comput.Aided.Molec. Des.* **1993**, *7*, 127-153
[15] P.Mata, V.J.Gillet, P.Johnson, J.Lampreia, G.J.Myatt,S.Sike, A.L.Stebbings, *J.Chem.Inf.Comput.Sci.* **1995**, *35*, 479-493
[16] S.H.Rotstein, M.A.Murcko, *J.Comput.Aided Molec.Des.* **1993**,*7*, 23-43
[17] P.A.Bartlett, G.T.Shea, S.J.Telfer, S.Waterman, in 'Molecular Recognition:Chemical and Biological Problems', S.M.Roberts (Ed.), Royal Society of London, **1989**, pp.182-196
[18] D.A.Pearlman, M.A.Murcko, *J.Comput.Chem.* **1993**, *14*, 1184-1193
[19] R.A.Lewis, P.M.Dean, *Proc.Roy.Soc.Lond.* **1989**, *B236*, 125-140
[20] R.A.Lewis, P.M.Dean, *Proc.Roy.Soc.Lond.* **1989**, *B236*, 141-162,
[21] P.L.Chau, P.M.Dean, *J.Comput.Aided Molec. Des.* **1992**, *6*, 385-396
[22] H.J.Böhm, *J.Comput.Aided Molec.Des.* **1992**, *6*, 61-78
[23] H.J.Böhm, *J.Comput.Aided Molec.Des.* **1992**, *6*, 593-606
[24] R.E.Babine, T.M.Bleckman, C.R.Kissinger, R.Showalter, L.A.Pelletier, C.Lewis, K.Tucker, E.Moomaw, H.E.Parge, J.E.Villafranca, *Bioorg. Med. Chem. Lett.* **1995**, *5*, 1719-1724
[25] M.T. Pisabarro, A.R. Ortiz, A. Palomar, F. Cabre, L. Garcia, R.C. Wade, F. Gago, D. Mauleon, G. Carganico, *J.Med.Chem.* **1994**, *37*, 337-341
[26] M.Melnick, S.H.Reich, K.K.Lewis, L.J.Mitchell, D.Nguyen, A.J.Trippe, H.Dawson, J.F.Davies, K.Appelt, B.W.Wu, L.Musick, D.K.Gehlhaar, S.Webber, B.Shetty M.Kosa, D.Kahil, D.Andrada, *J.Med.Chem.* **1996**, *39*, 2795-2811
[27] The Available Chemicals Directory (ACD) is distributed by Molecular Design Ltd., San Leandro, 2132 Farallon Drive, CA 94577
[28] I.Ugi, S.Lohberger, R.Karl, in 'Comprehensive Organic Synthesis: Selectivity for Synthetic Efficiency', Vol. 2, (Eds. B.M.Trost, I.Fleming, C.H.Heathcock), Pergamon Press, Oxford, **1991**, pp.1083-1109
[29] L.Weber, S.Wallbaum, C.Broger, K.Gubernator, *Angew.Chem. Int.Ed.Engl.* **1995**, *34*, 2280-2282
[30] D.W.Banner, P.Hadvary, *J.Biol.Chem.* **1991**, *266*, 20085
[31] P. Dauber-Osguthorpe, V.A. Roberts, D.J. Osguthorpe, J. Wolff, M. Genest, A.T. Hagler, *Proteins* **1988**, *4*, 31-47

[32] H.J. Böhm, *J.Comput. Aided Molec. Des.* **1994**, *8*, 623-632
[33] H.J. Böhm, *J.Comput.Aided Molec.Des.* **1996**,*10*, 265-272
[34] G.Klebe, Th.Mietzner, *J.Comp.Aided Molec.Des.* **1994**,*8*, 583-606
[35] H.J.Böhm, *J.Comput.Aided Molec.Des.* **1994**,8, 243-256
[36] F.C. Bernstein, T.F. Koetzle,, G.J.B. Williams, E.F., Jr. Meyer, M.D. Brice, J.R. Rodgers, O. Kennard, T. Shimanouchi, and T. Tasumi, *J. Mol. Biol.* **1977**, *112*, 535-542
[37] R.C.Thompsom, E.R.Blout, *Biochemistry* **1973**, *12*, 51-57
[38] F.J.Brown, D.W.Andisik, P.B.Bernstein, C.B.Bryant, C.Ceccarelli, J.R.Damewood, P.D.Edwards, R.A.Earley, S.Feeney, R.C.Green, B.C.Gomes, B.J.Kosmider, R.D.Krell, A.Shaw, G.B.Steelman, R.M.Thomas, E.P.Vacek, C.A.Veale, P.A.Tuthill, P.Warner, J.C.Williams, D.J.Wolanin, S.A.Woolson, *J.Med.Chem.* **1994**, *37*, 1259-1261
[39] C.A.Veale, P.B.Bernstein, C.B.Bryant, C.Ceccarelli, J.R.Damewood, R.A.Earley, S.Feeney, B.C.Gomes, B.J.Kosmider, G.B.Steelman, R.M.Thomas, E.P.Vacek, J.C.Williams, D.J.Wolanin, S.A.Woolson, *J.Med.Chem.* **1995**, *38*, 98-108
[40] H.J.Böhm, Th. Pfeiffer, W. Höffken, unpublished results

PART II. **LEAD FINDING STRATEGIES**

Chapter 9

Quantification of Molecular Similarity and Its Application to Combinatorial Chemistry

by **Richard A. Lewis**[a]), **Andrew C. Good**[a]) and **Stephen D. Pickett**[b])

[a]) Rhone-Poulenc Rorer, New Lead Generation, Rainham Road South, Dagenham, Essex, RM10 7XS, United Kingdom
[b]) Rhone-Poulenc Rorer, Modelisation Moleculaire, Centre de Recherches de Vitry-Alfortville, 13, quai Jules Guesde, 94403 Vitry-sur-Seine Cedex, France

Quantification of Molecular Similarity and Its Application to Combinatorial Chemistry

Richard A. Lewis, Andrew C. Good

Rhone-Poulenc Rorer, New Lead Generation, Rainham Road South, Dagenham, Essex, RM10 7XS, United Kingdom

and **Stephen D. Pickett**

Rhone-Poulenc Rorer, Modelisation Moleculaire, Centre de Recherches de Vitry-Alfortville, 13, quai Jules Guesde, 94403 Vitry-sur-Seine Cedex, France

---

The concept of molecular similarity is an essential part of modern computer-aided drug design methods, and has been successfully applied many times in the optimisation of lead series. Molecular diversity calculations are becoming an increasingly important tool in the field of combinatorial chemistry. For many research groups, this new chemical technology is and will continue to produce a huge increase in the number of compounds available for screening. In many cases, the number of molecules that could be constructed greatly exceeds potential synthesis and screening capacity; it will become vital to design libraries based on the properties of compounds already in existence, if the diversity of each new molecular collection is to be maximized. The central requirements of such an approach are the construction of a product-based descriptor which can be applied to an existing compound data set, and a robust method for optimising the choice of reagents to maximise diversity. This presents a major problem for many of the paradigms currently used in such calculations, as they are reagent-based and thus not suitable as descriptors of whole molecules. In this paper, we will discuss briefly some of the approaches and metrics used to describe molecular similarity, and techniques for combinatorial library design. We will then describe a novel product-based method for determining molecular diversity, comparing it critically to the ChemDiverse software. The new method includes direct reagent selection, sophisticated diversity measurement functions and improved pharmacophore keys. The potential of this tool to aid in diversity profiling is illustrated in three studies. We will finish with a discussion of the experimental validation of similarity measures, with the conclusion that this goal is not readily attainable given current information.

Molecular Similarity Concepts. - *Molecular Similarity.* Research into molecular similarity has been a very fruitful field, and it is not the intention of this paper to review in detail the large body of published work in this area. The reader is referred to more comprehensive sources for this information [1,2]. We intend to look at descriptors of molecular similarity from the view point of pragmatism and relevance to the problem in hand. The overall thesis of this work is to ***design combinatorial libraries***. As we discuss later, this is a computationally intensive task, so our descriptors must be easily computable and manipulable: the most precise description of a molecule is probably its complete Hamiltonian, but this is not the form most drug designers find easy to handle or understand. The second factor to bear in mind is the purpose of the libraries, which is, in most cases, for high-throughput screening against a protein target. This implies that the descriptors should be strongly biased in favour of those that are most relevant to intermolecular interactions. Naturally, other factors are important when a ligand is being optimised into a drug, but this is not the key concern here, when we are trying to discover the lead series for optimisation.

Rigorous Descriptors of Molecular Similarity. These describe the shape and/or the electronic surface of a molecule in a highly detailed way [3-7]. Their drawbacks are that they are computationally expensive to compute, and that they have to be computed for each conformer of a molecule. Conformational flexibility is a big issue when thinking about molecular similarity: a molecule is not a static entity, but a thermal ensemble of conformers, only one of which might be the bioactive conformation. We should try to ensure that we have included a representation of the bioactive conformation in our description of the molecule. For these reasons, we have not used rigorous descriptors in our work here.

Approximate Descriptors of Molecular Similarity. As the heading implies, these descriptors are more approximate, sacrificing rigour and detail for a coarser but more compact and manipulable descriptor [8-10]. Because of the simplification of information, conformers can be included in the description, but only as multiple instances of same compound, that is, in a database one would have, for example, 100 entries for a particular compound, each one describing one conformation.

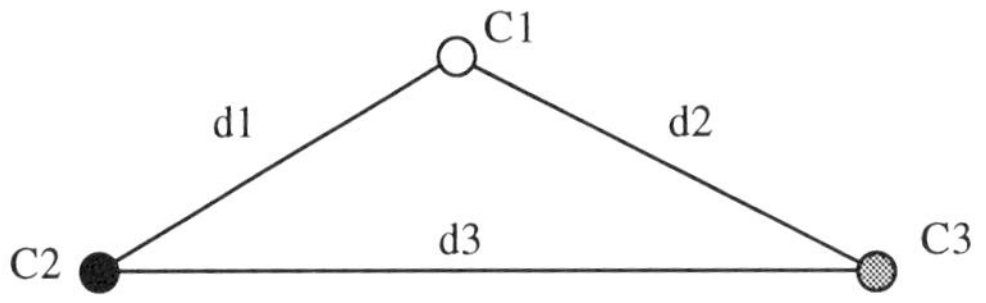

Fig. 1. *A schematic representation of a generic pharmacophore made up of three centres and three distances.*

Pharmacophore Descriptors of Molecular Similarity. Pharmacophore descriptors [11-13] are the next stage in the trade-off between rigour and pragmatism. The philosophy is to abstract only the essential information, allowing multiple conformer data contained within one descriptor. Because this is the descriptor we will be using most extensively for this work, we shall describe pharmacophores a little more fully than the other descriptors. A good working definition of a pharmacophore has been given by Wermuth and Langer [14] as 'the ensemble of steric and electronic features which are necessary to insure supramolecular interactions with a specific biological target structure'. The ensemble also contains distance information, that is, the geometric pattern of the features. The number of features necessary to describe a pharmacophore is case- and definition-dependent: for the purposes of this work, we make the pragmatic and arbitrary assumption that three features (C1, C2 and C3 in *Fig. 1*) and three distances (d1, d2 and d3) will suffice for most cases (relying on Ariens' hypothesis for justification). We freely acknowledge that for particular cases, the pharmacophore may be considerably more complex, but we note that a four feature pharmacophore can be formed by the combination four three-feature pharmacophores. We are currently investigating the use of higher level pharmacophores in our work. The features, which we also refer to by the term 'centres', are defined as hydrogen-bond donors, acceptors, aromatic ring centres, hydrophobes, bases, acids and so on. The chemical nature of the groups is less important than the way in which they interact with their partners during the molecular recognition process.

It should be clear for the above discussion that a pharmacophore is a necessary but not sufficient requirement for binding. Because of its approximation nature, the pharmacophore does not carry a detailed level of shape description, and it is well known that lack of shape complementarity will wreck binding. It is also hypothesised that high affinity arises mainly from shape complementarity, the enthalpic contribution of forming a hydrogen-bond being balanced by the cost of desolvation. The arguments in favour of the validity of the pharmacophore concept is that it has been used successfully to explain a great deal of diverse information within a concise hypothesis, and that it has been used to enhance demonstrably the efficiency of lead generation [15]. A comparison of a number of molecular descriptors has been made by Bath *et al.* [16].

Substructural and Physicochemical Descriptors of Molecular Similarity. These represent an even greater level of abstraction, as they focus on a description of a molecule in purely chemical terms or in terms of its macroscopic physicochemical properties. We have used both approaches in the past [17,18] to analyse our corporate database, but we feel now that the detachment of these descriptors from intermolecular interaction processes was too great to sustain, now that we can use pharmacophoric descriptors. There are also pitfalls associated with the use of substructural descriptors, which we deal with in the discussion section.

Combinatorial Libraries. - There are many different strategies for performing combinatorial chemistry, and the most appropriate technology to use is dependent on the resources and goals of the project. Several reviews have been written on this subject [19]. In its broadest sense, combinatorial chemistry can be defined as the process of making all possible combinations of some given sets of reagents and a given reaction. For the purposes of this work, we make the assumption that the combinatorial libraries are made as spatial arrays in 96-well plates, with less than ten, preferably one, components per well. Given this model, it is easy to see how the number of possible reaction products can vastly exceed available resources for all but the most spendthrift institution. If we consider the simple example of an amide condensation, we can look in a commercial catalogue and find about 3000 amine synthons, and 3000 acid synthons. Now, 3000 acids x 3000 amines = 9 million products; given a screening cost of $0.10 per compound, then the cost of screening the library would be $1 million, ignoring the fact that one would have to store 10000 plates, and the fact that a considerable amount of radioactive waste would be generated by the screening and would require to be disposed of. This forces us to adopt a strategy of designing a 'reasonable' subset. This begs the question of what is meant by the term 'reasonable', and we return to our earlier comments on similarity measures for our answer: pharmacophores are the most relevant and appropriate descriptors to use. We freely acknowledge that other workers in this area will adopt different similarity metrics to ours, but the principle is the same. It should however be noted that different similarity metrics will result in different library designs, and it is very difficult to say, *a priori*, which metric is the better one to use.

Reagent Selection in Isolation. Strategy: Select a subset of reagents by reference to the set of REAGENTS.
Reagent selection can be performed in two contexts: the reagents can be taken in isolation, ignoring the chemistry of the final product, or they can be assessed in terms of the set of final products. The principal application of this strategy is reducing the size of the reagent sets down from thousands to hundreds or tens of molecules. The key deficiency is in the assumption that the sites of variation in the core scaffold are independent. Experience within RPR would suggest that this assumption is not really justifiable, especially for 3D diversity measures like pharmacophores. The set of reagents can be clustered into homogenous subfamilies using a similarity measure, and one representative of each cluster selected for inclusion into the library. In terms of *Fig. 2*, the diversity of the library = Diversity(B,D,H) ⊕ Diversity(2,5,8,12).

	A	B	C	D	E	F	G	H
1	●	●	●	●	●	●	●	●
2	●	*	●	*	●	●	●	*
3	●	●	●	●	●	●	●	●
4	●	●	●	●	●	●	●	●
5	●	*	●	*	●	●	●	*
6	●	●	●	●	●	●	●	●
7	●	●	●	●	●	●	●	●
8	●	*	●	*	●	●	●	*
9	●	●	●	●	●	●	●	●
10	●	●	●	●	●	●	●	●
11	●	●	●	●	●	●	●	●
12	●	*	●	*	●	●	●	*

Fig. 2. *A schematic showing a subselection of 12 compounds (marked by* * *) from a possible 96. Only 7 reagents (2,5,8,12,B,D,H) are required to make the combinatorial sublibrary. Design methods are used to select the optimal choice of 7 reagents.*

This stage generally responds well to hierarchical clustering using Daylight fingerprints [20] or physicochemical measures [21]. These methods can be thought of as 2D or 1D metrics respectively. For the Daylight fingerprint, A molecule is described in terms of substructural fragments, e.g., the presence/absence of a ring, acid group, etc. Molecules from congeneric series will be highly similar in this metric, which has been shown to be well suited for producing representative screening sets. The key advantage of reagent selection in isolation is pragmatic, in that very large (up to 15K compounds) datasets can be clustered in a few hours.

Reagent Selection in Context. Strategy: Select a subset of reagents by reference to the set of PRODUCTS.
The most rigorous way to select reagents is to consider the final products of the reaction. In this case, using *Fig. 2*, the diversity of the library = Diversity({B,D,H} ⊗ {2,5,8,12}). The similarity measure will then measure the differences between two

products rather than two reagents. It should also be noted that one reagent will give rise to several products so that a poor score between two products does not mean that a reagent should be rejected; rather it is the total score of all the products of that reagent that must be considered. Conversely, the selection of some products automatically means that other products must also be selected, because of the implied set of reagents that must present to create the first set of products. This is obviously a more complex procedure, as there are many more products than reagents, but can be handled using optimisation techniques such as simulated annealing or genetic algorithms. The perceived disadvantages to reagent selection from products are the potential memory requirements required to store and process many thousands of products and the CPU requirements required to process the similarity scores. These obstacles have been overcome with present technology.

Methodology. - *Pharmacophore Similarity Descriptors.* The pharmacophore descriptors [13, 22] we have used are built up from combinations of three features/pharmacophore centres (hydrogen-bond donors, hydrogen-bond acceptors, aromatic ring centres, hydrophobes, bases, acids) in a triangle. The distances allowed for the triangle sides are partitioned into distances of between 2 and 25 Å. The distances are divided into bins e.g. 2-10 Å at 1 Å intervals to divide space up into a finite set of pharmacophore triangles.. Each triangle combination of features and distances is used to set a single bit in a binary key; it is this key that is the pharmacophoric descriptor. To get the key for a molecule, we perform a full conformational analysis for each molecule (using a strategy that will hopefully ensure that we include bioactive conformations in our analysis), and check to see which pharmacophores the molecule, in any one of its conformations, can express. For each pharmacophore expressed, we turn on the appropriate bit in the molecule key.

Library Design - Finding Available Reagents. The first stage in library design should be to understand the scope of the chemical reaction being used to generate the combinatorial library, and this can only be done by careful experimentation with a trainer set. From this exercise, one can then identify the functional groups (and their local environment) that need to be present, and interfering groups that need to be absent, for the reaction to proceed cleanly. In most cases, the products of a combinatorial reaction are not worked up to full purity (in contrast to more traditional synthetic chemistry), so the reaction must be high-yielding with few side-reactions. Examples might be the need to have no acidic protons for a Grignard reaction, or to have only one amine group in an amide condensation to prevent regioisomer formation. We code this information as a series of Daylight [23] SMARTS statements, and then search our reagents database (for example, the Available Chemical Directory [24]) using our own SMARTS parser [25].

Hierarchical Clustering Using a 2D Measure. The next step is optional, depending on the number of reagents that have been found, and the intended purpose of the library. Product-based design is limited by a necessary trade-off between pragmatism (CPU time, memory) and the rigour of similarity measure. It has been observed within RPR that in cases where there are many reagents, several of them are highly similar in a 2D substructural sense, and would by most criteria be considered redundant. Hierarchical clustering using a fingerprint measure is an excellent method for identifying such redundancies. The library should also be considered in context of the biological screen(s) for which it is intended. This may provide some limits on molecular weight or log P; it is a fairly simple matter to factor this into the reagent selection procedure. Arguably, cost and availability can also be included as a constraint. This stage of reagent selection is appropriate for filtering many thousands of reagents down to hundreds.

Library Design using ChemDiverse. The ChemDiverse software [26] for library profiling provides a number useful tools for diversity assessment, and is rapidly becoming an industry standard for this aspect of combinatorial chemistry. The ChemDiverse protocol for molecular diversity is based on trying to obtain the maximum coverage of pharmacophore space by potential combinatorial chemistry products. However, the protocol used by ChemDiverse for assessing pharmacophore diversity suffers from a number of technical drawbacks which are being addressed:

Selection of compounds requires on-the-fly conformational analysis of the products. Any pharmacophores found are added to a single pharmacophore key, which describes the whole growing ensemble of molecules. Compounds are only added to the ensemble if the set of pharmacophores that they express overlaps less than a certain extent with the ensemble key, that is, the molecule contains several new pharmacophores. As a consequence the searches are dependent on the order in which the molecules are extracted from the database. This approach is frequently referred to as cherry-picking, as the selection of the products does not take into account the reagents needed to make them, ignoring a fundamental feature of combinatorial chemistry. Instead ChemDiverse must be used repeatedly to select what it considers to be the best set of products. After each search the user must intervene to determine the most frequently occurring reagents, so that subsequent searches can be paired down until the final reagent choice is reached.

It is not possible to modify the search criterion to include other molecular properties such as shape. This is because the function used to determine diversity can not include any non pharmacophoric properties. It is possible to remove products based on placing upper and lower bounds for given properties.

The ChemDiverse pharmacophore keys are potentially extremely useful tools for directing library design. However, they are limited in the current incarnation, as the ensemble key only registers whether or not a particular pharmacophore either exists in the ensemble, and not how many times it is found. This makes the key prone to saturation, even when artificially small distance bins are applied. This binary description can also give a very distorted picture of the pharmacophoric space covered by the library.

Improvements to the ChemDiverse Method. If pharmacophore keys are to be used to their full potential, they should measure the absolute number of pharmacophores present within a given data set, rather then just presence/absence; this is a central part of the PDQ approach [13]. It would also be desirable for the molecules that express a particular pharmacophore to have enough variety of shapes to yield a lead. It would therefore be best if libraries were designed so as to provide a large coverage of pharmacophores in regions where the current in house databases show a dearth of pharmacophore coverage. As we have already mentioned, the ChemDiverse pharmacophore keys saturate too rapidly and do not show how many pharmacophores are present. It would be preferable if pharmacophore keys for each molecule could be stored on-line and accessed as and when required, rather than continually recalculating them on the fly, it. It is not possible to use standard ChemDiverse keys to store this, since each key requires > 300 Kbytes of hard disk space.

To get around the first and last problems, a ChemX PCL script has been written which profiles each molecule in a data set, extracts the ChemDiverse key, decodes it and writes out the individual pharmacophores found for the molecule. If most of the bits in the key for an individual structure are not set (the pharmacophore coverage is sparse), a large saving in space can be made with such an approach, since the absent pharmacophores are ignored, with only 4 bytes of space required to store each pharmacophore found. The number of 3D key distance bins was also adjusted to 16 instead of 32; this increased coarseness was felt to be justified when the large rotational increments used by the rule based analysis in ChemDiverse are considered. The zeroth bin (0-2Å) was also removed from consideration, since it allowed pharmacophores of the type aromatic - lipophilic - X in which the aromatic and lipophilic centres are actually on top of each other Calculation times for pharmacophore keys are around 10000 compounds/day on an SGI R4400 using our PCL script. It is now a fairly simple matter to write optimisation functions, and to manipulate the information in the key.

Combined Measures of Molecular Diversity. The central part of our measure for molecular diversity is based on pharmacophores. The first step is to compute and store the pharmacophores for all molecules in the potential set of products. It is then simple, each time a new product subset is generated, to add up the number of hits found for each pharmacophore, and to optimise some function of the number and spread of pharmacophores. Additional constraints on centre and distance type can be applied to constrain the profile to selected pharmacophores. A number of additional properties have been used as function modifiers:

(a) Number of heavy atoms
(b) Largest triangle perimeter present for all pharmacophores found
(c) Largest triangle area present for all pharmacophores found
(d) Number of accepted conformations found.

The function modifiers (a), (b) and (c) are used to provide a crude measure of shape distribution. The minimum and maximum values for each properties are determined for the entire data set. The properties are then divided into equal partitions between these limits. Each new product subset is split up into these partitions, and the number of molecules in each partition is compared the number expected for a perfectly even distribution (other distribution functions could be used). The occupancy function is a maximum when each bin is equally occupied, which forces the generated product subset to have a good distribution of shapes while still maximizing pharmacophore diversity.

The function modifier (d) is given a value inversely proportional to the square root of the number of conformations. In this way the number of unique pharmacophores per conformation can be maximized.

Library Design by Simulated Annealing. It is clear that, if we want to have a design procedure that is useful to the combinatorial chemist, we must perform our design in reagent space, so that no further analysis of products required. This we have done in the program Rpick, which we have implemented using simulated annealing [27] and genetic algorithms [28]. The flow chart for the simulated annealing algorithm is shown in *Fig. 3*.

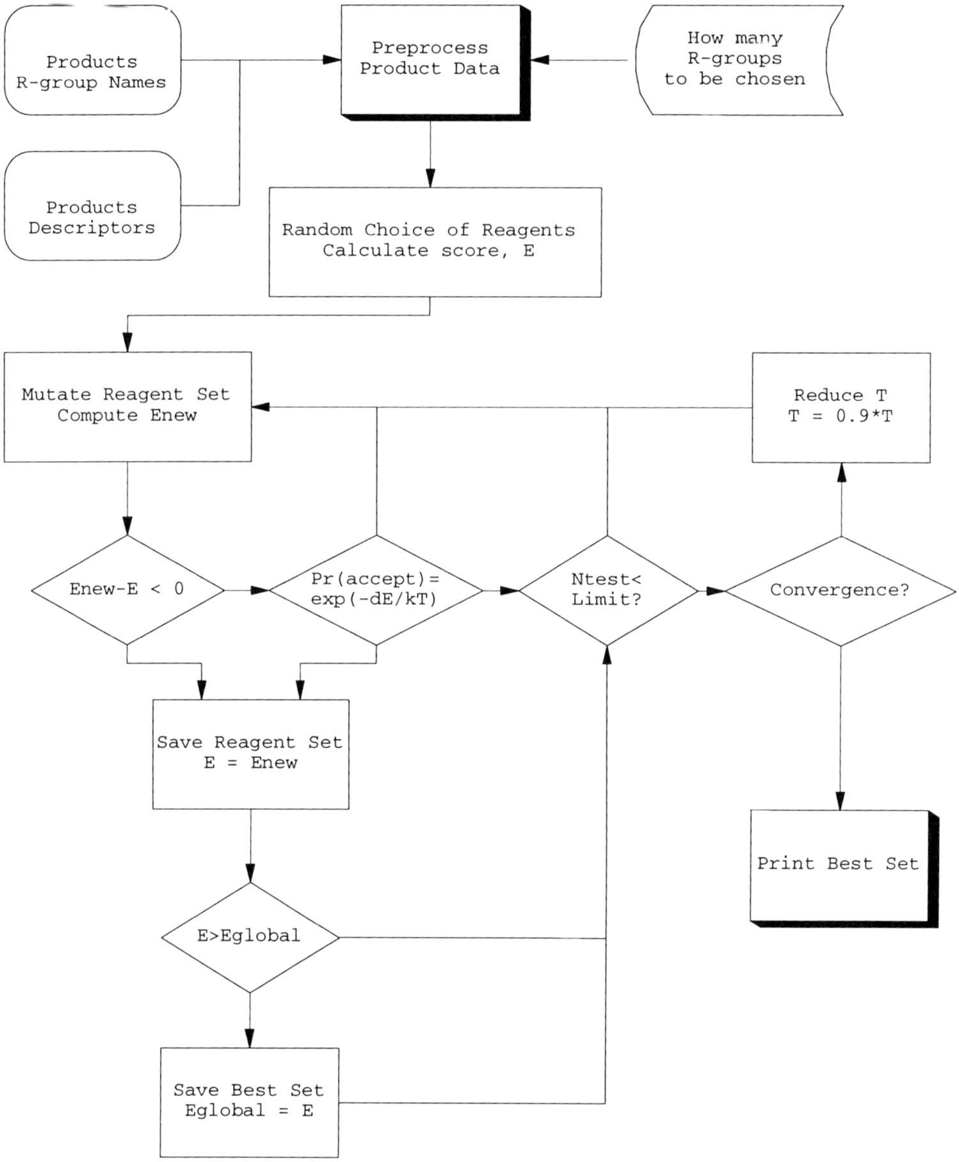

Fig. 3. *A flow chart for the Simulated Annealing version of Rpick.*

The data for the program consists of two parallel files of product data. The first file contains information on the product name and the names of the reagents that were used to make it. The second file contains the product name (for checking) and a list of product descriptors to be used later in the computation of molecular similarity. The first stage of the program checks that a complete and consistent set of reagents have been entered (the issue of missing entries is not dealt with in the current version), then asks the user to select how many reagents are required at each site of variation. This allows the run to be tailored to the actual chemistry. For example, for performing a bimolecular reaction in a 80/96 well plate, one would want 10 reagents in position 1 and 8 in position 2. The program now proceeds to the optimisation phase.The central feature of simulated annealing is the acceptance formula. In outline, an initial solution to the annealing problem is evaluated and the score, or fitness, of the solution becomes the current best solution. New solutions are proposed and evaluated; if a new solution has a better score, it is accepted and becomes the current best score. If the solution has a poorer score, it is accepted with a probability given by a Boltzmann distribution. This allows the program to climb out of local minima and thus to explore the solution space more thoroughly. The Boltzmann distribution is itself controlled by a parameter T that can loosely be thought of as temperature. At high values of T, many poorer solutions are allowed. As the program progresses, the value of T is decreased and the best solution is annealed into the optimal solution.

New solutions are proposed by the change_reagents routines, and evaluated by the Energy function. Change_reagents randomly swaps 1 reagent in the current solution for other possible reagents. More sophisticated strategies for mutation are possible; this will be the subject of future work. The Energy function plays two parts: the first part is to test quickly to see if the proposed solution is feasible, and then to evaluate the solution. The feasibility test allows hard constraints to be entered, for example that one reagent is incompatible with another, or that there are too many rotatable bonds in one product. Products that fail these tests are penalised by getting a diversity score of zero. The evaluation part computes a score, or energy, for the ensemble of molecules that can be created from the current reagent set. How this is done is up to the user, and depends on the molecular descriptors that were read in. The scoring function could reflect diversity within the product library, or diversity compared to some external standard; many different forms are possible. Once the optimisation has converged, the program then prints out its selection of reagents.

Pharmacophores are the key measure that we use, although mixed similarity metrics can be and are used. Although we do not discuss it here, runs can be constrained to enter space not filled by previous datasets: this has proved to be very useful for generating general screening sets and for filling holes in company registries. Running times of under 30 mins for typical problems are observed.

Reagent Selection by Genetic Algorithms. Genetic algorithms (GAs) use an evolutionary model to control the course of the optimisation. A flow chart is shown in *Fig. 4*. A solution is encoded as a single array, called a chromosome, which in our case contains the list of selected reagents. The GA maintains a population of these potential solutions. The initial population is selected at random from all possible solutions. Each solution is scored by generating the molecules associated with the selected reagents and calculating the score as described below. This score is often referred to as the fitness of the solution. The next population (generation) is chosen from the previous population by a randomized selection procedure that ensures that the number of times a solution is chosen is proportional to its fitness. Variation is introduced by applying genetic operators to the new population. These include cross-over and mutation. Cross-over takes two solutions (parents) and swaps portions of their binary representations to make children. Mutation randomly mutates one bit in a chromosome. The procedure is continued until a user-defined number of generations have been evaluated or the solutions have converged. As for the SA program, there is no constraint on the type of scoring function which can be used and the input is similar. The GA keeps track of the best solution(s) found during the course of the optimisation and these are output at the end of the run.

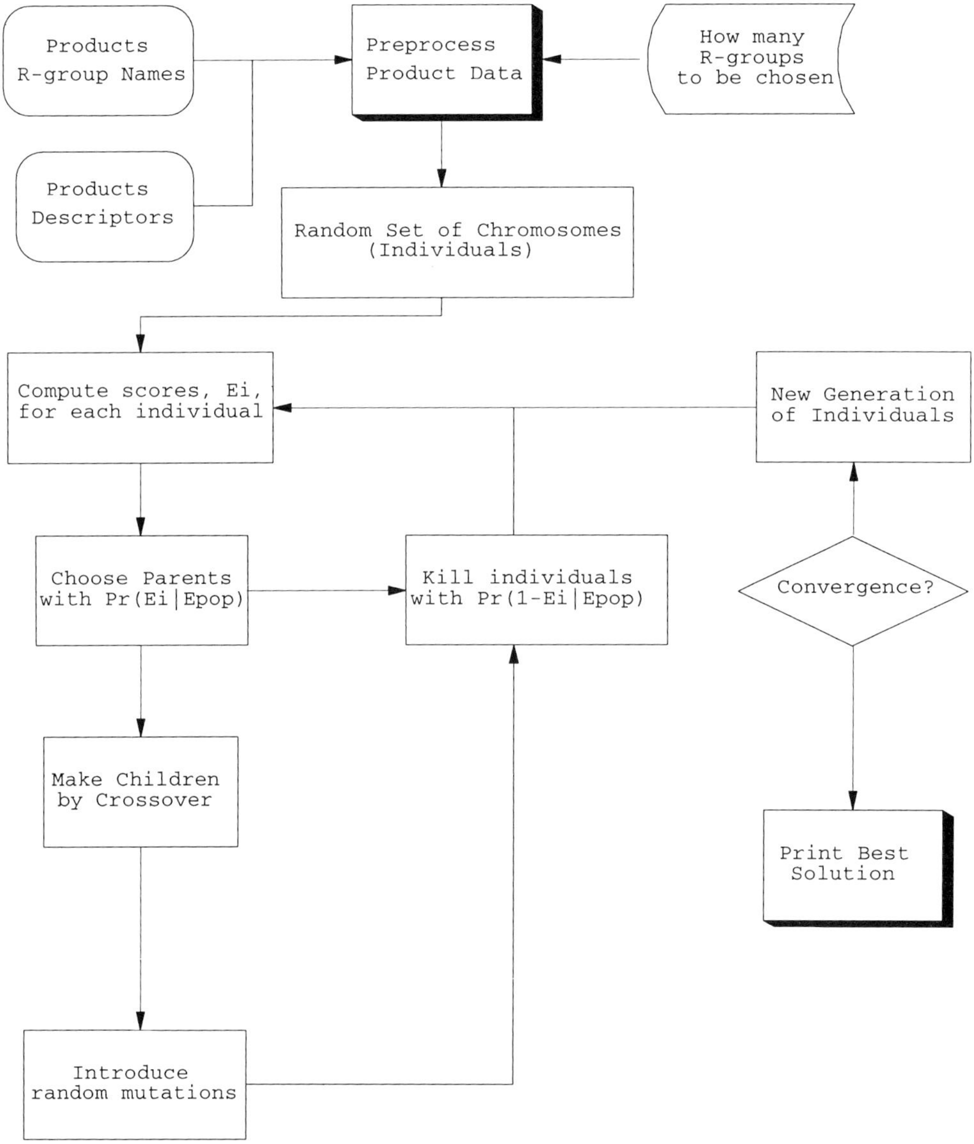

Fig. 4. *A flow chart for the Genetic Algorithm version of Rpick.*

Simulated Annealing vs Genetic Algorithms. The behaviour of the two methods for optimisation is very similar, and the correct method to choose will depend on several factors including the library itself and the scoring function used. In particular, SA techniques perform well on tightly constrained problems, whilst GAs are better suited to problems with discontinuous or noisy scoring functions. Both methods have a random selection procedure, so results will vary according the seed of the random number generator. We have therefore looked at both methods to try to build up our experience.

Results. - *Optimisation of ClogP Distribution for a Benzodiazepine Library.* ClogP [29] was chosen as a self consistent molecular descriptor that could be used purely for validation. The scoring function is derived from experimental design:

$$f = \mathrm{MIN}(Dij)$$

In plain language, we are trying to maximise the separation of the two closest neighbours in the product set. This will give an even sampling of space. For example, if we have an even spread of points with values between 0 and 4, selection of 2 representatives should give points at 0 and 4, 3 representatives should give points at 0, 2 and 4 and so on.

The maximum value of f is (Max(ClogP)-Min(ClogP))/(N-1).

To test the program, we used the benzodiazepine library of DeWitt *et al.* [30] (*Fig. 5*).

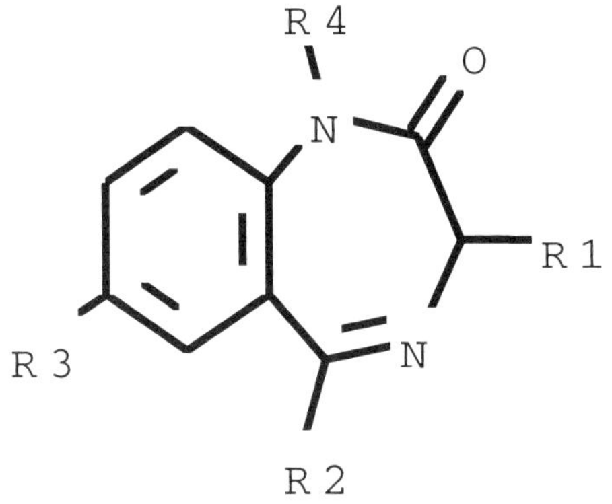

R1 from **Gly, Ala, Val, Phe, Trp, Asp, Asn, Glu, Gln, Thr, Lys**

R2 = Ph, 4-MeOPh, cHex, 2-thienyl, 4-NO_2Ph, 4-NH_2Ph

R3 = H, Cl, NO_2, OMe, NH_2

R4 = H, Me, benzyl, iBu

Fig. 5. *The Markush description of Benzodiazepine library used to test Rpick*

Two types of run were performed: (a) ignoring the combinatorial nature of the library and just selecting compounds (b) choosing reagents. The results are shown in *Table 1*. The Select column shows the number of reagents selected to appear at each position, so that 4.4.3.2 means 4 reagents at position R1, 4 at R2, 3 at R3 and 2 at R4, making 96 overall. The results of these runs can be compared with the theoretical maximum. All the results have been scaled up by a factor of 1000, to allow the calculation to be performed in integer rather than real values. Note that the type (a) runs are able to get far closer to the theoretical maximum, as expected, as a change in one reagent has a knock-on effect in all products. In particular, we would highlight the last two runs. The theoretical maximum is 94.6, cherry picking gives 46 and 4.4.3.2 product selection gives 13.

Table 1. *The Results of Running Rpick on a Benzodiazepine Library Using ClogP as a Descriptor*

Run Type	Select	Actual Max.	Poss. Max	R1	R2	R3	R4
a	16	552	567.5	-	-	-	-
b	2.2.2.2	331	567.5	Trp/Glu	Ph/Thien	NO2/NH2	Me/Bz
b	2.2.2.2	321	567.5	Trp/Asn	cHex/pNH2Ph	H/Cl	Me/iBu
a	96	46	94.6	-	-	-	-
b	4.4.3.2	13	94.6	Val/Phe/Glu/ Lys	Ph/cHex/Thien/ pNH2Ph	H/Cl/NO2	Me/Bz

Optimisation of Pharmacophore Coverage for a Benzodiazepine Library. The Pharmacophore-Derived Queries (PDQ) descriptor [13] describes a molecule by determining which of a set of 5916 three centre pharmacophores it expresses. In this development phase, a relatively simple scoring function was used for minimization:

$$f = 1 - (Npharm/Npmax)$$

where Npharm is the number of different pharmacophores contained in a solution and Npmax is the maximum possible pharmacophore coverage, i.e. the total number of different pharmacophores contained within the whole library.

This run was performed using the Genetic Algorithm version of Rpick. The goal is to find a subset of the library which covers as much of the available pharmacophore space as possible. The results are shown in *Table 2.* In this trial 96 compounds were requested made up of a selection of 4x4x3x2 reagents from each of R1, R2, R3 and R4 respectively. The table shows the best solution over 5 separate runs. This covers 83% of the pharmacophore space of all 1320 possible compounds in the library. The same minimum was obtained with different settings for the mutation and cross-over rates. Several other local minima were also obtained covering greater than 80% of the pharmacophore space.

Table 2. *The Results of Running Rpick(GA) on a Benzodiazepine Library Using PDQ as a Descriptor.*

Substitution Site	Reagents Selected
R1	Val, Trp, Glu, Lys
R2	4-MeOPh, c-Hex, 4-NO_2Ph, 4-NH_2Ph
R3	NO_2, OMe, NH_2
R4	H, i-Bu

For R1, a range of hydrophobic and polar amino acid sidechains have been chosen. For R2 and R3, a range of both hydrophobic and polar groups have been chosen. For R4, the selection of H generates a new H-bond donor group and the iBu is chosen to represent hydrophobic character. It should be noted that, because we are analysing the whole molecule, the phenyl in position R4 does not add many new pharmacophores because of the existence of aromatic rings in other positions. This highlights a crucial difference between reagent selection by reagent and by product. The running times of the Genetic Algorithm version of Rpick was much slower than the Simulated Annealing version.

A Comparison of ChemDiverse Against Rpick. A number of investigations have been carried out to compare ChemDiverse and Rpick. The data set chosen was derived from over 8000 amines present in the ACD [24]. These were whittled down to 2355 structures using an in-house SMARTS-based program to remove amines with low basicity (e.g. anilines with electron withdrawing substituents on the benzene ring) and a molecular weight below 250. These remaining amines were attached to an artificial diaminobenzene central group (*Fig. 6*) using ChemDiverse and then keyed using our molecular level pharmacophore storage procedure (5 CPU hours on an R4400). This simple system was chosen to allow a direct comparison between the two methods: Rpick can handle up to 4 different sites of combinatorial variation around the central fragment.

ChemDiverse was then run on the data set. Before execution the amines were ordered for keying using the Chem-X *sample* command. This procedure attempts to maximize the difference in 2D or 3D distance keys between structures at the top of the data set list, thus ensuring a more efficient sampling of structural diversity. The ChemDiverse calculation was executed using a maximum permitted overlap of 80% between the molecule being keyed and the ensemble pharmacophore key.

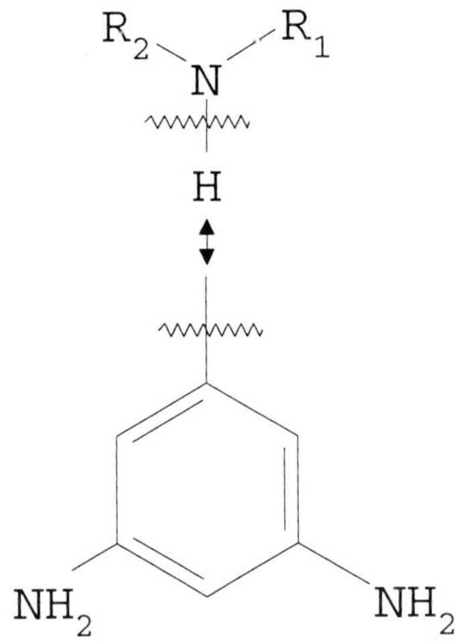

Fig. 6. *The structure of the diaminobenzene structure used to compare ChemDiverse and Rpick.*

The ChemDiverse run resulted in a set of 86 molecules being accepted (this highlights the ChemDiverse problem already mentioned, in that it is not possible to specify the number of hits required, rather you must try a range of calculation constraints to achieve the required number of structures. This can be done within a single run but is not the default). Three Rpick runs were then executed. For all three runs, 86 hits were requested to allow direct comparison with the ChemDiverse run. The three runs were set up as follows:

(1) Maximize the number of pharmacophores.
(2) Maximize the number of pharmacophores and also ensure an even distribution of shape properties (a), (b) and (c).
(3) Maximize the number of pharmacophores, ensure an even distribution of shape properties (a), (b) and (c), and minimize the number of accepted conformations for the structures chosen.

The calculation statistics determined from these investigations are shown in *Table 3* as general results, together with graphs showing partition information for the ChemDiverse set, Rpick (2) set and full amine data set (*Figs. 7 and 8*). In run Rpick(1), 86 products were specified, subject to maximising the number of pharmacophores. The selected set contains 18088 pharmacophores. However, it also contains a poor spread of shape, and a large number of accepted conformations. This means that a few highly flexible molecules are contributing a lot of pharmacophores. This can be misleading, as highly flexible molecules are not ideal screening candidates (the debate as to how flexible or rigid a library should be is unresolved). In Rpick(2), we try to force an even shape distribution in addition to maximising pharmacophore coverage; this has the side-effect of removing some of the most flexible molecules and reducing the number of pharmacophores.

Table 3. *The Results of Running Rpick on a Diaminobenzene Library Using Pharmacophores as the Principal Descriptor. Note that the Total Number of Unique Pharmacophores present in the Entire Data Set is 19136.*

Calculation	Number of Unique Pharmacophores	Number of accepted Conformations in Dataset	Partition function scores Property			Approximate Calculation time (min.s)
			a	b	c	
ChemDiverse	16532	41516	0.25	0.33	0.38	80
Rpick (1)	17848	75994	0.25	0.37	0.38	18
Rpick (2)	15107	63724	0.50	0.86	0.75	8
Rpick (3)	13889	15681	0.57	0.70	0.64	11

We can see how successful Rpick has been by comparing the shape distribution (as measured by the area of the largest pharmacophore expressed by each molecule) obtained by Rpick to the distribution obtained using the entire library and the set of molecules obtained using ChemDiverse (*Fig. 7*). We can see that even though the distribution in the whole library is skewed (this is reflected in the ChemDiverse results), Rpick has managed to spread the distribution as much as is possible. If we look at the distribution of heavy atoms in the molecule sets (*Fig. 8*), we again see that Rpick has produced a more even distribution. The peak at the higher end is caused by the interplay of the constraint functions and the need to maximise the number of pharmacophores; more flexible molecules tend to express more pharmacophores. The correct relative weights to give the various functions is an area of active study. In run Rpick(3), in which we penalise on the number of conformations, the reduction in the number of pharmacophores is even more pronounced, emphasising the relationship between pharmacophore coverage and flexibility.

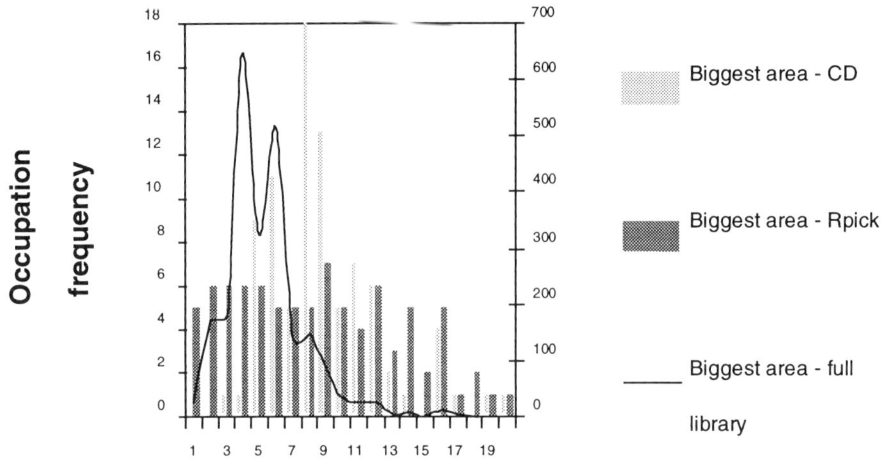

Fig. 7. *A graph showing the distribution of the area of the largest pharmacophore expressed by each molecule for the whole library (solid line), ChemDiverse (light grey bars; CD) and Rpick (dark grey bars).*

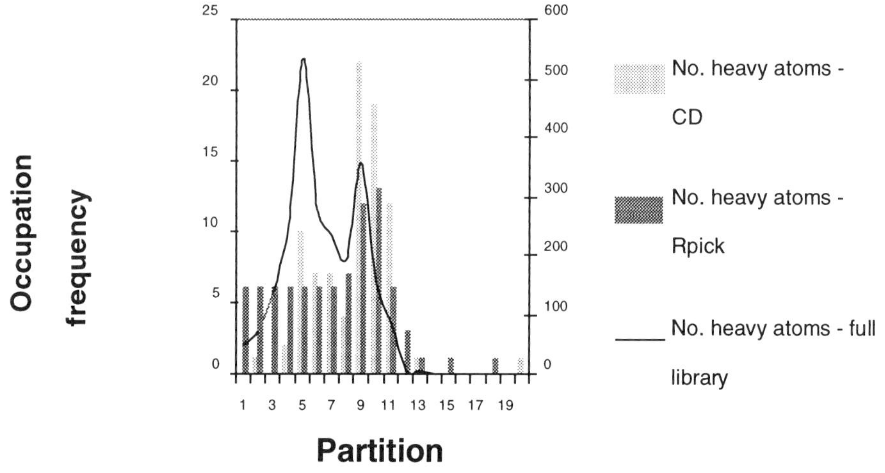

Fig. 8. *A graph showing the distribution of the number of heavy atoms in each molecule for the whole library (solid line), ChemDiverse (light grey bars; CD) and Rpick (dark grey bars).*

Discussion. - The results of the calculations undertaken clearly illustrate the advantages of the Rpick program over Chem-Diverse: Rpick is able to select an exact number of requested reagents for each substitution site. When attempting to just maximize the number of unique pharmacophores present, Rpick run (1) is able to find over 1000 more for the same number of structures. The Rpick optimisation function can be easily modified: Rpick runs (2) and (3) show how other properties can easily be incorporated, significantly changing the makeup of the chosen data set, while maintaining good pharmacophoric diversity.

The length of the simulated annealing run within Rpick is extremely dependent on the parameters (e.g. temperature); the runs undertaken here converged rapidly to excellent solution. Around 97% of the CPU time of Rpick is spent in the energy calculation routine. Around 20 energy calculations per R4400 CPU second are achieved when choosing 100 from 2355 structures. Limited tests suggest that the number of energy calculations per CPU second varies as approximately the square root of the number of products to be processed (6 calculations/sec when choosing 1000). This of course will also depend on the promiscuity of the molecules under consideration, since more pharmacophores require more processing.

Experimental Validation of Similarity Metrics. When drug designers are using any similarity measure, they should ask themselves a few questions, for example: How do we know if our similarity measure is meaningful? Have we achieved diversity within our libraries? Do we want to achieve diversity using this measure? Unfortunately, there are no easy answers, especially with respect to how meaningful a measure is. Most studies in the literature [31-34] have looked at separating actives from inactives in compound *collections*. The method is to take a number of reference structures with known biological activity (β-blocker, for example), to trawl through a database looking for similar compounds, using the defined metric and analysing the number of compounds with that activity missed due to insufficient similarity (false negatives), and the number of extraneous hits (false positives), and the enrichment ratio of actives found vs inactives discarded. There are a number of possible pitfalls to this method: (i) Biological activity classes can be too broad e.g. antihypertensives, so that a range of possible mechanisms are included leading to apparently a large number of false negatives. (ii) the cut-off between activity and inactivity can be crucial (*Fig. 9*), possibly excluding all but close analogues, which will favour the substructural similarity metrics. It should also be pointed out that many corporate compound collections contain diverse families but large familial groupings (steroids, penicillins, benzodiazepines), whereas combinatorial libraries contain a family of diverse compounds. These methods also make the assumption is that there is true SAR in the data [33]. Does this reflect a universal truth or a universal method of optimisation?

Another methodological pitfall comes from the comparison of binary keys, using, for example, the Tanimoto coefficient. In our hands, the Tanimoto coefficient does not work very well for sparse binary keys, for instance pharmacophore keys. In an attempt to cluster RPR corporate database of over 150K compounds using binary pharmacophore descriptors, a very sharp division was observed between the similarity level that gave 1 cluster of 150K members and the level that gave 150K singletons. Better results have been obtained with denser keys.

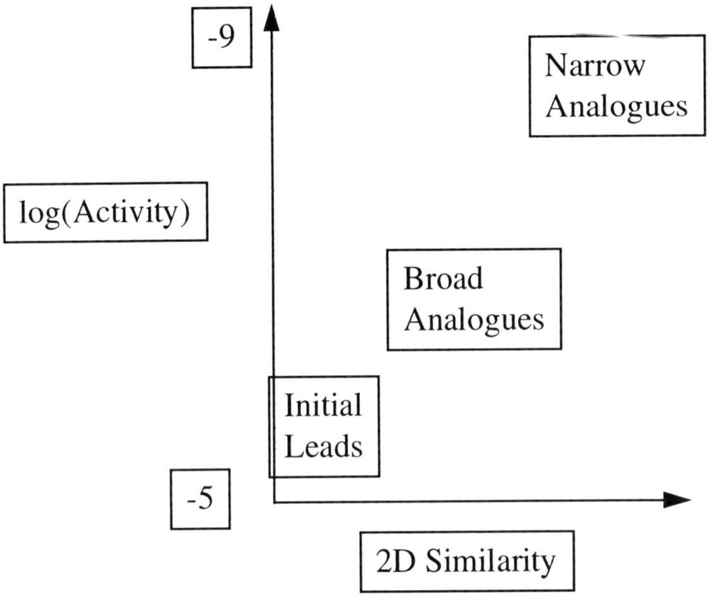

Fig. 9. *A schematic graph showing the relationship between activity and optimisation strategy. As a lead progresses through from low to high activity, a chemist will first make broad analogues, then narrower ones, as the SAR becomes more defined.*

We question the validity of these and other studies because in our experience, most data sets containing chemical structures and biological data are not complete and contain errors. We do run into a fundamental sampling problem, that we have no external reference against which to demonstrate completeness or incompleteness, that is, how can we know if the data sets we examine are a representative sample of the complete chemical universe? We have also observed that most data sets contain a bias that reflects the historical development of medicinal chemistry, in that certain therapeutic classes are over-represented, whereas molecules that influence more recently discovered targets are under-represented. There are also a large number of analogues, which reflects the standard practice of medicinal chemistry and drug development over the course of many years. It is for these reasons that we believe that company databases will respond well to substructural measures: based on our experiences using pharmacophores for lead generation [15], we think that combinatorial libraries will behave better with a pharmacophore metric. We also note that different test cases and screens give different answers, reflecting the many different mechanisms of molecular recognition.

Challenges for The Future. The quantification of molecular similarity and its application to combinatorial chemistry is not a solved problem. Three centre pharmacophore measures are a good starting point, but need to be expanded to include more centre, shape and chirality information. We need to be able to combine different similarity metrics in a sensible way, so that each descriptor receives the appropriate weight in the design procedure. The maximum possible diversity that can be achieved within a particular combinatorial reaction is constrained by reagents available. For common functional groups, this is not generally a problem, but for the more esoteric functions like isocyanates (used in the Ugi reaction among other), there are only 12 reagents available for purchase. Can we work from desired diversity of the products back to missing reagents? These are the outstanding challenges for future research in this area.

Conclusions. - Combinatorial Chemistry offers the temptation of being able to make more compounds than we can afford to synthesise, store and screen. Some form of design and prioritisation is therefore highly desirable. Previous methods have focused on selecting reagents in isolation, due to the complexities of product-based strategies. We have overcome this block, through the development of two programs, based on Simulated Annealing and Genetic Algorithms, that can be used to select automatically **reagents** which will produce the most desirable characteristics (according to the user) in the final set of synthesised **products**. The programs have been validated with several simple functions, and the results are consistent with theoretical expectations. The running times are also very quick, so that design is not an expensive option. We now have the opportunity to develop more complex, but more realistic, design criteria. Pharmacophore Similarity may not be the most exact descriptor to use for similarity, but it may well be the most appropriate to use for lead generation and library design

Acknowledgements. - We gratefully acknowledge helpful comments and criticism from our colleagues, E. Surcouf, A. Laoui, C. Luttmann, C. Carrez, A. Morgat, J. Mason, P. Menard, D. Cheney, I. Morize, P. Bamborough, D. Clark and C. Newton.

REFERENCES

[1] Dean, P.M. (Ed.), *Molecular Similarity in Drug Design*, Blackie Academic and Professional Glasgow, 1995

[2] Martin Y.C., Brown R.D., Bures M.G., *Quantifying Diversity.* In Kerwin J-F. and Gordon E.M.(Eds) 'Combinatorial Chemistry and Molecular Diversity', in press.

[3] Carbo, R., Leyda, L., Arnau, M.,. *Int. J. Quantum Chem.* **1980**, *17*, 1185- 1189.

[4] Good A.C., Hodgkin E.E., Richards W.G., *J. Chem. Inf. Comput. Sci.* **1993**, *33*, 112-116.

[5] Kuntz, I.D., Blaney, J.M., Oatley, S.J., Langridge, R., Ferrin, T.E., *J. Mol. Biol.* **1982**, 161, 269-288.

[6] Bemis, G.W., Kuntz, I.D., *J. Comput.-Aided Mol. Design.* **1992**, *6*, 607-627.

[7] Mezey, P.G., *Methods for molecular shape similarity and topological shape design.* In Dean, P.M. (Ed.), Molecular Similarity in Drug Design, Blackie Academic and Professional Glasgow, 1995, pp 241-268.

[8] Perry N.C., Van Geerestein V.J., *J. Chem. Inf. Comput. Sci.* **1992**, *32*, 607-616

[9] Nilakantan, R., Bauman, N. and Venkataraghavan, R., *J. Chem. Inf. Comput. Sci.* **1993**, 31, 527-530.

[10] Good, A.C., Ewing, T.J.A, Gschwend, D, *J. Comput.-Aided Mol. Design.* **1995**, 9, 1-12.

[11] Murrall, N.W., Davies, E.K., *J. Chem. Inf. Comput. Sci.* **1990**, 30, 312-316.

[12] Mason, J.S., *Experiences with searching for molecular similarity in conformationally flexible 3D databases.* In Dean, P.M. (Ed.), Molecular Similarity in Drug Design, Blackie Academic and Professional Glasgow, **1995**, pp. 151-154.

[13] Pickett S.D., Mason J.S., McLay I.M., *J. Chem. Inf. Comput. Sci.*, in press.
[14] Wermuth C., Langer T. *Pharmacophore identifcation.* In Kubinyi H. (Ed.) '3D QSAR in Drug Design', Escom, Leiden, 1993, pp. 117-136.
[15] Mason J. S., McLay I. M. and Lewis R.A. *Applications of Computer-Aided Drug Design Techniques to Lead Generation.* In Dean P.M., Jolles G. and Newton C.G. (Eds) 'New Perspectives in Drug Design', Academic Press, London, **1995**, pp. 225-253.
[16] Bath, P.A., Poirette, A.R., Willett, P., Allen F.H., *J. Chem. Inf. Comput. Sci.* **1994**, 34, 141-147.
[17] Lewis R.A., McLay I.M. and Mason J.S., *Chemical Design Automation News* **1995**, *10*, 37-38.
[18] Menard P.R., Lewis R.A., Mason J.S., *Rational Screening Set Design and Compound Selection: Cascaded Clustering*, manuscript in preparation
[19] Gordon E.M., Gallop M.A., Patel D.V., *Acc. Chem. Res.* **1996**, 29, 144-154.
[20] Barnard J.M., Downs G.M., *J. Chem. Inf. Comput. Sci.* **1992**, *32*, 644-649.
[21] Downs G.M., Willett P., *J. Chem. Inf. Comput. Sci.* **1994**, *34*, 1094-1102.
[22] Good A.C., Kuntz I.D., *J. Comput.-Aided Mol. Design* **1995**, *9*, 373-379
[23] Daylight Chemical Information Systems, Santa Fe, NM87501, USA
[24] Available Chemicals Database, developed and distributed by Molecular Design Ltd., San Leonardo, CA.
[25] This program is available on request from RAL
[26] ChemDiverse, developed and distributed as part of the Chem-X modeling package by Chemical Design Ltd., Roundway House, Cromwell Business Park, Chipping Norton, OXON, UK.
[27] Kirkpatrick S., Gellatt Jr. C.D., Vecchi M.P, Science **1983**, 220, 671-690.
[28] Clark D.E., Westhead D.R. *J. Comp.-Aided Mol. Design,* **1996**, *10*, 337-358.
[29] Hansch C., Leo A. In '*Substituent Constants for Correlation Analysis in Chemistry and Biology*'. Wiley Interscience, New York., pp. 18-43.
[30] DeWitt S.H., Kiely J.S., Stankovic C.J., Schroeder M.C., Cody D.M.R., Pavia, M.R., *Proc. Natl. Acad. Sci. USA.* **1993**, *90*, 6909-6913.
[31] Brown R.D and Martin Y.C., *J. Chem. Inf. Comput. Sci.,* in press.
[32] Sheridan R.P., Kearsley S.K., *J. Chem. Inf. Comput. Sci.* **1995**, *35*, 310-320.
[33] Cummins D.J., Andrews C.W., Bentley J.A. and Cory M. *J. Chem. Inf. Comput. Sci.* **1996**, *36*, 750-763.
[34] Patterson D.E., Cramer R.D., Ferguson A.M., Clark R.D., Weinberger L.E., *J. Med. Chem.* **1996**, *39*, 3049-3059.
[35] Taylor R. *J. Chem. Inf. Comput. Sci.* **1995**, *35*, 59-67.

PART II. **LEAD FINDING STRATEGIES**

Chapter 10

Exploiting Molecular Diversity: Pharmacophore Searching and Compound Clustering

by **Vincent J. van Geerestein***, **Hans Hamersma** and **Steven P. van Helden**

Department of Computational Medicinal Chemistry, N.V. Organon, P.O. Box 20, 5340 BH Oss, The Netherlands

Exploiting Molecular Diversity: Pharmacophore Searching and Compound Clustering

by **Vincent J. van Geerestein***, **Hans Hamersma** and **Steven P. van Helden**

Department of Computational Medicinal Chemistry,
N.V. Organon, P.O. Box 20, 5340 BH Oss, The Netherlands

Pharmacophore searching and compound clustering are important concepts of computer based methods for lead finding strategies. These concepts are very much linked to modern developments in drug discovery and they are the direct response of computer chemists to the current achievements of combinatorial chemistry (CC) and high throughput screening (HTS). Pharmacophore searching is now an established technique which has a proven track record in many discovery programs of pharmaceutical companies. The outstanding issue here is probably the problem of dealing with conformational flexibility. In the paper we will share our experiences using conformationally flexible pharmacophore searching and give an outline of the strategy used in our group. In the hit list of a pharmacophore search, we typically obtain a factor of 10 of enrichment of the biological activity of concern. Compound clustering is now regularly used for rational selection of compounds for acquisition for in-house compound collections used for screening. However, also here challenges and dilemmas remain. Clustering is typically done on the basis on 2D fingerprints, derived from techniques used for fast substructure searching by scientists working in the field of chemical information. We show that comparable results may be obtained with 3D fingerprint derived from 3D pharmacophores. Compound clustering in lead finding appears to be of value as a method of prioritising the test order or in case the throughput of the assay prevents testing all compounds. In the same way it may be used to select and prioritise compounds for acquisition.

Introduction. - Molecular modelling and chemometrics for the study of quantitative structure activity relationships (QSAR) are now standard tools within the pharmaceutical pre-clinical research. Many of these techniques are particularly suitable for assisting the synthetic chemist in optimising a given series of lead compounds. Nevertheless, rational drug design methods do not turn the process of lead optimisation into a trivial endeavour. Although increased potency may be readily obtained, the optimisation process is often seriously hampered by adverse ADME[1] problems. In reality the development of a suitable drug development candidate entails the simultaneous fulfilment of all the necessary criteria with a single compound. For that reason, the whole emphasis of lead optimisation should be on defining levels of acceptable performance rather than identifying properties associated with optimal performance in one of the relevant criteria. Consequently, a set of different starting points often increases the chance of success. A sufficient supply of leads

[1] Absorption, distribution, metabolism and elimination.

is thus of decisive importance to this multidimensional optimisation process. The leads may either be taken from the literature, originate from a structure based drug design project or be obtained from a mass screening program.

The advent of high-throughput screening (HTS) has given pharmaceutical companies the ability and possibility to test very large numbers of compounds in a short time [1]. HTS is expected to give a sufficient number of lead compounds for further optimisation. The probability of success in HTS depends on factors that include assay throughput, number and diversity of compounds available for testing but also the ability to discriminate meaningful activity from false positives and assay noise. Given a robustly designed HTS assay, the rate determining step in the discovery of new leads may to a large extent depend on the diversity of the compound collections tested [2]. For that reason, diversity of compound collections is currently receiving a high level of attention.

The concept of molecular diversity has already been exploited in number of different ways. Nowadays, large compound libraries may be synthesised with combinatorial chemistry or parallel synthesis technology. However, rather than focusing solely on achieving huge numbers of compounds, the discovery efficiency may be improved by using experimental design methods to maximise molecular diversity for a given synthetic library size or to bias the synthetic library with key features for a specific receptor [3]. An alternative source of compound diversity is the acquisition of compound collections through brokers. In this instance concepts of diversity may be used to decide which compounds will enrich existing compound collections with the maximum relevant diversity. Notwithstanding the fact that molecular diversity has become a quite popular paradigm in drug discovery, no single "best" definition of molecular diversity can possibly be given [4]. This is a situation that is similar to the paradigm of molecular similarity, where the actual measure also depends on its particular application [5].

This paper tries to give some idea about the strategic values of two computational methods for lead generation: three dimensional (3D) database searching and compound clustering. The former approach is now an almost traditional tool in drug discovery research whereas the latter received quite some interest lately because of the recent HTS developments. We would like to assess the efficiency and effectiveness of these methods as strategic approaches in drug discovery in particular in comparison to random screening. Unfortunately, we have to perform these studies on non-ideal validation sets because large sets with broad screening results are currently not publicly available.

Test Database. - The test database for the present studies is our in-house DrugsFile containing molecular structures and other data of selected drugs in development or on the market. All information on biological and therapeutic activities has been gathered from the literature during the last 13 years[2]. Drugs can be searched by biological activities, therapeutic classes, manufacturers, stage of development or chemical (sub)structures. Nearly 900 items are in use for labelling compounds with biological activities varying from broad pharmacological actions to specific receptor or enzyme nomenclature. The database holds currently about 13,000 drugs for which 36,000 biological activities have been reported. We maintain a 3D database version of the DrugsFile with single molecular conformations generated with CORINA [6, 7]. For the current studies we used 11686 single-compound drugs with 865 different activity codes. The DrugsFile has proven to be very valuable for training of pharmacophores and other validation purposes.

Pharmacophore Searching

In case any structure-activity relationships are known, computational screening of 3D databases may effectively be used as a pre-screen to *in vivo* or *in vitro* screening [8, 9]. If a 3D model of the putative binding pocket of the molecular target is available, automatic docking procedures [10] or *de novo* design methods [11] may be put to use. Unfortunately, for these methods the necessary high quality experimental information about the receptor is often lacking. For that reason, 3D databases are frequently screened on the basis of pharmacophore models derived from information of active and inactive analogues. In this paper we attempt to give some estimate of the effectiveness and efficiency of pharmacophore searching for a small diverse set of pharmacophores from the literature. We have also looked at the effect of using different protocols for database preparation and searching. In order to be able to comprehend the various options in conformationally flexible pharmacophore searching, quite some detail of the procedures followed has to be given.

Keying and Conformational Flexibility. - At Organon, 3D database pharmacophore searches are performed using the Chem-X software [12]. Publications about the use of this system include those from the groups of Milne [13] and Mason [9]. In Chem-X terminology a pharmacophore may consist of 4 different so-called centre types: hydrogen bond donors, hydrogen bond acceptors, positively charged centres and aromatic ring centres. Negatively charged centres are not used explicitly because they coincide with hydrogen bond acceptors and are thus implicitly taken into account. Recently, a fifth centre type (so called hydrophobes) has been introduced but this type is not included in the present study. Chem-X not only stores the 3D molecular structure in the database but also "distance keys" holding information about distances between pharmacophore centres. Since 4 different types of pharmacophore centres are defined in Chem-X, 10 different

[2] Primary sources are Drugs of the Future, Drug Data Report, Martindale, Pharmaprojects, and USP Dictionary of USAN and International Drug Names.

distance keys are stored in the database corresponding to distances of 10 different centre-centre combinations. Each distance key consists of 32 so-called bins each corresponding to a certain distance range. The set of 320 distance bins thus indicates which centre-centre distances occur within a certain molecule. For each molecule a 32-bin information key and a 64-bin bond key are set as well [14].

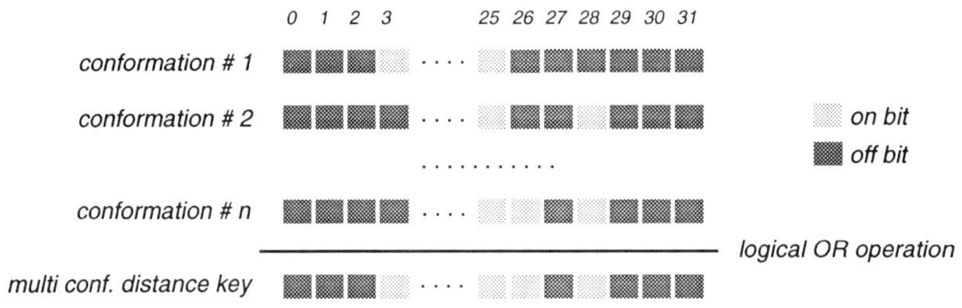

Fig. 1. *Schematic representation of the generation of multi-conformational distance keys by Chem-X. Single distance keys of low energy conformations are merged into a multi-conformational distance key.*

Conformational freedom is taken into account by applying a rule-based approach for generating a representative set of low energy conformations. For each conformation the centre-centre distances are added to the distance key (see *Fig. 1*). After completion of this procedure, the distance key indicates which distances may occur in any of the conformations considered for that molecule. This process of calculating the combined distance keys for all compounds in the database is called "database keying". At search time, the key of the query is compared with the key of each database compound to check whether that molecule might have a conformation with the proper distances. Only molecules passing this key screen are subjected to the time-consuming regeneration of conformations during which the conformations are tested individually for a possible match with the query.

The keying of the database is a rather time consuming process because it involves a conformational analysis for all compounds in the database[3]. In principle, conformations are generated by rotating in pre-defined steps about all rotatable bonds. Rules are applied for discarding conformations that are unlikely to be energetically feasible. By default Chem-X rotates sp^3-sp^3 bonds in 3 steps of 120°, sp^3-sp^2 bonds in 6 steps of 60°, conjugated bonds in 2 steps of 180° and double bonds also in 2 steps of 180°. The keying protocol may be customised to suit the specific needs of the user. At Organon we normally keep double bond rotations fixed. Dihedral angles around sp^3-sp^2 bonds do not in fact have a clear preference for any particular value and the choice of number of rotation

[3] Keying 10,000 drug-like compounds takes about 20 hours on a Silicon Graphics Power Challenge using a single 190 MHz MIPS R10000 CPU. All pharmacophore searches were done on this computer.

steps about sp^3-sp^2 bonds is thus more or less dictated by the balance between level of precision and economics of time spent on conformational analysis. As an alternative to the conformationally flexible keying method, Chem-X also offers a much faster, but less accurate, keying method where distance bins are set on the basis of upper and lower bounds for distances between pharmacophore centres which are directly estimated from the chemical graph by counting the number of bonds between the pharmacophore centres. These estimated distance keys are called 2D distance keys as opposed to the 3D distance keys mentioned above which are obtained from real distances taken from a set of (conformationally analysed) 3D structures. The 2D distance keys will usually have more bits set than in the corresponding 3D distance keys, but all bits set the 3D distance key should also be set in the 2D distance key.

Table 1. *Combinations of Protocols for Database Keying and Database Searching.*

protocol	conformational analysis during keying	conformational analysis during searching
sc/sc	single conformation	single conformation
sc/a6	single conformation	6 x 60° steps for sp^3-sp^2 bonds
a6/a6	6 x 60° steps for sp^3-sp^2 bonds	6 x 60° steps for sp^3-sp^2 bonds
a4/a4	4 x 90° steps for sp^3-sp^2 bonds	4 x 90° steps for sp^3-sp^2 bonds
2D/a6	distances estimated by path length	6 x 60° steps for sp^3-sp^2 bonds
no/a6	none	6 x 60° steps for sp^3-sp^2 bonds
a6/flex	6 x 60° steps for sp^3-sp^2 bonds	flexi-fit

At search time conformations are normally regenerated using the same rule based protocol as used for the database keying. Alternatively, a flexi-fit algorithm may be applied for constrained torsional optimisation to match the database molecule onto the pharmacophore query. We have performed a comparative study of seven different combinations of database keying and searching protocols with the objective to determine the best protocol for general application in pharmacophore searching (see *Table 1*).

Effectiveness Measures. - In our study we have used the following measures for qualifying the results of a pharmacophore search:

$$\%\text{ACTFOUND} = \frac{\text{number of actives in hit list}}{\text{number of actives in DrugsFile database}} \cdot 100\%$$

$$\%\text{ACTHITS} = \frac{\text{number of actives in hit list}}{\text{number of compounds in hit list}} \cdot 100\%$$

$$\text{ENRICHMENT} = \frac{\%\text{ACTHITS}}{\text{percentage of actives in DrugsFile database}}$$

The ideal pharmacophore query for a given biological activity should in principle return all known actives (%ACTFOUND = 100). However, the descriptive power of a pharmacophore is frequently inadequate in capturing a whole series of different molecules displaying a similar activity. This may obviously be rationalised by compounds exhibiting different binding modes and thus in fact requiring a multiple pharmacophore description. On the basis of the %ACTFOUND and %ACTHITS measures, the balance between the number of false positive hits and number of false negative hits may be optimised by adjusting the pharmacophore. The lowest possible number of false positive hits will be obtained by maximising the %ACTHITS value. However, this poses a basic problem because not all activities will necessarily have been reported for all the compounds in the DrugsFile. Consequently, some of the false positives found by the search may actually be true positives with yet unreported activity of interest. For that reason, the value of %ACTHITS, as returned from a database search, should be considered as a lower bound. ENRICHMENT gives the increase in probability of finding a hit as a result of a pharmacophore search over finding a hit by random selection. ENRICHMENT in combination with %ACTFOUND is a good measure for determining the effectiveness of 3D database search.

Flowchart. - A search in Chem-DBS 3D is usually presented as a two-step process: first the screening of the keys followed by the 3D structure matching. A compound may fail to pass the second step for a number of different reasons which we have analysed in detail using the Chem-X Programming Control Language (PCL). The various stages of the search process are depicted in a flowchart shown in *Fig. 2*. For example we used the results for a query for histamine (H1) antagonists as originally defined by Moock [15]. The "total" and "actives" columns indicate how many of all DrugsFile compounds and how many known histamine H1 antagonists, respectively, pass each step. Out of the 211 reported H1 antagonists in the database only 48 were identified with the Moock query, resulting in a %ACTFOUND value of only 22.7%. We may differentiate various reasons for rationalising this poor result. The 49 H1 antagonists rejected by the key screen are mostly compounds that lack the two requisite aromatic rings. These compounds thus may have an other mechanism of action. A mismatch in atom types caused the rejection of 4 further H1 antagonists having a secondary amine instead of a tertiary amine as defined in the query. Finally, 105 known H1 antagonists were dropped because they violated the distance limits on the geometry. We then used the data from this flowchart for optimising the query. In this particular case, relaxing the tolerances on the centre-centre distances and changing the tertiary amino group into a generalised N sp^3 atom gave a significantly larger number of identified H1 antagonists, namely 157 instead of 48, whilst maintaining the values for the other quality measures.

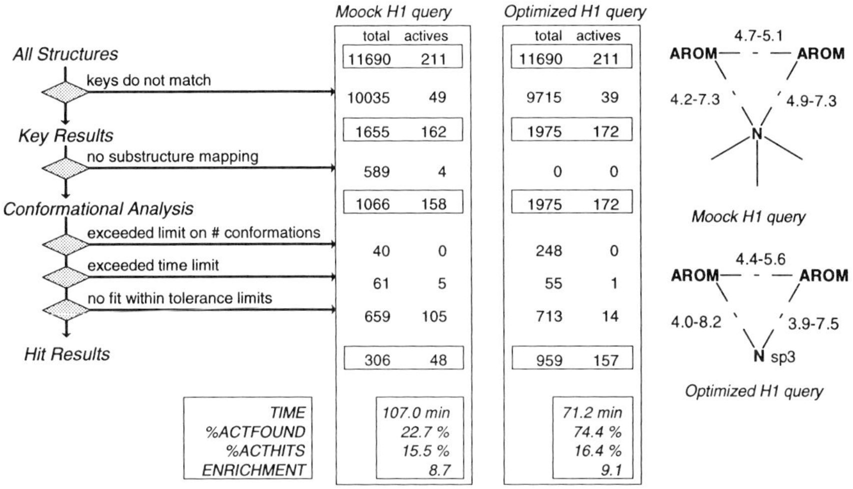

Fig. 2. *Flow chart of a Chem-X pharmacophore search used for query optimisation. Compounds from the DrugsFile database are rejected by different criteria at the various stages of the matching process. Result data are shown for the histamine H1 antagonist pharmacophore taken from Moock [15]*

An interesting aspect is the occurrence of biological activities in the compound hit list other than the required biological activity for which the pharmacophore has been mapped. The bar chart in *Fig. 3* shows the distribution of the 20 most frequently present activities of the compounds in the hit list that was obtained with the optimised H1 antagonist query. The total height of the bars gives the enrichment for each activity class found. For example, the 959 compound hit list contains 35 reported 5-HT_6 antagonists (code S1). Since the total database holds 41 compounds with 5-HT_6 antagonistic activity, this activity has been enriched by a factor of (35/959)/(41/11690) = 10.4. Out of these 35 reported 5-HT_6 antagonists, 11 compounds also have the desired antihistaminic H1 activity (code HU). The grey part of bar thus shows intrinsically present biological activity and anticipates the probable side effects of H1 antagonists. The black part of the bar may be interpreted as the part of the enrichment that is due to false positive compounds, and this information may be used for further training of the query in order to reduce the incidence of undesired activities.

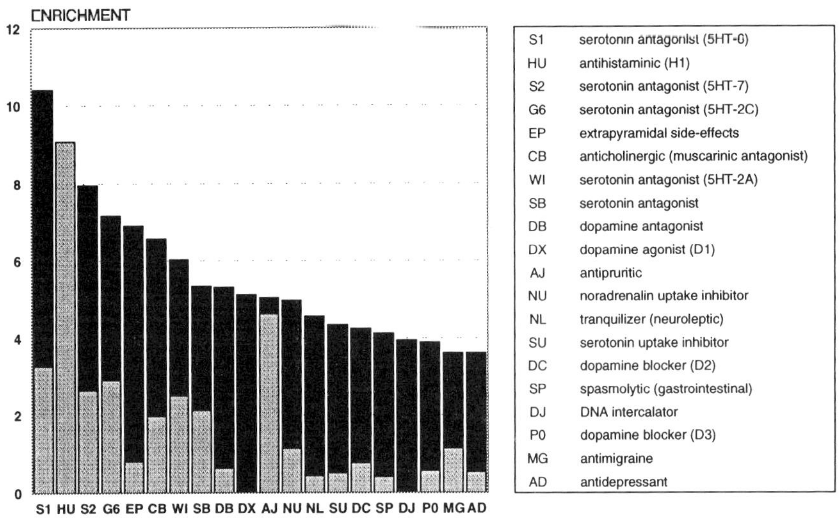

Fig. 3. *Spectrum of reported biological activities of compounds identified with the optimised histamine H1 antagonist pharmacophore. The total bars show the enrichment of a particular activity. The grey part of the bars represent the fraction of the enrichment originating from known histamine H1 antagonists.*

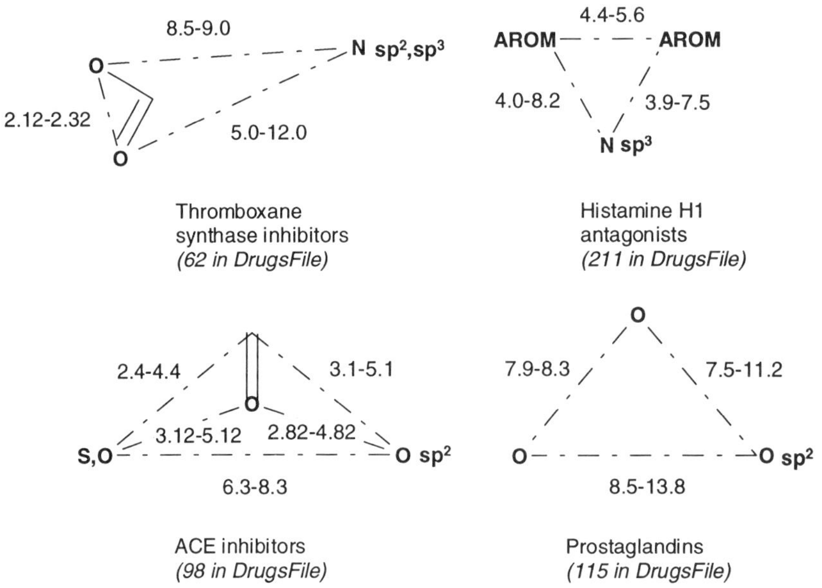

Fig. 4. *Optimised pharmacophore queries used in the present study. Dashed lines indicate distance (Å) ranges between pharmacophore centres and atoms.*

Evaluation of Effectiveness. - In order to evaluate the efficiency of pharmacophore searching techniques, four diverse queries were selected from the literature [15,16]. Each of the queries was used to search the DrugsFile database by means of the default **a6/a6** protocol and the results were analysed using the method depicted in the flowchart. In a few cases this led to slight modifications of the original queries. *Fig. 4* shows the optimised queries that were used to search the DrugsFile database with each of the seven different combinations of protocols listed in *Table 1*. The results for %ACTFOUND, %ACTHITS and ENRICHMENT effectiveness measures are given in *Table 2*. *Fig. 5* and *Fig. 6* give the corresponding bar charts for %ACTFOUND and ENRICHMENT data respectively.

Table 2. *Results for %ACTFOUND, %ACTHITS and ENRICHMENT for Chem-X Pharmacophore Searches obtained for 4 Queries with 7 Combinations of Protocols for Database Keying and Searching.*

protocol:	**sc/sc**	**sc/a6**	**a6/a6**	**2D/a6**	**no/a6**	**a4/a4**	**a6/flex**
%ACFOUND							
Tromboxane synth.inhibitor	14.5	50.0	50.0	50.0	50.0	48.4	25.8
Histamine H1 antagonist	72.0	73.5	74.4	74.4	74.4	73.5	73.5
ACE inhibitor	25.5	61.2	61.2	61.2	61.2	62.2	46.9
Prostaglandin	13.9	55.6	57.4	57.4	57.4	59.1	19.1
%ACTHITS							
Tromboxane synth.inhibitor	2.9	3.7	3.7	3.7	3.7	3.7	3.5
Histamine H1 antagonist	23.6	19.7	16.4	16.0	16.0	16.9	21.1
ACE inhibitor	3.3	7.9	7.8	7.8	7.8	8.0	5.0
Prostaglandin	3.8	8.7	10.8	10.8	10.8	10.9	3.3
ENRICHMENT							
Tromboxane synth.inhibitor	5.5	7.0	7.0	7.0	7.0	7.0	6.7
Histamine H1 antagonist	13.1	10.9	9.1	8.9	8.9	9.3	11.7
ACE inhibitor	4.0	9.4	9.4	9.4	9.3	9.5	6.0
Prostaglandin	3.8	8.9	11.0	11.0	11.0	11.1	3.3

The **sc/sc** protocol does not apply any conformational analysis but uses the single conformation that is stored in the database. We anticipated that the **sc/sc** protocol would give mediocre results which is indeed confirmed by the low percentage of identified known actives (see *Fig. 5*) for 3 out of 4 queries. Only for the histamine query, the **sc/sc** protocol gave most of the known antagonists which are apparently stored in the database in a conformation that matches the query. The **a6/a6**, **2D/a6** and **no/a6** combinations all use the same search protocol and should thus give identical values for %ACTFOUND,

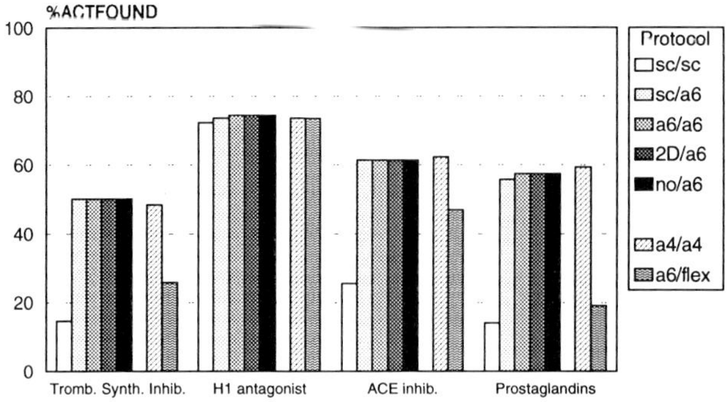

Fig. 5. *Percentage of identified known actives (%ACTFOUND) for Chem-X pharmacophore searches obtained for 4 queries with 7 combinations of protocols for database keying and searching.*

%ACTHITS and ENRICHMENT, which was established indeed, except for a few minor deviations of unknown origin. The **2D/a6** combination, using estimated distance keys, avoids the time-consuming conformational analysis during the database keying phase. However, database searches using **2D/a6** are expected to take longer since more compounds will pass the initial screen because of the fuzzy nature of 2D distance keys. In **no/a6** the initial screen is skipped completely and all compounds are submitted directly to the conformational regeneration which may take a considerable amount of time to complete. For **sc/a6**, which takes conformational flexibility into account during database searching, we expected quite a few compounds to be wrongly rejected by the initial screen, specifically all those compounds that have a stored conformation different from the one that fits the pharmacophore query. However, in fact remarkable similar %ACTFOUND values were obtained for **sc/a6** and **a6/a6** which may either indicate that stored conformations are indeed matching the pharmacophore or that the initial key screen is not very discriminative which may have been caused by relatively wide distance tolerances.

Although the **a4/a4** combination gives a different set of conformations compared with **a6/a6**, the results for the two combinations are very similar (see *Table 2*). In theory, **a6/flex** should give similar or better results for %ACTFOUND than **a6/a6**. In practice, however, **a6/flex** was found to perform quite badly because the flexi-fit procedure was taking up large amount of CPU time and as a consequence often exceeded the time limit before a good match could be obtained.

The effect of the different combinations of protocols on the ENRICHMENT measure is somewhat difficult to appreciate and should always be seen in combination with %ACTFOUND. For example, the ENRICHMENT values for **sc/sc** are higher than expected because it produces smaller hit list than when conformationally flexible search protocols like **a6/a6** are used. However, for reasons already pointed out, **sc/sc** is definitely not the best choice. The enrichment factor for **a6/a6**, which is the protocol combination of choice, varies from 7 to 11 depending on the pharmacophore query (see *Table 2*).

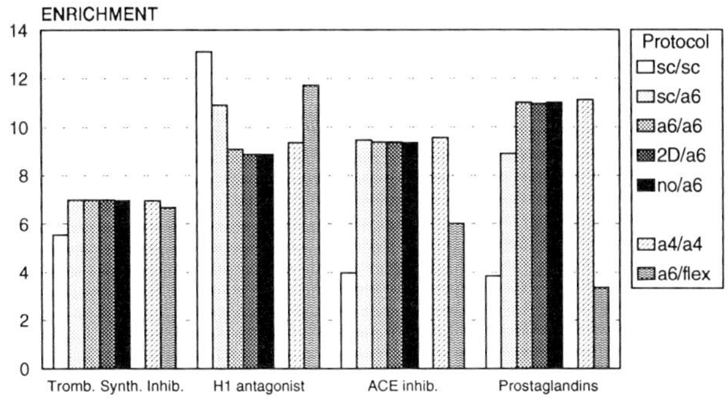

Fig. 6. *Improvement in the probability of finding a hit by a pharmacophore search over finding a hit by random selection (ENRICHMENT) for Chem-X pharmacophore searches obtained for 4 queries with 7 combinations of protocols for database keying and searching.*

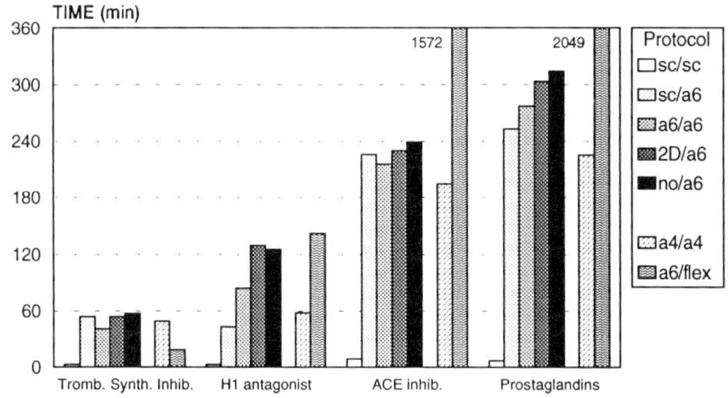

Fig. 7. *CPU time (minutes on 190 MHz R1000) for Chem-X pharmacophore searches obtained for four queries with seven combinations of protocols for database keying and searching.*

Evaluation of Efficiency. - The CPU times required for the pharmacophore searches are shown in *Fig. 7*. Large differences between the queries are observed due to the difference in conformational flexibility of the matching compounds. As expected the **sc/sc** protocol is by far the fastest of the protocols tested, but it leads to poorer results as discussed above. Results for the relative efficiency of conformationally flexible search protocols are difficult to generalise, but a number of observations can be made. Except for the histamine query, the **2D/a6** protocol is only marginally slower than the **a6/a6** protocol indicating no prohibitive reasons for using the estimated 2D distance keys and skipping the time-consuming rule-based keying for obtaining 3D distance keys. An unexpected result was obtained with the **no/a6** protocol because it gave the CPU times comparable to those obtained with the **2D/a6** protocol. In the **no/a6** protocol up to 10 times more compounds are subjected to substructure mapping and one would thus expect a significant increase in search times for the **no/a6** protocol. Apparently, the Chem-X pharmacophore matching algorithm implements a number of efficient heuristics for rejecting non-matching compounds at an early stage of the substructure mapping process. On the other hand, it remains peculiar that the percentage of compounds rejected by the very fast initial screen has little effect on the overall efficiency. The **a4/a4** protocol is somewhat faster than the **a6/a6** protocol because a smaller number of conformations is being considered. The current implementation of the flexi-fit algorithm may lead to excessive search times and does not give any advantage in terms of effectiveness.

Conclusion. - The results show that in our study the keying protocol has surprisingly little effect on the search performance as long as conformational flexibility is taken into account by a rule-based algorithm at search time. The **sc/sc** protocol is "quick and dirty" and may be used as such. We feel that the choice between **a6/a6**, **2D/a6** and **no/a6** is almost a matter of personal preference. We would recommend using **2D/a6** because it does not require a full conformational analysis for database keying making the database preparation very straightforward. Our findings for **no/a6** need further examination before we would recommend using this in general. The choice between **a6/a6** and **a4/a4** is also almost a matter of taste, **a6/a6** being somewhat slower but generally giving more realistic conformations than **a4/a4**. Usage of the current Chem-X implementation of flexi-fit is not recommended. We have shown that an ENRICHMENT factor of circa 10 can typically be achieved for a skilfully optimised pharmacophore.

Compound Clustering

Although a particular HTS assay may provide the means for screening many or all of the compounds available, this may be undesirable for reasons of costs and time because resources may be wasted if this large scale effort results in the production of redundant information. So rather than testing all compounds, the idea is that a representative subset may be tested instead which, in principle, should result in similar information on structure-activity relationships [17]. A representative subset is usually obtained by means of compound clustering and selecting one compound from each cluster. Obviously, this concept will only work well if the biological activity of interest is truly correlated to a molecular representation that can be used in the clustering. Unfortunately, the validity of this concept is rather hard to prove because a good validation set of compounds with a complete set of biological activities is often not available.

In a recent study, Brown and Martin [18] have evaluated a variety of structure based clustering methods and they considered a number of different ways for characterising compounds. These authors reported some gain of cluster-based over random-based strategies but did not report large benefits. Cluster information may also be put to use in other ways. For example, even when the entire collection is eventually to be tested, Taylor [19] showed that when priority for order of testing is assigned to compounds from clusters for which a member was already found to be active, the most active compounds within the collection are identified at an earlier stage than when a random testing order is used. Cluster information may also be used to select compounds for a follow-up screen after initial hits with the representative subset have been obtained.

In this paper we will not go into the technical details of clustering [20] but we will try to evaluate the validity of the "representative subset concept" in terms of the spectrum of biological activities displayed by a subset compared to the one displayed by the whole compound set. Here we introduce an initial study aimed at assessing how well a subset, rationally selected from a large compound collection, represents the spectrum of biological properties present within the whole set. Although it is quite accepted that the 3D attributes of a drug molecule determine its biological activity, 2D descriptors have been found to perform at least as well as 3D descriptors in separating active and inactive molecules [18]. We have verified whether the choice between a small set of different 2D and 3D molecular "fingerprint" descriptors affects the number of different activities presented by the subset.

Table 3. *Fingerprint Descriptors used for Compound Clustering*[4].

descriptor	origin	conformations	number of bins
BCI	topology dictionary	N/A.	2467
3D/SC	pharmacophore distances	single	320
2D+3D/SC	info key + bond key + pharmacophore distances	single	416
3D/MC	pharmacophore distances	multiple	320
2D+3D/MC	info key + bond key + pharmacophore distances	multiple	416

Methods. - For the present study we confined ourselves to Ward's method of clustering as implemented by Barnard and Downs (BCI) [21]. In other studies we have found Ward's method to be most rewarding and this clustering method was also the preferred choice of cluster methods tested by Brown and Martin [18]. The BCI implementation of the program first produces a list of Reciprocal Nearest Neighbours

[4] Pharmacophore distances refer to the above mentioned 10 Chem-X distance keys with 32 bins each. The single conformation has been built with CORINA and multiple conformations have been generated with the **a6** protocol for conformational analysis (see Table 1).

(RNN) and subsequently merges the most similar entities into a single point until the required number of clusters has been reached [20]. The rate-determining step, the generation of the RNN table, took about 45 CPU minutes for 11686 structures on a 190 MHz R10000. The fingerprint descriptors used are presented in *Table 3*. For the 2D fingerprints we chose the (sub)structure fingerprints implemented in the BCI clustering programs [22]. In this method, non-hashed fingerprints are generated by comparing every molecule with a dictionary of structure fragments. Our present customised dictionary contains 2467 fragments comprising atom and bond sequences, atom pairs, ring composition and ring fusion sequences [22]. For the DrugsFile a sparse set of fingerprints was obtained with an averaged number of bits set of 136 and a maximum of 513. As an alternative, the pharmacophore keys implemented within Chem-X, as described in the previous section, were used as descriptors. These 3D fingerprints can be generated either for a single conformation or for a set of multiple conformations (see *Fig. 1*). The single-conformation fingerprints **3D/SC** have fewer bits set than multiple-conformation ones **3D/MC** as indicated by an average number of 25 versus 55 respectively.

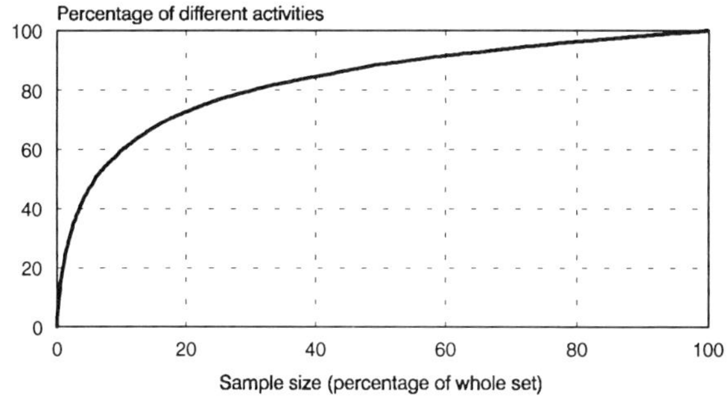

Fig. 8. *Percentage of biological activities displayed by randomly selected subsets. The graph shows the percentage of different biological activities (K_{RAN}/K_{TOT}) versus the relative size of the subset (N/M).*

We have taken the following approach for our validation procedure. For a given set of M compounds and N clusters, a representative subset of N compounds was obtained by selecting the centroids of the N clusters. We then counted the total number of different activities reported in the DrugsFile for the subset of N representatives and compared this number with the total number of different activities reported for a random selection of N compounds. We will call these sum totals K_{REP} for the representative and K_{RAN} for the random subsets. A rationally selected subset of representatives will exhibit a larger diversity in activity than a random selection of the same size if K_{REP} is larger than K_{RAN}.

We first obtained a reference curve for K_{RAN} versus N. One by one the compounds were randomly taken from the DrugsFile whilst recording the cumulative total of different activities represented by the total set of compounds taken. Fig. 8 shows the averaged curve of ten independent runs, where the X and Y axes have been scaled by the total number of compounds (M = 11686) and the total number of different activities (K_{TOT} = 865) respectively. For a given sample size, the standard deviation for the number of different activities is within 5% which is in pictorial sense close to the thickness of the curve shown in Fig. 8. On the one hand, it can readily be seen that a random sample of 1% of the whole set displays 20% of all activities and 10% already displays 70% of activities. On the other hand, about 10% of the activities are still not encountered when over 50% of the whole set is sampled. The curve does not reach an asymptote thus a first, quite obvious, conclusion is that only after sampling 100% of the database all different activities will have been encountered. In other words, certainty about the presence or absence of a hit can only be obtained by screening the whole library.

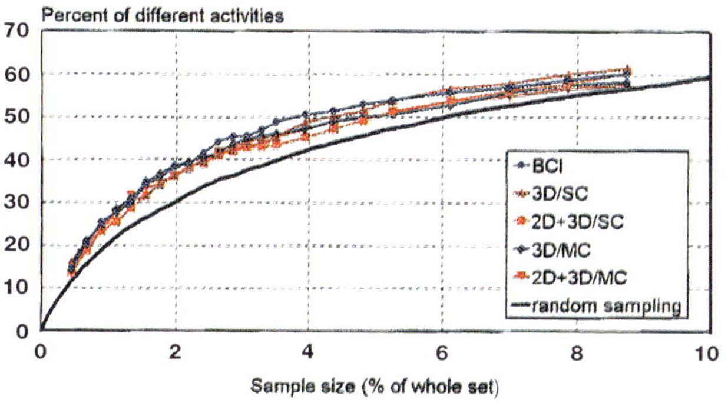

Fig. 9. *Comparison of percentage of biological activities displayed by representative subsets and randomly selected subsets (bottom curve). The graphs show the percentage of different biological activities (K_{REP} /K_{TOT}) versus the relative size (N/M) for the 5 different fingerprint descriptors in Table 3.*

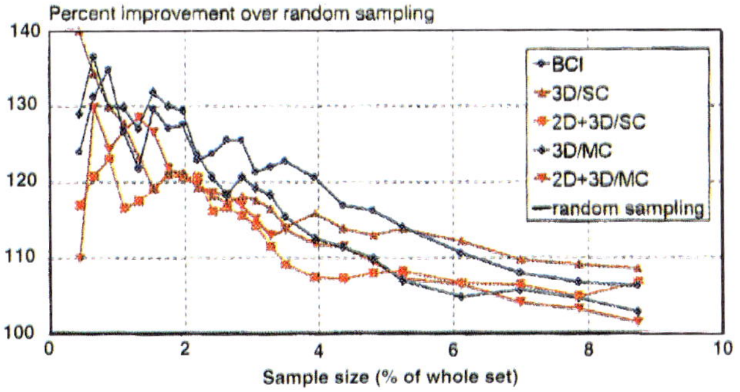

Fig. 10. *Comparison of percentage of biological activities displayed by representative subsets and randomly selected subsets (bottom curve). The graphs for the representative subsets shown in Fig. 9 are expressed relative to the graph for the randomly selected subsets which has been put at 100%.*

Results and Discussion. - We performed cluster analyses using the different fingerprints listed in *Table 3*. For each of the different descriptors the following procedure was followed. The RNN list was processed in order to obtain subsets for a given sequence of numbers of clusters, e.g. 25, 50, ... ,1000. Representative subset at each level of clustering was obtained by selecting the centroid of each of the clusters. Finally, for each set of representatives the total number of different activities was calculated by the activity information in the DrugsFile. In this way, for each of the 5 different fingerprint descriptors, a discrete set of K_{REP} values for a sequence of different N values was obtained. By connecting the points for a particular descriptor a curve of K_{REP} versus N is obtained which may be compared to the reference curve shown in *Fig. 8*. The final results are shown in *Fig. 9* and also here the X and Y axes have been scaled. *Fig. 10* shows the same data as in *Fig. 9* but here the percentage of different activities K_{REP} is expressed relative to K_{RAN} which has been put at 100%.

Visual inspection of the structures of the resulting clusters conforms to chemical intuition, as both for topological and pharmacophore keys clusters of largely analogous structures are produced, especially when the number of clusters N exceeds 10% of the number of structures M in the whole set.

The results show that use of representative subsets samples a larger number of activities than does random selection. The difference is most pronounced between 1 and 5% sampling of the whole set where a relative improvement over random sampling is obtained varying from roughly 130% to 110% (see *Fig. 10*). When more than 25% of the whole set is selected, the effect becomes marginal. The 2D substructure fingerprint descriptor **BCI** show the best overall performance. The differences between the 4 types of fingerprints derived from Chem-X pharmacophore keys are relatively small. Including the information and bond key does not seem to have any benefit, probably because of the very limited number of bits used, 32 for the information key and 64 for the bond key [14]. Small differences are found between the single- and multi-conformational fingerprints: **3D/MC** or **2D+3D/MC** select more properties at low sampling density (up to 4% of sample size), whereas **3D/SC** or **2D+3D/SC** give better results when a larger set of representatives is being used (5-10% of sample size).

The strategy of using representative subsets obviously works best if each particular activity is concentrated in as few clusters (preferably just one) as possible and if the clusters largely consist of compounds with that particular activity. We will use the term "active cluster" to indicate a cluster that includes at least one compound with a specific biological activity. To study this matter, we determined the partitioning of twelve biological activities at various levels of clustering. *Table 4* shows the results obtained with **BCI**, **3D/SC** and **3D/MC** fingerprints. It can readily be seen that, with rare exceptions, the use of **BCI** fingerprints concentrates the actives in the lowest number of active clusters, and includes fewer inactives in those clusters. Unsurprisingly, a larger number of clusters concentrates the actives in a (slightly) larger number of smaller and more homogeneous clusters. The choice of how many clusters to use is thus very important for the effectiveness of using the concepts of compound clustering.

Table 4. *Number of Active Clusters and the Total Number of Compounds present in the Active Clusters (between Brackets) for 2D and 3D Fingerprint Descriptors and Number of Clusters (N).*

	BCI	**BCI**	**BCI**	**3D/SC**	**3D/MC**
activity (# in database)	N =3000	N =1500	N =1000	N =1000	N = 1000
5HT 1A agonist (101)	50 (292)	45 (508)	39 (661)	46 (1001)	47 (684)
5HT 1D agonist (24)	15 (65)	14 (151)	14 (243)	15 (265)	15 (141)
5HT agonist (104)	54 (346)	49 (624)	45 (836)	50 (1129)	48 (843)
5HT antagonist (201)	123 (711)	151 (1118)	94 (1539)	98 (1887)	120 (2050)
ACE inhibitor (98)	43 (191)	35 (345)	30 (486)	47 (1000)	65 (1463)
Antidepressant (409)	231 (1414)	186 (2227)	157 (2849)	150 (3745)	184 (3629)
Antihistamine H1 (211)	92 (565)	75 (893)	66 (1224)	72 (1612)	97 (1966)
Antiprogestagen. (33)	8 (50)	6 (93)	5 (104)	8 (522)	8 (275)
Cytostatic (1093)	577 (2785)	423 (4140)	352 (5094)	39 (6105)	386 (6560)
Progestagen (108)	21 (189)	14 (259)	12 (281)	11 (725)	12 (827)
Prostaglandin (115)	25 (146	14 (173)	13 (202)	27 (511)	32 (635)
Thromboxane inhib.A2 (63)	30 (111)	25 (172)	21 (262)	28 (394)	35 (459)

What, then, are the chances that, when selecting a compound from each cluster, an active one will be picked? This obviously depends on the ratio actives/inactives in these clusters. However, these chances are higher than simple mathematics would seem to indicate. *Fig. 11* shows that the majority of cluster sizes lies below the arithmetical average N/M, with rapidly dwindling occurrence of larger clusters. Clearly, this improves the chances of finding one od the actives when randomly selecting a representative from each active cluster. In addition, the ratio of actives/inactives is also skewed, as is shown in *Fig. 12* for the four activities on which the pharmacophore searching protocols have been tested. That is, of the active clusters, most have either a very high or a very low proportion of inactives, which greatly increases the probability of selecting actives. Indeed, for most activities at least one cluster had been formed consisting exclusively of actives. Thus, chances are quite good that actives will be found. Moreover, the advantage over random sampling is larger than apparent from these examples, as clustering gives the opportunity to examine the active clusters and potentially identify further active compounds.

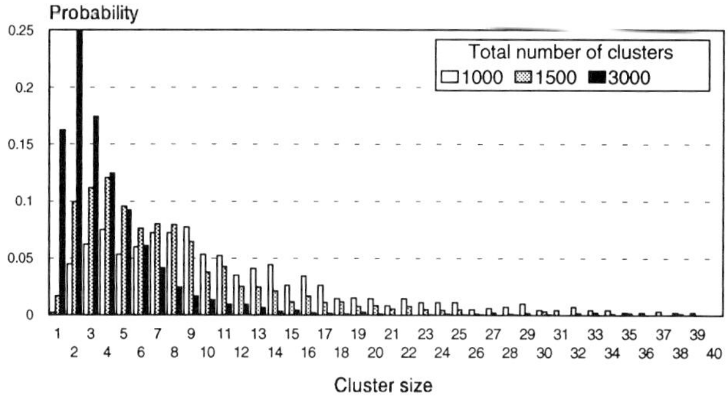

Fig. 11. *Cluster size distribution for three different levels of clustering. 11686 compounds were clustered using Ward's method bases on their 2D fingerprints.*

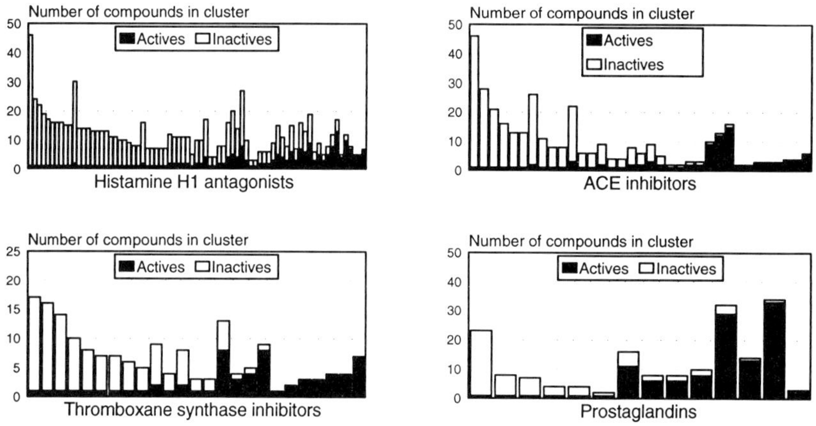

Fig. 12. *Distribution of actives/inactives for active clusters containing at least one active. 11686 compounds were clustered into 1500 groups using Ward's method bases on their 2D fingerprints.*

Conclusion. - We have shown that using cluster representatives gives a significantly better sampling of the activity space than using randomly selected subsets. However, this advantage decreases rapidly for subsets larger than 10% of the total set. Although the 2D descriptors worked best in general, we think that we have shown that also using 3D pharmacophore descriptors can be used advantageously. However, the current limitations of Chem-X pharmacophores as already shown in the previous part, give rise to only partly homogeneous groups with respect to activity profiles. Therefore, the relatively good performance of these low dimensional pharmacophore fingerprints is maybe even quite remarkable.

The validity of the concept of using a representative subset depends on the assumption that similar structures display similar properties. This assumption seems not unfounded as the property of interest, the biological activity, generally is linked with the structure, i.e. closely related structures tend to display similar activity, although any given activity may be present in several, sometimes quite unrelated, chemical series. The question, however, is whether our results really indicate that this assumption is true. The fact that the predominant structural features differ from cluster to cluster indicates that already testing cluster representatives should give a broader sampling of properties than would be expected from random selection. Further studies will be needed to understand these matters in more detail than we do now. For that purpose we need good validation sets which hopefully will also become available to the academic community.

Acknowledgement. - We would like to express our sincere gratitude for the continuous effort of Dr. Jan Kelder for maintaining the DrugsFile. His relentless effort over more than a decade has helped us in many beneficial ways.

REFERENCES

[1] J.S. Major, *Pharm. Manufact. Int.* **1995**, 37-41.

[2] M.S. Ashton, M.C. Jaye, J.S. Mason, *Drug Discovery Today* **1996**, *1*, 71-78.

[3] E.J. Martin, J.M. Blaney, M.A. Siani, D.C. Spellmeyer, A.K. Wong, W.H. Moos, *J. Med. Chem.* **1995**, *38*, 1431-1436.

[4] C.J. Stankovic, *Pharm. Manufact. Int.* **1995**, 19-24.

[5] D.H. Rouvray, *J. Chem. Inf. Comput. Sci.* **1992**, *32*, 580-586.

[6] J. Sadowski, J. Gasteiger, *Chem. Rev.* **1993**, *93*, 2567-2581.

[7] J. Sadowski, J. Gasteiger, G. Klebe. *J. Chem. Inf. Comput. Sci.* **1994**, *34*, 1000-1008.

[8] Y.C. Martin, *J. Med. Chem.* **1992**, *35*, 2145-2154.

[9] A.C. Good, J.S. Mason, in 'Reviews in Computational Chemistry', Eds. K.B. Lipkowitz and D.B. Boyd, VCH Publishers, Inc. New York, 1996, Vol. 7, pp. 67-117.

[10] I.D. Kuntz, *Science*, **1992**, *257*, 1078-1082.

[11] V.J. Gillet, W. Newell, P.Mata, G. Myatt, S. Sike, Z. Zsoldos, A.P. Johnson, *J. Chem. Inf. Comput. Sci.* **1994**, *34,* 207-217.

[12] N.W. Murrall, E.K. Davies, *J. Chem. Inf. Comput. Sci.* **1990**, *30,* 312-316.

[13] G.W.A. Milne, M.C. Nicklaus, J.S. Driscoll, S.Wang, D. Zaharevitz, *J. Chem. Inf. Comput. Sci.* **1994**, *34,* 1219-1224.

[14] Chem-X user manual, Chemical Design Ltd., Roundway House, Cromwell Park, Chipping Norton OX7 5SR, UK.

[15] T. E. Moock, D. R. Henry, A. G. Ozkabak, M. Alamgir, *J. Chem. Inf. Comput. Sci.* **1994**, *34*, 184-189.

[16] D.E. Clarck, P. Willett, P.W. Kenny, *J. Mol. Graphics* **1993**, *11*, 146-156.

[17] G.M. Downs, P. Willett, in 'Advanced Computer-Assisted Techniques in Drug Discovery', Ed. H. van de Waterbeemd, VCH, Weinheim, 1995, Vol 3, pp 111-130.

[18] R.D. Brown, Y.C. Martin, *J. Chem. Inf. Comput.* **1996**, *36*, 572-584.

[19] R. Taylor, *J. Chem. Inf. Comput. Sci.* **1995**, *35*, 59-67.

[20] B. Everitt, Cluster Analysis, Gower, London 1986.

[21] Barnard Chemical Information (BCI) Ltd., 46 Uppergate Road, Stannington, Sheffield S6 6BX, UK.

[22] J.M. Barnard, G.M. Downs, *J. Chem. Inf. Comput. Sci.* **1992**, *32*, 644-649.

PART II. **LEAD FINDING STRATEGIES**

Chapter 11

On the Benefits of Attractive Pseudo-Potentials in a Genetic Algorithm Approach for Structure-Based Library Screening

by **Jochen Antel**[a]), **Ingmar Reuter**[b]) and **Dietmar Schomburg**[b])

[a]) Solvay Pharma Research, Hans-Böckler-Allee 20, D-30173 Hannover
[b]) Gesellschaft für Biotechnologische Forschung mbH, Mascheroder Weg 1, D-38124 Braunschweig

On the Benefits of Attractive Pseudo-Potentials in a Genetic Algorithm Approach for Structure-Based Library Screening

by **Jochen Antel**

Solvay Pharma Research, Hans-Böckler-Allee 20, D-30173 Hannover

and **Ingmar Reuter** and **Dietmar Schomburg**

Gesellschaft für Biotechnologische Forschung mbH, Mascheroder Weg 1, D-38124 Braunschweig

A novel, tailorable database docking program using a Genetic Algorithm (GA) approach as the final mesh in a „*Protein Structure Driven Database Screening*" procedure, starting with UNITY as the widest mesh, is described. For test runs a small database containing co-crystallized inhibitors of the metalloprotease Thermolysin was used. First results from database searches and subsequent GA runs, influenced by the active site geometry showed that the involvement of attractive pseudo potentials improves significantly the accuracy of the docking procedure and looks as an ideal „secondary screening" regime leading to a much saver set selection.

Introduction. - During the past few years a rush into the new, chemical counterpart of HTS - as there are Combinatorial Chemistry [1,2] and Automated Parallel Chemistry [3] either in solution phase or solid support - could be observed. However, despite the ability to generate now thousands of compounds in much shorter periods, first thoughts about the necessity to cut down the astronomic numbers of „possible compounds" to the still enormous numbers of „realizable compounds" arose [4].

One reasonable approach in this direction is to screen - via database searches - libraries of existing or still virtual compounds. While similarity/diversity considerations are one side of the medal, we concentrate here on „*Protein Structure Driven Database Screening*" as the other side.

First results convinced us about the obtainable benefits of this approach, however it was also soon obvious, that additional software development is required for an improved cut down of false negatives and false positives.

A tailorable database docking program using a Genetic Algorithm (GA) approach is being in development [5] as the final mesh in the database screening procedure, starting with UNITY [6] as the widest mesh.

Methodology. - The genetic algorithm is currently written in SPL (SYBYL Programming Language) the inherent programming language of SYBYL [6]. Though SPL is very powerful and convenient because of its integration, it bears the obvious disadvantages of a macro language, the lack of speed. Nevertheless, the features of the GA can be easily optimized in this programming environment and first results are very encouraging for further investigations.

The GA is currently implemented as follows:
The genetic information is represented by torsion angle values (real type numbers). The whole conformational space of a molecule can be evaluated. However, aiming at the examination of databases we currently restrict the „allowed“ rotatable bonds to acyclic single bonds (amide bonds are identified and restricted to cis/trans).
The initial population is created by random rotation of the dihedrals of each start structure (hitlist members of a previous UNITY search).
Since every part of the molecule - dissected by the chosen rotatable bond - can be rotated relative to the remaining one, the resulting structures are at first fitted to a user defined set of anchor atoms (e.g. an amide moiety).
The fitness of this initial population (generation 0) is then examined by a scoring function which takes into account:

- the van der Waals interaction energy of the ligand and the site („SYBYL energy function“).
- the intramolecular energies of the generated conformations of the ligands („SYBYL energy function“).
- additional „pseudo-energies“ (see below)

For the breeding of subsequent generations („Reproduction“), the genetic operators „Mutation“ and „Crossover“ are applied.
Mutation is implemented by random selection of one individual, random choice of one rotatable bond and assigning a random value for its torsion angle.
Crossover is implemented by a roulette wheel selection of the two individuals, choice of one rotatable bond and mutual exchange of the all torsional angles of two individuals for the smaller part of the molecule as determined by the chosen rotatable bond.
The fitness of the resulting „children“ is evaluated (see above) and an ordered list of the fitness of the individuals of the parent generation and the fitness of the structures of the new children is generated.
The fittest individuals are used for the next reproduction step. The new generation has the same number of individuals as the previous one (steady state).
As an additional feature, attractive pseudo energy terms are applicable for customizable sets of atoms or functional groups, allowing to define additional interaction sites (e.g. hydrogen bonding facilities, „hydrophobic interactions). These features can be considered as „or“ type features since these interaction sites need not necessarily to be occupied but will contribute to the fitness if they are.

Discussion. - For test runs the 14 co-crystallized inhibitors of the metalloprotease Thermolysin (derived from the PDB [7]) were investigated and for various reasons (e.g. occurence of a „Zn-ligand, P1'-P2' amide bond, P2' residue) we restricted ourselves to a very small database containing 7 co-crystallized inhibitors with a more or less common backbone (**Fig.1**).

In order to use one „common active site“ **4TMN-INH** and **1TMN-INH** were used as template to define a 600 pm sphere. Every enzym residue within this sphere was than considered to belong to the site (subsequently called site). The „common co-crystallized orientation“ of the ligands was obtained by fitting all His142, His146, and Zn atoms of the various enzyme structures to the „site“ and extracting the co-crystallized ligands.

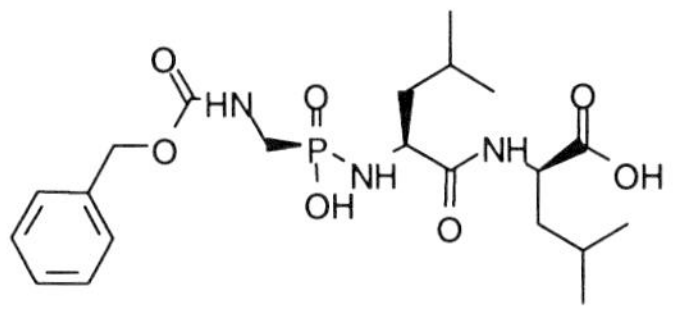

5TMN-INH

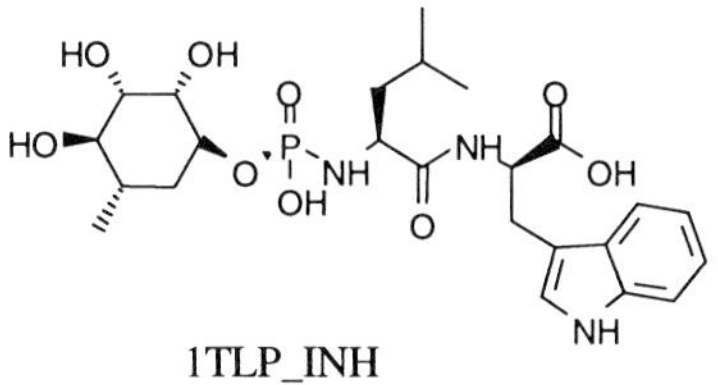

1TLP_INH

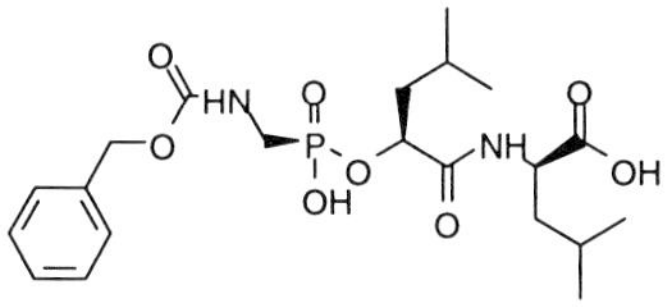

6TMN_INH

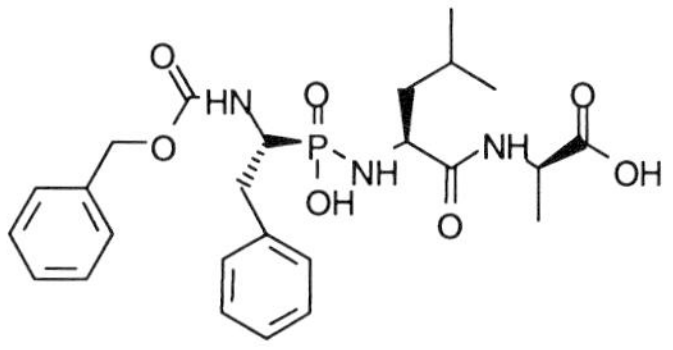

4TMN_INH

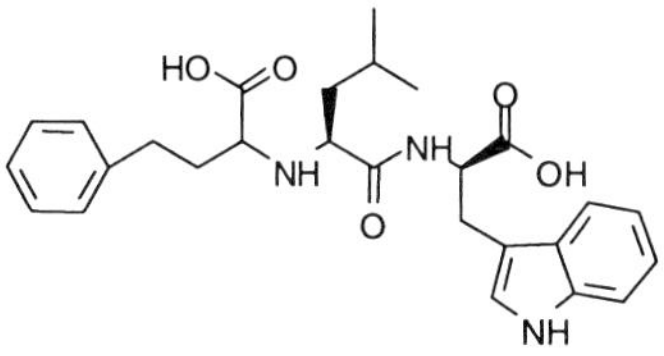

1TMN_INH

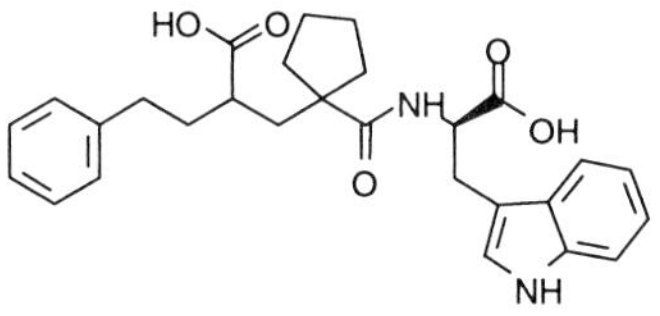

1THL_INH

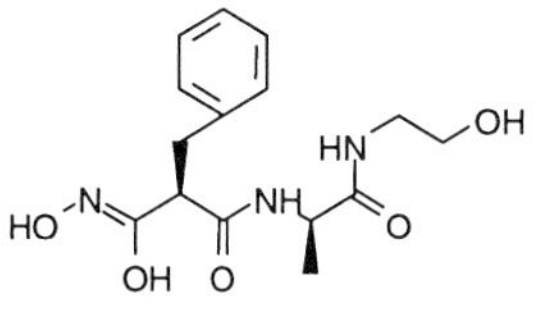

5TLN_INH

Fig. 1. *Structures of the test set of 7 co-crystallized inhibitors. The indicated names are corresponding to the PDB codes.*

For test purposes two small UNITY databases were built up, one with the unmodified structures (**Fig. 3.** and **Fig. 4.**) in order to validate the queries and a second (subsequently called test-database) with randomly chosen and modified dihedral angles (**Fig. 5.**).
Three queries were built up for 3D searching:

1. Query 1 (subsequently called **Q1**): The query consists of the P1'-P2' amid moiety, the centroid of 3 carbon atoms of any atom type and an oxygen atom as „Zn-ligand" (see **Fig. 2.**). The distance constraints and allowed tolerances were chosen to satisfy the observed distances for all co-crystallized ligands (veryfied with static 3D search).
2. Query 2 (subsequently called **Q2**): The atom definitions and distance constraints are identical to **Q1**. In addition in order to describe the shape of the „site" the UNITY feature „excluded volume" was applied to all Cα-atoms of the „site" (36 „excluded volumes", default settings used). Again the query setup was validated by a successful static 3D search for the unmodified co-crystallized ligands.
3. Query 3 (subsequently called **Q3**): **Q3** is an extension of **Q2** by 10 more excluded volumes (see discussion below). Again the query setup was validated by a successful static 3D search for the unmodified co-crystallized ligands.

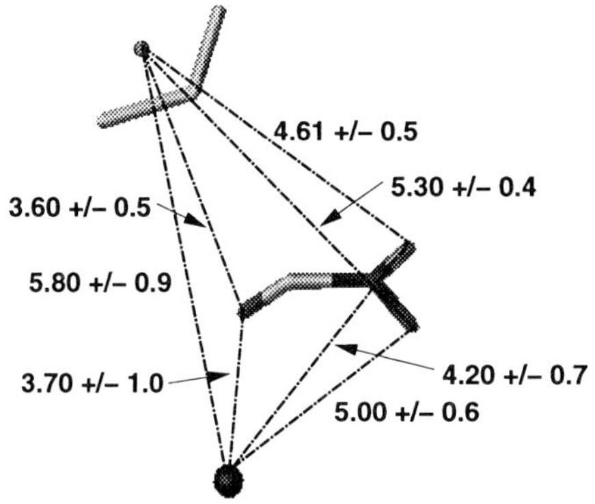

Fig. 2. *3D-query: **Q1**. The 4 amide atoms are also used as anchor atoms in the GA runs.*

While the static UNITY-3D searches with **Q1**, **Q2**, **Q3**, within the database of the unmodified structures yielded always 7 hits (100%), the situation differed for searching within the database of the modified structures: **Q1** yielded 7 hits with structures aligned to the query but - because of the lack of volume constraints in **Q1** - misaligned with respect to the „site". **Q2** yielded no hits for the static UNITY-3D search and 6 hits for flexible searching (standard setup plus: maxtime=300, vdW=yes). **Q3** yielded no hits for the static UNITY-3D search and 5 hits for flexible searching (standard setup plus: maxtime=300, vdW=yes). The structures of the hits of **Q2** and **Q3** satisfied the query conditions, did not violate the site, but were not optimal aligned within the site.

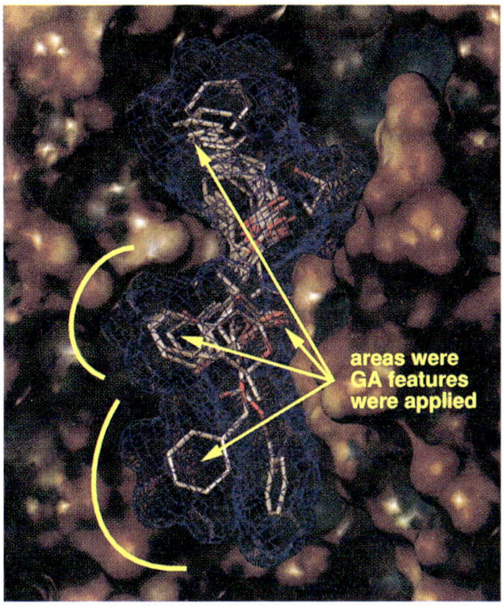

Fig. 3. *Plot of the 7 co-crystallized ligands (unmodified) within the „site" and indication of spatial areas where additional GA features were applied.*

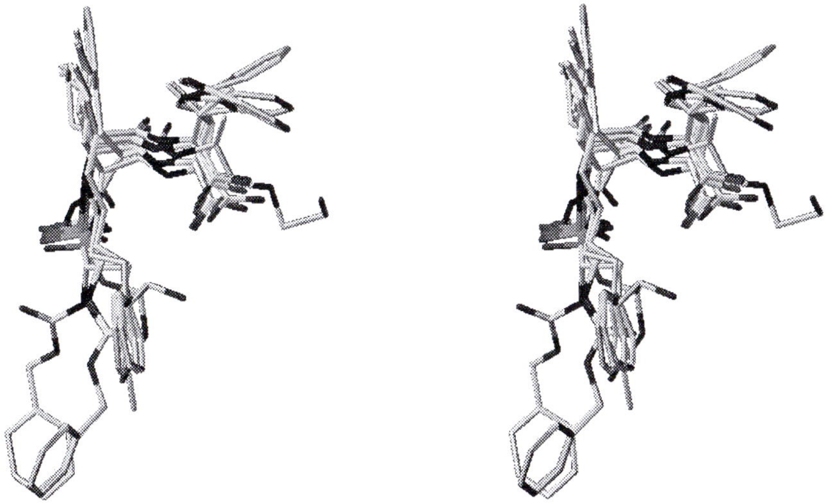

Fig. 4. *Stereoview of the unmodified ligands.*

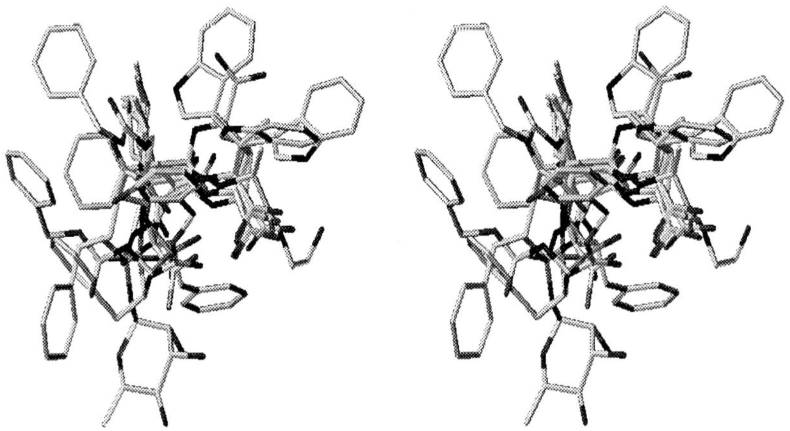

Fig. 5. *Stereoview of the modified ligands (content of test-database).*

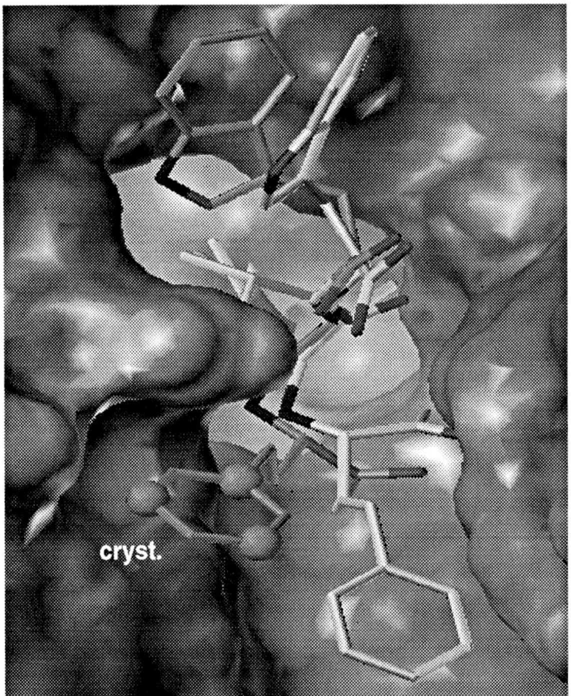

Fig. 6. *Comparison of the co-crystallized conformation of 1TMN_INH (cryst.) and the conformation resulting from Q2 (light grey) within the site.*

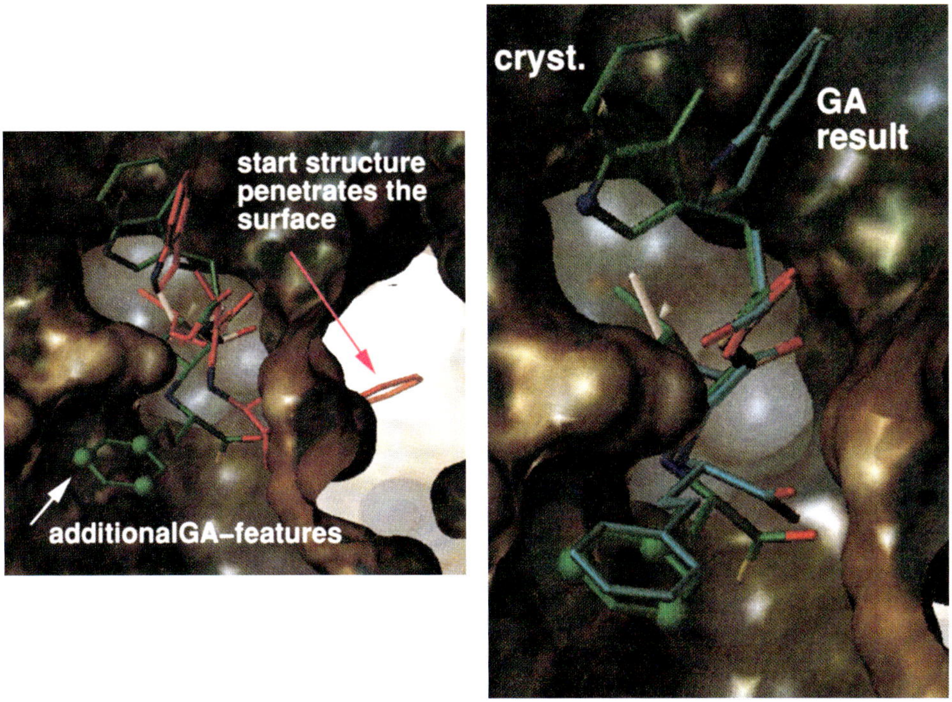

Fig. 7. *Comparison of the X-ray conformation of 1TMN_INH with the start conformation (left) and the GA result (right) obtained by additional attractive pseudo energy terms.*

Subsequently GA runs were performed with different settings. In all cases the 4 atoms (H-N-C=O)of the P1', P2' amide moiety were used as anchor atoms. The remaining 4 atoms from **Q1** - 3 carbon type atoms to address bulk in P1' and 1 oxygen type atom as the „Zn-ligand" - were kept in all trials. Attractive pseudo-energy values were always attributed to these 4 atomic positions which rank individuals that occupy these positions higher in their fitness and resulted finally in a good match to the query and based on our setup (evaluation of van der Waals interaction energy between ligand and site) the fittest individuals do not violate the site.

The involvement of additional attractive pseudo potentials improves significantly the accuracy of the docking procedure. As one example the results obtained with N-(1-carboxy-3-phenylpropyl)-L-leucyl-L-tryptophan (from „1TMN") are shown below. The addition of attractive pseudo potentials for a hydrogen bond at P2' and „carbon-type atoms" at P2, resulted for the fittest individuals in a conformation close to the co-crystallized.

Conclusion. - Molecular recognition of a ligand to its receptor or enzyme is a complex process in three dimensions where mutual complementary molecular properties contribute to the strength of the resulting complex. The whole process is additionally influenced by entropic terms. Taking only distance constraints into consideration for flexible 3D searching is not sufficient for a reliable selection of compounds, because too many false positives will result. Therefore volume/shape constraints need to be applicable as a second filter.

However, while these type of constraints are in principle available in UNITY their use lead - at least in the actual version (V.2.5) - to false negatives.

Applying our GA approach to the hitlist of a 3D flexible search yielded all of the 7 hits and resulted especially after the application of additional „attractive pseudo potentials“ in conformations close to the co-crystallized ones. This allows for a quick assessment of the quality of the hits and a ranking of the hitlists.

We therefore conclude: While the results from the initial UNITY database searches are obtained quickly, the somewhat slower but more accurate GA approach looks as an ideal „secondary screening“ regime leading to a much saver set selection.

REFERENCES

[1] E. M. Gordon, R. W. Barett, W. J. Dower, S. P. A. Fodor, and M. A. Gallop, *J. Med. Chem.* **1994**, *37*, 1385-1401.

[2] L. Weber, *Nachr. Chem. Techn. Lab.* **1994**, *42*, 698-702.

[3] E. Peisach, D. Casebier, S. L. Gallion, P. Furth, G. A. Petsko, J. C. Hogan jr., D. Ringe, *Science* **1995**, *269*, 66-69.

[4] E. J. Martin, J. M. Blaney, M. A. Siani, D. C. Spellmeyer, A. K. Wong, W. H. Moos, *J. Med. Chem.* **1995**, *38*, 1431-1436.

[5] J. Antel, I. Reuter, D. Schomburg, A genetic Algorithm Program for the second step in a Structure Based Virtual Library Screening Approach, in preparation.
See also: G. Jones, P. Willett, R. C. Glen, *J. Mol. Biol.* **1995**, *245*, 43-53 et. cit. lit.

[6] Unity (2.5) Chemical Information Software and SYBYL Molecular Modeling System 6.22, Tripos Associates: St. Louis, Mo.

[7] Bernstein, F. C.; Koetzle, T. F., Williams, G. J. B.; Meyer, E. F.; Brice, M. D.; Rogers, J. R.; Kennard, O.; Shimanouchi, T.; Tasumi, M.; J. Mol. Biol. **1977**, *112*, 535-542.

PART II. LEAD FINDING STRATEGIES

Chapter 12

Intelligent Combinatorial Libraries

by **Torbjörn Lundstedt**[a]), **Sergio Clementi**[b]), **Gabriele Cruciani**[b]), **Manual Pastor**[b]), **Nouna Kettaneh**[c]), **Per M. Andersson**[d]), **Anna Linusson**[d]), **Michael Sjöström**[d]), **Svante Wold***[d]) and **Bo Nordén**[e])

[a])Pharmacia-Upjohn, Structure-Property Optimization Centre, Malmö, Sweden
[b])Laboratorio di Chemiometria, Dipartimento di Chimia, Universita di Perugia, Italy
[c])MDS Inc., Winchester, MA, USA
[d])Research Group for Chemometrics, Institute of Chemistry, Umeå University, Sweden
[e])Astra-Hässle, Computational Chemistry Group, Mölndal, Sweden

Intelligent Combinatorial Libraries

by **Torbjörn Lundstedt**
Pharmacia-Upjohn, Structure-Property Optimization
Centre, Malmö, Sweden

and **Sergio Clementi, Gabriele Cruciani, Manual Pastor**
Laboratorio di Chemiometria, Dipartimento di Chimia,
Università di Perugia, Italy

and **Nouna Kettaneh**
MDS Inc., Winchester, MA, USA

and **Per M. Andersson, Anna Linusson, Michael Sjöström, Svante Wold***
Research Group for Chemometrics, Institute of Chemistry, Umeå
University, Sweden

and **Bo Nordén**
Astra-Hässle, Computational Chemistry Group, Mölndal, Sweden

A chemical (combinatorial) library is a, usually large, set of compounds made with the purpose to contain all possible compounds of a certain type. The library is often made in order to find a lead compound for a specific drug action or for the optimization of a lead. Because of the large number of synthesized compounds in the library, their biological activity is usually measured by rapid and simple tests, "high throughput screening" (HTS), giving crude answers, for instance "active" or "not".

Except for very simple libraries, however, it is impossible to synthesize all possible compounds. Hence a selection must be made, either according to synthetic feasibility, or according to some other principles. Moreover, the risk is substantial to miss interesting compounds due to the crude biological testing. Finally, there is a large risk for many false positives due to the large number of tested compounds and their individual evaluation.

We here present a strategy for constructing a library with optimal information while still taking synthetic feasibility into account. The objective is to provide high (optimal) chemical diversity with a moderate number of compounds, plus adequate depth and width of the biological testing.

The strategy is based on a multivariate characterization of the synthesis starting materials (building blocks, BB.s), principal component analysis, multivariate design, and multivariate QSAR. The strategy applies both to solid phase synthesis as well as libraries in solution.

Introduction. — In the initial phases of a drug design project, fairly large numbers of compounds, usually in the order of a couple of hundreds to a couple of thousands, are synthesized and tested [1-3]. With combinatorial chemistry (**CombC**) these numbers may now be increased by several magnitudes of ten; it is not unusual to make and test millions of compounds in a single project [4]. CombC started about 10 years ago with the work of *Geysen* and *Houghten* [5, 6], who showed that, in the synthesis of peptides, one may make all or almost all of the possible compounds, and then find the ones that are most active by using sensitive and selective tests. This generated an avalanche of efforts to use the same principles to generate and test candidate structures (or mimics) for the purposes of drug design. More recently this has been extended to the synthesis of ordinary organic compounds of the type A-B-C, where diverse sets, libraries, of building blocks (**BB**) A_i, B_j, and C_k (i=1,2,..,I; j=1,2,..,J; k=1, 2,.., K) can generate a large set of combinations A_i-B_j-C_k with the help of automatized parallel organic syntheses. The number of building block libraries (**BBL**) is not restricted to three; the same approach can be used to make compounds of the type A-B-C-D, A-B-C-D-E, etc. This approach of rapidly and semi-automatically creating and testing large "libraries" of compounds has been labelled "combinatorial chemistry", here abbreviated **CombC**.

Put somewhat crudely, the philosophical basis of CombC is the belief that making and testing a larger number of compounds leads to a higher chance to find an active compound than if a smaller number of compounds is investigated. In the beginning CombC was believed to be a route to all possible compounds of a certain structural type. This would, with appropriate biological testing, of course, maximize the probability to find the most active compound. However, we are now starting to realize that making *all* compounds is not possible, neither in principle nor in practice, chemistry is open ended. Also, large numbers of compounds present administrative and testing problems.

Hence, continuing to use CombC as a route to very many compounds now corresponds to the assumption that each additional compound provides some additional information about the structural requirements for a successful drug candidate. A million compounds are presumed to give substantially more information than, say, a thousand compounds.

It is easy to understand, however, that there is no direct and strong relation between the number of investigated compounds, **N**, and the chance to find a compound with substantial biological activity. Such a relation holds *only* when (**a**) all compounds are well spread out over the pertinent abstract structural space, **S-space** (*Fig. 1*), and (**b**) the compounds are adequately tested. Another selection of **N** compounds may be distributed over the S-space with much less information content (*Fig. 2*).

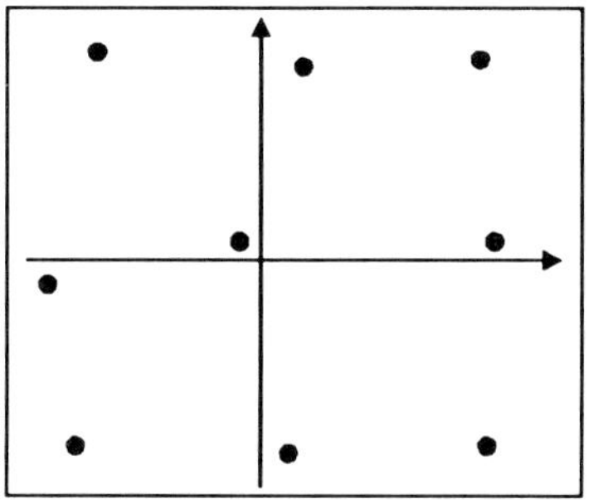

Fig. 1. *A compound can be represented as a point in an abstract structural space (S-space) where the axes corresponds to pertinent structural features, e.g. lipophilicity, size, polarity, polarizability, hydrogen bonding, steric shape, etc. Each of these properties may need a multitude of axes to describe, e.g. the distribution of lipophilicity over different regions of the molecule, the distribution of polarity over the same regions, etc.*

Given a perient description, a set of compounds can be represented as a set of points in the corresponding S-space. The information content of the set is closely related to how well the set is spread out over the available region of the space. The compounds shown here constitute an informative set; each compound is far from the others, and together they well span the S-space. The two coordinate axes may, for example, here represent lipophilicity and molecular weight.

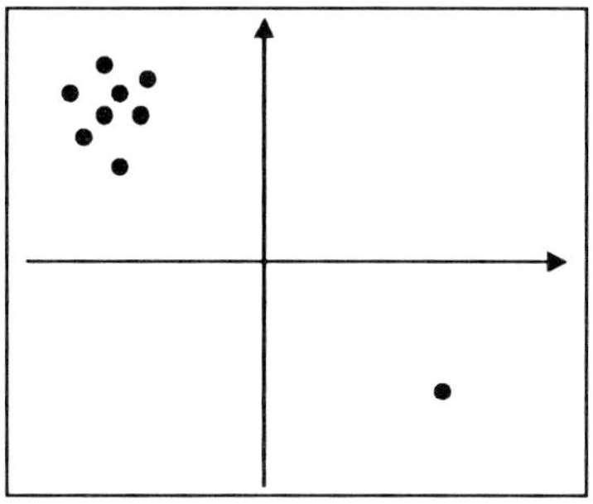

Fig. 2. *The same number of compounds as in Figure 1, in the same S-space, but here constituting of a uninformative set. The information content in this set of compounds corresponds approximately to the same information as two compounds.*

Spreading compounds over the available S-space corresponds to the selection according to a good statistical molecular design (**SMD**) for selecting combinatorial libraries and BBL.s [7-9]. Thus, it matters a lot *which* compounds that are made. It may be more informative to make and test 25 well selected compounds than 25000 poorly selected ones.

Of course, when all compounds of a certain type can be made, they together represent the best possible coverage of the S-space. Even if this were possible, however, it may be advantageous to select a subset that is almost as well covering the S-space, since this subset can be better tested for biological activity. The selection of a subset would also be preferable because it will simplify the identification of the chemical structure.

Like most complicated processes, CombC involves a number of compromises. The more compounds that are made, the larger is the possible resulting information, particularly if the compounds are well selected. However, the larger the number of compounds, the less deeply they can be biologically tested. Hence, if SMD can reduce the number of compounds needed to give a certain amount of information, the compounds can be tested in more depth, thereby improving the total, chemical + biological, information.

After making and testing a set of compounds the resulting data should be properly evaluated to find which combination of structural features that are required for high activity, low side effects, metabolic stability, etc. Some kind of quantitative modelling is necessary in this evaluation, and we will here discuss some requirements for this modelling, particularly in relation to CombC. Also, for the growth of knowledge in a project it is necessary to have a quantitative framework for connecting and representing results and helping their interpretation. Chemometrical or other models provide the needed tools for these purposes.

The present article proposes an *efficient* approach to CombC. This includes how to obtain a representative set of compounds using the methods of CombC and parallel organic synthesis, followed by the measurement of an informative profile of "biological activities", and the efficient use of these measured data. Hence, we shall emphasize the use of SMD and modelling with CombC. The purpose of SMD in this context is to find an appropriate set of compounds that can be synthesized by the fast and automatized methods of CombC at the same time as this set is optimally informative about the investigated biological activity and its structural requirements. The purpose of the modelling is to utilize this information efficiently for understanding and interpreting the relationship between structure and activity for interesting classes of compounds and using the resulting models for predicting the activity of new, yet untested, compounds, thus resulting in compounds with improved activity profiles.

Two Types of CombC, Searching for, and Optimizing a Lead. — The greatest interest in CombC has been as an approach for finding lead compounds. This is done by making large libraries of "diverse" compounds. These are then subjected to a fast biological testing, so called high through-put screening (**HTS**). This interest in CombC reflects the difficulty in drug industry to find leads, even with the availability of crystallized receptors, 2D and 3D-NMR, 3D molecular modelling, super-computers, and all.

Lead finding by any "rational" strategy is difficult, because we do not have a firm idea what we are looking for. Hence, it is difficult both to construct a S-space, because we do not know what structural features are important, and to specify a search strategy. Still, CombC can, if appropriately used, be used to improve the chances for finding leads. And, if we know a little more what we are looking for in terms of structural classes, the chances improve even further.

The second use of CombC, to optimize a known lead compound, is rather straight forward. Here we can use our experience of how to investigate congeneric series, develop Quantitative Structure Activity Relationships (**QSAR**), optimize structures, and so on. Here CombC is a practical approach to rapidly generate substantial sets of compounds that well cover the S-spaces of limited structural variation. The investigation and statistical modelling of such a set is needed for the purpose of optimization. Of course SMD comes in here as an integral part for the selection of these sets, as well as statistical modelling for connecting the results of all the tests of all the compounds, and how to do this is fairly well known and understood [2,3].

Interestingly, our understanding of statistical modelling and its philosophical and theoretical foundation shows us what we can do, parametrize, design, and model sets of chemically and biologically similar compounds, and thereby also shows us what we can and cannot do in the search for leads. This will guide us to what possibly can be done in this first, more difficult problem of lead finding, and it turns out that this is better than nothing, but perhaps not as much as we had dreamt of.

The CombC Approach and the Steps Involved. — We here consider CombC in a broad sense as an approach to find active compounds for the subsequent purposes of drug design and optimization. This involves more than just making many compounds, namely (1) How to quantitatively describe the chemical structures of the compounds, (2) how to select which compounds to make, (3) how to make the biological testing, and (4) how to evaluate the test results.

These problems will be discussed below, illustrated by two hypothetical examples.

Examples. — To illustrate the proposed strategy, we shall use two "simulated" examples. The objective in both cases may be to find a compound that cures hair loss or some other troublesome ailment.

The first example is, for historical and educational reasons, based on small peptides -- hexa-peptides. *Houghten et al.* [4] used hexa-peptides to illustrate the principles of CombC making **all** possible hexa-peptides; 64 millions = 20^6, if we restrict the BB.s to the 20 coded amino acids (AA). *Hellberg, Jonsson, Sandberg, et al.* [10-12] characterized free amino acids by a number of physicochemical measurements, and derived scales -- z_1, z_2, and z_3 -- that may be used to describe each position in a peptide. Hence, a hexa-peptide can approximately be described by 6 x 3 = 18 variables, and a dipeptide can be described as 2 x 3 = 6 variables. *Table 1* shows the **z**-values of the 20 coded AA [10].

Table 1. *The Three PC Scores (Principal Property Values) of the 20 Coded AA from ref. [10]. Values for Additional Non-Coded AA are found in Ref. [11, 12].*

No.	Name	z_1	z_2	z_3
1	ALA	0.07	-1.73	0.09
2	VAL	-2.69	-2.53	-1.29
3	LEU	-4.19	-1.03	-0.98
4	ILE	-4.44	-1.68	-1.03
5	PRO	-1.22	0.88	2.23
6	PHE	-4.92	1.30	0.45
7	TRP	-4.75	3.65	0.85
8	MET	-2.49	-0.27	-0.41
9	LYS	2.84	1.41	-3.14
10	ARG	2.88	2.52	-3.44
11	HIS	2.41	1.74	1.11
12	GLY	2.23	-5.36	0.30
13	SER	1.96	-1.63	0.57
14	THR	0.92	-2.09	-1.40
15	CYS	0.71	-0.97	4.13
16	TYR	-1.39	2.32	0.01
17	ASN	3.22	1.45	0.84
18	GLN	2.18	0.53	-1.14
19	ASP	3.64	1.13	2.36
20	GLU	3.08	0.39	-0.07

The second example is based on ketones as starting materials. We may imagine that we wish to make a library from A_i = ketones (R_1COR_2) reacting with B_j = Grignard reagents (R_3MgCl), then further reacting with C_k = acid chlorides (R_4COCl), to give the final library $A_iB_jC_k$ = $R_1(R_2)(R_3)C\text{-}OCO\text{-}R_4$. A number of ketones were multivariately characterized and modelled by *Carlson, Prochazka* and *Lundstedt* [13]. The structural variation is approximately captured by two principal properties (denoted by t_1 and t_2) as indicated in *Figs. 3 and 4*. The Grignard reagents and acid chlorides, both being of the generalized structure R-X, are assumed to be well characterized also by two principal properties, which will not be further considered in any detail.

Fig. 3 shows the distribution of the ketones in the two dimensional score space (t_1,t_2), resulting from the principal component analysis (**PCA**) of the table of 78 ketones described by the 11 structure descriptor variables. The figure also shows 9 compounds selected by a D-optimal design to well span this score space, as will be discussed below. *Fig. 4* shows the same score space but with another selection of 12 compounds, claimed below to be superior.

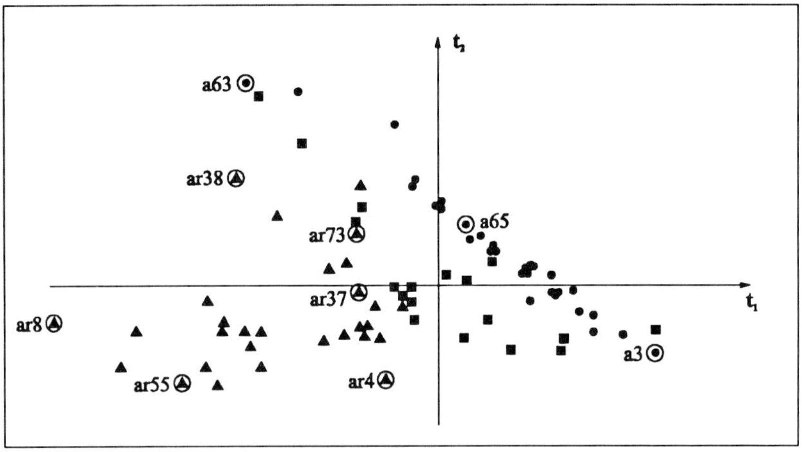

Fig. 3. *An S-space defined by the two first principal component scores (principal properties t_1 and t_2) of the 78 ketones according to Carlson, Lundstedt, et al. [14]. The rings indicate the selection of 9 compounds according to a D-optimal design and a quadratic model in t_1 and t_2. The design has selected three aliphatic ketones (prefix **a**, filled circles), and six aromatic ketones (prefix **ar** filled triangles). No alicyclic ketones (filled squares) were selected by the design.*

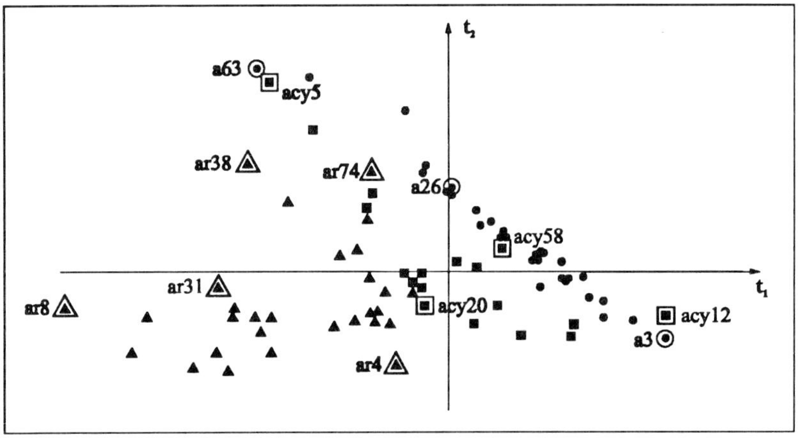

Fig. 4. *The same S-space as in Figure 3, with aliphatic (prefix **a**, filled circles), aromatic (prefix **ar**, filled triangles), and alicyclic ketone (prefix **acy**, filled squares). The selection of compounds was made according to three separate D-optimal designs for the three classes, with three, five, and for compounds for **a**, **ar** and **acy**, respectively.*

Chemometrical Aspects of CombC. — The CombC approach implicitly or explicitly involves several types of **modelling**. *Biological* models are used for the biological testing, and *chemical, statistical* and *chemometrical* models are used for selecting compounds, and for relating the variation of the biological data to the variation of chemical structure. These models are employed even if we do not **explicitly** mention them.

The use of modelling is based on a number of assumptions. The most important relate to the "similarity" of the modelled set of phenomena or events or cases (here chemical compounds). Any mathematical / statistical modelling approach must either be based on fundamental "ab initio" theory without many adjustable parameters, or be derived as a "Taylor or other serial expansion" of such fundamental models. In the latter case the model is valid only "locally", i.e., for moderate changes in its variables. In the present context this corresponds to having only a moderate variation in chemical structure between the compounds included in one model. If there is a wide variation, this necessitates the use of several models, one for each "structural type" of compounds.

In CombC there is little hope that models and approaches can be based on fundamental theory, particularly when we are searching for leads. This because interactions between chemical compounds and parts of living systems are complex and the details are rather poorly understood. Hence, CombC must be based on semi-empirical models with only local validity. This corresponds to the requirement that any model based procedure used in a given CombC application — for instance biological testing, SMD, and QSAR — is applied only to compounds that are structurally and biologically similar.

When searching for leads this seems absurd; we do not know what we are looking for, so how can we assure structural and biological similarity? In practice this is not so hard, however. We know fairly well what structural variation that can be considered "moderate", and we can use profiles of measured properties — say, lipophilicity, pKa, IR, UV and NMR spectra — plus functional groups, structural "class", and molecular weight, for "diagnostics of similarity". In addition, a profile of multiple biological tests measured on each investigated compound form an excellent basis for assessing the biological similarity of the compounds.

Consequently, if we wish to investigate a structurally diverse set of compounds, it is necessary to divide it into subsets of similar compounds. One objective with the data-analysis then is to investigate how wide structural variation a single model can tolerate, and then divide the total data into domains, "clusters", that can be approximated by single models. Again this is not too hard in practice; we have a rather good understanding of chemical similarity. With the ketones, for example, we may distinguish the three classes of aliphatic, alicyclic, and aromatic compounds. In addition, there are good clustering procedures available for these purposes provided that the structures are reasonably well parametrized.

Since drug design and CombC involve the modelling of the relation between chemical structure and biological activity, it is essential that the investigated compounds, in a reasonable way, span the pertinent space representing the structural variation. This leads to the use of SMD for the selection of these compounds, as further discussed below.

Specifying the Search Space -- the Structure Descriptors. To be able to quantitatively compare chemical compounds with respect to similarity and diversity, and to model the relation between their structural variation and biological activity, we need to translate the chemical structure (including conformation, tertiary structure, etc.) to an array of numbers. This is an area where recently great progress has been made -- see e.g., ref.'s [2, 3, 7, 9] -- with computational chemistry and physical organic chemistry providing the essential tools and concepts. Usually we can arrange this array of numbers as a vector, where each vector element of one compound corresponds to the same element of another compound. We call these vector elements structure descriptors, or structure descriptor variables.

This quantitative description has a number of requirements to be suitable for CombC. It has to be fast to derive or calculate. It has to be relevant, i.e., capture the essential structural properties that influence the biological activities of interest. The structure descriptor variables have to be chemically interpretable. If possible, the description should be reversible, i.e., be possible to also translate backwards from descriptor values to structure. The description should be consistent with similarity—dissimilarity, so that compounds that are closely similar have very similar values of the descriptors, and dissimilar compound have widely different values of at least most of the descriptors.

There are two choices for this structure quantification, either the structural description of the BB.s, or the description of the whole "final" library molecules. The former has several advantages; describing the separate BB.s -- A_i, B_j, and C_k -- automatically yields also a description of their combinations, $A_iB_jC_k$. An advantage with the separate description of A_i, B_j, and C_k is that this also can be used for the selection of the BB.s according to an SMD. We shall therefore consider only this alternative.

Physical and computational chemistry provide us with good approaches for quantification of structure. These apply differently for the two cases of a BBL being a fixed backbone (scaffold) plus "substituents", or when the BBL has a more complicated composition.

(a). Classical Substituents. Over the last six decades physical organic chemists have developed a number of structural descriptors -- substituent scales -- for the parts of organic compounds that can be seen as varying around a fixed backbone, the organic substituents. These scales describe electronic (Hammett, Taft), steric (Taft, Verloop), lipophilic (Hansch, Rekker), and hydrogen bonding effects. For recent compilations, see [2, 14, 15]. The "principal property" scales (z_1, z_2, z_3) of the AA discussed above represent continuations in this physical organic chemistry tradition [10-12].

(b). Other Fragments. Often BB's cannot be described in terms of a common backbone with varying substituents. One then needs a way to calculate structure descriptors that still capture "effects" such as steric, lipophilic, and polar. Here the GRID approach [16] and other "3D approaches" [reviewed in ref. 3] seem to work well, accounting both for intrinsic properties such as volume, shape, atom type, etc., as well as interactions with the surroundings of the molecule (van der Waals, hydrophobic, Coulombic, and hydrogen bonding). A recent example of this approach is given by the GRID characterization of heteroaromatics [17].

For the description of large numbers of molecules needed for selecting a BBL, however, 3D and other computational chemistry methods tend to be slow and cumbersome unless they are automatized and connected to other software for the automatic generation of structural coordinates. The development of such systems is underway in several laboratories, but they are not yet routinely available.

Preprocessing. In the common case when the K structural descriptors are not independent, but are strongly correlated, an initial principal component analysis (PCA) should be made of the descriptor data to find the "real" dimensionality of the data. Thereafter the design is made in the resulting principal components (**PC**) scores -- so called multivariate design [18, 19]. This has been done in both the present examples. The three z-scales of the AA are derived from a PCA of 29 variables, including measured properties from chromatography, NMR, pK_a, log P, PI, plus calculated van der Waals volume and molecular weight [10]. Analogously, the two t-scales of the 78 ketones are the first two PC scores from a PCA of 11 descriptors derived from IR and NMR spectra and other properties such as density, molecular weight, etc. [13].

Commonly this PCA indicates that a BBL is clustered -- exemplified by the ketones (example 2), and the heteroaromatics (10 clusters [17]). Then a second separate PCA for each cluster may be needed to derive good principal properties, design variables, for the clusters. This was not done here in example 2, however.

Building Block Selection by Statistical Molecular Design (SMD). — Whenever non-exact data are used in modelling, this becomes at least partly statistical. In the present context, neither the biological test data, nor the structural descriptors or their summaries, the principal properties, are "exact". Although one can exactly reproduce theoretical calculations of, for instance, a molecular volume, this does not make the resulting number exact. All theoretical calculations in computational chemistry are based on approximate models, and different implementations use different approximations, and hence give different results.

Consequently, the selection of compounds needs to be made by taking noise and uncertainties into account. Regardless of the size of these uncertainties, this makes two structurally similar compounds -- sitting close to each other in S-space -- provide less information than two dissimilar compounds, sitting far from each other in S-space. Hence, the number of compounds in a set is not directly related to the information of the set, and we need statistical design (in the present context called statistical *molecular* design, SMD).

The principle of SMD is basically to select compounds so that they are representative of the structural variation one wishes to investigate. The issue of representativity is then translated to the mathematical property of adequate coverage of the abstract space formed by the structure descriptors, the S-space. In this K dimensional space any potential compound, quantified by the pertinent values of the K structure descriptors, is represented by a point.

In the context of CombC one may make design of the BBL.s, or of the final library. It is clear that if we make SMD.s of the BBL.s, the combination of the designed BBL.s will result also of a designed final library. For practical reasons it is much easier to apply SMD to the BB.s, since they are smaller than the final molecules. We shall therefore consider only this approach, namely SMD of the BBL.s, followed by a designed combination of the BBL.s. The latter is discussed in a special section below. One important point to remember is that any selection, including that of an SMD, is based on some kind of implicit or explicit model. Since we do not know everything, and hence all our models are incomplete, we should complement the designed selection of compounds with some compounds "randomly" selected from the BB candidate set.

Provided that, for a given type of BB.s, we are interested only in a fairly small region of the S-space, and provided that the K descriptors are statistically independent, traditional statistical designs such as factorial designs, fractional factorial designs, and composite designs [20], or "tailor-made" designs, such as D-optimal designs [21], are adequate. The type of design and the required number of compounds (N) depend on the underlying model (e.g., linear or quadratic), and on the number of descriptors, K.

If, however, the desired structural variation of a BBL is large, as is common in "lead finding", the use of more dense designs is necessary. Two possibilities can be seen. In the rare case when the structural variation is moderate and homogeneous, so called space filling designs are appropriate. In the typical case that the structures are clustered, however, "cluster based" designs are needed, as discussed below.

As was discussed above (structure descriptors) an initial multivariate analysis is needed when the structural descriptors are not independent. This will give "principal properties" that are ideal as design variables because they are few, statistically independent, and constitute an optimal summary of the original descriptor data.

The application of SMD is usually different in the two cases -- (a) optimizing a set of similar compounds, and (b) searching for leads in a set of diverse compounds.

(a) Lead Optimization. Since lead optimization is less complicated than lead finding, we start with this situation. Here we can usually assume the structural variation in each BBL to be fairly small. Also, we can assume that this variation can be seen as moderate variations of some structural features on a common, fixed, scaffold.

The hexa-peptides provide an example of this situation. The "scaffold" is the chain of six repeated -NH-CHR-CO- units, and the variation of the side chains R defines the structural variation. The syntheses required to give a small library of hexa-peptides -- up to, say, N=300 -- is nowadays simple and needs not be considered here. We shall consider dipeptides as BB.s, and use SMD to construct a BBL of representative dipeptides.

The sequences of small peptides can rather well be parametrized by describing each position by the three z-scales of *Hellberg, Jonsson, Sandberg et al.* [10-12]. Hence, considering dipeptides as BB.s, each BB is characterized by means of six descriptors; z_1, z_2, and z_3, for AA position 1 and 2.

When optimizing a system, such as a chemical compound interacting with a biological "target", at least quadratic models are usually needed, because one is searching for a maximum or a minimum. In drug design, optimal biological activity corresponds to an optimal size, lipophilicity, hydrogen bonding, etc. Designing the BBL for a quadratic model means that the linear description is extended by quadratic terms and interactions, and sufficiently many compounds must be selected for the BBL to support this extended model. Thus, if the BB.s are described by K descriptors, the quadratic model has $\mathbf{p} = \{\ 2K + K(K-1)/2\ \}$ terms, and the BBL needs to contain at least, say, p+5 compounds.

The selection of these p+5 compounds is easily done by means of a composite design [20] in the case that the distribution of the available BB.s (the candidate set) in S-space is fairly regular, or a D-optimal design [21] in the general case.

For a BBL of dipeptides, for lead optimization, K = 6, and p = 12+15 = 27. This gives around 32 dipeptides, selected from the candidate set by D-optimal design.

Adding a "random complement" of, say, about half this number, consisting of 16 dipeptides randomly selected from the candidate set, we end up with a BBL of around 50 dipeptides. If we restrict ourselves to the 20 coded AA, the candidate set has 20*20 = 400 members, but if we wish to use also non-coded AA, the candidate set is larger. We here refrain from showing the resulting (N=50) BBL, since its actual composition is of little interest for the present discussion.

(b) Lead Search. SMD, i.e., selection of representative compounds (here BB.s), is difficult when there is a large structural variation (diversity). This because we do not know what we are looking for (we do not have a lead), and hence we do not know which structural features that are important. This may be remedied by including many descriptors, but then the search space becomes high-dimensional, with the consequence that very many compounds are needed for the resulting BBL to be adequate. An additional complication is that the structure descriptors may not mean the same thing over the S-space. This problem may be dealt with by using many levels in the designs, but again this will result in very many compounds in the BBL. If, for example, we manage to develop 20 independent descriptors that fairly adequately describe a diverse library, and we feel that five levels are needed of each descriptor, we arrive at an SMD of 5^{20} » 100 billion compounds. This number seems to be rather much both for synthesis and testing.

Interestingly however, the CombC approach based on making libraries as combinations of BB.s, provides a practical solution to the problem. Using tools of multivariate analysis such as cluster analysis and principal component analysis (PCA), we can investigate whether a BBL is homogeneous -- consisting of chemically similar compounds -- or whether it is clustered. If a BBL is homogeneous, it can be well represented by an SMD containing rather few compounds, selected by, for instance, a quadratic D-optimal design. If we, for a moment, consider dipeptides as BB.s, this would be done as shown above.

If a BBL is clustered, this corresponds to that the S-space containing dense regions, clusters, with empty space in between. Each cluster will, if things are done right, contain similar compounds. In this case, a separate SMD made for each cluster will provide an adequate representation of the whole S-space [28]. If there are compounds between the clusters, a separate designed selection of these should be made to complement the BBL.

In example 2, the ketones are indeed clustered. The 11 dimensional space shows three strong groups that are chemically interpretable as aromatic, alicyclic, and aliphatic compounds. This is seen in *Figs. 3 and 4* which show the two significant PC scores (R^2 = 0.74) of the (78 x 11) descriptor matrix.

A design in these PC scores (N=9, D-optimal with center point, for a linear model) made *without* the recognition of the grouping is shown in *Figure 3*. Although the design covers the space of t_1 and t_2, it completely misses the alicyclic compounds. Also, the aromatic group is over-represented due to its greater area in the PC plot. The selection resulting from three separate designs, one for each cluster, is shown in *Figure 4*. These designs were made with five points for the aromatic group, four for the alicyclic, and three for the aliphatic group. This selection is much better than the first, since it has an adequate coverage of the three types of ketones, with about the same total number of compounds. Adding, say, 6 randomly selected compounds gives a BBL of ketones containing N=18 compounds.

Assuming (without solid motivation) that the Grignard reagents and acid chlorides have 4 and 5 clusters, similar reasoning would give BBL.s of around 25 and 30 compounds, respectively, for these classes.

We call this approach cluster design, where we first divide each set of BB.s into subsets (clusters) of similar compounds, then making BBL.s by separate SMD.s for each cluster. The approach has positive consequences: Simple designs result, because structural variation in each subset is limited. Relevant descriptors are not too hard to find because subset can often be seen as backbone + substituents, and we have lots of good substituent descriptors for size, lipophilicity, polarity, polarizability, etc. Finally, QSAR.s of each cluster are possible to construct for the same reasons, making results interpretable, predictions possible, etc.

When applying cluster analysis for finding groups, we must remember that clusters rarely are round in S-space, but elongated as seen in *Figures 3 and 4*. Also the different clusters have different variance-covariance structure, i.e., different orientation in S-space. Cluster analytical methods that are consistent with this non-roundness and differences in orientation include those of fuzzy clustering [22] and SIMCA [23]. Also hierarchical structures are common in chemistry. This corresponds to grand clusters divided in sub clusters divided in sub clusters, etc.

Design of the Final Library -- Combining the Building Blocks. — As discussed above, the selection of compounds in one BBL, say A_i, will normally be made according to a combination of clustering and SMD in each cluster, plus a "random complement". The combination of compounds from the BBL.s to form the final library should, of course, also be made according to an SMD. This design would almost always be made using a D-optimal selection (plus a few center points) from the candidate sets comprising all possible combinations of the BBL compounds. In addition, a "random complement" of the final library is made by combining parts of the random complements of the BBL.s.

The SMD is made based on the total number of descriptors of the BB.s, i.e., ($K_A + K_B + K_C$) if we are making the final library ABC, and the BB.s are characterized by K_A, K_B, and K_C descriptors, respectively. Depending on the stage of investigation -- lead search or lead optimization -- different models would normally be used as the basis of the designs. Linear models, or linear + selected interactions models would typically be used with lead search, and full quadratic models with lead optimization. Let us first consider the latter, with hexa-peptides as the example.

A hexa-peptide is approximately characterized by 18 (6 x 3) numbers (z-scales). For linear models, the S-space thus has 18 dimensions, and for a model including quadratic terms and interactions, the S-space gets 189 dimensions (18 + 18 + 18*17/2). Hence an SMD with around 200 hexa-peptides will be well representative of all 64 millions possible hexa-peptides. These 200 hexa-peptides are selected from the candidate set with 125000 members (50*50*50) defined by the 50 membered BBL of dipeptides. Hence, the 200 selected compounds can be made as combinations of these 50 selected BB.s (dipeptides). Including a "random complement" to this design of another 100 hexa-peptides gives a total library of around 300 compounds that in many ways is informationally optimal, and at the same time small enough to allow for a thorough biological testing (MTMMS, see below).

Turning now to "lead search", let us first consider a more difficult peptide problem of, for example, 21-peptides. Describing each position with the three z-scales, we arrive at 63 parameters. If, as is usually done in designed screening [20], we restrict ourselves to a linear model in these 63 variables, plus neighbor interactions (12, 23, 34, etc., totally 20 position pairs times 9 variable pairs), and "helix" interactions (14, 25, 36, etc, totally 18 position pairs times 9 variable pairs), we get totally 405 design variables. Hence we need a final library of around 450 to 500 compounds, which is feasible from both synthetic and testing aspects.

With clustered BBL.s, we can use several designed approaches for their combination. The simplest is to consider each BBL as a qualitative variable with n_m levels (the number of BB.s in the m.th BBL). In example 2 we have BBL.s with n_1=18, n_2=25, and n_3=30 members. A linear model in three qualitative variables has 70 (18+25+30-3) free parameters, which would give between 70 and, say, 150 compounds in the final "minimal" library. A more realistic model with interactions AB, AC, and BC would add 1597 (17*24 + 17*29 * 24*29) parameters, however, necessitating at least 1700 compounds in the final library.

However, recognizing the clustering of the three BBL.s, we can reparametrize the problem as one of three qualitative variables for the clusters (with 3, 4, and 5 levels), plus the position in each cluster described by two qualitative variables with three levels (low, middle, high), assuming two principal properties per cluster. This gives totally nine qualitative variables. The number of independent linear parameters is 21, there are 26 cluster interaction terms (2*3 + 2*4 + 3*4), and 162 cluster--position interactions, totally 209. A final library with 225 members would well span the corresponding S-space, and provide information about all effects likely to be the most important.

Table 2 shows a table of parameters and resulting sizes of final libraries for up to 15 clusters per BBL, showing the somewhat surprising result that if the BBL.s are clustered, and if each cluster can be represented by 2 PC.s, we can do well with small libraries of between 150 and 1500 compounds. These assumptions seem realistic according to studies by *Carlson et al.* reviewed in [19] and *van de Waterbeemd et al.* [9].

Table 2. *The cluster based screening SMDs as function of the average number of clusters (M) of the BBL.s. Each cluster is assumed to be modeled by a two component PC model. The models underlying the designs contain linear terms, plus cluster-cluster interactions (col. 3) and cluster-position interactions (col.4).*

M (avg number of clusters)	Linear terms 3*(M-1+4)	Cluster interactions 3*(M-1)^2	Clus -- pos interactions 3*18(M-1)	Total parameters B+C+D	Total library	Comment
3	18	12	108	138	150	
4	21	27	162	210	225	example 2
7	30	108	324	462	500	
10	39	243	486	768	800	
15	54	588	756	1398	1500	

Adding "random selections" of compounds to each BBL increases the size of the final library, but not catastrophically. We can include these random selections as a couple of additional clusters in the design calculations, pushing the average number of

BBL clusters in *Table 2* up by two or three, leading to at most a few hundred additional compounds in the final libraries.

Making the Final Library — the Chemical Synthesis. — The first act in the chemical synthesis is to investigate the scope and limitations of the synthetic reactions [29], since different clusters of compounds, or even compounds within a cluster, may require different experimental settings. Even with automatized solid state syntheses using polymer beads as carriers of compounds, the synthetic steps need to be optimized so that reasonable yields are obtained for all compounds in the library. Otherwise unbalanced test data will result, with subsequent loss of information. Such optimization is easier accomplished with few compounds. Recently, robots that optimize organic syntheses on the basis of statistical modelling and experimental design have been developed [26]. For a moderate number of compounds, say up to 1000, the robots can be programmed to perform all the synthesis, including the optimization of each individual step, in a reasonable time. This possibility is a further strong argument for limiting the number of compounds and just make as many as needed from the information point of view.

Defining the Compass -- the Biological Testing (MTMMS preferable to HTS). — The biological testing is the most important step of the CombC as well of any other strategy of drug design. Without good and relevant measures of the biological activity, no rating of compounds can be made, and nothing is achieved regardless of how many compounds are made. Hence we consider the biological testing to be an integral part of CombC, not a separate step glued on afterwards. Since the information content of the biological test data is so important, we strongly recommend MTMMS (moderate throughput multiple model screening) instead of HTS (high throughput screening) which is usually favored in CombC applications.

In the pharmaceutical industry the objective is to develop drugs with desired clinical effect(s). To achieve this it is necessary to have access to tests which well model the clinical situation. Hence these test systems are often called pharmacological models. These models have very different degrees of sophistication, from simple and fast *in vitro* models to very complex and time consuming *in vivo* models. They have, as well, a large variation in relevance and adequacy, from being good to very poor approximations of the clinical situation.

A fast biological screening will necessarily be only a crude approximation of the clinical situation. To still get useful biological data it is necessary to focus on the (often hypothetical) biochemical mechanism explaining the reasons for a disease. The mechanism has to be expressed in biological activities which can be measured in HTS, i.e., receptor affinities, enzyme inhibition, etc.

In CombC it is common to apply HTS in order to find compounds with a specific activity, e.g., compounds which are selective enzyme inhibitors. The HTS is then run only to trace active compounds (not selective) by just measuring active/non active towards the target enzyme. However, to find and to have a possibility to "optimize" the selectivity it is necessary to measure on a wider spectrum of fast screening models. Hence, in order to obtain the best possible information from CombC, several HTS models must be used in parallel.

Fast shallow testing of many compounds involves two types of risks, that of false positives (over-optimistic evaluation), and that of false negatives (missed opportunities).

Risk for False Negatives. With a large library of compounds the only biological testing that can be afforded is quick and shallow "high throughput screening", HTS. The "hits" in this screening process are then taken to further testing with more informative, but also more time consuming, tests. This strategy involves great risks. It works only if the initial HTS results are strongly correlated to the "real" biological activity. And, of course, this can rarely be investigated beforehand, one just has to assume that simple test of receptor or enzyme binding is informative about this "real" activity.

The risk for false negatives, i.e., that an actually "active" compound is classified as inactive is a property of the actual biological test, and the only way to decrease it is to use more extensive, more deep biological testing. This takes time and resources, and hence the number of compounds in the library must be weighed against the demands on resources for the biological testing.

Risk for False Positives. The larger the number of tested compounds, the greater is the risk of false positives for a given biological test. This arrives because each compound is evaluated separately, with a given probability that the test result is falsely positive. Hence, CombC with many compounds evaluated by HTS may give a large number of false positive compounds.

The risk for false positives when evaluating a large set (with N members) ***one compound at a time*** increases as $\{1 - 0.95^N\}$ or $\{1 - 0.99^N\}$, or .... if we consider compounds having an activity larger than 2 SD.s (standard deviations of the testing variability), 3 SD.s, ... With the usual testing noise of 0.3 log units (a factor of two in potency) the Bonferroni correction should be applied to avoid too many false positives, i.e., testing on the probability level of 0.05/N or 0.01/N instead of 0.05 or 0.01 respectively. With large values of N this leads to very small probability levels, and absurd requirements on activity levels to be accepted as a "real hit". This is another strong argument for (1) having moderate values of N, and (2) evaluating all compounds together with respect to some kind of **model** — distribution or/and structure related — to get the problems of false positives under control.

Another more problematic type of a false positive is a compound with real activity in the target system, but which for some reason is useless for further development. This may be, for example, a compound that binds well to a receptor, but which is metabolized by other enzymes in the cell or on the way to the "target". The best way to decrease this risk is to test fewer compounds and to have a more extensive biological testing with a smaller risk for a false positive result. Again, we see the need to make a compromise between the number of compounds and the depth of the biological testing.

A similar argument applies to false negatives. Since any single test value has a non-negligible measurement error (noise), there is a certain risk to miss real positive results. A good way to decrease the total risk is to analyze ***all*** data together by means of a model; all activity values support each other and the consequences of noise and measurement errors are minimized.

Depth of Biological Testing. To avoid or reduce these problems with false negatives and positives, the following is suggested;

1. Run several HTS models in parallel.
2. If possible, investigate all of the biological models with a set of compounds with known clinical activity, designed in accordance with a multivariate design regarding the clinical action as well as the chemistry. By doing this it is possible to make a multivariate analysis of the results and obtain information on the discrimination power of the different HTS models.

Hence there are strong arguments for an as extensive biological testing as possible. This is, in turn, a strong argument for the use of SMD which may greatly reduce the number of compounds needed to give a certain amount of information. Hence, with appropriate SMD the compounds may be tested in more depth, thereby improving the total, chemical + biological, information. We shall call this testing strategy MTMMS for moderate throughput multiple model screening.

Efficient Use of the Biological Test Results by Modelling (QSAR). — When compounds are selected according to SMD, this necessitates the adequate description of their structures by means of quantitative variables, "structure descriptors". This description can then be used *after* the compound selection, synthesis, and biological testing to formulate quantitative models between structural variation and activity variation, so called quantitative structure activity relationships (QSAR). For extensive reviews, see ref.'s [2, 3]. With multiple structure descriptors and multiple biological activity variables (responses), these models are necessarily multivariate in their nature, making the Partial Least Squares Projections to Latent Structures (**PLS**) approach [14,15] suitable for the data analysis.

The use of QSAR.s for the final evaluation of the test data further decreases the risk for both false positives and false negatives. This because *all* data are analyzed together, and thereby the random variation of individual test results are averaged out. Moreover, outliers and erroneous test results are identified by model diagnostics, which even further stabilizes the results.

Summary and Conclusions — A Recommended Strategy for CombC. — The selection of compounds to be synthesized to form a library is made with the purpose to make the library as "informative" as possible with respect to a certain type of biological activity. The library compounds will be tested with either the hope to find a "lead" for a certain pharmaceutical application, or for the optimization of a certain class of compounds with respect to, usually, a profile of biological and physico-chemical measurements.

We have emphasized that the chances to have a real positive hit in at least one library compound are not directly related to the number of compounds in the set, but rather to how well the library covers the pertinent S-space.

Based on our present understanding of the mathematical modelling of complicated systems, such as that of a chemical compound in a living organism, we can devise a strategy for generating "informative" compounds in sufficient numbers for either of these two objectives, lead search, or lead optimization.

We point out that lead search can be based on the same firm foundation as the "easier" lead optimization, namely perturbation theory applied to a complex relationship. This is achieved by dividing the final library into groups, clusters, of chemically and biologically fairly similar compounds. This grouping may be derived from the structural clusters found in the preceding BBL.s or it may be estimated by a cluster analysis of the Y-profiles of the final library itself, or preferably from both.

For each cluster in the BBL, a separate SMD is made to select compounds that span this structural cluster. Analogously, a separate model of the relation between structural variation and activity variation is developed for each cluster.

We note that this use of multiple designs and multiple models does not lead to any loss of information or understanding. It is just a consequence of our knowledge that a model to have any validity must be based on, and derived from, a set of homogeneous data. Here this means a set of compounds and their biological activities where the interaction between structure and "biology" remains essentially the same. The same multivariate and interpretational tools that are used to find the "clusters" can be used to classify new compounds according to which cluster they are most similar. Hence predictions and optimization is achieved as easily with multiple models as with a single one. Moreover, since the separate cluster models are much simpler than any "global" single model, the clustering also simplifies the interpretation and the understanding of the results.

A great advantage with the present approach is that the final libraries contain fairly few compounds, usually well below 1000. This is well within reach of automatized parallel synthesis combined with automatized multiple model biological testing [26], making CombC an interesting approach to speed up the initial drug development stages.

An important part of the philosophy behind the proposed approach is the recognition of the importance of the biological test data. Without relevant and fairly extensive biological test data nothing is achieved at all. This is a weakness with HTS data; the correlation between HTS results and real clinical activity is often weak or even absent. This makes CombC based on casual HTS equivalent to throwing dice. The biological testing must be an integral part of CombC, not a separate phase that can be glued on afterwards.

Additional advantages when the libraries contain fairly few compounds are that the knowledge about the chemical syntheses is better. This will simplify the structural analysis for the chemical identification.

Comparing our proposed strategy for CombC with what actually is done in most laboratories today there are differences in (a) the selection of compounds — we recommend the use of design, SMD — and (b) the evaluation of the results — we recommend a formal multivariate modelling similar or equivalent to QSAR. Design is used by *Martin, Blaney et al.* at Chiron [7], by *Young et al.* at Glaxo [8], and others. The design part is difficult to do well without the recognition of possible clusters in the data. In related areas such as selecting compounds for QSAR, for chemical synthesis, etc., we have long stressed the necessity to model (and therefore design) separate clusters separately [24, 25].

Acknowledgements. - Financial support by the Swedish Natural Science Research Council to the Umeå Chemometrics Group is gratefully acknowledged.

REFERENCES

[1] B. Spilker. 'Multinational Drug Companies'. Issues in Drug Discovery and Development. Raven Press, New York, 1989.

[2] H. van de Waterbeemd (Ed.), 'QSAR: Chemometric Methods in Molecular Design. Methods and Principles in Medicinal Chemistry', Vol 2., Verlag Chemie, Weinheim, Germany, 1995.

[3] H. Kubinyi. (Ed.), '3D QSAR in Drug Design; Theory, Methods and Applications'. ESCOM Science Publishers, Leiden, Holland, 1993.

[4] R. A. Houghten, C. Pinilla, S. E. Bondelle, J. R. Appel, C. T. Dooley, J. H. Cuervo. *Nature*, **1991**, *354*, 84-86.

[5] H. M. Geysen, R. H. Meloen and S. J. Barteling. *Proc. Natl. Acad. Sci. U.S.A.* **1984**, *81*, 3998-4002.

[6] R. A. Houghten. General method for the rapid solid-phase synthesis of large numbers of peptides: Specificity of antigen-antibody interaction a the level of individual amino acids. *Proc. Natn. Acad. Sci. U.S.A.* **1985**, *82*, 5131-5135.

[7] E. J. Martin, J. M. Blaney, M. A. Siani, D. C. Spellmyer, A. K. Wong, W. H. Moos. *J. Med. Chem.* **1995**, *38*, 1431-1436.

[8] S. S. Young, D. M. Hawkins. *J. Med. Chem.* **1995**, *38* 2784-2788.

[9] H. van de Waterbeemd, G. Constantino, S. Clementi, G. Cruciani, R. Valigi. 'Disjoint Principal Properties of Organic Substituents'. In QSAR: Chemometric Methods in Molecular Design (Ed. H. van de Waterbeemd), Methods and Principles in Medicinal Chemistry, Vol 2., Verlag Chemie, Weinheim, Germany, 1995.

[10] S. Hellberg, M. Sjöström, S. Wold. *Acta Chem. Scand.* **1986**, *B40*, 135-140.

[11] J. Jonsson, L. Eriksson, S. Hellberg, M. Sjöström, S. Wold. *Quant. Struct.-Act. Relat.* **1989**, *8*, 204-209.

[12] M. Sandberg, J. Jonsson, S. Wold. Principal Properties for an extended set of 77 amino acids. In manuscript, 1996.

[13] R. Carlson, M. P. Prochazka, T. Lundstedt. *Acta Chem. Scand.* **1988**, *B42*, 145-156.

[14] C. Hansch, A.J Leo. Substituent Constants for Correlation Analysis in Chemistry and Biology, Wiley, New York, 1979.

[15] B. Skagerberg, D. Bonelli, S. Clementi, G. Cruciani, C. Ebert. Principal properties of organic aromatic substituents. *Quant. Struct-Act.Relat.* **1989**, *8*, 32-38.

[16] P. J. Goodford. *J. Med. Chem.* **1985**, *28*, 849-57.

[17] S. Clementi, G. Cruciani, P. Fili, D. Riganelli, R. Valigi, G. Musumarra. *Quant. Struct-Act. Relat.* **1996**, *15*, 108-120.

[18] S. Wold, M. Sjöström, R. Carlson, T. Lundstedt, S. Hellberg, B. Skagerberg, C. Wikström, J. Öhman. *Anal. Chim. Acta* **1986**, *191*, 17-32.

[19] R. Carlson, Design and optimization in organic synthesis. Elsevier, Amsterdam, 1992.

[20] G. E. P. Box, W. G. Hunter, J. S. Hunter Statistics for experimenters. Wiley, New York, 1978.

[21] M. Baroni, S. Clementi, G. Cruciani, N. Kettaneh-Wold, S. Wold. *Quant. Struct.-Act. Relat.* **1993**, 12, 225-231.

[22] R. W. Gunderson, K. R. Thrane, R. D. Nilson. *Chemometrics and Intell. Lab. Syst.* **1992**, *3*, 119-131.

[23] S. Wold, K. Andersson. *J. Chromatogr.* **1973**, *80*, 43.

[24] S. Wold, E. Johansson, M. Cocchi. 'PLS -- Partial least-squares projections to latent structures'. In H. Kubinyi, Ed., 3D QSAR in Drug Design; Theory, Methods and Applications. ESCOM Science Publishers, Leiden, Holland, **1993**, 523-550.

[25] S. Wold. 'PLS for Multivariate Linear Modeling'. In QSAR: Chemometric Methods in Molecular Design (Ed. H. van de Waterbeemd), Methods and Principles in Medicinal Chemistry, Vol 2., Verlag Chemie, Weinheim, Germany, **1995**, 195-218.

[26] Scitec Laboratory Automation SA, Av. de Provence 18, CH-1007 Lausanne, Switzerland.

[27] H. M. Geysen, S. J. Rodda, T. J. Mason. *Molec. Immun.* **1986**, *23*, 709-715.

[28] T. Lundstedt, R. Carlson, R. Shabana. *Acta Chem. Scand.* **1987**, *B41*, 157-163.

[29] R. Carlson, T. Lundstedt. *Acta Chem. Scand.* **1987**, *B41*, 164-173.

PART II. **LEAD FINDING STRATEGIES**

Chapter 13

Molecular Dynamics Simulations As a Tool to Investigate the Three-Dimensional Diversity of Peptide and Pseudopeptide Libraries

by **Gérard Grassy**[a]), **Abdelaziz Yasri**[a])[e]), **Pierre Sans**[a])[e]), **Anne-Marie Armbruster**[b]), **Roger Lahana**[c]) and **Jean-Luc Fauchère***[d])

[a]) Centre de Biochimie Structurale, UMR CNRS 9955, INSERM U414, Université de Montpellier I, F-34060 Montpellier Cedex
[b]) IBM France, Tour Septentrion, F-92081 Paris La Défense Cedex 9
[c]) Oxford Molecular Group, X-Pole, Ecole Polytechnique, F-91128 Palaiseau Cedex
[d]) Institut de Recherches Servier, 11 rue des Moulineaux, F-92150 Suresnes
[e]) Present address: SYNT:EM, Parc scientifique G. Besse, F-30000 Nimes

Molecular Dynamics Simulations As a Tool to Investigate the Three-Dimensional Diversity of Peptide and Pseudopeptide Libraries

by **Gérard Grassy, Abdelaziz Yasri**[1]), and **Pierre Sans**[1])
Centre de Biochimie Structurale, UMR CNRS 9955, INSERM U414,
Université de Montpellier I, F-34060 Montpellier Cedex

and **Anne-Marie Armbruster**
IBM France, Tour Septentrion, F-92081 Paris La Défense Cedex 9

and **Roger Lahana**
Oxford Molecular Group, X-Pole, Ecole Polytechnique, F-91128 Palaiseau Cedex

and **Jean-Luc Fauchère***
Institut de Recherches Servier, 11 rue des Moulineaux, F-92150 Suresnes

The three-dimensional exploration of the conformational space by di-, tri-, tetra-, penta-, hexa- and decapeptides is evaluated using mainly molecular dynamics, and compared to that of the corresponding ψ[CH(OH)-NH] pseudopeptides and to a series of rigid steroid compounds. The dynamic data are expressed by the autocorrelation vector on which multivariate statistics can be applied. It is concluded that while steroids and, to a less extent, dipeptides are poor explorers of the 3D space, hexapeptides or longer peptides homogenously span the conformational space, nearly as well as their pseudopeptide counterparts, and that hexapeptide libraries are suitable sources of ligands to macromolecular targets.

Introduction. - Peptide libraries are a convenient source of molecular leads for optimisation towards peptide or non peptide drugs. Whether obtained by chemical synthesis on solid phase [1] or by genetic engineering on the surface of phages [2], they consist of mixtures of large number of individual peptides, e.g., 64 millions of all possible hexapeptides made out of the 20 natural amino acids.

1) Present address: SYNT:EM, Parc scientifique G. Besse, F-30000 Nimes (France)

Chemical libraries are generated in a combinatorial way by a "mix and divide" procedure after each incorporation of a new amino acid, before they are screened in a convenient assay with the peptide substrate still attached on the bead [3] or free in solution [1].

Several reasons are in favour of a limitation of the sequence length to the hexapeptide size in peptide libraries. Firstly, this length is believed to be sufficient for proper epitope recognition [4], at least for a free peptide in solution (as opposed to a bead-linked peptide). Even shorter segments are recognized by G protein-coupled receptors, as suggested by Ariens [5], or are convenient leads (substituted dipeptides) for chemical optimization [6]. Secondly, calculations based on a Poisson distribution of the amount of resin needed to insure proper representation of each individual peptide after the mix and divide procedure, result into gram quantities in each reactor for hexapeptides, while unpracticably higher quantities of resin would be needed for hepta- or longer peptides [7].

Besides amino acids in peptides, several other building blocks have been envisaged and partially used for the synthesis of libraries. These include D- and other non natural amino acids, N-alkylglycines, chiral aminocarbonates, phosphonate dipeptide surrogates or organic building blocks such as benzodiazepine, hydantoin or steroid units (review [8])· The resulting oligomers may of course display three-dimensional features very different from those of true peptides. Hence, pseudopeptides with throughout reduced peptide bonds ψ (CH_2NH) or ψ(CH(OH)-NH) are likely to be more flexible, N-methylated peptides more rigid than the corresponding parent peptides [9].

Since peptide libraries are used as sources of new leads and since no a priori limitation is put on the biological target used in screening, a maximum coverage of the available conformational space is the aim, at least in the first discovery step. The incorporation of conformational constraints into flexible lead molecules is expected to occur in a second optimisation step as a strategy to increase potency, selectivity and stability of the peptidomimetic ligand [10].

The question arises whether random peptides, as contained in synthetic libraries of up to hexapeptides, are representative collections of all possible conformations and therefore of potential ligands to any complementary receptor site. Although sophisticated methods have been put forward to estimate molecular diversity in compound collections [11, 12, 13], no such systematic study has been published so far.

The purpose of our work was therefore to investigate the exploration of the conformational space by libraries of di- tri- tetra- penta- and hexapeptides compared to pseudopeptide and non peptide libraries.

In the present first study, we offer the results of molecular dynamic simulations of representative peptides and pseudopeptides in function of size and flexibility, using the spatial autocorrelation vector for representation and principal components analysis for interpretation. We show that the rigidity of the peptide bond has less marked effects when the peptide size increases and that full libraries of hexapeptides sample the conformational space nearly as well as the corresponding rotation-free pseudopeptides.

Methods. - ***Molecular Mechanics and Dynamics.*** Preliminary investigations using molecular mechanics as implanted in the software MAD (Oxford Molecular Ltd., Magdalen Center, Stanford-on-Thames, Oxford OX4 4GA, U.K., version 2.20, 1994) allowed tree search and Monte Carlo simulations to be performed on model di- and tripeptides and on the corresponding pseudopeptides. The coverage of the conformational space by dipeptides was found to be relatively low compared to pseudopeptides (not shown), due to the rigidity of the peptide bond. Since these studies could not be extended up to the hexapeptide size, because of too long computing times, we decided to turn to molecular dynamics as a means to simulate the peptide conformations and to choose the method of spatial autocorrelation to simplify the description of the peptide dynamics [14, 15, 16, 17].

The starting conformation of each of the peptides or analogs was obtained from extended peptide models using in a first step, the molecular mechanics software MM2 (implanted in MAD) [18] and in a second step, the semiempirical quantum method AM1 implanted in the MOPAC software [19]. The resulting energy-minimized 3D structure was then subjected to molecular dynamics simulation using AMBER 4.0 [20]. Energy minimization and molecular dynamics calculations were carried out on a cluster of HP 735 workstations and molecular display was obtained on a Silicon Graphics Iris 4D 30 station using the Prosimulate graphics program. For each simulated peptide, the total simulation time was 1 ns, resulting in a computation time of about 1 min per ps. Calculations were performed at the constant temperature of 300°K. Water solvation was simulated using a distance-dependent dielectric constant $\varepsilon = 1 \times r$. All trajectories were calculated with a non-bonded cut-off distance equal to 15 Å, and the non-bonded pair list was updated every 25 fs. The molecule coordinated were collected every 1 ps during walking over the trajectories.

The 3D autocorrelation vector was then generated for each retained conformation as already described [16, 21]. Briefly, this vector represents the distribution of interatomic distances of the molecule grouped in discrete arbitrarily defined intervals. While the first component of the vector is the total number of atoms, the other components *A(i)* are defined by the number of atom pairs separated by a distance between *i* x *dx* and (*i+1)*x *dx* where *i* is an integer and *dx* an arbitrary distance between zero and d_{max} *(dmax* = maximum interatomic distance in the molecule). The 3D autocorrelation vector is endowed with outstanding properties, since it achieves a substantial reduction of the conformational data and is easy to calculate, it is translationally and rotationally invariant but sensitive to conformational changes and can be subjected to multivariate statistics such as principal component and cluster analysis [16]. The vector components were obtained with the software TSAR (Oxford Molecular Group) with a distance increment of 0.5 Å.

Principal Component Analysis (which reduces large number of descriptors to a smaller number of orthogonal variables with decreasing information content [22]) was used to visualize the components of the autocorrelation vector in a 2D or 3D representation which contains as much of the original information as possible. To this end, the original variables were normalized and the covariance matrix diagonalized using the classical Jacobi transform routine. Hence, e.g., the projections of the autocorrelation vectors in the plane of the two principal components are representative of the exploration of the conformational space by the investigated molecule during the dynamic simulation.

Cluster Analysis.- In order to detect possible fluctuations around finite sets of privileged conformations, cluster analysis [23] was performed, which helped to retrieve the observed conformation classes and to calibrate the dynamic simulation.

Similarity Indices. In order to complement the qualitative visual comparison of the point clouds in the projection planes after principal component analysis, the data were compared quantitatively using two exploration/similarity indices: the hypervolume ratio I(n) of the clouds and the cloud interdistance D(n), where n indicates the polymer size (n=3 for tripeptide or pseudotripeptide). These quantities were estimated as follows:

$$I(n) = V'(n)/V(n) \qquad (1)$$

where V'(n) and V(n) are the hypervolumes (dimension i = number of components of the autocorrelation vector) of the conformational space explored by the reduced and natural peptide, respectively, while D(n) was simply the distance in the hyperspace of the centers of gravity of the two clouds. The lower these indices, the higher the similarity. These two joint informations allow one to estimate the extent and the homogeneity of the coverage of the conformational space.

Results. - ***Ala-Ala s-cis and s-trans Conformational Spaces.*** - The 3D autocorrelation vector was first applied to the analysis of the dipeptide Ala-Ala, in order to check the potency and sensitivity of the method. The s-cis and s-trans isomers of the peptide bond were chosen as the starting conformations, of which only the s-trans isomer is generally found in natural peptides and proteins. The minimized 3D-structures of the two isomers obtained by the molecular mechanics (tree method) and the quantum mechanics (MOPAC 6.0 using the Hamiltonian AM1 and the option PRECISE) treatments, are shown in *Fig. 1*.

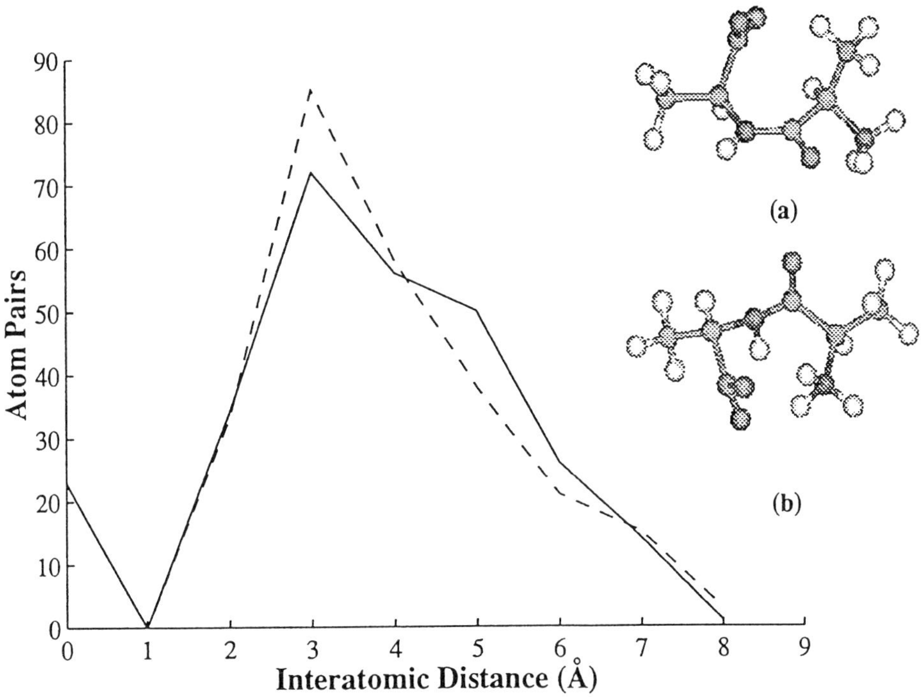

Fig. 1. *Autocorrelation vector of the s-cis (--) and s-trans (—) configurations of Ala-Ala as represented in balls and sticks (a) and (b), respectively.*

Despite the high similarity of the interatomic distance distribution of both s-cis and s-trans isomers (3D autocorrelation vectors using a 0.5 Å step), the conformational spaces of the two isomers projected in the principal plane (PC1 versus PC2, the two major principal components explaining 53% of the variance), occupy different portions of space in a symmetrical manner (*Fig. 2*). This expected result for the two isomers demonstrates that principal component analysis applied on the autocorrelation vector is a powerful tool to display the conformational landscape of flexible molecules.

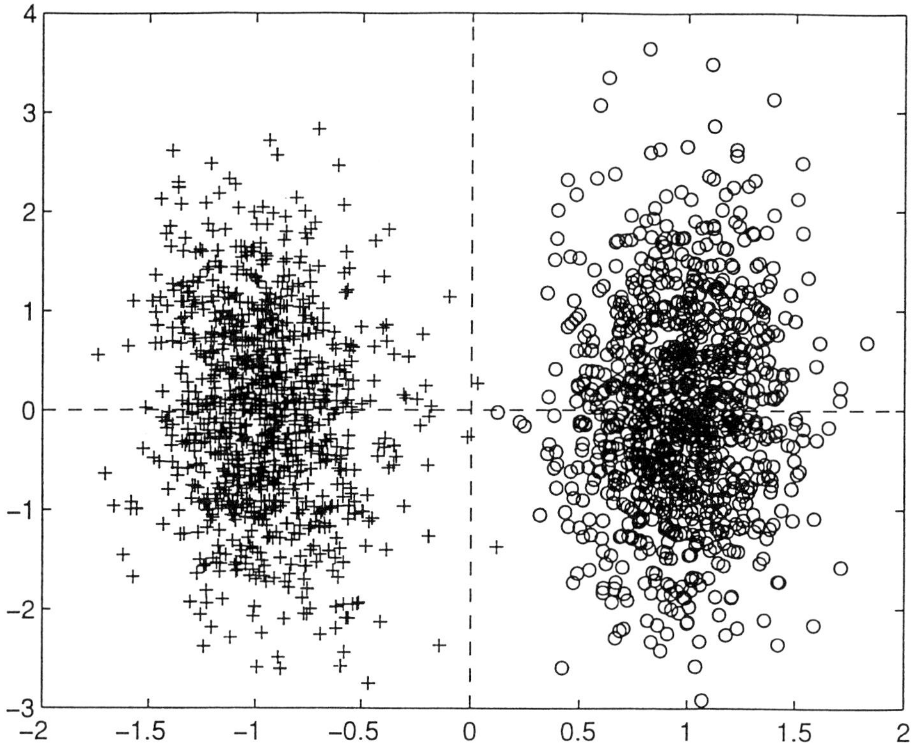

Fig. 2. *Projections on the PC1/PC2 plane of the conformations of the s-cis (+) and s-trans (o) isomers, after principal component analysis of Ala-Ala.*

Conformational Spaces of Peptides and Pseudopeptides. - In order to compare the conformational space spanned by similar compounds such as those found in peptide libraries, we have considered two series of peptides and pseudopeptides of increasing length listed in *Table 1*.

In the pseudopeptides, the peptide unit -CO-NH- was replaced throughout by its reduced counterpart -CH(OH)-NH-, which in contrast to the planar rigid peptide bond, exhibits free rotational behaviour around the C-N bond. Using the

Table 1. *Sequence of the Investigated Peptides or Pseudopeptides* [a])

number of residues	sequence
2	Ala-Ala
	Ala -Tyr
3	Ala-Ser-Ala
4	Ala-Ser-Ala-Ser
5	Ala-Ser-Ala-Ser-Tyr
6	Ala-Ala-Ala-Ala-Ala-Tyr
	Ala-Ser-Ala-Ser-Ala-Tyr
10	Ala-Ala-Ala-Ala-Ala-Ala-Ala-Ala-Ala-Tyr

[a]) pseudopeptides are throughout ψ[CH(OH)-NH]

same protocol for the two types of compounds, and following the same molecular mechanics and dynamics procedures (see methods), the full dynamic simulation was performed for each couple of molecules. In order to avoid an only partial description of the conformational space, a calibration was applied which ensured that the observed conformational clusters were walked over several times. The conformations produced during the equilibration phase were then discarded and the autocorrelation vectors calculated, retaining the same number of conformations for each couple peptide/pseudopeptide. The principal component analysis was then applied on the joint ensemble of autocorrelation vector components of the peptide and pseudopeptide couple. The results were then evaluated from visual inspection of the projections in the principal planes (*Fig. 3*) and from the values of the similarity indices I(n) and D(n).

For the calculation of I(n), the conformations generated for each molecule were enclosed in a parallelepiped of the same dimension as the autocorrelation vector and its hypervolume evaluated. I(n) was then calculated according to equation 1 for each couple peptide/pseudopeptide of the same size n. The value of I(n) tends to one when the conformational spaces explored by the peptide and dipeptide tend to be identical.

The values of the similarity index I(n) based on the volume ratio, as displayed in *Fig. 4*, first increase from the di-to the tetrapeptide, then decrease from the tetra- to the hexapeptide and stabilize for larger peptide sizes (only one decapeptide investigated).

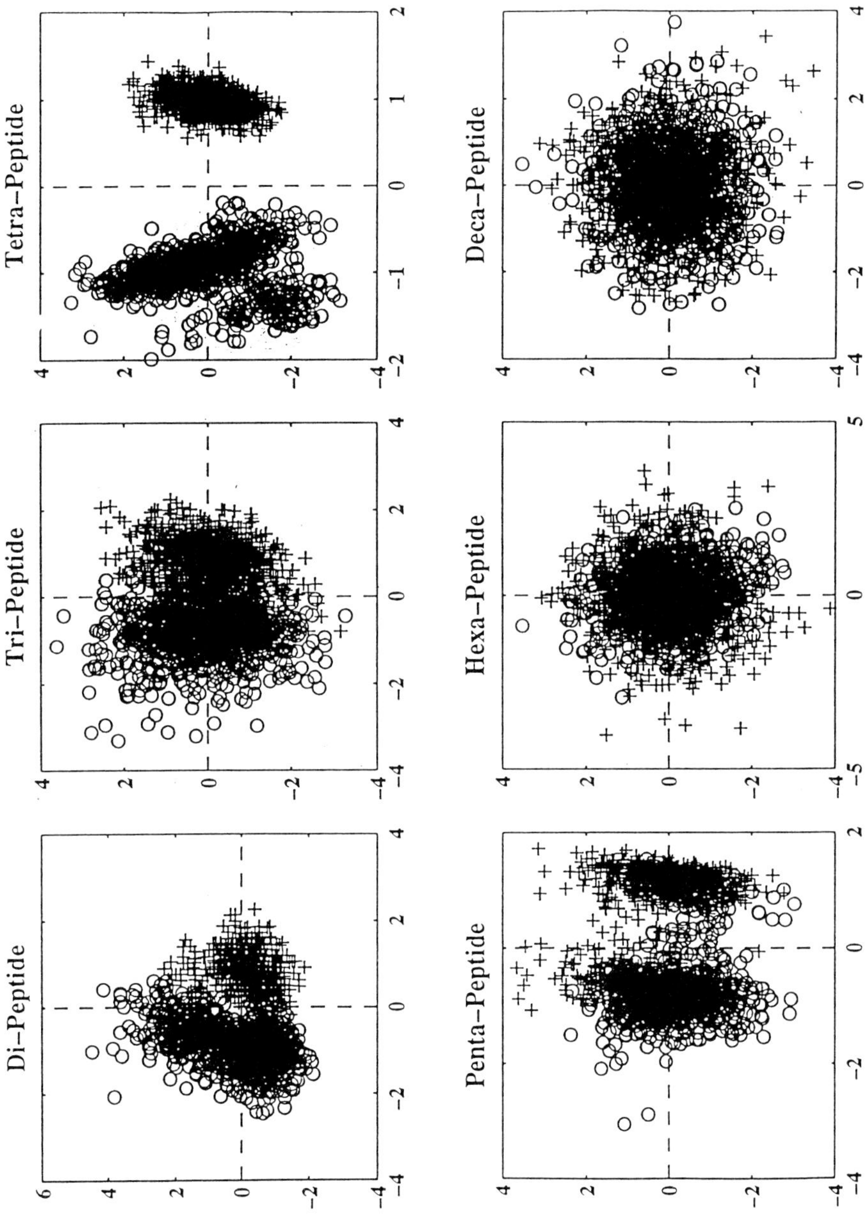

Fig. 3. *Projections on the PC1/PC2 plane of the conformations of the peptides (+) and pseudopeptides (o) after principal component analysis of the dynamic data.*

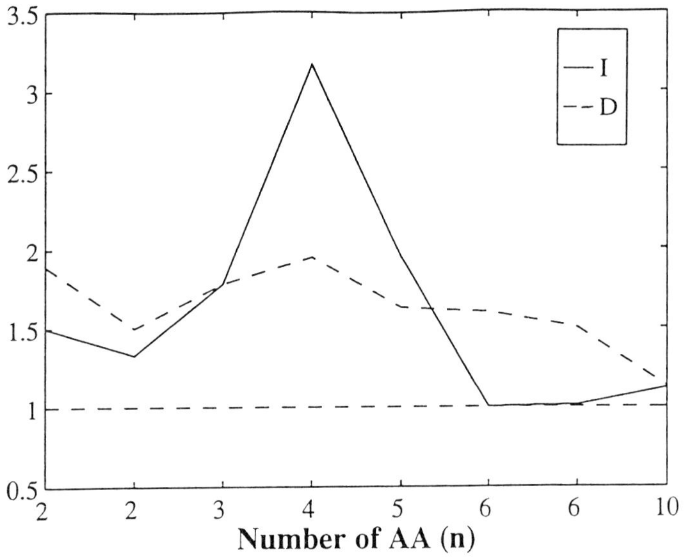

Fig. 4. *Variation of the exploration/similarity indices I(n) and D(n) in function of peptide size n.*

Strikingly, the ratio I(n) nearly reaches unity at the level of the hexapeptide. Similarly, the values of the index D(n), the distance of the centers of gravity of the conformational cloud (in the hyperspace of the same dimension as the autocorrelation vector), walk over a maximum for the tetramers, decrease to dipeptide/pseudodipeptide values for hexamers and reaches a low value (around one) for the decapeptide (*Fig. 4*).

In contrast to peptides and pseudopeptides, libraries of compounds based on heterocyclic scaffolds decorated with combinations of a number of substituents, do not cover homogenously the available conformational space. The above method applied to benzodiazepine and steroid librairies leads to very discrete distributions of the conformations in spite of relatively number of analogs considered in each series. As an example, a series of 156 steroid-like compounds acting on the same receptor [24, 25] have been evaluated. *Fig. 5* shows that the normalized coverage of the 3D space by the 156 analogs is extremely low compared to that of one single dipeptide Ala-Ala.

Discussion. - The purpose of the present study was to evaluate the three-dimensional diversity of peptide libraries compared to similar collections of more flexible pseudopeptides and more constrained polysubstituted heterocyclic compounds. To this end the dynamic behaviour of series of peptides and pseudopeptides from the di- to the decapeptide size was investigated. The

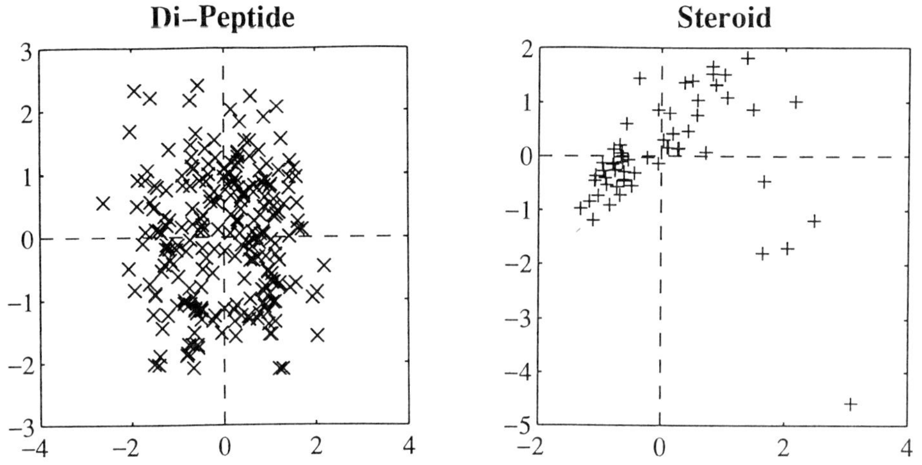

Fig. 5. *Conformational space explored by the dipeptide Ala-Ala compared to the total space explored by a series of 156 parent steroid compounds.*

dynamic trajectories were expressed by the autocorrelation vector for each retained conformation and the results made readable as projections in two- or tridimensional maps after principal component analysis. Comparison between compound classes was further quantitized by two exploration/similarity indices. From a practical point of view, these methods appear as powerful, since they strongly reduce the complexity of the dynamic data and greatly improve their analysis and representation.

The results indicate that 1) the investigated natural dipeptides strongly differ from their pseudopeptide counterparts since, despite comparable size, they cover a considerably smaller portion of the conformational space. Dipeptide libraries are likely to be poor collections in terms of three-dimensional diversity, mainly due to the peptide bond rigidity; 2) the investigated natural hexapeptides or longer peptides homogenously explore the conformational space (covered by their most extended form) as seen from the projections in the principal planes. Hexapeptide libraries are likely to display a large three-dimensional diversity, a desirable property in view of the unlimited number of biological targets used in screening; 3) differences between peptides and pseudopeptides become less apparent with increasing size and become negligible at the hexamer level, as seen from the exploration/similarity indices.

Further investigations will be needed to evaluate the sequence dependence of our results and to complement the present purely steric study by an estimation of the spatial explorative power of compound collections in terms of lipophilicity and electrostatic potentials. However, our observations seem to confirm that full libraries of hexapeptides, (which in contrast to pseudopeptides are easily synthesised in an automated combinatorial way [26]), exhibit in addition to a relatively large fingerprint and functional diversity, a large spatial exploratory power compared to scaffold-based heterocyclic libraries and are convenient sources of ligands for a variety of receptor or enzyme targets.

REFERENCES

[1] R.A. Houghten, C. Pinilla, S.E. Blondelle, J.R. Appel, C.T. Dooley, J.H. Cuervo, *Nature* **1991**, *354*, 84.

[2] J.K. Scott, G.P. Smith, ***Science*** **1990, *249***, 386.

[3] K.S., Lam, S.E. Salmon, E.M. Hersh, V.J. Hruby, W.M. Kazmierski, R.J. Knapp, *Nature* **1991**, *354*, 82.

[4] H.M. Geysen,., Rodda, S.J., Mason, T.J., Tribbick, G., Schoofs P.G., *J. Immunol. Meth.* **1987,** *102*, 259.

[5] E.J. Ariëns, in "Trends in Drug Research" , Ed. V. Claassen, Pharmacochemistry Library, Ed. H. Timmerman, Elsevier, Amsterdam, 1990, Vol. 13, pp. 1-10.

[6] D.C. Horwell, W. Howson, G.S. Ratcliffe, D. Rees, *Bioorg. Med. Chem.Lett.* **1993**, *3*, 799.

[7] A.L. Fauchère, J.A. Boutin, *Trends Pharmacol. Sci.* **1996**, *17,* 8.

[8] E.M. Gordon, R.W Barrett,., W.J. Dower, S.P. Fodor, M.A. Gallop, *J. Med. Chem.* **1994**, *37*, 1385.

[9] A.F. Spatola, in: "Chemistry and Biochemistry of Amino Acids, Peptides and Proteins". Ed. B. Weinstein, M. Dekker, New York, 1983, Vol. 7, pp. 267-357.

[10] J.L. Fauchère, in " Computer-Aided Drug Design in Industrial Research", Eds. E.C. Herrmann, R. Franke, Springer, Berlin, 1995, pp. 129-161.

[11] E.J. Martin, J.M. Blaney, M.A. Siani, D.C. Spellmeyer, A.K. Wong, W.H. Moos, *J. Med. Chem.* **1995**, *38*, 1431.

[12] D.E. Patterson, R.D. Cramer, A.M. Ferguson, R.D. Clark, L.E. Weinberger, *J. Med. Chem.* **1996**, *39*, 3049.

[13] R.D. Cramer, R.D. Clark, D.E. Patterson, A.M. Ferguson, *J. Med. Chem.* **1996**, *39*, 3060.

[14] Broto, P., Moreau, G., Vandycke, C. *Eur. J. Med. Chem.* **1984**, *19*, 66.

[15] G. Grassy, R. Lahana, in "Trends in QSAR and Molecular Modelling 92", Ed. C.G. Wermuth, Escom, Leiden, NL, 1993, pp. 216-219.

[16] J. Haiech, T. Koscielniak, G. Grassy, *J. Mol. Graphics* **1995**, *13*, 46.

[17] Wagener, M., J. Sadowski, J. Gasteiger, *J. Am. Chem. Soc.* **1995**, *117*, 7769.

[18] N.L. Allinger, *Adv. Phys. Org. Chem.* 1976, *13*, 1.

[19] J.J Stewart, *J. Comp.-Aided Mol. Des.* **1990**, *4*, 1.

[20] P.K. Weiner, P.A. Kollman, *J. Am. Chem. Soc.* **1984**, *106*, 765.

[21] A. Yasri, L. Chiche, J. Haiech, G. Grassy, *Protein Engineering* **1996**, in press.

[22] P.H. Weiner, E.R. Malinowski, A.R. Levinstone, *J. Phys. Chem.* **1970**, *74*, 4537.

[23] B. Everitt, in "Cluster analysis", Wiley, New York, 1974.

[24] G. Grassy, P. Trape, J. Bompart, B. Calas, G. Auzou, *J. Mol. Graphics* **1995,** *13*, 356.

[25] M. Claire, H. Faraj, G. Grassy, A. Aumelas, A. Rondot, G. Auzou, *J. Med. Chem.* **1993**, *36*, 2404.

[26] J.L. Fauchère, J.M. Henlin, J.A. Boutin, *Canad. J. Physiol. Pharmacol.* **1997**, *75*, in press.

PART II. **LEAD FINDING STRATEGIES**

Chapter 14

CATALYST Pharmacophore Models and Their Utility As Queries for Searching 3D Databases

by **Peter W. Sprague** and **Remy Hoffmann**

Molecular Simulations Inc., 13 Harbourton Ridge Rd.,
Pennington, NJ 08534-371213, USA

CATALYST Pharmacophore Models and Their Utility As Queries for Searching 3D Databases

by **Peter W. Sprague** and **Remy Hoffmann**

Molecular Simulations Inc., 13 Harbourton Ridge Rd., Pennington, NJ 08534-371213, USA

The experimental approach of screening large collections of compounds for activities in new biological assays has also been effective historically in identifying new leads [3]. Experimental methods are expensive and time consuming so considerable effort has been invested in developing systems to accurately express molecules electronically in organized databases [4] that are suitable for rapid searching. Most recently, these efforts include the ability to search large 3D electronic databases with abstract pharmacophoric queries [5]. Our purpose in this chapter is to review the technology developed at MSI [6] to enable the computation of abstract pharmacophoric queries, build large indexed 3D databases of compounds, and provide filtering and searching capabilities with these constructs.

Introduction. - Pattern recognition has long been employed in a qualitative sense by medicinal chemists in search of new lead compounds. The "chemical intuition" that so often plays a decisive role in the creative process of drug discovery is rooted in a chemist's ability to recognize similarities between structurally different molecules. While our native ability to recognize patterns is quite powerful, it is easily saturated when the numbers of compounds and possible interrelating patterns becomes large. Thus, the need for an automated pattern recognition system suitable for the analysis of structure activity data encompassing tens or hundreds of compounds. Such systems can be viewed as extensions of the traditional QSAR methods [1] that go beyond attempts to relate physical properties of compounds with activity by incorporating more abstract computed properties such as pharmacophores [2] into the analysis.

The Catalyst^{TM} software suite consists of three modules, a conformational search program (catConf/conFirm), a pharmacophore, or hypothesis, generation program (catHypo/hipHop), and a database building program (catDB). CatConf computes an electronic representation that emphasizes coverage of conformational space [7]. CatHypo computes chemical feature-based [8] pharmacophore models from a training set of compounds with activity data. HipHop computes models that contain all features in common across a training set of molecules without activity data. CatDB builds and administers compound databases and indices that can be searched with several kinds of queries including catHypo pharmacophore and compound shape models.

Catalyst Models. - Catalyst models, termed hypotheses [9], are constructed from abstract definitions of five kinds of ligand-receptor binding interactions. These are, hydophobic, hydrogen bond donating (HBD) and accepting (HBA), salt bridges and aromatic ring pi stacking interactions. These "features" are located relative to each other in coordinate space as points surrounded by a tolerance sphere. Each sphere represents the region in space that should be occupied by chemical functionality capable of the kind of interaction specified by the feature type. Thus, a positive ionizable sphere would translate to a functional group such as a basic amine that would be expected to be protonated at physiological pH. A collection of these spheres models the attributes of a molecule that are most important for binding of a ligand to its receptor. This model will be more or less accurate depending on how the relative positioning of each sphere is computed and on the relative importance (weighting) assigned to each sphere. Construction of a Catalyst hypothesis is accomplished either manually, using positioning data from Xray crystallography or NMR experiments, or automatically using catHypo which attempts to infer feature types and relative positions from structure-activity data. Examples of both kinds of procedures will be presented in the following section.

From Structure-Activity Data. - Automatic computation of a chemical feature-based model starting with structure-activity data can be successful provided that two assumptions about the data can be made. First, that all of the active compounds in the set bind to the same receptor in roughly the same fashion. Second, compounds that can have more binding interactions with the receptor are more active than those with fewer. This second assumption also means that less active compounds are expected to be smaller and possess fewer chemical features than more active compounds. In many structure-activity data sets, there may be large, perhaps even feature rich, molecules that are not very active. This could be caused by an adverse steric interaction between the receptor and a portion of the molecule that prevents actual mapping to the pharmacophore. Current versions of Catalyst do not consider adverse steric interactions during automatic pharmacophore generation and will always overestimate the activities of these compounds so they are better left out of the analysis. If information about receptor constraints is available, it can be expressed in the model by adding exclusion spheres manually after the automatic generation step.
Training set selection is key to a successful application of the catHypo analysis. Basically, one attempts to populate the training set with structurally diverse examples that vary in activity over a range as close to four orders of magnitude as possible. Each compound is selected to provide as much unique information in terms of structure or activity as is possible. It is important to avoid structural redundancy because this confounds the statistical apparatus used to correlate predicted and measured activities during the analysis. Hypotheses from the analysis are validated mathematically and visually. The best hypotheses will estimate the activities of the training set members correctly to within a order of magnitude, and will pass standard statistical tests for non-randomness [10]. Catalyst provides interface tools for visualizing how compounds fit to the hypothesis, it selects the best fitting conformer and displays this overlaid on the hypothesis. Using this tool, the medicinal chemist can decide which fits are meaningful, and thus which hypotheses best model what is known about the structure-activity. Finally, the predictivity of the best hypotheses can be established by determining how well activities of non-training set members are estimated. Two examples of hypotheses derived in this way are presented below.

ACE Training set. The angiotensin converting enzyme (ACE) model was derived from a training set of di and tripeptides that were active over a range of four orders of magnitude [11]. The ACE hypothesis shown in *Fig. 1* was found to be reasonably predictive for compounds outside the training set. For example, this hypothesis correctly predicts that captopril and enalapril will be active ACE inhibitors and that the enantiomer of captopril will not. This hypothesis is thus a suitable query for discovering potential ACE inhibitory compounds in 3D compound databases.

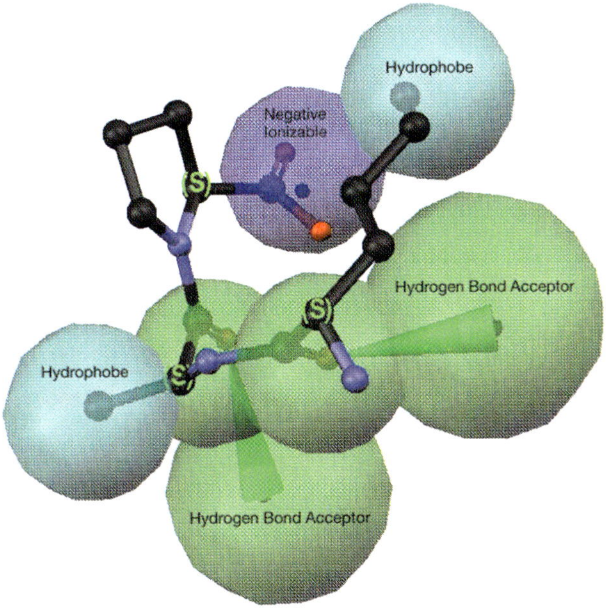

Fig. 1 *The ACE hypothesis mapped to nLeuAlaPro*

PFT Training Set. The protein farnesyl transferase (PFT) model was deduced from a training set of 20 CaaX box tetrapeptides [12] that also varied in activity over a range of four orders of magnitude. The hypothesis in *Fig. 2* predicts that of the three potential hydrogen bond forming carbonyl oxygens in CVFM, only the Phe-amide is required for activity. The peptide-related inhibitor L-731,735, which lacks all other carbonyl oxygens (*Fig. 3*), provides experimental verification for this prediction [13]. Compounds that can present chemical features in the relative orientation suggested by this hypothesis should therefore exhibit PFT inhibitory activity. Note that the hypothesis does not require any specific backbone connection, only the presence of appropriately positioned chemical features. This independence from connectivity makes the model useful for finding potentially active compounds with quite different overall structures when used as a database search query.

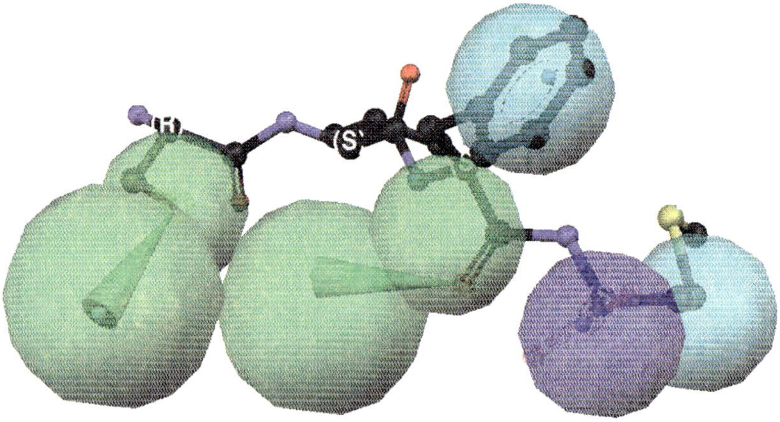

Fig. 2 *The PFT hypothesis mapped to CVFM*

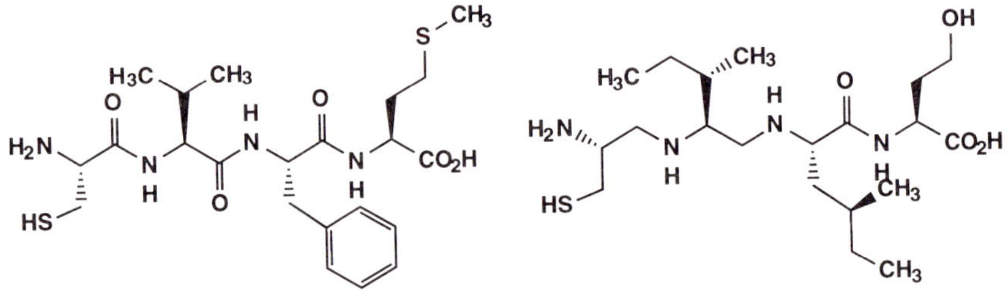

CVFM I_{50} = 10nM   L-731,735 I_{50} = 18nM

Fig. 3 *Structures of CVFM and L-731,735*

From Target Structure Data. - If one is fortunate enough to have Xray crystallography structure data for a target receptor containing a bound ligand, translation of this rich information into a Catalyst database query is straight forward. If only the receptor structure is known, or partially known, it is still possible to extract relevant information that can be used to build database queries. Both procedures will be described with examples below. Information about the receptor can also be transferred to Catalyst queries by appending excluded volumes that represent regions where ligands are not allowed. In this way, the query can express important attributes of both the ligand and its receptor.

Exact ligand-receptor data (PHIV Xray). The HIV protease research area is blessed by much Xray crystallography data including many examples of bound inhibitors [14]. We used the bound conformer of SB-203386 from one such study (1sbg.pdb) as a template for preparing a Catalyst database search query [15]. The coordinates for all heavy atoms were imported in mol format and tools within the Catalyst interface were used to manually position chemical features along the molecule. The intermediate result, that contains all possible binding interactions (*Fig. 4A*), can be converted directly to a query (*Fig. 4B*), or modified to include an interesting subset of the features (*Fig. 4C*).

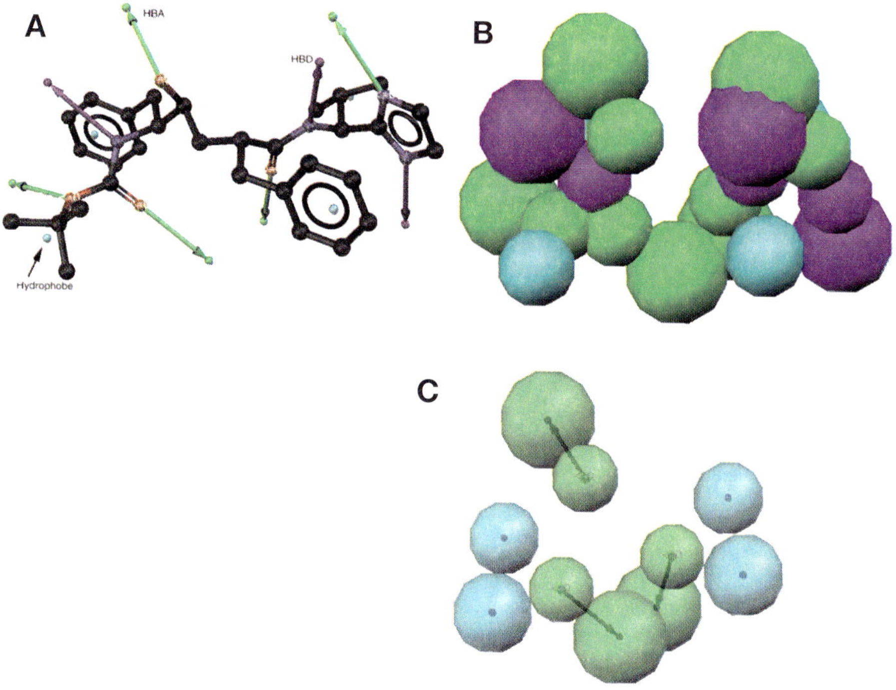

Fig. 4 ***A***, *SB-203386 with positioned features;* ***B***, *The corresponding Catayst feature-based model;* ***C***, *A selected subset of chemical features.*

LUDI Models of Receptor Data. Proteins also constitute a source of information from which a feature-based query can be derived. Human Leucocyte Elastase (HLE: EC 3.4.21.37) is a serine protease produced by polymorphonuclear leukocytes upon inflammatory stimuli that has been implicated in the development of emphysema and rheumatoid arthritis. In this enzyme, the amino acids Ser195, His57 and Asp102 constitute the catalytic triad [16]. The crystal structure of HLE at a 2.3 Å [17] resolution was used as starting point for analysis within Insight 95 [18]. LUDI [19] was designed to automatically construct ligands for enzymes, where the 3D structure is known, from molecular fragments. This program was used to search the ACD database [20] for all the possible fragments and/or molecules that could interact with hydrophobes, HBAs or HBDs in the protein within a 10 Å sphere around the centroid of the active site. During this procedure, a map of all the possible interaction sites was created and visualized. Within the 10 Å sphere, the different residues show a variable density of possible interaction sites depending on the number of fragments found in the database that interact with each of these residues. The experiment suggests that HBA interactions with Ser195 and Val216, HBD interactions with Cys191, His57 and Phe41 and hydrophobic interactions with His57 and Phe192, are the most significant interactions (*Fig. 5a*). For each of these residues, the x,y,z coordinates of the average interaction site was used (considering also the x,y,z coordinates of the atom on the protein for hydrogen bond donating or accepting interactions) to derive, using Catalyst, the seven feature hypothesis shown in *Fig. 5b*.

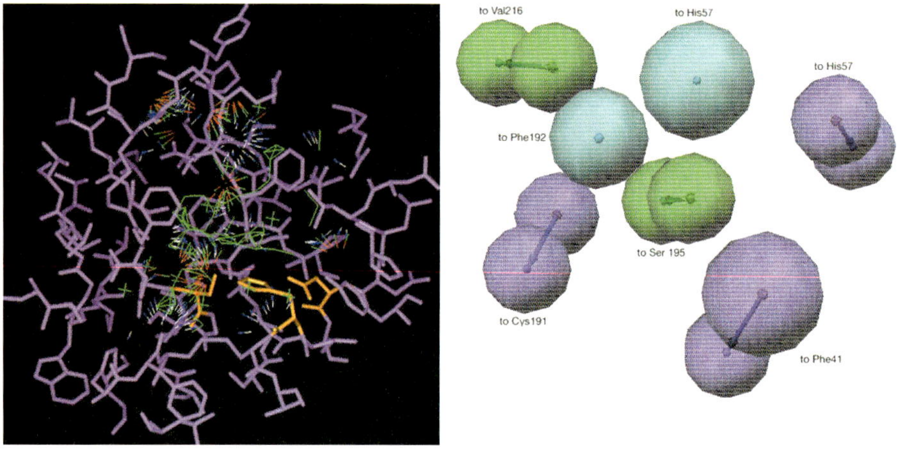

A B

Fig. 5 *From Interaction sites to Chemical Feature based models*
***A** Interaction map (Green: Hydrophobic Red: HBA Blue: HBD). The catalytic amino acids are colored in orange. **B** Seven feature hypothesis derived from Ludi interaction map (Cyan: Hydrophobic, Green: HBA, Magenta: HBD).*

This hypothesis was used as a 3D query to search the Derwent World Drug Index database [21]. Because a molecule must fit all the features of a Catalyst query to be retrieved as a hit, this query was far too complex and returned no hits. The complexity of the query was reduced by creating new hypotheses that are subsets of the initial hypothesis. In these hypotheses, interactions with the Ser195 and His57 side-chains from the catalytic triad were conserved. Two six-feature hypotheses, one missing the HB Donor interaction with Cys191 (**HYPO1**), and one missing the HB Acceptor interaction with Val216 (**HYPO2**), were constructed. In addition, two five-feature hypotheses, one missing the HB Donor interactions with the backbone carbonyls of Phe41 and His57 (**HYPO3**), and one missing the HB Donor interaction with Cys191 and the HB Acceptor interaction with Val216 (**HYPO4**), were prepared. Derwent DB search results are shown in *Table1*.

Table1. *Number of Hits found with the Subset Hypotheses*

Hypothesis	Hits [a]	Hits with MW<600
HYPO1	145	51
HYPO2	19	3
HYPO3	769	319
HYPO4	1887	699

[a] The Fast Flexible Search method was used to search this database.

In order to focus on drug-sized molecules, we considered only hits with a molecular weight lower than 600. An example of a hit for each query is shown in *Fig. 6.* Note that none of these hits are peptides and therefore represent interesting candidates for biological testing.
These hits can be tested as potential HLE inhibitors in an appropriate assay. If any prove active, they can be used as starting points for lead development and evaluated against the full seven-featured hypothesis to look for chemical modifications that could be made in order to improve their fit.

The Chemical Feature Abstraction. - The chemical feature-based model promotes a new way of thinking about molecules and ligand - receptor interactions. Rather than attempting to keep track of all atoms and bonds in a molecule, one focuses on only those regions that are likely to be most important for intermolecular binding interactions. Because these regions, chemical features, are abstract definitions, they can represent many different kinds of molecules. This property makes chemical features very useful for dealing with the important drug development problem of how to translate activity from one chemical class of compounds to another.

Medicinal chemists are often faced with active lead compounds that exhibit undesirable properties that limit their drug development potential. The difficulties encountered with poor bioavailability in the case of renin inhibitors is a good example. Many research groups were able to identify very potent inhibitors of the renin enzyme that were structurally similar to the renin substrate. Most of these compounds were structurally similar to peptides and, perhaps for this reason, did not develop high enough blood levels after oral administration to be effective *in vivo* [22]. The time honored approach for working around such problems is to find active compounds that are as structurally different from the original leads as possible. An excellent tactic for executing this approach, is to construct a chemical feature-based model, representing the functionality that is important for activity, and use this model to search databases for new structural types.

What is the likelihood that a chemical feature-based construct truly models the types and relative positions of functional groups needed in a ligand for effective binding to a receptor? The answer to this question depends on the quality of information available to construct the model. If an Xray crystallography study fixing the position of a ligand to the receptor is available, then the probability is relatively high. On the other hand, if one has only structure-activity data available, even if it is an ideal data set, the probability is clearly much less because of the assumptions that make application of the Catalyst procedure possible. However, many modern drugs, for example ACE inhibitors, have been successfully developed in the absence of detailed ligand receptor binding information. If used as an adjunct to intuition, the Catalyst modeling method can provide important guidance and is a powerful tool for inexpensively testing ideas.

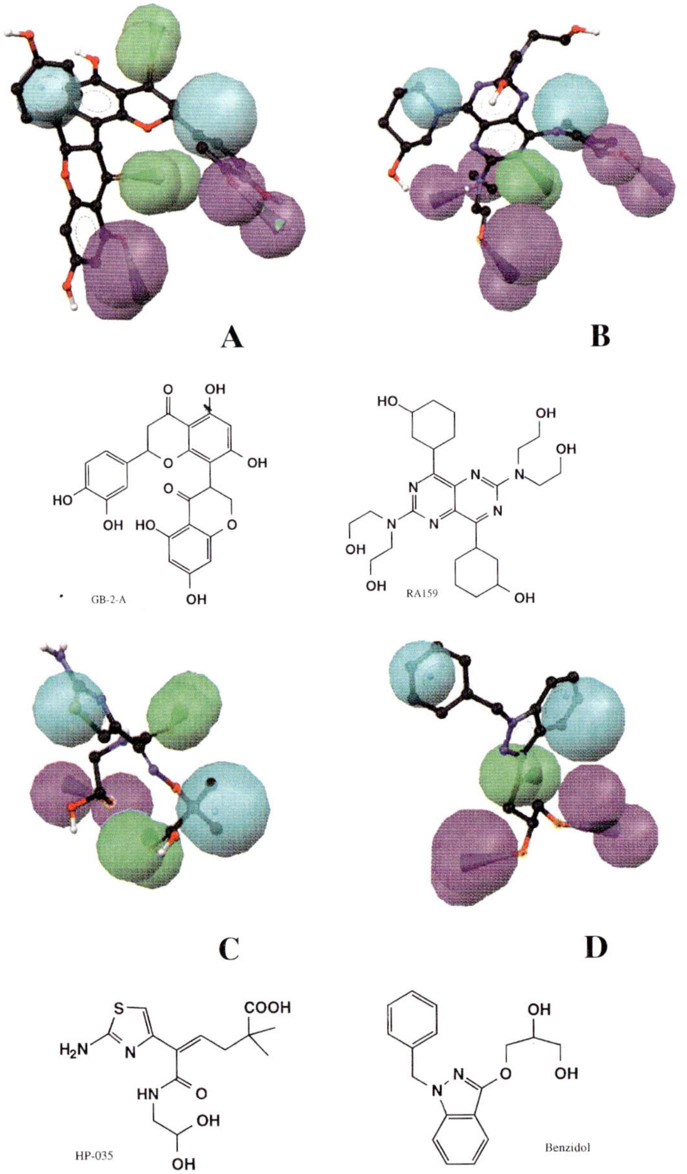

Fig. 6. *Different low molecular weight candidates for HLE inhibition.*
***A**. GB-2-A mapped on HYPO1, **B**. RA-159 mapped on HYPO2, **C**. HP-035 mapped on HYPO3, **D**. Benzidol mapped on HYPO4.*

Database Query. - A Catalyst hypothesis, consists of chemical feature objects that are defined in terms of molecular groups and fragments. A negative ionizable feature, for example, includes groups that are ionized at physiological pH such as carboxylic, phosphorus, and sulfur-based acids and more rare features like trifluoroacetylamides. The definitions are based on a list of examples that can be modified to include whatever special features the user has need of. Because Catalyst hypotheses are automatically database queries, this flexibility means that custom built queries can be focused broadly, or very narrowly, depending on the desired result. It is also possible to include in a query specification any one dimensional (1D) constraint for which there is information stored in the database. These constraints can be any alphanumeric data that is of interest, for example, molecular weight, CLOGP, biological activity, and compound cost. A 1D property unique to Catalyst is the Fit number, a computed value that represents how well the chemical features in a molecule superimpose to the features of a hypothesis. This value is computed by finding the mapping in a conformer that maximizes the following function:

$$Fit = \sum_{i=1}^{|F|} w_i \max\left[0, 1 - \left(\frac{\varepsilon_i}{t_i}\right)^2\right]$$

where $|F|$ is the number of features in the hypothesis, w_i is the weight of feature i, e_i is the error to feature i (the distance between the location of the hypothesis feature and the relevant part of the molecule), and t_i is the tolerance of feature i.
Individual, or combinations of such constraints are very useful for limiting database hit lists as will be discussed.

3D Database Searching. - Recently, construction of virtual databases containing high quality three dimensional representations of molecules has become possible and methods for searching using chemical feature-based queries have been perfected [23]. It is now practical to sort through several hundred thousand compounds with sophisticated 3D queries in a matter of minutes. This capability has generated new tactics for identifying drug discovery leads and promises to have a major impact on how drug discovery programs are carried out in the future [4]. In this section, we discuss one way in which such databases can be prepared and provide examples of their use.

Database Construction. The major problems that must be addressed in 3D database construction are conformational flexibility and search performance. The first impacts on the effectiveness of a search experiment; if the molecules flexibility is inadequately expressed, no really relevant output can be expected. The second, while not critical for scientific accuracy, is extremely important for the success of a searching experiment. If one cannot get results within a reasonable period on the human time scale, the experiment may never even be attempted.

Conformational Flexibility Problem. There are two fundamentally different approaches to covering conformational flexibility that have appeared in contemporary 3D databases. These are storage of molecules as sets of different conformers that represent the conformational space open to the molecule [7], and on-the-fly conformational search of molecules stored as individual conformers [24]. Both methods have strengths and weaknesses and both require scientific compromises in order to be practical.

Regardless to the approach, a conformational search must be done. Most contemporary search algorithms require generation of large numbers of similar conformations followed by a clustering and selection operation to remove close duplicates [25], This would seem to favor the on-the-fly paradigm because conformational search could be focused on only those compounds that appear interesting and database storage requirements would be minimized. On the other hand, if conformational search could be made efficient, comprehensive, and rapid, storage of multiple conformers of molecules in a database could be practical. With multiple conformer storage, extensive indexing becomes possible and that can make searching very efficient. Also, in contrast to the on-the-fly method, extra computational time can be invested to ensure good conformational coverage.

Work at BioCAD [26] has shown, that for drug sized molecules with no more degrees of freedom than found in tetrapeptides, conformational space can be represented quite well [7] with a few hundred conformers generated using a poling algorithm [27]. Furthermore, the storage requirements for large numbers of molecules modeled in this way can be reasonable. For compound collections like the All Chemicals Directory (ACD), the average number of conformations required is about 33 and the storage space per molecule is about 4.6 Kbytes. A Catalyst multiple conformer ACD database containing 166,705 compounds (5.53 million conformers) and including indices and 1D data occupies 1.45 Gbytes of disk storage; quite practical on todays inexpensive multi-gigabyte disks. Construction time for such a database is extensive, but not forbiddingly so. Current versions of Catalyst build 7-10,000 multi-conformer compound models per cpu day using a SGI R4400 equipped machine running at 150 MHz. Thus, the ACD database, including indices, can be built in one week on a four processor SGI Challenge machine.

In order to achieve practical build times, the quality of conformational coverage is adjusted to the minimum necessary to ensure reasonable performance. The use of tolerance spheres around the central coordinate for each chemical feature in a Catalyst query ensures that searches will succeed in finding relevant hits among these very terse conformational models [5].

Databases with higher quality conformational coverage are possible but require much longer build times. For example, the best quality conformational models that Catalyst can produce require build times 1400 times longer than the default database build method.

Performance Problems. The amount of time available for experiment is always limited and may even dictate the types of experiments that are attempted. Therefore, it is important to design into any database system as much efficiency as possible Database build times are only part of the problem, there are search time requirements that are perhaps even more critical. Catalyst databases were designed from the outset to provide very fast search retrieval times. This is made possible by extensive indexing of all stored properties, including chemical feature types and locations. As a result, search times for four or five featured queries are typically less then ten minutes for databases as large as the ACD.

Hits lists from pharmacophore searches can be very large, perhaps tens of thousands, so there is an obvious need for automatic filtering of these lists. This is most conveniently done by attaching 1D constraints to the queries. In adddition to fixed constraints like molecular weight ranges, one can select fit numbers that express how well a molecule must fit to the query in order to be selected as a hit. We will illustrate the effect of various 1D constraints in the examples section.

Kinds of Databases. Compound collections typically encountered in the pharmaceutical environment range from small sets (less than 1000) of compounds of immediate interest, to large archival (100,000 or more) sets that are of occasional interest. We make no distinction between these types, all are amenable to the same construction and searching treatment.

The recent focus on combinatorial chemistry methods has defined another kind of compound database, the virtual library. Virtual libraries are simply databases containing, in electronic form, all of the compounds that would result from applying a specified chemistry to lists of reagents and core structures. For any given chemistry and reagent set, this number can be so large that actual synthesis of the entire library is impractical, even with modern automated methods. Because synthesis of a random subset might well result in missing large areas of compound space, some logical means for subsetting the theoretical virtual library into a representative and manageable experimental library is necessary. The use of a pharmacophore model that represents desired chemical functionality is clearly a logical way to filter theoretical libraries. In the Catalyst paradigm, this reduces to building the multi conformer database from the output of a compound enumerator program such as the Cerius^2 Analog Builder [28], and searching this library with a chemical feature-based query. The filtered hit list becomes the experimental library.

Searching Strategies. When a large database containing over 100,000 compounds is searched with a simple pharmacophore query that might contain three or four simple features, the most likely outcome will be very large hit lists of 10,000 or more compounds. The user is then faced with a generally unpleasant task of evaluating these huge lists. There are two basic next step procedures that can be employed. One can look at each compound, a truly Herculean task that is beyond the patience of most, or one can modify the query to find only a subset of the most interesting compounds. This later approach is well supported by tools within Catalyst. The number of hits necessary to guarantee that no interesting compounds are missed in a search can easily conflict with the number that is practical to evaluate. What is a reasonable compromise? To help answer this question, we searched the Derwent World Drug Index (48,405 compounds) with our ACE inhibitor hypothesis knowing beforehand how many distinct ACE inhibitors were in the database to be found [29]. The results are presented in *Table 2*.
In this table, we compared how well the ACE query performs in finding known ACE inhibitors in the database depending on how different query attributes are varied. Using the query shown in *Fig. 1* and default attribute settings, 55 % of a standard set of ACE inhibitors known to be in the Derwent World Drug Index are present in the hit list after performing a Best Flexible search. The ACE inhibitors make up 2.9% of the hit list, representing a 17 fold enrichment over what is in the Derwent database. We used the enrichment factor to measure the selectivity of a given query compared to a browse of the entire database (row 1 of table). Because this query does not find 45 % of the list compounds, we experimented with two approaches for increasing the number of hits returned from a search. In the first, we increased the size of the tolerance spheres for each feature from 1.6 Å to 2.0 Å for heavy atoms, and from 2.2

Å to 2.6 Å for projected points; an increase of 20%. This tactic results in a distinct improvement in the number of ACE inhibitors found (compare query ace1.1F0 with ace1.1EF0). In the second method, we removed the projected points from the hydrogen bond acceptor features of the query. This, removes the directional constraint from these features and makes the overall query easier to satisfy. This tactic also results in more ACE inhibitors found (compare ace1.1F0 with ace1.1NPF0). The two tactics are somewhat additive and, when both are employed simultaneously, more ACE inhibitors are found (compare ace1.1F0 with ace1.1ENPF0). A few more of the standard list of ACE inhibitors are found when the tolerance sphere tolerance for heavy atoms is further increased by 40% to 2.4 Å, (compare ace1.1F0 with ace1.1E2NPF0).

Table 2 *ACE Query Search Results*

Query[1]	**Total Hits**[2]	**ACE Hits**[3]	**% ACE**[4]	**% Hit list**[5]	**enrich**[6]
MW<700	43089	87	100	0.2	1
ace1.1F0	1681	48	55	2.9	14.5
ace1.1F3	765	27	32	3.5	17.5
ace1.1F6	78	6	7.2	7.7	38.5
ace1.1EF0	2852	61	70	2.1	10.5
ace1.1EF3	1833	52	60	2.8	14
ace1.1NPF0	2614	55	63	2.1	10.5
ace1.1NPF3	1843	47	54	2.6	13
ace1.1ENPF0	3540	64	74	1.8	9
ace1.1ENPF3	2910	60	69	2.1	10.5
ace1.1E2NPF0	4161	66	76	1.6	8
ace1.1E2NPF3	3649	65	75	1.8	9

[1] Query, name of the query used showing origin (ace1.1), fit number minimum (F), non default tolerance sphere size (E, heavy atom = 2.0 Å, projected point = 2.6 Å; E2, heavy atom = 2.4 Å), absence of projected point in hydrogen bond acceptor feature (NP). all queries include a 1D filter to reject any hit with molecular weight > 700.
[2] Number of hits found [3] Number of ACE inhibitors found. [4] Percentage of standard list.
[5] Percentage of total hits found. [6] Enrichment factor.

To shorten the hit list, we increased the fit number parameter attached to the ACE query. The fit number refers to how well a hit conformer must map its chemical features to the query chemical features. The default setting is 0 meaning that any calculated fit not less than 0 will qualify, and is intended to provide hit lists where the number of false negatives is minimized at the expense of some false positives. Each Catalyst chemical feature-based query has a calculated maximum fit number. This is the value returned when all chemical features in a molecule superimpose exactly on the chemical features in a query, and is equal to the sum of all query feature weights. For example, the maximum fit number for the ace1.1 query is 9.5. We used a value approximately one third of this maximum for query ace1.1F3. The effect of this maneuver was to increase the selectivity of the query at the expense of finding as many of the possible legitimate hits. This effect increases as the fit number parameter is increased (compare ace1.1F3 and ace1.1F6). The user can thus tune a query to provide hit lists that bias the search toward thoroughness or selectivity.

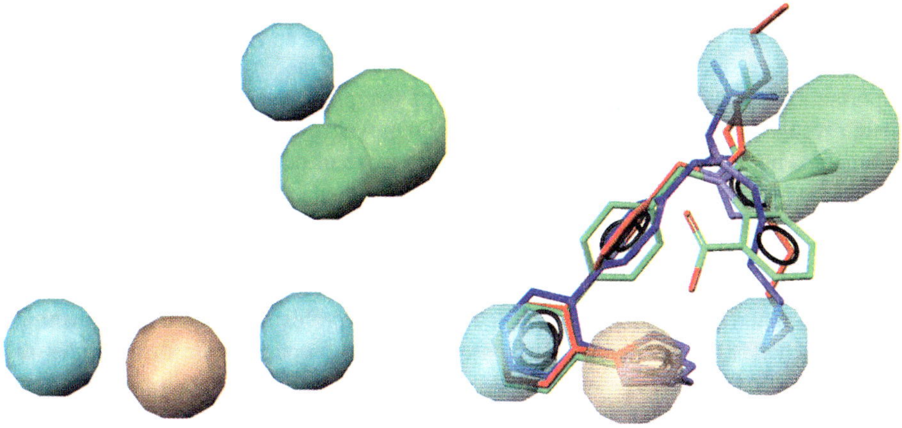

Fig. 7 *The angiotensin II antagonist query at left mapped to candesartan (green), valsartan (blue), and forasartan (red) on the right.*

Table 3 *Angiotensin II Antagonist Query Results*

Query[1]	Total Hits[2]	AII Hits[3]	% AII[4]	% Hit list[5]	enrich[6]
MW<700	48405	154	100	0.32	1
AII4.32F0	1004	134	85	13.3	37
AII4.32F5	462	100	65	21.6	60
AII4.32EF0	1725	149	97	8.6	24
AII4.32EF5	997	137	89	13.7	38

[1] Query, name of the query used showing origin (AII4.32), fit number minimum (F), non default tolerance sphere size (E, heavy atom = 2.0 Å, projected point = 2.6 Å), all queries include a 1D filter to reject any hit with molecular weight > 750. [2] Number of hits found [3] Number of AII antagonists found. [4] Percentage of standard list. [5] Percentage of total hits found. [6] Enrichment factor.

We present results from a search of the Derwent World Drug Index (*Table 3*) with an angiotensin II antagonist query 30 shown in *Fig. 7*. These results are somewhat better than in the ACE inhibitor experiment. The default query AII4.22F0 found 85% of the non-peptide antagonists known to be present in the database. This percentage was increased to 97 by expanding the query feature tolerance spheres by 20%. As before, applying a fit number constraint (maximal fit number for the AII

[25] M. Lipton, W. C. Still, *J. Comput. Chem.* **1989,** *9*, 343-355; R. A. Dammkoehler, S. F. Karasek, E. F. Gerkley-Shands, G. A. Marshall, *J. Comp. Aided Mol. Des.* **1989**, *3*, 3-21; M. Saunders, H. A. Jiminez-Vasquez, *J. Comput. Chem.* **1993,** *14*, 330-348.
[26] Conformational Analysis in Catalyst, Technical White Paper, 1991, BioCAD Corporation (contact Molecular Simulations Inc. 9685 Scranton Road, San Diego, CA 92121-3752).
[27] A. Smellie, S. L. Teig, P. Towbin, *J. Comp. Chem.* **1995,** *16*, 171-187.
[28] Cerius^2 Diversity, Molecular Simulations Inc. 9685 Scranton Road, San Diego, CA 92121-3752.
[29] DB searches were done to find all compounds with the word 'angiotensin' in the mode of action field, and all compounds with 'pril' in the name field. The hits from these searches were evaluated visually and only those compounds recognizable as non-peptide ACE inhibitors were retained.
[30] P. W. Sprague Building A Hypothesis for Angiotensin II Antagonism, Application Note, BioCAD Corp., Sunnyvale, CA, 1994 (contact Molecular Simulations Inc., San Diego, CA).

PART II. **LEAD FINDING STRATEGIES**

Chapter 15

The Nature and Geometry of Intermolecular Interactions: Combination of 3D-Database Information with Theoretical Tools

by **Jos P.M. Lommerse**[a]), **Robin Taylor**[a]),
Frank H. Allen[a]) and **Sarah L. Price**[b])

[a]) Cambridge Crystallographic Data Centre, 12 Union Road, Cambridge CB2 1EZ, UK
[b]) University College London, Department of Chemistry, 20 Gordon Street, London WC1H 0AJ, UK

preferred geometries. For example, they might need to know the strength of a hydrogen bond to a π-ring system, or the relative strengths of a dipolar carbonyl-carbonyl interaction compared with a hydrogen bond to a carbonyl oxygen, or whether hydrogen bonds to ether oxygens prefer to form along the lone pair directions of the oxygen. This sort of information is important not only in designing a ligand for a known protein structure, but also in structure-activity studies of a series of compounds that bind to an unknown protein active site. Finally, it is important to understand why a particular intermolecular interaction is strong, or has a preferred geometry. When questions like 'do long-range electrostatic forces play the key role in this particular interaction?', or 'why do ether oxygens usually accept hydrogen bonds whereas ester oxygen hardly ever do so?' can be answered satisfactorily, then a powerful tool is available to assist the design and optimisation of protein ligands which have excellent binding properties.

In this paper, an approach is described that might satisfy these needs in rational drug design, and will also be very useful in other fields of chemistry where intermolecular interactions play a key role. It is based on the extraction from crystal structures of the three-dimensional coordinates of contacts between small organic molecules. These experimentally based, geometrical data are further explored by theoretical energy calculations. Using examples, it will be shown that these two independent sources of information are complementary and consistent, and can be used to rationalise protein-ligand binding interactions in complexes that occur in the Brookhaven Protein Databank (PDB) [1].

Brief Description of Methodology. — *Extraction of data from the Cambridge Structural Database.* Experimentally derived data on intermolecular interactions are obtained from the Cambridge Structural Database, which contains coordinates of about 150.000 crystal structures of small organic molecules [2]. Chemical functional groups are defined in terms of atom types, bonds and limiting three-dimensional geometries. Using the CSDS QUEST 3D software [3], the CSD can be searched for (i) the presence of two particular fragments (called here the "central-group" and the "contact-group") in the same crystal structure, and subsequently, (ii) the occurrence of intermolecular contacts or short distances between the central-group and contact-group atoms. When such a contact is found, the coordinates of both groups are stored. In post-search processing, the central groups are superimposed and the distribution of contact groups around the central group is symmetrised according to the central group's structure. The resulting three-dimensional scattergrams are stored in MOL2 format [4] and visualised in the RasMol molecular viewer [5].

Table 1. *Overview of the Central and Contact Groups in the Library of 3D Scattergrams.*

Central Groups[a]					**Contact groups**[b]	
General:			*Nucleic Acids & Proteins*	*Solvates*		
Terminal	*Acyclic links*	*Rings*				
C only (3)	C,N (5)	C only (1)	Bases (5)	Water (1)	N-H	C=O
C,N (7)	C,O (10)	C,N (18)	Terminl (9)		O-H	C=O (amide)
C,O (6)	C,N,O (5)	C,O (3)	Links (2)		N-H + O-H	X=O
C,N,O (6)	C,P,... (4)	C,N,O (5)	Rings (5)		N-H (amide)	X=S
C,P,... (3)	C,S,... (15)	C,S,... (4)			C-H	any O
C.S,... (5)		Frags (10)			X-N=X	any S
C,Hal[c],..(7)					X-O-X	N (aromatic)
					X-S-X	

[a]) In between brackets the number of central groups of a particular type.
[b]) X=any atom.
[c]) Hal=halogen atom.

The result is a database or library of 3D scatterplots. A summary of some of the central and contact groups that are currently included in the library is given in *Table 1*. The library contains a scatterplot for each combination of central and contact group, there being in excess of 2300 such plots. In general, the central groups can be viewed as representatives for protein ligands, whereas the contact groups usually relate to amino acid functional groups.

Theoretical energy calculations. In addition to the crystallographically-derived data, theoretical energy calculations are performed on various model systems that correspond to particular interactions in the library. The model compounds chosen to represent the contact groups usually contain the functional groups of amino acids. For example, *N*-methylacetamide, imidazole, and methanol are used to represent a peptide bond, a histidine, and a serine or threonine side group, respectively. A wide range of molecules is used to model the various central groups. Using these model compounds, calculations are first performed to search the potential energy hypersurface of a dimer (central-group ... contact-group) system, so as to locate low-energy orientations. Then, subsequently, higher quality ab initio based intermolecular energies are calculated for these low-energy dimer orientations. This exercise thus provides theoretical intermolecular geometries and energies that can be compared with the preferred geometries in the 3D scattergrams.

Because the calculations are based on model compounds, the obtained energies can only be used as approximates to the strengths of the intermolecular interactions.

The *ab initio* package CADPAC 6.0 [6] is used to calculate the wave functions of individual monomers and to optimise their geometries at the 6-31G** basis set level. The charge distributions are corrected for electron correlation by the MP2 method [7] and then a distributed multipole analysis (DMA) is carried out on the wavefunctions to obtain atomic multipoles up to hexadecapole.

Calculation of minimum energy orientations for dimers are carried out using the ORIENT 3.0 program [8]. The program can efficiently minimise the interaction energy of a given dimer starting-orientation using a model intermolecular potential. The model potential describes the long-range electrostatic energy using the DMA multipoles by including all terms of the multipole expansion up to and including r^{-5}, where r is the intermolecular separation of the pair of atomic sites, and all other energy contributions by an empirical, isotropic *exp*-6 repulsion-dispersion term [9]. Long-range interactions, viz. involving molecules not in (van der Waals) contact, are dominated by electrostatic forces (which have only a reciprocal dependence on distance r, compared with other intermolecular forces, which have dependencies on higher reciprocal powers of r). The charge distribution of the wave function of an isolated molecule is accurately described by the atomic multipole expansion. The *exp-6* model potential (FIT [9]) reproduces a wide range of polar organic crystal structures, and so will provide a reasonable first estimate of the intermolecular separations, but ignores any anisotropy in the atomic repulsive wall. The methodology described is adequate to compute long-range interactions and, therefore, results in well orientated monomer positions for energy minimised dimers. Since the method is fast, it can be used to scan and minimise quickly the full intermolecular hypersurface. Typically, starting with 100-500 initial orientations, the global and 5-20 local minima are obtained for any particular dimer.

For selected minimum-energy dimer orientations, the intermolecular interaction energies are calculated using the Intermolecular Perturbation Theory (IMPT) [10] method available within the CADPAC 6.0 package. This method is free of basis set superposition error, a major problem within *ab initio* supermolecule calculations [11]. It provides an estimate for each significant contribution to the total interaction energy: at first order, the electrostatic energy (E_{es}) term describes the attractive or repulsive classical Coulombic interaction. The exchange-repulsion term (E_{er}) is the sum of (i) an energy lowering due to exchange of electrons of parallel spin between the molecules, and (ii) a repulsive term arising from the Pauli exclusion principle. The repulsive E_{er} increases exponentially with decreasing separation between the wavefunctions. At second order, the polarisation or induction energy (E_{pol}) term accounts for the energy gain caused by the change of the intramolecular wave function of one molecule due to the presence of the undistorted charge distribution of the second molecule; the charge-transfer energy (E_{ct}) term is the attractive energy due to actual charge transfer between molecules. The dispersion energy (E_{disp}) is calculated at the second order double excitation level and is the result of instantaneous correlation of fluctuating electron density distributions.

Due to the limited quality of the empirical repulsion-dispersion parameters used in the ORIENT program (*e.g.* only one set of isotropic parameters is used for all types of nitrogen atoms), it is necessary to optimise the energy as function of the distance between the model molecules maintaining fixed mutual orientations using the IMPT method. The energies of the resulting distance-optimised dimers are stored in the library and the corresponding dimer orientations can be viewed using the RasMol molecular viewer.

Discussion of Some Nonbonded Interactions. — One of the most important nonbonded interactions in protein-ligand complexes is the hydrogen bond, because of its high frequency of occurrence and relatively large strength. Therefore, in the design of drugs hydrogen bonding is usually an important parameter to be accounted for, and can be used to predict, *a priori*, orientations of ligands in protein active sites. The library of nonbonded interactions covers most of the potentially important hydrogen bonds, as well as other intermolecular interactions.

Browsing through the library of 3D scatterplots, a number of interesting or unexpected interactions can quickly be discovered, as well as many well-known ones. A few examples will be discussed below, including some hydrogen bonded and non-hydrogen bonded interactions.

Taking the ester and ketone hydrogen bond acceptor functionalities as central groups and hydroxyl as contact group, it is seen at a glance in a 3D scatterplot that the carbonyl group of the ester has similar acceptor characteristics to the carbonyl of a ketone group (*Figs. 1a,b*). In both cases, hydrogen bonding occurs preferentially in or near to the lone pairs of the carbonyl oxygen atom. It is striking that the (bridging) ester oxygen hardly accepts hydrogen bonds at all (*Fig. 1b*), whereas, by way of comparison, ether oxygens accept many hydrogen bonds (*Fig. 1c*: note that in this case the preferred hydrogen bond geometry is *not* in the lone pair direction of the ether oxygen, but rather all positions near and in between the lone pairs are equally favourable). This observation was further explored using theoretical energy calculations, with Z-methylacetate, 2-propanone, methoxymethane and methanol as model compounds. *Table 2* lists the intermolecular energies for the hydrogen bonded dimers of these model compounds in

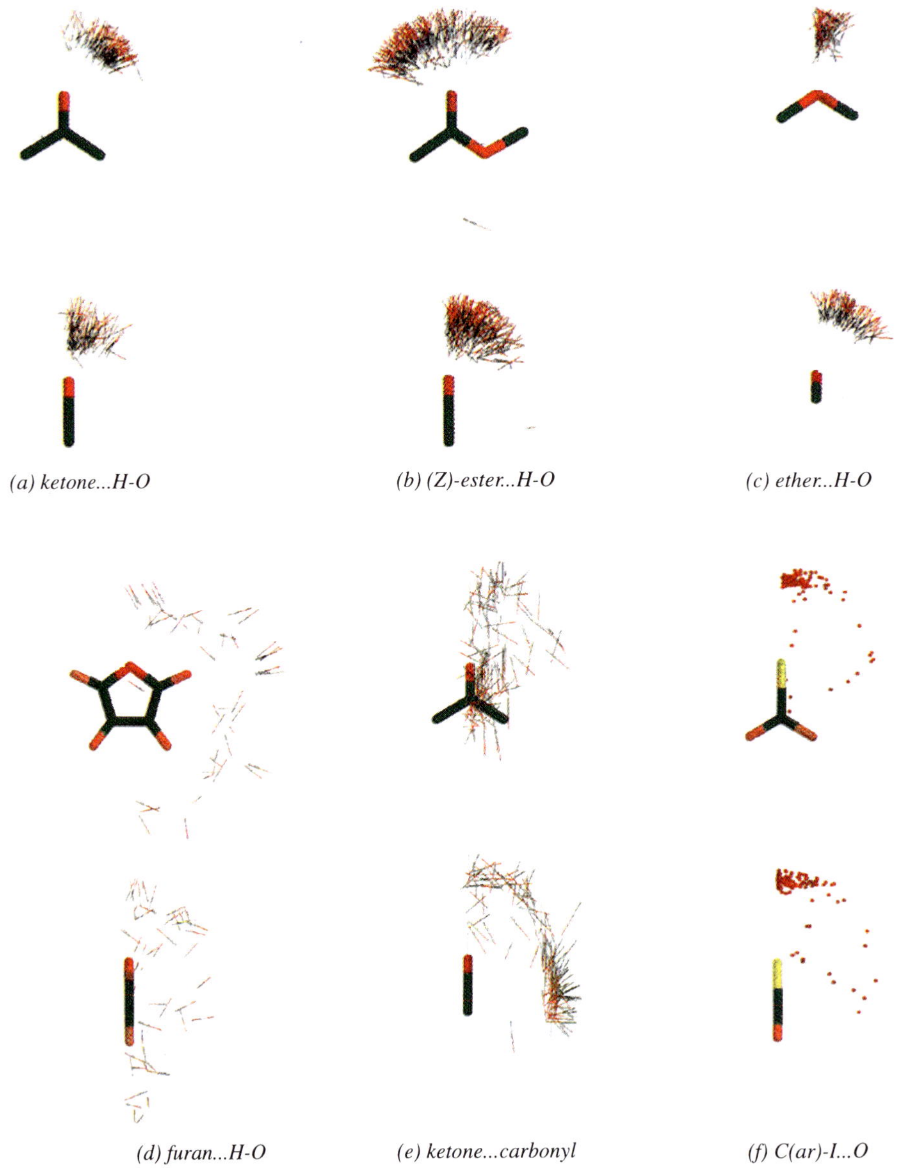

Fig. 1. *3D scattergrams derived from CSD searches showing the distributions of contact groups around superimposed central groups. For all contacts shown, at least one intermolecular distance between a contact group atom and a central group atom is less than v, where v = sum of van der Waals radii (in a, b and c) or v = sum of van der Waals radii plus 1.5 Å (in d, e and f). Top and edge view projections of the central group are given for each scattergram. Shown are: O-H contact group distributions around (a) ketone groups, (b) (Z)-ester groups, (c) ether groups and (d) furan rings; (e) carbonyl contact groups around ketone groups; and (f) oxygen atoms around iodine atoms which are covalently bonded to aromatic carbon atoms.*

Table 2. *IMPT Energies of Some Intermolecular Interactions [kJ/mol].*

System of model compounds	E_{tot}	E_{es}	E_{er}	E_{pol}	E_{ct}	E_{disp}	r^e
2-propanone...methanol	-24.7	-36.3	+28.0	-3.8	-3.9	-8.6	1.94
Z-methylacetate[a]...methanol	-24.7	-36.5	+27.9	-3.8	-3.7	-8.6	1.94
methoxymethane...methanol	-21.1	-32.4	+24.8	-3.1	-3.4	-7.0	1.94
Z-methylacetate[b]...methanol	-15.0	-22.7	+20.2	-2.2	-2.4	-7.8	2.00
furan[c]...methanol	-14.4	-20.5	+16.4	-1.9	-1.8	-6.7	2.08
furan[d]...methanol	-10.1	-10.5	+8.3	-1.0	-1.2	-5.8	2.72[f]
2-propanone...2-propanone	-22.2	-19.6	+14.0	-2.4	-1.2	-13.0	3.13[g]

[a]) Carbonyl oxygen acceptor, accepting lone pair *trans* with respect to the ester bridging oxygen

[b]) Ester oxygen as the acceptor

[c]) Furan ring oxygen as acceptor

[d]) Furan π-system as acceptor

[e]) H...O hydrogen bond length (in Å)

[f]) H...centroid of the furan ring

[g]) between centroids of C=O bonds

their 'distance-optimised' IMPT orientations (see methods).

Very similar interaction energies are obtained for hydrogen bond formation to the carbonyl groups of the ester and the ketone functions, which are calculated for the global energy minima of the model systems and agree with the preferred geometry according to the CSD data. The electrostatic energy dominates the attractive contributions to the total energy (E_{es} = -36.5 and -36.3 kJ/mol, respectively). In contrast, the electrostatic contribution for hydrogen bonding to the ester oxygen is much less than for H-bonding to the ether oxygen (E_{es} = -22.7 and -32.4 kJ/mol, respectively). It was concluded that an unfavourable dipole-dipole interaction between the methanol O-H and the ester C=O lead to the strongly reduced electrostatic energy [12]. Obviously, this has a dramatic effect on the total interaction energy, and the optimum distance for hydrogen bonding to the ester oxygen will naturally be longer than to an ether oxygen (2.00 Å and 1.94 Å, respectively). The small number of hydrogen bonds to the ester oxygen in the 3D scatterplot is therefore likely to be the result of a competition between the strong carbonyl and the weaker ester oxygen acceptor atom.

In this context, it is interesting to observe the hydrogen bond acceptor characteristics of furan (*Fig. 1d*). In this heterocycle, the oxygen atom only rarely accepts hydrogen bonds, which is reflected in the low number of O-H...O hydrogen bonds in the relevant library scatterplot (3%; 3 O...H-O distances less than 2.4 Å, out of 89 CSD

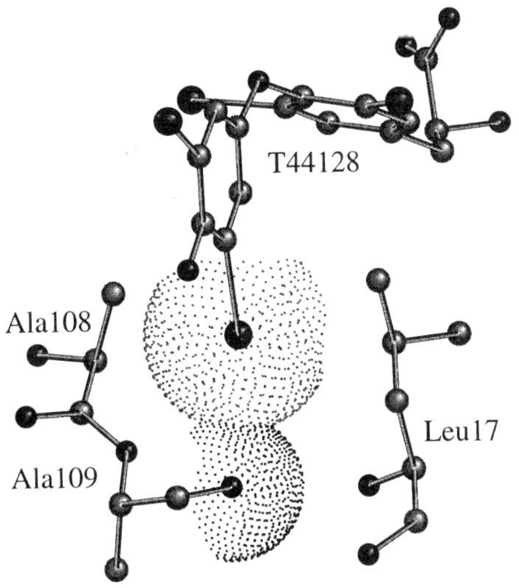

Fig. 2. *Short I...O contact between the iodine of 3,5,3',5'-tetraiodo-L-thyronine and carbonyl oxygen of Ala109 in trans-thyretine (PDB entry 1ETA [17]).*

easily polarisable, 'soft', halogen atom as well as having an isotropic charge distribution. When the iodine is attached to an electron withdrawing aromatic system, as represented in the 3D scattergram, the iodine...oxygen attraction should be rather strong. This is in agreement with the CSD data, because very short distances are present in the scattergram (about 3.1 Å, which is 0.4 Å shorter then the sum of the conventional van der Waals radii of iodine and oxygen). This interaction also occurs in a protein-ligand complex in the PDB (*Fig. 2*). In the tetraiodo ligand, the iodines are attached to aromatic rings. One of the iodines interacts with a carbonyl oxygen of a protein peptide linkage at a distance of only 3.1 Å and with a fairly linear C-I...O angle (161^{0}), as would be expected from the CSD data.

Concluding Remarks. — The described library is a useful tool for exploring and studying intermolecular interactions. It will provide the molecular modeller an overview of intermolecular interactions relevant to protein-ligand binding, both qualitatively (which interactions occur?) and quantitatively (what are their strengths and preferred geometries?). Although the CSD contains the crystal structures of small, organic molecules, not of protein-ligand complexes, there is substantial evidence that the entries that contain both an O-H group and a furan ring). The theoretical calculations show that the oxygen is a rather weak acceptor (hydrogen bonding with hydroxyl group, E_{tot} = -14.4 kJ/mol) and that the π-ring system can be viewed as another weak acceptor (E_{tot} = -10.1 kJ/mol, *Table 2*). Indeed, in the 3D scattergram, an O-H...π hydrogen bond is present. In both the ester and the furan systems, the oxygen atoms are bonded to sp^2 hybridised atoms. The weakness of this type of oxygen as a hydrogen bond acceptor has been studied extensively by Böhm *et al.* [13].

Other strong intermolecular interactions occur between neutral molecules apart from hydrogen bonds. A nice example is the carbonyl...carbonyl interaction. The carbonyl dipoles have a strong preference to form an anti-parallel orientation (*Fig. 1e*). The energy calculations demonstrate that this anti-parallel dipole...dipole stacking is a strong interaction (E_{tot} = -22.2 kJ/mol) which is comparable with the strength of a hydrogen bond to a carbonyl group (e.g. methanol...propanone -24.7 kJ/mol, see *Table 2*). It is obvious that electrostatic attraction plays a major role (E_{es} = -19.6 kJ/mol), but the dispersive forces are considerable (-13.0 kJ/mol; note also that the 6-31G* basis set level of the IMPT calculations will result in an underestimation of this energy term [14]). Although no hydrogen bonding is involved, it is clear that the (C=O)...(C=O) interaction is potentially important. Indeed, it has been calculated that the carbonyl...carbonyl interaction contributes 45% of the favourable attraction energy holding main chain peptide groups together [15].

A less obvious example of an intermolecular attraction that does not involve hydrogen bonding is the halogen...oxygen/nitrogen interaction (e.g. for iodine in *Fig. 1f*). In a recent paper, this interaction was analysed for carbon-bonded halogens (C-X) [16]. It was shown that the interaction is primarily the result of an anisotropic charge distribution around the halogen nucleus. This results in (i) a slightly non-spherical shape for the halogen atom, viz. a shorter van der Waals radius along the extension of the C-X axis than perpendicular to this axis, and (ii) an anisotropy in the electrostatic interaction, which can be attractive for orientations of the (partially negatively charged) oxygen or nitrogen atom. Both effects can lead to a close approach of the halogen atom to an oxygen or nitrogen positioned approximately on the extension of the C-X bond. In favourable circumstances, there may be a reasonably strong attraction energy: *e.g.* -10 kJ/mol as calculated by the IMPT method for a case where a chlorine atom is in a strong electron withdrawing environment and interacts with nitrogen [16].

Unfortunately, these *ab initio* based calculations are not feasible for heavier atoms than bromine, due to the lack of adequate basis set functions. Therefore, one has to rely on the CSD data when studying, say, the iodine...oxygen interactions (*Fig. 1f*). Iodine is an observed intermolecular contacts in both types of structures are similar [18,19]. Nevertheless, scattergrams derived from PDB protein-ligand complexes are currently included in the library as a separate source of information. The number of central groups and contact groups will be soon extended and a total of more than 5000 3D scatterplots will finally be stored. Software will make it possible to have easy access to the information included in the library.

REFERENCES

[1] F.C. Bernstein, T.F. Koetzle, G.J.B. Williams, E.F. Meyer, M.D. Brice, J.R. Rodgers, O. Kennard, T. Shimanouchi, M. Tasumi, *J. Mol. Biol.* **1977**, *112*, 535.
[2] F.H. Allen, J.E. Davies, O.J. Johnson, O. Kennard, C.F. Macrae, E.M. Mitchell, G.F. Mitchell, J.M. Smith, D.G. Watson, *J. Chem. Inf. Comp. Sci.* **1991**, *31*, 187.
[3] Cambridge Structural Database System Users Manual, Getting Started with the CSD System, Cambridge Crystallographic Data Centre, 12 Union Road, Cambridge, U.K., **1994**.
[4] Tripos Associates, St Louis, USA.
[5] R.A. Sayle, GlaxoWellcome Group Research.
[6] R.D. Amos, CADPAC6.0: The Cambridge Analytical Derivatives Package, Issue 6.0, University of Cambridge, Cambridge, U.K., **1996**.
[7] C. Møller, M.S. Plesset, *Phys. Rev.* **1934**, *56*, 618.
[8] A.J. Stone, P.L.A. Popelier, D.J. Wales, ORIENT Version 3.0: A Program for the Calculation of Electrostatic Interactions between Molecules, University of Cambridge, Cambridge, U.K., **1994**.
[9] D.S. Coombes, S.L. Price, D.S. Willock, M. Leslie, *J. Phys. Chem.* **1996**, *100*, 7352.
[10] I.C. Hayes, A.J. Stone, *Mol. Phys.* **1984**, *53*, 83.
[11] J.H. van Lenthe, J.G.C.M. van Duijneveldt-van de Rijdt, F.B. van Duijneveldt, *Adv. Chem. Phys.* **1987**, *69*, 521.
[12] J.P.M. Lommerse, S.L. Price, R.Taylor, *J. Comp. Chem.,* accepted for publication.
[13] H.-J. Böhm, S. Brode, U. Hesse, G. Klebe, submitted for publication.
[14] S.L. Price, A.J. Stone, J. Lucas, R.S. Rowland, A.E. Thornley, *J. Am. Chem. Soc.* **1994**, *116*, 4910.
[15] P.M. Maccallum, R. Poet, E.J. Milner-White, *J. Mol. Biol.* **1995**, *248*, 361.
[16] J.P.M. Lommerse, A.J. Stone, R. Taylor, F.H. Allen, *J. Am. Chem. Soc.* **1996**, *118*, 3108.
[17] J.A. Hamilton, L.K. Steinrauf, B.C. Braden, J.J. Liepnieks, M.D. Benson, M., *J. Biol. Chem.* **1993**, *268*, 2416.
[18] J.P.M. Lommerse, R. Taylor *J. Enzyme Inh.*, accepted for publication.
[19] G. Klebe, *J. Mol. Biol.* **1994**, *237*, 212.

PART II. **LEAD FINDING STRATEGIES**

Chapter 16

Non-supervised Neural Networks: A New Classification Tool to Process Large Databases

by **Dmitri B. Kireev***[a]), **Frédéric Ros**[a]), **Philippe Bernard**[a]), **Jacques R. Chrétien***[a]) and **Natalia Rozhkova**[b])

[a])Laboratoire de Chimiométrie, Université d'Orléans, B.P. 6759, F-45067 Orléans Cedex 2, France
[b])Plant Protection Chemicals Research, Ugrezhskaya 31, 109088 Moscow, Russia

Non-supervised Neural Networks: A New Classification Tool to Process Large Databases

by **Dmitri B. Kireev***, **Frédéric Ros**, **Philippe Bernard**, **Jacques R. Chrétien***
Laboratoire de Chimiométrie, Université d'Orléans, B.P. 6759, F-45067 Orléans Cedex 2, France

and **Natalia Rozhkova**
Plant Protection Chemicals Research, Ugrezhskaya 31, 109088 Moscow, Russia

Establishing QSAR in large databases which contain structural and biological information has qualitative differences compared to QSAR studies on compact series of chemical homologues. Indeed, classical QSAR tools often work not so well with databases which count thousands of compounds. The Self-Organising Maps (SOM) suggest a new solution to the problem. A key difference between SOM, also known as Kohonen's neural network, and many other neural networks is that SOM learns without supervision. SOM is a projection technique which reduces the descriptor multidimensional space into a space of any given dimensionality. To adopt the method for QSAR purposes quality estimates for the SOM model evaluation have been developed by the authors. A case study involving SOM is also presented. The processed database contains more than 2000 organo-phosphorous compounds with various pesticide activities from the STRAC database.

Introduction. - Exploratory analysis of a data distribution in the multidimensional descriptor space plays an important role in QSAR. This kind of analysis is indispensable in the initial stage of a QSAR study, *i.e.* experimental design, and it should also be done before regression methods are applied to the training set.

The projection techniques constitute a principal tool which allows to reveal compact clusters of chemical compounds in descriptor space. The data are usually projected from the multidimensional descriptor space onto a two or three dimensional space by means of principal component analysis (PCA) or hierarchical cluster analysis (HCA). Both PCA and HCA are capable of essential reducing of the data dimensionality. The both possess however of some significant limitations. For instance, very often the reduced principal component space though keeping a major part of the data variability has the dimensionality higher than three. Hence it may not be visually examined on presence of compact clusters. The cluster analysis dendrograms are easily computable, even for large amounts of data, but then hardly interpretable since the information is strongly degenerated.

A projection technique which seems free from the aforementioned disadvantages has emerged a decade ago. It belongs to the artificial neural network (ANN) family [1] and is referred to as Kohonen's ANN or self-organising map (SOM) [2]. In contrast to very popular in chemistry supervised back-propagation networks [3], SOM is a neural network with non-supervised learning. Kohonen's ANN are able to project the data onto the reduced space of any given dimensionality. This capability of projecting the data distributions of any complexity onto two or three dimensional space is due to the non-linear nature of ANN which is inspired by brain processing.

The implication of SOM in QSAR is especially interesting in the context of processing huge amounts of data. Indeed, because of the high structural diversity in databases which count thousands of compounds, PCA does not yield a reduced principal component space of proper dimensionality. The expert classification of compounds by their belonging to chemical classes is often impossible because of its arbitrariness and significant time expenses. In view of growing interest to modelling the chemical diversity as an emerging new lead searching tool, such a robust projecting technique as the Kohonen's SOM should play an important role in the automated data processing in chemical diversity libraries.

First applications of SOM to chemical problems have already appeared. The Kohonen's ANN have been used in modelling protein sequences, in QSAR analysis and in some applications related to classification [4-6].

However using SOM as a chemometrical tool in QSAR envisages some additional methodological work to be done. For example, we have recently introduced an estimate for the evaluation of the SOM quality and robustness [7]. A systematic investigation of which SOM architecture and initial parameters are the most appropriate in QSAR is also necessary.

In order to stimulate an interest to SOM applications in QSAR and make it more accessible we give here not only our own contribution to the development of the SOM methodology but also some basic principles of SOM functioning. Then new indices for the SOM model evaluation are presented. In the next two sections we demonstrate the applications of SOM. The first one is a model example which shows how our estimates of the SOM model quality work and the second one is an application of SOM to the real life QSAR problem The latter concerns the processing of information in the database of organophosphate pesticides from the Moscow Plant Protection Institute which counts more than two thousand compounds.

Self-Organising Maps. - *Basic Principles.* Kohonen's ANN is a topology-pre-serving technique. It means that the chemical compounds which are close to each other in the multidimensional descriptor space are still neighbours on the Kohonen's map after the learning is finished. Though the Kohonen's map may take any dimensionality we shall refer below only to bidimensional maps.

The SOM consists of Kohonen and input layers. The two-dimensional Kohonen layer contains M×M nodes and its size does not depend on the number of the molecular descriptors. After training, the Kohonen layer represents a map with compounds of the training set distributed on its surface. In the input layer each node corresponds to a molecular descriptor. Hence, the number of nodes in the input layer is equal to the number of descriptors (N). Each node of the Kohonen layer is connected to each node in the input layer. Both the descriptors and the connection weights are normalised to 1. Hence, each node of Kohonen layer is characterised by N weights varying between 0 and 1 and each chemical compound is characterised by N descriptors varying as well between 0 and 1. Thus, Euclidean distance may be measured between any node in Kohonen layer and any compound. When the descriptor values of the first compound are submitted onto the input layer the distances between this compound and each node of the Kohonen layer are calculated. A node having the shortest distance to this first compound is called the winner and the compound is placed on this node of the map. The weights of the winner node are modified in order to make this node even closer to the compound according to:

$$X_{ij}(t+1) = W_{ij}(t) + \alpha^*(X - W_i(t)) \quad \textbf{(1)}$$

where $\alpha \ll 1$ is the learning rate, X is the input vector and W_i the weight vector of the winner and t is the iteration number.

The weights of the nearest neighbour nodes of the Kohonen layer are also modified to become closer to the winner and, hence, to the current compound. Then the same procedure is repeated for all the compounds. Finally each compound is placed on a node in the Kohonen layer and the compounds which are close to each other in the descriptor space should occupy the same or neighbouring nodes of the Kohonen layer. If one accepts that the Euclidean distance is a measure of the similarity between molecules, then one may conclude that Kohonen's ANN reveals the chemically similar compounds.

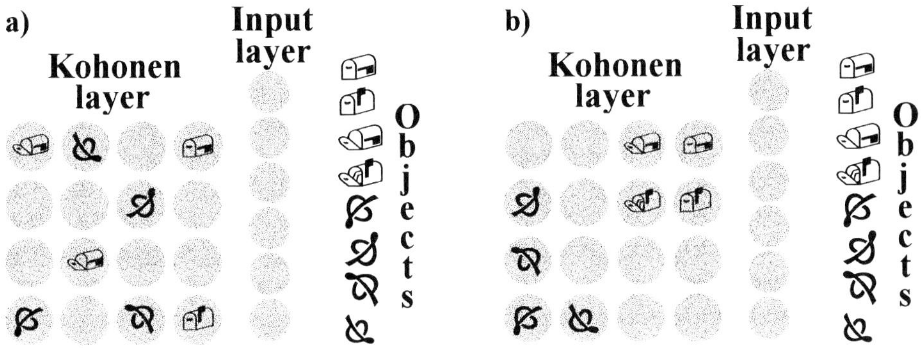

Fig. 1. *a) The learning of SOM at an initial phase: the Kohonen layer nodes are randomly populated; b) the learning at a final phase: similar objects are grouped together.*

An important improvement has recently been proposed to increase efficiency of the algorithm. This feature is referred to as the conscience mechanism [8]. It favours the losers in order to make the self-organisation more robust. This mechanism is implicated in form of an additional term to the *Eq. 1*:

$$B_i = \gamma\,(N \cdot F_i - 1) \quad \textbf{(2)}$$

where γ is a constant and F_i is the frequency of winning by the i-th node.

One more feature concerns the Kohonen layer configuration. It may be processed as wrapped or flat (see *Fig. 2*).

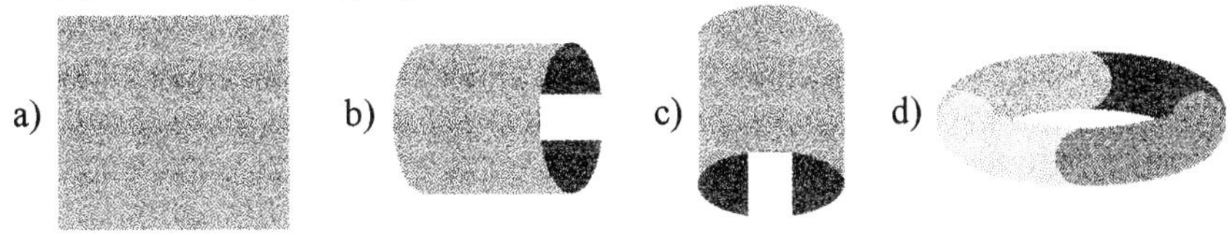

Fig. 2. *The Kohonen layer configurations: a) flat, b) horizontal wrap, c) vertical wrap, d) double wrap.*

It should be noted as well that due to the very nature of the Kohonen ANN the image of SOM is always dependent on the initial weights and on the renumbering of compounds.

Model Evaluation. While the supervised learning methods may be validated by comparing the actual and estimated values of the modelled property, this can not be applied to the non-supervised techniques. In PCA, which is a linear non-supervised method, the model quality may be estimated by the quantity of the original information kept in the principal components. The recalculation of the descriptors from the principal components can be made and then the comparison between the original and reconstituted values is possible. This approach may not be applied to SOM as well.

As it was already mentioned, the optimised image of the Kohonen's map depends on the starting weights of connections between the input and Kohonen layers. These are randomly initiated before each run of the algorithm. This means that each run will produce a different image. However, if compounds are clustered in their multidimensional descriptor space, the Kohonen's ANN should always yield the clusters on the map. Hence the repeatedly reproduced elements of the image indicate a robust data structure. But if the distribution of compounds on the map is always different it means that the compounds represent a single compact cluster and may not be classified whatever is the classification technique.

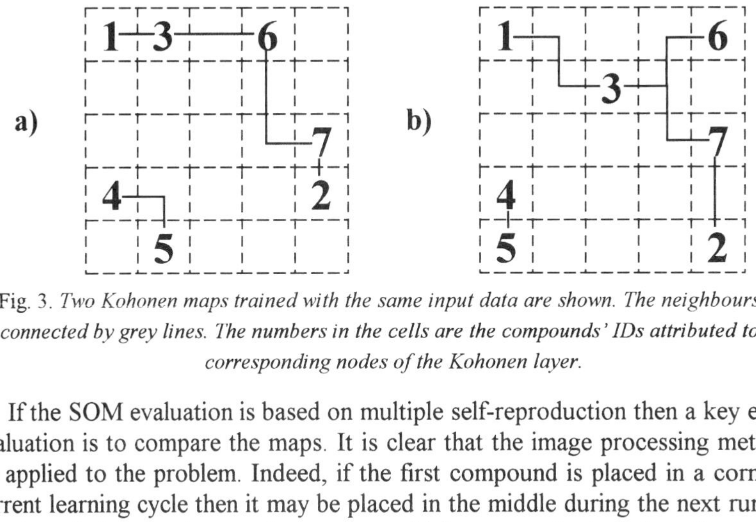

Fig. 3. *Two Kohonen maps trained with the same input data are shown. The neighbours are connected by grey lines. The numbers in the cells are the compounds' IDs attributed to the corresponding nodes of the Kohonen layer.*

If the SOM evaluation is based on multiple self-reproduction then a key element of the evaluation is to compare the maps. It is clear that the image processing methods may not be applied to the problem. Indeed, if the first compound is placed in a corner during the current learning cycle then it may be placed in the middle during the next run that will change the whole map image. A property which may really be taken for comparison purposes is a neighbourhood ordering. Compounds taking neighbouring Kohonen layer nodes during the first run should always keep this neighbourhood. For example, one may call neighbours any pair of compounds if the distance between them on the map is not greater than three. The « block-city » distance may be taken as a suitable distance type. Then, an appropriate measure of the similarity between two maps may be defined as the number of the pairs of compounds which remain neighbours on both maps. It should be noted that only the neighbours are suitable objects to apply this similarity criterion because the distances between the distant compounds may vary in wide range.

In *Fig. 3* an example is shown of two similar maps calculated from the same input data. If the neighbourhood distance is not greater than three, then the map (a) counts 6 pairs of neighbouring compounds and the map (b) counts 5 neighbouring compounds. The compounds **1** and **6** are the neighbours in the map (a), but they have lost their neighbourhood in the map (b).

A normalised index SI_{ij} quantifying the similarity between two maps may be introduced here:

$$SI_{ij} = NN_{ij} / NN_i \quad \textbf{(3)}$$

where NN_{ij} is the number of pairs of compounds who are the neighbours on both maps *i* and *j* and NN_i is the number of the neighbouring pairs on the map *i*. This index may take a value less or equal to 1. It should be noted that the SI_{ij} value may be different from the SI_{ji} value. This normalised index allows to compare the similarities of the maps which have been issued from different training sets and the similarities of the maps of different dimensionality in the same manner as one is able to compare the quality of two different regression models by their correlation coefficients.

When a similarity measure is found we may formulate two following basic statements which underlie the SOM model evaluation.

- if there is no any similarity between the Kohonen maps issued from the same input data then the data are not clustered.
- if different Kohonen maps issued from the same input data are similar then the best map for use in further data analyses is the one which is similar to as many other maps as it is possible.

For any series of *L* maps trained from the same input data a *L*×*L* matrix may be composed whose elements are the similarity indices for each pair of the maps. The mean value over all non-diagonal elements of such a matrix (*SI*) may be used as an indicator of how many compounds of the training set are clustered.

$$SI = \sum_{i,j(i \neq j)}^{L} SI_{ij} / L(L-1) \quad \textbf{(4)}$$

The mean value over the *i*-th column (SI_i) quantifies the similarity of the *i*-th map to the other maps.

$$SI_i = \sum_{j(i \neq j)}^{L} SI_j / (L-1) \quad \textbf{(5)}$$

The first of the two estimates (*SI*) should be calculated for each series of the maps generated from the same input data to verify if the data have an intrinsic structure. The second one (SI_i) is useful to decide which of the generated maps is the most suitable for the visual analysis.

Applications. - *A Model Example*. The training sets used in this study have been artificially generated in order to examine whether the Kohonen's ANN is able to reveal the structure of the data studied and to validate our quality estimates. One of these two training sets consists of 30 hypothetical objects each of which is described by 5 descriptors. Each descriptor is a series of random values normally distributed around 10 with a standard deviation of 2.5. Another training set differs from the first one in that the descriptors are normally distributed around three numbers (10, 20, 30) with the same standard deviation of 2.5. Hence, the objects from the first training set are randomly distributed in the whole descriptor space. The objects of the second series of data form three distinct clusters in the five dimensional descriptor space.

Twelve maps have been generated from each of the training sets. The Kohonen layer have consisted of 10×10 nodes. An improvement to the visualisation of the Kohonen's map has been used in order to increase the resolution of the maps. According to the improvement the co-ordinates of each object on the map are calculated as the mean values over three nearest nodes of the Kohonen layer. The co-ordinates calculated in such a way are the real values. Examples of the maps obtained from the different series are shown in *Fig. 4*.

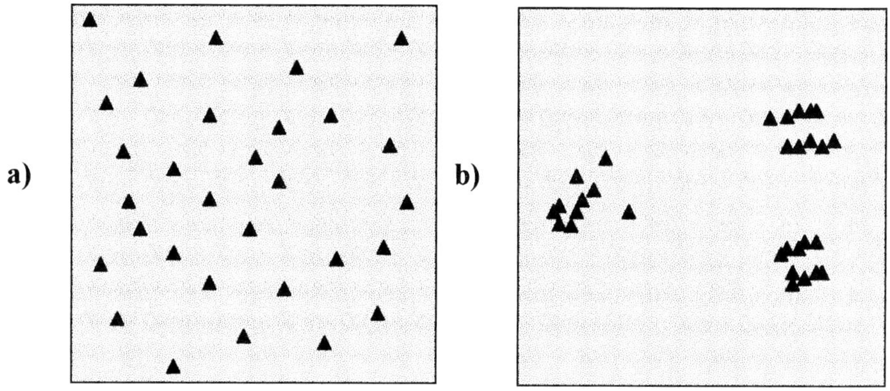

Fig. 4. *Two Kohonen maps generated from two different training sets. The map (a) has been issued from five dimensional space in which all the points are randomly distributed. The map (b) is calculated from the data set consisting of three distinct clusters. This map clearly shows the structure of the corresponding data.*

The 12×12 similarity matrices have been calculated for both training sets. The better matrix is presented in *Table 1*. This better matrix corresponds to the structured data of the second training set. The values of individual similarity coefficients clearly show the better reproductivity of the maps for the second training set consisting of three spatially distinct data series.

The best three maps are shadowed in the *Table 1*. The best SI_i value has the first map (SI_1=1.0). The total SI value for the maps corresponding to the clustered data is 0.98. The SI value for another series of data is less than 0.5. A low SI value does not means that the objects are clustered in such a complicated manner that SOM is not able to reveal this data structure. It only means the absence of any structure. Hence, if the classification is the ultimate goal of the study, the low SI value would mean that the descriptors are not relevant to the problem.

Table 1. *The Similarity Indices Calculated for a Series of Maps Corresponding to Structured Data.*

	1	2	3	4	5	6	7	8	9	10	11	12
1	1,0	0,98	0,99	1,0	0,96	0,99	0,94	1,0	0,96	0,99	0,98	0,98
2	1,0	1,0	1,0	1,0	0,98	1,0	0,98	1,0	0,93	1,0	1,0	0,98
3	1,0	0,92	1,0	0,98	0,94	0,93	0,89	0,99	0,96	0,97	1,0	0,93
4	1,0	0,89	0,97	1,0	0,95	1,0	0,95	0,99	0,94	0,98	0,94	0,98
5	1,0	0,98	1,0	1,0	1,0	1,0	0,98	1,0	1,0	1,0	1,0	1,0
6	1,0	0,96	0,93	0,99	0,95	1,0	0,95	1,0	0,94	0,98	0,98	0,94
7	1,0	0,96	0,99	1,0	1,0	1,0	1,0	1,0	0,97	1,0	1,0	1,0
8	1,0	1,0	1,0	1,0	1,0	1,0	1,0	1,0	0,94	1,0	1,0	1,0
9	0,97	0,84	0,95	0,94	0,90	0,92	0,87	0,94	1,0	0,90	0,94	0,90
10	1,0	0,86	0,97	1,0	1,0	1,0	0,99	1,0	1,0	1,0	1,0	1,0
11	1,0	0,97	1,0	1,0	1,0	1,0	1,0	1,0	1,0	1,0	1,0	1,0
12	1,0	0,92	0,98	0,99	0,99	0,99	0,99	0,99	0,93	1,0	1,0	1,0

A Case Study. - The subject of the study is to establish QSAR in a pesticide database *STRAC* from the Moscow Plant Protection Institute. This database counts more than 50,000 entries. At the first step of the study we have limited our research area to ~2,000 organophosphorous compounds with various pesticide activities. A high degree of structural diversity of the training set (see sample structures below) makes it impossible to use traditional linear tools.

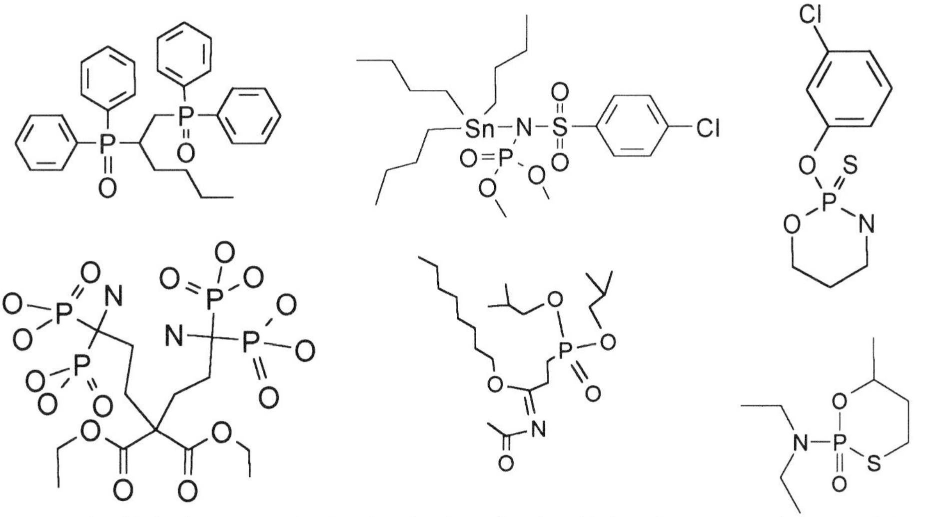

The biological diversity in the database is also high. Five types of the pesticide activity on different insect species have been measured for these compounds. Some compounds are active simultaneously against two or three species. The statistics on the activity distributions in the database is shown in *Table 2*. The figures in this table concern only very active compounds which cause at least a 95% level of mortality in the tested species.

Table 2. *Statistics on the Biological Diversity of the STRAC Database.*

Species	Active compounds
Aphis **(Green-Fly)**	**135 compounds**
Caland.**(Weevil)**	**138 compounds**
Musca Domestica **(Fly)**	**45 compounds**
Aphis-Caland.	**60 compounds**
Aphis-Musca D.	**27 compounds**
Caland.-Musca D.	**36 compounds**
Aphis-Caland.-Musca D.	**22 compounds**

At first, classical QSAR analyses has been applied to the data. More than 100 physico-chemical, electronic, geometric and topological descriptors have been calculated. It has been found out that the insect toxicities from the database can not be modelled by means of multiple linear regression or PLS. Moreover, PCA have shown that at least 10 principle components (PCs) are necessary to explain 95 % of the data variation. This result is not surprising because of the high structural diversity in the database. Hence, the data can not be visualised in a two or three dimensional hyperspace. As to the hierarchical clustering (HC) then it has yielded hundreds of clusters whose role and relative positions is hard to interpret. Hence, no reasonable classification and modelling can be made on the basis of these models.

The application of the Kohonen's ANN to this training set addresses two independent issues. The first one is applied, *i.e.* to derive activity models, and the second issue is methodological, *i.e.* to investigate which network parameters and configurations are the most appropriate for such kind of problems. In both cases we have used our similarity indices for validating the models and for comparing the results.

The importance of the conscience and wrapping was studied in the methodological framework. After numerous network trainings it was found out that the most stable maps have been issued by the networks with the double wrapped Kohonen layer. The conscience was only necessary during an initial training phase. Finally, the network architecture used in the QSAR analyses was the following : double wrapped

In the applied part of our study, the problem we addressed was to delineate the regions on the Kohonen map corresponding to the different types of activity. It was found that the compounds having the same kind of activity occupy a compact area on the map. The examples of the regions occupied by the compounds active against the weevil and the green-fly are shown in Fig. 4 a,b. Moreover, the compounds which are active against both the weevil and the green-fly are mostly situated in the region corresponding to the overlap of the weevil and green-fly regions (Fig. 4 c). It should be noted again that the shape and the placement of the areas are different from one map to another. It is the similarity index which made it possible to establish that the active compounds are always the neighbours.

If a series of stable maps is generated, the best map is chosen and the regions corresponding to different activities are delineated then this best map may be used for prediction. For any hypothetical compound the co-ordinates on the map can be easily calculated. The following steps look simple: one has only detect in which region of the map the new compound is placed and then to register whether the new compound is active against one or several species or it is inactive.

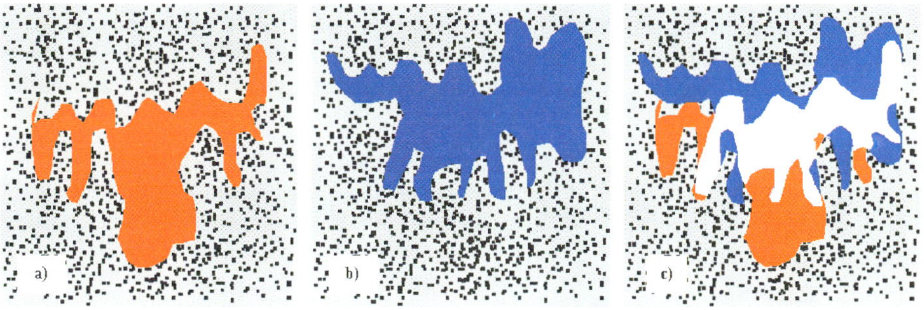

Fig. 5. *« Active » regions on the maps: the regions corresponding to the activity against the weevil (a) and the green-fly (b). The third map (c) shows the overlap (white) of the weevil and green-fly regions.*

Yet the reality is more complex. For example the 135 compounds with activity against the weevil make up about 5% of the training set. Their region on the map occupies however at least 15-20% of the map surface. Taking into account that the compounds of the training set are distributed more or less homogeneously by the surface of the map one may conclude that there are some inactive compounds which also occupy the active region. This superposition of the active and inactive compounds on the map is due to the asymmetric data structure [9]. It means that in contrast to the active compounds which have a specific data structure the data of the inactive compounds are unstructured. The reason of this asymmetry is the following: the structural diversity of the active compounds is limited by rather small number of similar compounds possessing of given biological properties while the diversity of the inactive compounds is infinite.

The absence of a structure in the data corresponding to the inactive compounds may help however to discriminate the *actives* from the *inactives*. Indeed, we have already shown that the Kohonen maps of structured data are always similar one to each other in terms of our similarity indices. On the contrary, the values of the similarity indices for unstructured data are very low. These observations may be applied to the prediction of activity for a new compound. The new compound should join the training set and a series of Kohonen maps should be generated. Then, if the compound was systematically placed in the neighbourhood of the compounds with a given activity then this compound is supposed to have this activity. In this respect, the SOM may be considered as some kind non-linear discriminant analysis.

Conclusion. - It has been demonstrated that the Kohonen's self-organising map (SOM) is an effective new tool in QSAR. Being a non-supervised learning machine based on the neural network (NN) methodology, the method helps to reveal the clusters of compounds in a multidimensional descriptor space by visualising the data in two or three dimensions. It has numerous advantages as compared with the classical non-supervised learning techniques as the principle component analysis and hierarchical cluster analysis. Kohonen's NN is always able to reduce the dimensionality of complex and multidimensional data to any given dimensionality.

We have introduced the similarity indices (*SI*) which have demonstrated their efficiency as estimates of the quality and robustness of the SOM analyses. These *SI* have helped to turn a drawback of SOM, *i.e.* a dependence of the result on the uncontrolled, initial network conditions, into an advantage which increases the predictivity of the model. Indeed, an automated similarity analysis of numerous maps obtained from the same input data but from different initial conditions allows to distinguish the irrelevant inactive compounds from the active ones.

The possibility of an automated analysis is especially important when working with huge databases. The compounds in the large databases can not be individually classified by experts. Hence, the Kohonen NN may become an indispensable tool for classifying significant amounts of data.

REFERENCES

[1] P.K. Simpson, 'Artificial Neural Systems: Foundations, Paradigms, Applications and Implementations', Pergamon Press, New York, 1990.

[2] T. Kohonen, Self-Organization and Associative Memory, Springer-Verlag: Berlin, Heidelberg, New-York, Tokyo, 1988.

[3] D.E Rumelhart, J.L McClelland, 'Parallel Distributed Processing', MIT Press:Cambridge, MA, 1986.

[4] J. Gasteiger, J. Zupan, *Angew. Chem. Int. Ed. Engl.*, **1993**, 32, 503-527.

[5] V.S. Rose, I.F. Croall, H. MacFie, *J. Quant. Struct.-Act. Relat.* **1991**, 10, 6-15.

[6] J. Zupan, J. Gasteiger ,'Neural Networks for Chemists - An Introduction'; VCH Publishers, Weinheim, 1993.

[7] F. Ross, D.B. Kireev, Ph. Bernard, J.R. Chrétien, *J. Chem. Inf. Comp. Sci.*, submitted.

[8] D. DeSieno; Proc. ICNN, IEEE Press. NY **1988**, v1, pp. 117-124.

[9] W.J. Dunn, S. Wold, In 'Chemometric Methods in Drug Design'; Ed. H. van de Waterbeemd, VCH, Weinheim, 1995, pp. 49-62.

PART III. COMPUTATIONAL APPROACHES IN DRUG DISPOSITION

Chapter 17

Physicochemical Properties in Drug Metabolism and Pharmacokinetics

by **Dennis A. Smith**

Department of Drug Metabolism, Pfizer Central Research, Sandwich, Kent CT13 9NJ, UK

Physicochemical Properties in Drug Metabolism and Pharmacokinetics

by **Dennis A Smith**
Department of Drug Metabolism, Pfizer Central Research, Sandwich, Kent CT13 9NJ, UK

The classical pharmacokinetic phase/pharmacodynamic phase model of dose response can be represented physiologically by considering each component of the phases being separated by lipid membrane barriers that may or may not possess aqueous channels (pores) and which divide essentially aqueous environments. By studying only steady state conditions, the only interactions that need to be considered are how drug molecules traverse the compartments and how they interact with the proteins that trigger pharmacological activity (target) or metabolise and excrete them. The free concentration of drug in the plasma is in direct equilibrium with the interstitial fluid bathing most cells of the body, since the capillary wall contains sufficient pores to allow the rapid passage of relatively small molecules, regardless of physicochemistry. Such aqueous channels are much fewer in number in the capillaries of the brain (blood brain barrier) and rapid transfer into the brain fluids requires molecules to traverse the lipid cores of the membranes. Actual passage into cells requires the molecules to also transit the lipid core of the membrane. Distribution to the target for a drug, for instance whether a cell membrane receptor or an intracellular enzyme, therefore determines the range of physicochemical properties available for the drug discoverer to exploit.

The concentration of free drug in plasma is governed by the rate of delivery and the rate of clearance. Clearance of drug normally occurs from the liver and kidneys and it is an important assumption that only the free drug is available for clearance. Lipophilic drugs are metabolised by intracellular membrane bound enzyme systems (e.g. cytochrome P450's) and glucuronyl transferases) to more water soluble derivatives. The active sites of the major forms of the cytochrome P-450 superfamily rely heavily on hydrophobic interactions with their substrates although ion-pair and hydrogen bonding interactions also occur. Exit from the hepatocyte may be by simple passive diffusion back into the plasma or via cannicular active transport systems which excrete drugs and their metabolites, again with wide ranging physical properties, into the bile. Glomerular filtration by the kidney, of free drug in plasma, does not pose physicochemical barrier, so that all small molecules, with relatively low molecular weight, will appear in the filtrate. The secretion of drugs into the urine can also occur through tubular carrier systems similar to those present on the sinusoid face of the hepatocyte. The aqueous concentration processes that occur in the kidney mean that for drugs capable of travelling through the lipid core of the tubule membrane, significant reabsorption back into the plasma will occur. This process has the end result of only hydrophilic molecules being voided in the urine to any substantial degree. Our understanding of the various processes suggest that expert systems could be developed to predict dose response from minimal *in vitro* data.

Introduction. - The subject of physicochemical properties and their relationship to the many facets of drug metabolism and pharmacokinetics is complex. The situation is not helped by there being no unifying concept that aids the understanding of the events occurring between the administration of a drug and its effect. This paper outlines a simple model of drug action which relates to physicochemical properties and one that ultimately could form the basis of an expert system. The model is developed along the lines that the two most important parameters to predict for a compound during drug discovery are the administered dose and the frequency of dosing and that these should be predicted with the minimum of experimentation.

The classical pharmacokinetic phase/pharmacodynamic phase model (PK/PD) of dose response can be represented pictorially by considering each component of the phases as being separated by lipid membrane barriers that may or may not possess aqueous channels (pores) and which divide essentially aqueous environments (*Fig. 1*).

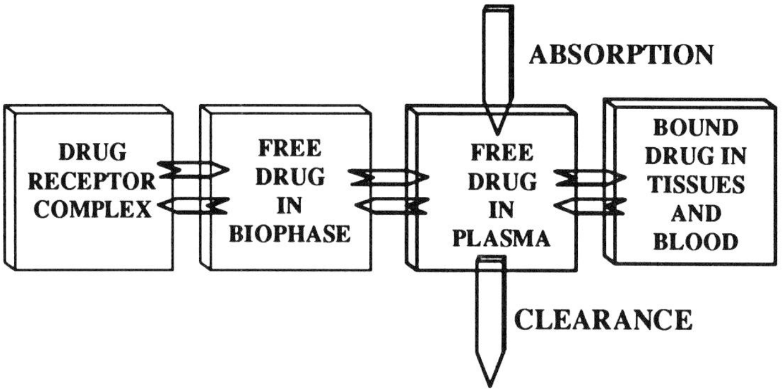

Fig.1. *Conceptual pharmacokinetic/pharmacodynamic model illustrating the dose response of a drug as a series of interconnecting compartments.*

The proceeses of absorption have been discussed elsewhere [1] and this paper looks at the behaviour of a drug following access to the systemic circulation. By studying only steady state conditions of the free drug in solution in plasma, the processes of reversible drug binding to plasma or cell constituents is essentially negated. Steady state concentrations of free drug (Cp_{free}) depend on the input rate of drug and the free drug clearance,

(Cl_{free}) : dose per unit time = Cl_{free} x Cp_{fre}e (1).

Reversible binding of drug is important in determining the frequency of administration to obtain the steady state conditions, and not the overall rate. Under steady state conditions, the only interactions that need to be considered are how drug molecules traverse the compartments and how they interact with the proteins that trigger pharmacological activity (target) or metabolise and excrete them. These interactions will often have a component of hydrophobicity in them and published log $D_{7.4}$ (octanol) values have been used to clarify them These have been obtained for the compounds of interest from the MedChem database (Thor, Daylight Chemical Information Systems Inc.), and are referred to in text as log D. It is easiest to consider the concentration of free drug in plasma as the pivotal point of the model and consider how this firstly drives the pharmacodynamic phase and secondly how this is controlled by the administered dose and the drug clearance processes.

Pharmacodynamic Response to Drug. - The free concentration of drug in the plasma is in direct equilibrium with the interstitial fluid bathing most cells of the body, since the capillary wall contains sufficient pores to allow the rapid passage of relatively small molecules, regardless of physicochemistry. Most receptor targets are accessed extracellularly. We can expect therefore that all drugs regardless of their physicochemistry will be in direct equilibrium, at these targets, with the free drug in plasma. For instance the

G-protein coupled receptors have a binding site which is accessible to hydrophilic molecules. This is exemplified by the endogenous agonists of these receptors which are usually hydrophilic by nature. Adrenaline, dopamine and histamine are representative and have log D values of -2.6, -2.4 and -2.9 respectively.

Figure 2 examines this concept of direct equilibrium further. The figure illustrates G-protein coupled receptor antagonists working at non-CNS sites (β-adrenoceptor antagonists, H_1 and H_2 antagonists and antimuscarinics) and compares potency from *in vitro* pharmacology (K_b, pA_2 etc.) with the steady state free drug concentration required for efficacy.

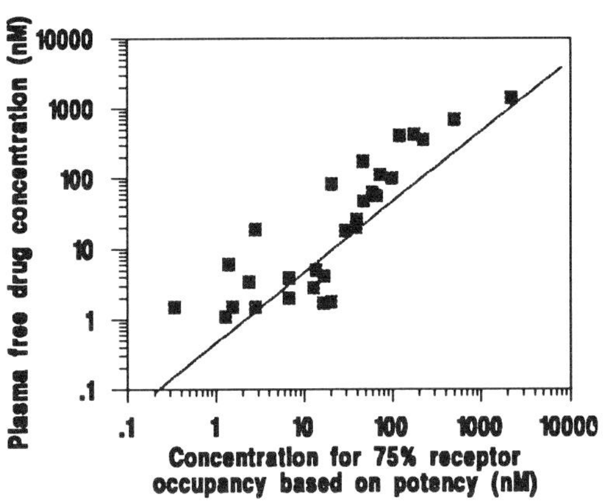

Fig. 2. *Correlation of in vitro measure of potency with plasma free drug concentration required for efficacy in a number of receptor antagonists.*

Direct comparison of these values gave a positive intercept, hence the data are presented recalculated as the free drug concentration required to achieve 75% receptor occupancy. The antagonists included in *Figure 2* range in log $D_{7.4}$ value, for example within the β-adrenoceptor antagonists the range is from -1.9 for atenolol to 0.8 for propranolol [1]. This range indicates the ease of passage from the circulation to the receptor site for both hydrophilic and lipophilic drugs. Aqueous channels are much fewer in number in the capillaries of the brain (blood brain barrier) and rapid transfer into the brain fluids requires molecules to traverse the lipid cores of the membranes. Actual passage into cells, like crossing the blood brain barrier, also requires molecules to transit the lipid core of the membrane due to the relative paucity of aqueous channels. Distribution to the target for a drug, whether a cell membrane receptor in the CNS or an intracellular enzyme or receptor, therefore critically determines the range of physicochemical properties available for the drug discoverer to exploit. The access of the CNS to drugs is illustrated by reference to a series of dopamine D_2 antagonists. Here receptor occupancy can be measured by the use of PET scanning and this "direct measure" of receptor occupancy can be compared with theoretical occupation calculated from the free drug plasma concentrations in the same experiment. *Figure 3* shows this comparison for the lipophilic antagonists remoxipride, haloperidol and thioridazine and the hydrophilic compound sulpiride [2,3,4,].

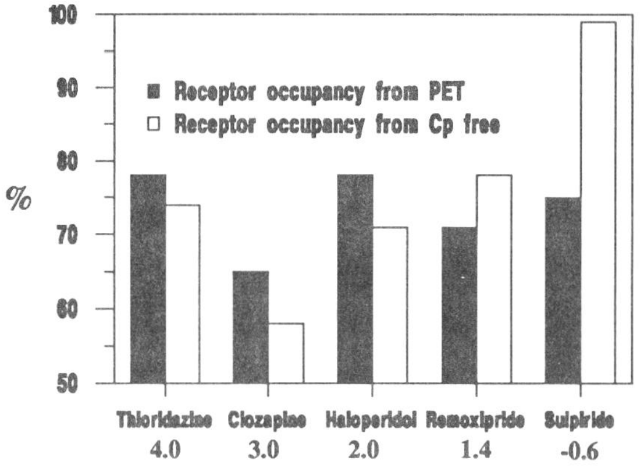

Fig. 3. *Receptor occupancy based on PET scanning and in vitro potency combined with free drug concentration (Cp free) and illustrating the ready equilibrium for lipophilic molecules and the lower equilibrium for hydrophilic compounds. Log D values are shown below compound names.*

As for the more comprehensive set of non CNS site antagonists, referred to above, efficacy is observed around 75% receptor occupancy. Noticeably the lipophilic antagonists, thioridazine, clozapine, haloperidol and remoxipride, are in direct equilibrium with the free drug concentration in plasma. However, the hydrophilic drug, sulpiride, has to be present at free concentrations in plasma over 50 fold greater than that required for simple direct equilibrium due to the poor penetration of this hydrophilic molecule across the blood brain barrier. As explained above, the reasons for this are the low number of aqueous channels or pores at the capillary wall of the blood vessels of the brain, thereby restricting entry of hydrophilic compounds to the extracellular fluid of the CNS. Similar observations can be made for intracellular targets, whether enzymes or receptors. Here the need to penetrate the lipid core is reflected in the physicochemistry of the endogenous agonists. For instance, steroid receptors are intracellular and steroids have log $D_{7.4}$ values such as 3.3, 1.7 and 2.3 for testosterone, cortisol and corticosterone respectively. These values make an interesting contrast with the values for the endogenous agonists of G-protein coupled receptors.

Drug Clearance. - The concentration of free drug in plasma is governed by the rate of delivery and the rate of clearance. The rate of delivery is a reflection of dose size and frequency. Clearance of drug normally occurs from the liver and kidneys and it is an important assumption that only the free drug is available for clearance. For hepatic clearance, passive diffusion through the lipid core of the hepatocyte membrane (available only to lipophilic drugs) is augmented by sinusoidal carrier systems, particularly for ionised molecules (either anionic or cationic) of molecular weights above 400. The presence of these carrier systems provides access to the interior of the hepatocyte to drugs with a wide range of physicochemical properties, ranging from hydrophilic to lipophilic. Lipophilic drugs are metabolised by intracellular membrane bound enzyme systems (e.g. cytochrome P450's and glucuronyl transferases) to more water soluble derivatives. The active sites of the major forms of the cytochrcme P450 superfamily rely heavily on hydrophobic interactions with their substrates, although ion-pair and hydrogen bonding interactions also occur. Exit from the hepatocyte may be by simple passive diffusion back into the plasma or via cannicular active transport systems which excrete drugs and their metabolites, again with wide ranging physical properties, into the bile. Glomerular filtration by the kidney, of free drug in plasma, can be likened to the passive diffusion into interstitial fluid with no physicochemical barriers to encounter. The result of which is that all small molecules, with relatively low molecular weight, will appear in the filtrate. The secretion of drugs into the urine can also occur through tubular carrier systems similar to those present on the sinusoid face of the hepatocyte. Unlike the liver, however, the aqueous concentration processes that occur in the kidney mean that for drugs capable of travelling through the lipid core of the tubule membrane, significant reabsorption back into the plasma will occur. This process has the end result of only hydrophilic molecules being voided in the urine to any substantial degree.

To further investigate the correlation between lipophilicity and these processes, the pharmacokinetic parameters of a number of drug were compared. Pharmacokinetic parameters were recalculated for free drug for the compounds listed by Benet *et al* [5]. *Fig. 4* illustrates the relationship between lipophilicity and free renal and non-renal clearance for neutral, basic and acidic drugs.

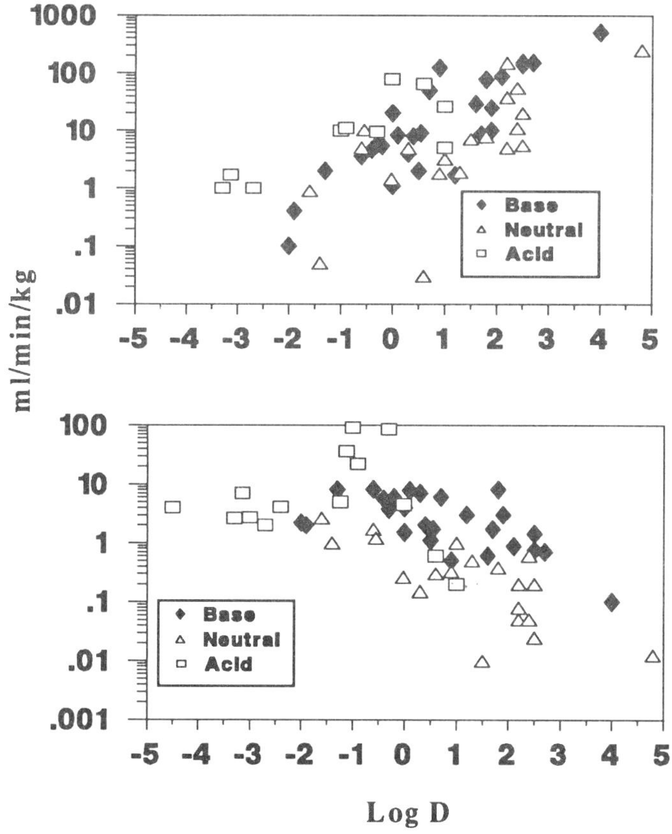

Fig. 4. *Relationship between lipophilicity and free metabolic (top) and free renal clearance (bottom) for different drug classes.*

In all cases there is a general trend to decreasing renal clearance and increasing non-renal clearance with increased lipophilicity. These trends can be rationalised by the descriptions of the processes outlined above. Closer examination of the data sets reveals further relationships. Renal clearance of neutral drugs (i.e. those not substrates for carrier systems) can be simulated by a model assuming glomerular filtration rate of free drug followed by a direct relationship between lipophilicity and reabsorption. The model assumes that a 200 fold concentration of the filtrate occurs during passage through the tubules (*Figure 5*).

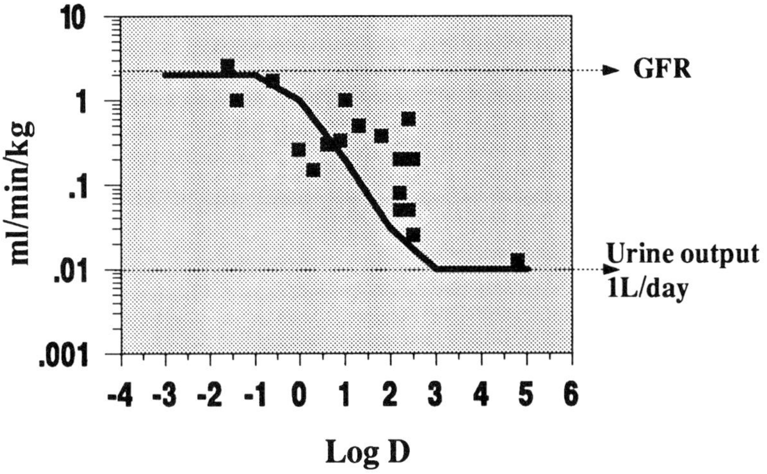

Fig. 5. *Relationship between lipophilicity and renal clearance for neutral drugs compared to a simple partioning model assuming concentration of the drug in the tubules.*

The relationships for acidic and basic drugs are more complex due to the presence of active transport systems, which increase the rate of clearance above GFR. Tubular pH also influences the reabsorption. Due to the acidic nature of urine in the tubules the reabsorption of acids is enhanced and bases decreased over that seen for neutral compounds. These shifts are due to the altered proportions of ionised drug for the two classes under the different pH conditions.

Free non-renal clearance of neutral drugs can be assumed to be metabolic in origin. Whilst important for many basic and acidic drugs active hepatic processes may also be involved, particularly when molecular weights exceed 400-450. Here the relationship to lipophilicity can be considered more complex since changes in lipophilicity are associated with increased or decreased functionality. It is hard to separate the increased lability to metabolism by cytochrome P450 seen in substitution of a molecule with say methyl, ethyl or isopropyl, with simple increased lipophilicity or with an increased lability to the relative ease of hydrogen abstraction from primary, secondary and tertiary carbon respectively [1]. In a single chemical series (benzodiazepines) the role of individual functionality can be dissected to give more general guiding rules. *Figure 6* shows that within this series, it is noticeable that the presence of an N-methyl function is associated with increased metabolic lability despite similar lipophilicities for the compounds shown.

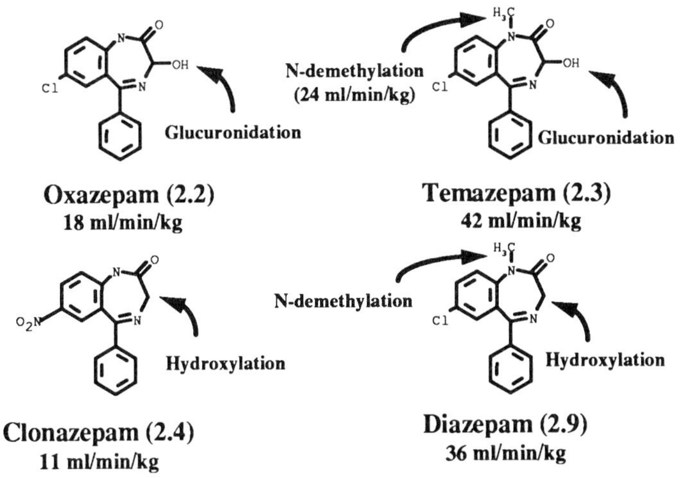

Fig. 6. *Partial free metabolic clearance rates for benzodiazepines (and log D values) illustrating the additive nature of the processes and the inportance of N-dealkylation.*

Exploration of the significance of the N-dealkylation reaction in a further subset of these compounds shows how increasing basicity of the nitrogen leads to enhanced rate of N-dealkylation (*Figure 7*), possibly due to a change in mechanism from hydrogen abstraction from the methyl to electron abstraction from the nitrogen [6].

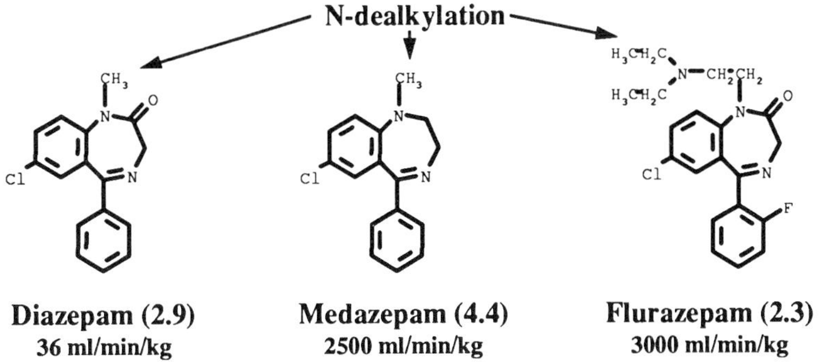

Fig. 7. *Partial free metabolic clearances (and log D values) for benzodiazepines showing increasing basicity and rate of N-dealkylation.*

By definition, such a trend leads to a breakdown in the relationship between log $D_{7.4}$ and free metabolic clearance. Thus, whilst general trends have been observed for free metabolic clearance and lipophilicity over a large data set, such trends are only global in nature and functionality is very important. *Figure 7* shows a series of Na^+ channel blockers in which N-dealkylation is the major clearance route in the more lipophilic compounds (lidocaine and bupricaine). Enhanced metabolic stability is exhibited by analogues with secondary or primary amine structures (flecainide and tocainide). The substitution of secondary or primary amine groups for tertiary amine functions, to enhance metabolic stability, has been described in detail elsewhere [1]. Alternatively overall reduction in lipophilicity by substitution of lipophilic regions distant from the site of metabolism (procainamide) also results in much reduced metabolic lability.

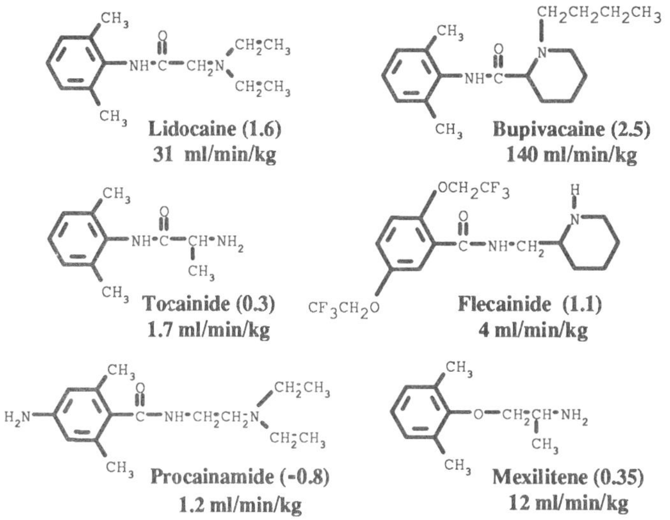

Fig. 8. *Free metabolic clearances of a series of Na^+ channel blockers (and log D values), the more lipophilic of which are subject to N-dealkylation.*

The rules governing the metabolic lability of particular functions are becoming better understood [1,7]. With cytochrome P450, metabolism of a substrate is governed by the topography of the active site, the degree of steric hindrance to access of the iron-oxygen complex to the possible sites of metabolism and the ease of electron or hydrogen abstraction from the various carbons or heteroatoms of the substrate [1]. Within this, it is readily possible to examine a structure and predict sites of metabolism. At present the actual prediction of rate is more complex. The subdivision of P450's into isoenzymes which rely heavily on substrate/ active site binding interactions, such as CYP2C9 and CYP2D6, and whose substrates thus exhibit low Km values, and those that that rely heavily on substrate chemistry (potentially high Vmax) such as CYP3A4 are important steps in this process.In the series of Na^+ channel blockers shown in *Figure 8*, the N-dealkylation reaction is principally catalysed by CYP3A4, an isoenzyme that relies principally on hydrophobic effects for substrate binding [7]. It is thus predictable that removal of functionality (possible sites of metabolism), or reduction in bulk lipophilicity should increase metabolic stability. Mexilitene, the non-amide containing analogue of the series falls within the substrate "requirements" for CYP2D6 [7] and the metabolic lability of this compound is due to aromatic and benzylic hydoxylation by this isoenzyme.

Dose Size. - Intrinsic metabolic clearance is calculated directly by the use of *in vitro* metabolism systems such as liver microsomes [8,9] and these data can be converted readily into other clearance estimates quite readily. Thus it is possible to calculate the daily dose requirements, for a metabolically cleared drug, purely from the anticipated therapeutic free drug concentration, based on receptor potency (*Figure 2*) and the predicted free metabolic clearance [1] using equation (**1**).

The model illustrated in *Figure 1* indicates that the free drug model does not discriminate between plasma and tissue binding. *Figure 9* illustrates how drug binding as indicated by free volume of distribution is influenced by lipophilicity.

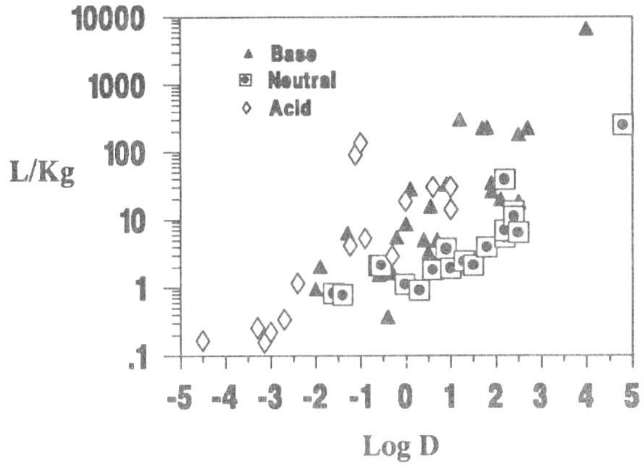

Fig. 9. *Free volume of distribution for different drug classes and its relationship with lipophilicity.*

The increase in free volume of distribution (Vd free) largely reflects association with plasma proteins (albumin) for acidic compounds and association with tissues for neutral and lipid compounds. The high association for acidic compounds with albumin is due to both ion-pair and hydrophobic interactions. Similar reasons apply for the affinity of basic compounds for tissues. Here the ion-pair and hydrophobic interactions are due to the head and tail groups of phospholipids present in cell membranes [1,10]. Similar to the prediction of intrinsic metabolic clearance, such associations can be studied with *in vitro* systems such as plasma protein and cell membrane binding [11,12]. The relationships in *Figure 8* also indicate that the volume of distribution of free drug is also predictable from physicochemical parameters alone. The volume term becomes important in understanding dose frequency since k_{el} = free Clearance/Vd free. Thus with the above knowledge of dose size, some estimate of frequency can be obtained from the predicted t½ (= 0.693/kel).

Conclusion. - The model presented forms the basis for a method of prediction of dose and effect based purely on physicochemical parameters augmented by simple *in vitro* systems. The assumptions that can be made for many compounds and targets is that there is a direct relationship between *in vitro* potency and the free drug plasma concentration required for efficacy. The free drug concentration is dependent on dose size and free drug clearance. The latter term is predictable from physicochemical and structural considerations, or from in vitro experimentation. The model could provide the basis for an expert system to maximise the use of existing physicochemical and structural data. This paper has used log D as the major physicochemical parameter. Modification of these values is needed, for further refinement, to take account of the numerous examples where H-bonding centres, in molecules, give a poor translation between the measured value and the ability of the molecule to traverse the lipid core of the membrane [1,13].

REFERENCES

[1] D.A. Smith, B.J. Jones, D.K. Walker, *Med. Res. Revs.* **1996**, *16*, 243.
[2] L. Farde, F-A Wiesel, C. Halldin, G. Sedrall, *Arch. Gen. Psychiatry* **1988**, *45*, 71.
[3] P. Seeman, *Acta Psychiatr. Scand.* **1990**, *82 (Suppl. 358),* 14.
[4] C. von Bahr, G. Movin, W-A, Yisak, K-G Jostell, M. Widman, *Acta Psychiatr. Scand.* **1990**, *82 (Suppl. 358)*, 41.
[5] L.Z. Benet, S. Oie, J.B. Schwartz, in 'Goodman and Gilmans' The Pharmacological Basis of Therapeutics', Eds. J.G. Hardman, L.E. Limbird, P.B. Molinoff, R.W. Ruddon, A.G. Gilman, McGraw-Hill, New York, **1995**, 9th ed. pp 1707-1792.
[6] F.P. Guengerich, T.L. Macdonald, *Faseb, J.* **1990**, *4*, 2453.
[7] D.A. Smith, B.K. Jones, *Biochem. Pharmacol.* **1992**, *44*, 2089.
[8] D.K. Walker, C.T. Alabaster, G.S. Congrave, M.B. Hargreaves, R. Hyland, B.C. Jones, L.J. Reed, D.A. Smith, *Drug.Met.Disp.* **1996**, *24*, 447.
[9] J.B. Houston, *Biochem. Pharmacol.* **1994**, *47*, 1469.
[10] R.P. Mason, D.G. Rhodes, L.H. Herbette, *J. Med. Chem.* **1991**, *34*, 869.
[11] J.J. Ramsden, *Experientia* **1993**, *49*, 688.
[12] P. Smejtek, S. Wang, *Arch. Environ. Contam. Toxicol.* **1993**, *25*, 394.
[13] D-C. Kim, P.S. Burton, R.T. Borchardt, *Pharm. Res.* **1993**, *10*, 1710.

PART III. COMPUTATIONAL APPROACHES IN DRUG DISPOSITION

Chapter 18

Experimental and Theoretical Predictions of Intestinal Drug Absoption

by **Katrin Palm**[a]), **Per Artursson**[a]) and **Kristina Luthman**[b])

[a]) Department of Pharmaceutics, Uppsala University, Box 580, S-751 23 Uppsala
[b]) Department of Organic Pharmaceutical Chemistry, Uppsala University, Box 574, S-751 23 Uppsala

Experimental and Theoretical Predictions of Intestinal Drug Absorption

by **Katrin Palm** and **Per Artursson**

Department of Pharmaceutics, Uppsala University, Box 580, S-751 23 Uppsala

and **Kristina Luthman**

Department of Organic Pharmaceutical Chemistry, Uppsala University, Box 574, S-751 23 Uppsala

Combinatorial chemistry and high throughput screening of pharmacological activity have resulted in a vast increase in the number of potential drug candidates in drug discovery programmes. This increases the demand for the development of screening methods for drug absorption and metabolism. In this review, we will discuss some general aspects of prediction of drug absorption and will focus in particular on two predictive methods for oral drug absorption under development in our laboratory: one experimental - the Caco-2 model, and one theoretical - based on molecular surface properties.

Introduction. – Many factors influence the oral absorption of drugs *in vivo,* e.g. drug solubility, chemical and metabolic stability, and permeability characteristics. Often, the most important factor influencing oral drug absorption is the permeability barrier of the monolayer of intestinal epithelial cells lining the gastro-intestinal tract. This barrier may be studied using intestinal segments mounted in Ussing-type chambers or intestinal cell monolayers (such as Caco-2 and HT29 grown on permeable filter supports) [1-6]. Most studies of drug transport in cell monolayers have been performed using Caco-2 cells and have been of a mechanistic nature. However, several independent correlations between cell monolayer permeability and oral absorption *in vivo* have also been established [7-9]. These studies have shown that cell monolayers can be used to identify drugs with potential absorption problems and possibly also to predict drug absorption *in vivo*. The advantages of the cell line models include a high throughput, simplified sample analysis, small amounts of substance needed, and minimal use of experimental animals. Disadvantages include the relatively low capacity of active transport systems and metabolic enzymes [1, 2, 4, 5, 10-12] and the inability to study regional differences in the gastro-intestinal tract [13].

Experimental methods based on physicochemical parameters have also been used to model drug absorption. The established relationship between passive epithelial transport and lipophilicity (octanol/water partition coefficients) is useful but imprecise [14], especially for new classes of drugs such as peptides [15, 16]. Other physicochemical parameters, such as hydrogen bonding potential reflecting desolvation energy, have been suggested to provide a better drug absorption model [17, 18]. The possibility of theoretically estimating oral drug absorption *a priori* is very desirable. Traditionally, such theoretical estimations have been based on fragmental constant methods to calculate lipophilicity. However, these methods are two-dimensional and to circumvent this drawback, new methods are being developed which are based on the three-dimensional shape. We are investigating the possibility of using dynamic molecular surface properties as predictors of epithelial permeability [19]. This method takes into account the 3D structure and the flexibility of the molecule. A combination of new *in vitro* techniques (such as the Caco-2 model) and molecular modelling can contribute to a better understanding of structure-absorption relationships.

The Epithelial Barrier of Drug Absorption. — Most drugs are transported across the intestinal epithelium by a passive diffusion process. Passive absorption can occur via the transcellular and the paracellular routes (*Fig. 1*). It is often generalised that rapidly transported lipophilic drugs are absorbed by the transcellular route, whereas slowly transported hydrophilic drugs are absorbed via the paracellular pathway. However, the paracellular pathway is restricted by the gating function of the tight junctions between the epithelial cells [20] and by the small surface area of the paracellular space, at least 1000-fold smaller than the surface area of the luminal cell membrane [21]. Therefore, many drugs exhibiting low lipophilicity may be absorbed not exclusively by the paracellular pathway but also transcellularly [22, 23].

A small number of drugs, usually nutrient-like, can utilise active carrier-mediated routes in the small intestine. It should be noted that there are also systems that mediate drug transport in a secretory direction, such as P-glycoprotein [24]. It has been speculated that large hydrophilic compounds (e.g. peptides and proteins) that are normally excluded from other pathways may be absorbed by transcytosis (*Fig. 1*). However, this pathway is of limited value in drug development due to its low capacity and the presence of proteolytic enzymes in the vesicles [25].

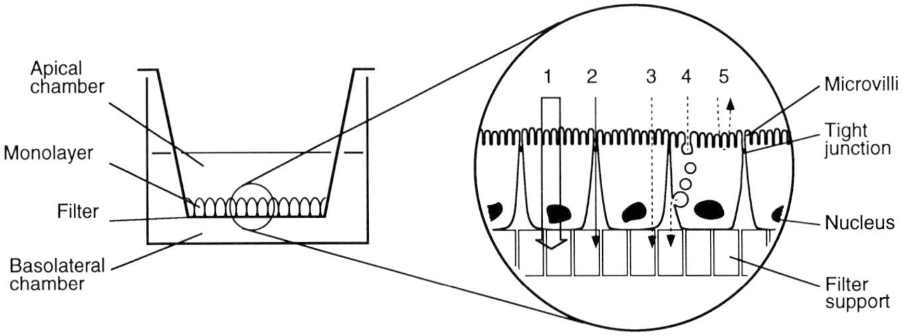

Fig. 1. *Drug transport across the intestinal epithelium may be studied in cell culture models. Caco-2 cells form confluent monolayers of well-differentiated enterocyte-like cells after 2-3 weeks in culture on porous filter supports. The enlargement shows a schematic drawing of a monolayer. The arrows indicate four absorption routes: 1 - passive transcellular, 2 - passive paracellular, 3 - active carrier-mediated, and 4 - transcytosis; and one secretory mechanism: 5 - the P-glycoprotein efflux system.*

Experimental Predictions. — Good screening tools for drug absorption should be rapid and easy to use, require small amounts of compound, be relatively inexpensive, and result in reliable predictions.

As *in vitro* models increase in complexity, more and more factors influencing oral drug absorption *in vivo* are accounted for. Unfortunately, the more closely the method approaches the *in vivo* situation, the more substance, time and money are usually required, and hence its suitability as a screening model decreases (*Fig. 2*). As the dominating barrier of intestinal drug absorption is the monolayer of epithelial cells lining the gastro-intestinal tract [26], models of this barrier should be suitable for screening/prediction of drug absorption.

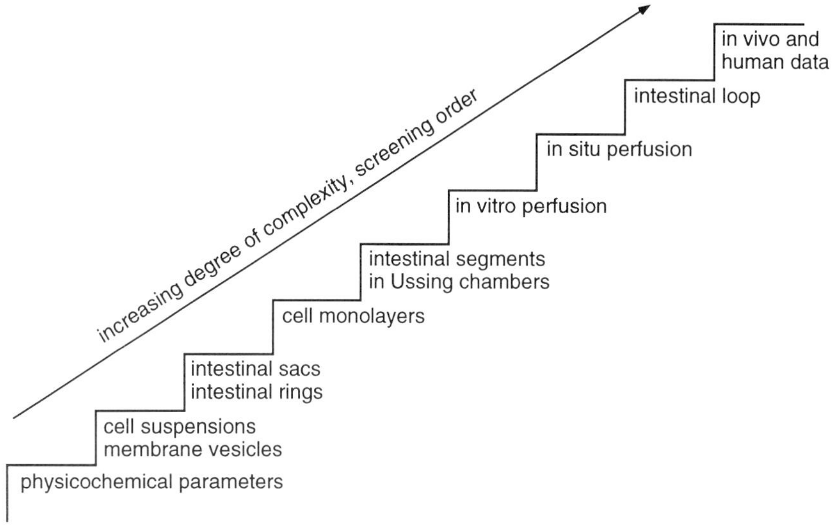

Fig. 2. *Different experimental methods for assessment of oral drug absorption. Redrawn from [27]*

Physicochemical Parameters. — Many physicochemical characteristics of drug molecules are believed to influence their ability to passively permeate the epithelium. These parameters reflect the lipophilicity, hydrogen-bonding potential, size and/or charge of drug molecules. Establishing prediction models based on physico-chemical parameters is attractive since such parameters are reasonably easy to determine experimentally and can sometimes be estimated theoretically.

Lipophilicity expressed as partition coefficients comprises two factors: hydrophobicity and polarity. The octanol/water partition coefficient (log P_{oct}) has long been used as a rule of thumb in estimations of oral drug absorption, but the relationship is usually impaired when structural diversity is introduced [7, 14, 28, 29]. Correlations between log P_{oct} and drug absorption often exhibit a sigmoidal shape which has been explained by paracellular transport of hydrophilic drugs and restricted transport of lipophilic drugs due to the unstirred water layer or extraction into cell membranes [30]. If very lipophilic drugs are included in the correlations, a parabolic relationship has been observed [31]. These models can often be improved by including a size descriptor [28, 32, 33]. Nevertheless, the predictive strength is best for very hydrophilic (log $P_{oct} < -1$) and lipophilic (log $P_{oct} > 1$) solutes, and the greatest uncertainty is between these limits [28]. Often, several physicochemical parameters are combined in a multiple linear regression analysis. This can be hazardous since physicochemical parameters (such as lipophilicity and molecular weight) are often interrelated and thereby incompatible with multiple linear regression [34, 35].

According to the pH-partition theory only the un-ionised form of the solute will partition into the cell membrane. The influence of charge has mainly been documented through the observation that the apparent distribution coefficients at the experimental pH (log D_{oct}) is better correlated to intestinal permeability than are log P_{oct} values [36]. The relationship between charge and permeability should intuitively be different for drugs transported transcellularly and paracellularly [37]. Rubas *et al.* have shown that in a series of cyclized RGD peptides assumed to be paracellularly transported, molecules with a net charge of -1 and -2 displayed the highest transport rates [38]. In contrast, a cationic selectivity has been observed in the paracellular transport of smaller drugs [39].

The hydrogen bonding potential of drug molecules has been proposed as an alternative predictor for passive transcellular transport [18]. The difference between log P_{oct} and log $P_{isooctane}$ (Δlog P), describing the desolvation energy, has been found to explain the epithelial permeation of a series of peptides more accurately than log P_{oct} [15, 16]. However, in a study by Buur *et al.*, log D_{oct} was found to be better correlated with permeability coefficients of 5-fluorouracil prodrugs in Caco-2 monolayers than Δlog P [40]. Apart from these studies, few comparisons of methods based on log P_{oct}/log D_{oct} and Δlog P have been published [36]. Generally, it can be expected that either a too low or too high lipophilicity, or a very high hydrogen bonding potential, will result in poor epithelial permeability [36, 41].

Caco-2 Cell Monolayers. — Caco-2 is a heterogeneous cell line derived from a human colon adenocarcinoma [42, 43]. Caco-2 cells form monolayers of well-differentiated polarised enterocyte-like cells with tight junctions when cultured on porous filter supports (*Fig. 1*). Despite their colonic origin, Caco-2 cells express many brush-border enzymes [e.g. 4, 5, 44] and active transport systems [e.g. 10-12].

Correlations between the fraction absorbed *in vivo* after oral administration and Caco-2 cell monolayer permeability have been established in several laboratories (*Fig. 3*) [7, 8, 11, 45]. These sigmoidal correlations are qualitatively similar but are shifted along the log permeability axis [22]. Therefore, each research unit wanting to use this model for prediction of oral drug absorption has to establish their own correlation between the fraction absorbed and cell monolayer permeability. This discrepancy among laboratories may be due to differences in experimental conditions and/or in the cell line itself [22].

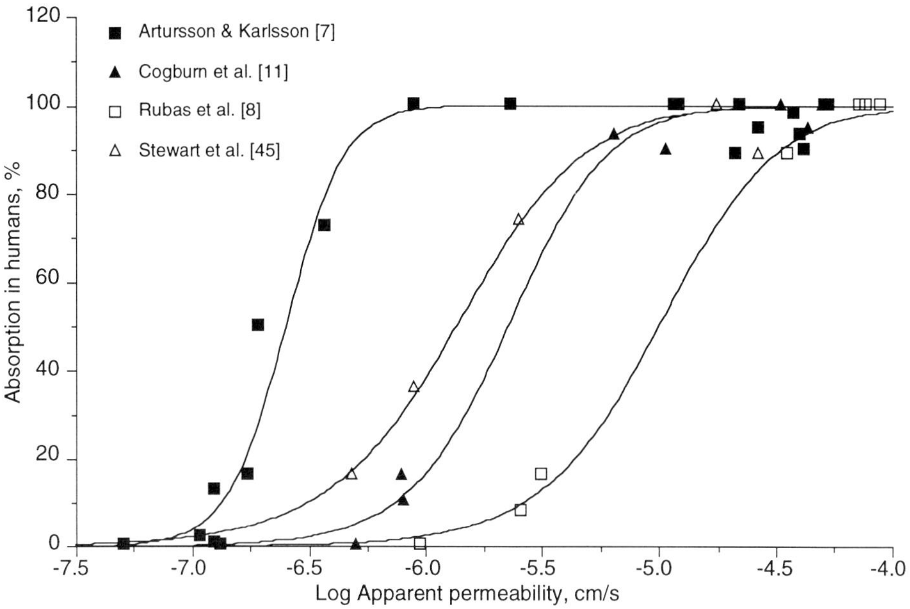

Fig. 3. *Correlation between the absorbed fraction of various drugs in humans after oral administration and Caco-2 monolayer permeability, obtained in four different laboratories [7, 8, 11, 45]. Reproduced from ref. [22] with permission from the publisher.*

How do Caco-2 cell monolayer permeability values compare with those in the human jejunum? This was recently investigated for three different classes of compounds, those undergoing rapid passive transportation, slow passive transportation, and active transportation [46]. Caco-2 cell monolayer permeability coefficients (P_c) [47] were compared with those obtained in the human jejunum *in vivo* with a double balloon perfusion technique (Loc-I-Gut) [48]. The permeability coefficients of the rapidly transported lipophilic drugs were quantitatively similar (2- to 4-fold difference) whereas the permeability coefficients of the slowly transported hydrophilic drugs were 30- to 80-fold lower in the Caco-2 model than in the human jejunum [46]. This shows that the Caco-2 monolayer is a good model for transcellular passive transport. The low paracellular permeability of the Caco-2 cells reflects the colonic origin of the cells and results in a large quantitative difference between the models. This discrepancy may be a result of differences in the paracellular pathway and in the absorptive surface area [22]. Previous studies indicate that the difference is quantitative rather than qualitative [49, 50]. New cell models with higher paracellular permeability, i.e. more similar to the *in vivo* situation, are therefore of interest. Two such cell lines displaying lower transepithelial resistance and higher paracellular permeability have been introduced: IEC-18, a normal but relatively undifferentiated cell line [51], and 2/4/A1, a conditionally immortalised intestinal cell line [52, 53]. The third class of compounds investigated by Lennernäs *et al.*, the actively transported compounds, was found to have a more than 100-fold lower absorption rate in the Caco-2 model than in the human jejunum [46]. This result can be explained by the low capacity of some active transport systems in Caco-2 cells [10, 12].

Theoretical Predictions. — *Calculated Physicochemical Parameters.* — More than twenty years ago, fragmental constant methods to calculate log P_{oct} were introduced [54, 55]. Since then, several similar methods have been developed [e.g. 56]. The accuracy of these calculations differs among classes of compounds, due to the quality of the fragmental constants used [57, 58]. The two-dimensionality of these methods will probably affect the log P_{oct} estimations negatively with increasing size and flexibility of the molecules. A promising new approach involves the calculation of the molecular lipophilicity potential (MLP) [59]. The lipophilicity of the molecule can be calculated from the lipophilicity potential mapped on the molecular surface, thus taking into consideration both the 3D shape and intramolecular interactions [60, 61].

Conradi *et al.* found a good correlation between the apparent Caco-2 monolayer permeability and the number of potential hydrogen bonds that small peptides form with water [15]. This provides a rapid method for the estimation of hydrogen bonding potential and desolvation energy. However, no correlation was found between the number of potential hydrogen bonds and penetration through the blood-brain barrier for a series of H_2-antagonists and other CNS penetrating compounds [62]. As pointed out by Conradi and co-workers, this method fails to account for intramolecular *vs.* intermolecular hydrogen bonding and steric or electronic influences on solute-solvent hydrogen bond strengths. A more accurate thermodynamic description, taking into consideration the acidity and basicity of different hydrogen bonding atoms, has been introduced by Raevsky *et al.* [63, 64].

Molecular Surface Properties. — The molecular surface reflects the 3D shape of the molecule and its properties determine how the compound is perceived by the surroundings. In a recent study, we therefore investigated whether molecular surface properties could model the passive epithelial permeation potential of drugs [19]. A series of six beta-receptor antagonists covering a wide range of lipophilicities (corresponding to the interval of log P_{oct} for which it is difficult to obtain reliable predictions of drug absorption [28]), but with similar pK_a values and molecular weights, was used as model compounds. We found a good correlation between the size of the dynamic polar van der Waals surface area and epithelial permeation in Caco-2 monolayers and rat ileal segments (*Fig. 4*) [19]. The polar surface area was defined as the area associated with nitrogen and oxygen atoms and the hydrogen atoms bonded to these. Since the surface properties of a molecule are influenced by its conformation, the dynamic average of the surface areas was calculated [65] (*Fig. 5*).

Similar correlations to those presented in *Fig. 4* were obtained using the water accessible surface areas. The linear shape of the correlation between the polar surface area and log P_c in Caco-2 monolayers can be explained by the use of permeability coefficients corrected for the unstirred water layer (P_c) [47]. Interestingly, despite the similar size of the drugs, poor correlations (r^2 = 0.54-0.69) were obtained between the nonpolar surface area and the permeability data. The epithelial permeability for these drugs was also correlated with log D_{oct} values calculated according to the method of Hansch and Leo [54, 66]. The calculated log D_{oct} values ranked the intestinal epithelial permeability values of the drugs in an erroneous order and resulted in linear correlations of decreased statistical quality (*Fig. 4*). With only six observations it is difficult to draw conclusions on whether the correlations are linear or sigmoidal, therefore, the interpretation of the r^2 values should be made with caution.

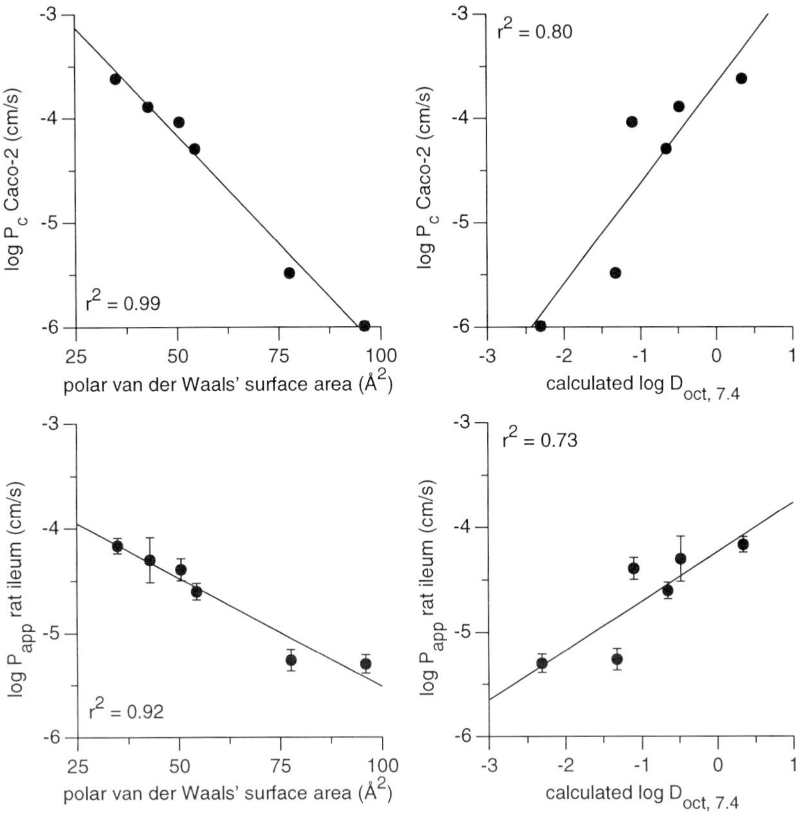

Fig. 4. *Correlations between structural descriptors of a series of beta-blocking agents (dynamic polar van der Waals area and calculated log D_{oct}) and the logarithm of the permeability coefficients determined in Caco-2 monolayers and rat ileal segments. The polar surface area was defined as the area associated with oxygen and nitrogen atoms and hydrogen atoms attached to these heteroatoms. The log D_{oct} values were calculated according to the method of Hansch and Leo [54, 66]. Reproduced with permission from the publisher [19].*

The conformational influence on molecular properties has also been discussed by others. In MLP calculations using esters of L-dopa it was shown that calculated log P values of different conformers suggested considerable differences in lipophilicity [61, 67].

What does the polar surface area describe? It has been found to be correlated with physicochemical properties such as lipophilicity [68, 69], solubility [70] and desolvation energy [71]. In a recent study by van de Waterbeemd *et al.* the relationship between Caco-2 permeability and several calculated molecular descriptors was investigated [72]. In this

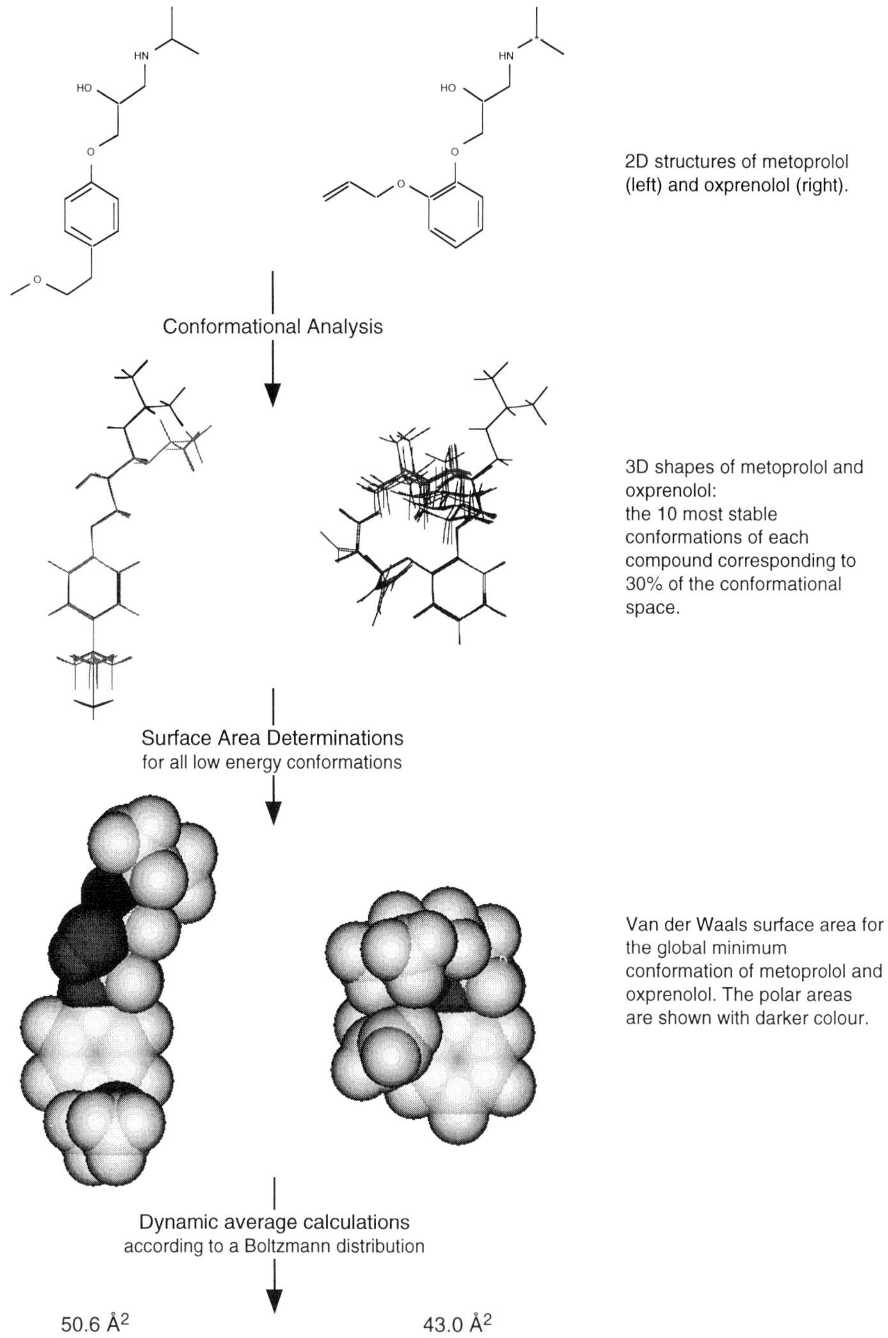

Fig. 5. *Strategy for dynamic surface area calculations [19].*

study the polar surface area was considered to be a factor of hydrogen bonding. The static polar surface areas were found to provide only a rough estimate of the total hydrogen bonding capacity and instead were more strongly correlated with the hydrogen bond accepting capability. The best correlations with cell monolayer permeabilities were found with calculated log P_{oct} and total hydrogen bonding capacity [72].

Conclusion. — Prediction of intestinal drug absorption is relevant in the early stages of drug development. As an experimental method, Caco-2 cell monolayers have been shown to be a useful screening tool, especially for passive transcellular absorption. In recent years the interest in theoretical predictions of drug absorption has increased [19, 36, 62, 68, 72, 73]. This can be attributed partly to the introduction of molecular modelling as a tool in this field and partly to an increased understanding of the physicochemical properties governing the drug absorption process. We are investigating a new theoretical model for the prediction of intestinal drug absorption using dynamic molecular surface properties, and good correlations have been found between the size of the molecular polar surface area and epithelial permeability for homologous series of drugs [19]. However, to improve the applicability of the model to more structurally diverse molecules, other factors such as size and charge should also be investigated. Studies of this nature are currently under way in our laboratory.

By combining new *in vitro* techniques and molecular modelling, it is hoped that a better understanding of the structural requirements for oral absorption and thereby better *a priori* prediction models can be obtained.

REFERENCES

[1] P. Artursson, *Crit. Rev. Ther. Drug Carrier Syst.* **1991**, 8, 305-330.

[2] K.M. Hillgren, A. Kato, R.T. Borchardt, *Med. Res. Rev.* **1995**, 15, 83-109.

[3] A.R. Hilgers, R.A. Conradi, P.S. Burton, *Pharm. Res.* **1990**, 7, 902-910.

[4] I.J. Hidalgo, T.J. Raub, R.T. Borchardt, *Gastroenterology* **1989**, 96, 736-749.

[5] M. Pinto, S. Robine-Leon, M.-D. Appay, M. Kedinger, N. Triadou, E. Dussaulx, B. Lacroix, P. Simon-Assmann, K. Haffen, J. Fogh, A. Zweibaum, *Biol. Cell* **1983**, 47, 323-330.

[6] G. Wilson, C.J. Hassan, I. Williamson, R. Shah, M. Mackay, P. Artursson, *J. Contr. Rel.* **1990**, 11, 25-40.

[7] P. Artursson, J. Karlsson, *Biochem. Biophys. Res. Commun.* **1991**, 175, 880-885.

[8] W. Rubas, J. Villagran, M. Cromwell, A. McLeod, J. Wassenberg, R. Mrsny, *S.T.P. Pharma Sci.* **1995**, 5, 93-97.

[9] P. Wils, A. Warnery, V. Phung-Ba, D. Scherman, *Cell Biol. Toxicol.* **1994**, 10, 393-397.

[10] I.J. Hidalgo, R.T. Borchardt, *Biochim. Biophys. Acta* **1990**, 1028, 25-30.

[11] J.N. Cogburn, M.G. Donovan, C.S. Schasteen, *Pharm. Res.* **1991**, 8, 210-216.

[12] M. Hu, R.T. Borchardt, *Pharm. Res.* **1990**, 7, 1313-1319.

[13] W. Rubas, N. Jezyk, G.M. Grass, *Pharm. Res.* **1993**, 10, 113-118.

[14] Y.C. Martin, *J. Med. Chem.* **1981**, 24, 229-237.

[15] R.A. Conradi, A.R. Hilgers, N.F.H. Ho, P.S. Burton, *Pharm. Res.* **1991**, 8, 1453-1460.

[16] R.A. Conradi, A.R. Hilgers, N.F.H. Ho, P.S. Burton, *Pharm. Res.* **1992**, 9, 435-439.

[17] R.A. Conradi, A.R. Hilgers, T.J. Vidmar, P.S. Burton, *Quant. Struct.-Act. Relat.* **1994**, 13, 4-10.

[18] R.C. Young, R.C. Mitchell, T.H. Brown, C.R. Ganellin, R. Griffiths, M. Jones, K.K. Rana, D. Saunders, I.R. Smith, N.E. Sore, T.J. Wilks, *J. Med. Chem.* **1988**, 31, 656-671.

[19] K. Palm, K. Luthman, A.-L. Ungell, G. Strandlund, P. Artursson, *J. Pharm. Sci.* **1996**, 85, 32-39.

[20] M. Furuse, T. Hirase, M. Itoh, A. Nagafuchi, S. Yonemura, S. Tsukita, S. Tsukita, *J. Cell Biol.* **1993**, 123, 1777-1788.

[21] J.R. Pappenheimer, K.Z. Reiss, *J. Membr. Biol.* **1987**, 100, 123-136.

[22] P. Artursson, K. Palm, K. Luthman, *Adv. Drug Deliv. Rev.* **1996**, in press.

[23] H.N. Nellans, *Adv. Drug Deliv. Rev.* **1991**, 7, 339-364.

[24] J. Karlsson, S.-M. Kuo, J. Ziemniak, P. Artursson, *Br. J. Pharmacol.* **1993**, 110, 1009-1016.

[25] L. Lazorová, P. Artursson, Å. Engström, A. Sjölander, *Am. J. Physiol.* **1996**, 270, G554-G564.

[26] M.J. Jackson in 'Physiology of the Gastrointestinal Tract', Eds. L.R. Johnson, Raven Press, New York, 1987, pp. 1597-1621.

[27] A.-L. Ungell, in 'Current Status on Targeted Drug Delivery to the Gastrointestinal Tract', London: Capsugel Library, 1993, pp 79-84.

[28] D.E. Leahy, J. Lynch, D.C. Taylor in 'Novel Drug Delivery and its Therapeutic Application', Eds. L.F. Prescott, W.S. Nimmo, John Wiley & Sons Ltd, 1989, pp. 33-44.

[29] S. Chong, S.A. Dando, K.M. Soucek, R.A. Morrison, *Pharm. Res.* **1996**, 13, 120-123.

[30] D. Winne, *Acta Pharm. Technol.* **1987**, 33, 53-59.

[31] P. Wils, A. Warnery, V. Phung-Ba, S. Legrain, D. Scherman, *J. Pharmacol. Exp. Ther.* **1993**, 269, 654-658.

[32] A. Wilschut, W.F. ten Berge, P.J. Robinson, T.E. McKone, *Chemosphere* **1995**, 30, 1275-1296.

[33] R.O. Potts, R.H. Guy, *Pharm. Res.* **1992**, 9, 663-669.

[34] R.O. Potts, R.H. Guy, *Pharm. Res.* **1993**, 10, 635-636.

[35] P.S. Burton, *Pharm. Res.* **1993**, 10, 636-637.

[36] A.M. ter Laak, R.S. Tsai, G.M. Donné-Op den Kelder, P.-A. Carrupt, B. Testa, H. Timmerman, *Eur. J. Pharm. Sci.* **1994**, 2, 373-384.

[37] N.F.H. Ho, T.J. Raub, P.S. Burton, C.L. Barsuhn, A. Adson, K.L. Audus, R.T. Borchardt in 'Transport Processes in Pharmaceutical Systems', Eds. K.J. Himmelstein, G.L. Amidon, P.I. Lee, Marcel Dekker, Inc., 1996, in press.

[38] W. Rubas, M. Cromwell, T. Gadek, D. Narindray, R. Mrsny, *Mat. Res. Soc. Symp. Proc.* **1994**, 331, 179-185.

[39] J. Karlsson, A.-L. Ungell, P. Artursson, *Pharm. Res.* **1994**, 11, S-248.

[40] A. Buur, L. Trier, C. Magnusson, P. Artursson, *Int. J. Pharm.* **1996**, 129, 223-231.

[41] C.A. Lipinski, in 'AAPS 10th Annual Meeting and Exposition. Accelerating Drug Discovery and Development', Miami Beach, Florida: American Association of Pharmaceutical Scientists, 1995, pp 20.

[42] P. Artursson, *J. Pharm. Sci.* **1990**, 79, 476-482.

[43] P.H. Vachon, J.-F. Beaulieu, *Gastroenterology* **1992**, 103, 414-423.

[44] E. Walter, T. Kissel, *Pharm. Res.* **1994**, 11, 1575-1580.

[45] B.H. Stewart, O.H. Chan, R.H. Lu, E.L. Reyner, H.L. Schmid, H.W. Hamilton, B.A. Steinbaugh, M.D. Taylor, *Pharm. Res.* **1995**, 12, 693-699.

[46] H. Lennernäs, K. Palm, U. Fagerholm, P. Artursson, *Int. J. Pharm.* **1996**, 127, 103-107.

[47] J. Karlsson, P. Artursson, *Int. J. Pharm.* **1991**, 71, 55-64.

[48] H. Lennernäs, Ö. Ahrenstedt, R. Hällgren, L. Knutson, M. Ryde, L.K. Paalzow, *Pharm. Res.* **1992**, 9, 1243-1251.

[49] P. Artursson, A.-L. Ungell, J.-E. Löfroth, *Pharm. Res.* **1993**, 10, 1123-1129.

[50] Y. Tanaka, Y. Taki, T. Sakane, T. Nadai, H. Sezaki, S. Yamashita, *Pharm. Res.* **1995**, 12, 523-528.

[51] T.Y. Ma, D. Hollander, D. Bhalla, H. Nguyen, P. Krugliak, *J. Lab. Clin. Med.* **1992**, 120, 329-341.

[52] V. Milovic, S. Olsson, J. Hochman, E.C.A. Paul, P. Artursson, *Gastroenterology Suppl.* **1996**, 110, A822.

[53] E.C.A. Paul, J. Hochman, A. Quaroni, *Am. J. Physiol.* **1993**, 265, C266-C278.

[54] A. Leo, P.Y.C. Jow, C. Silipo, C. Hansch, *J. Med. Chem.* **1975**, 18, 865-868.

[55] R.F. Rekker in 'The Hydrophobic Fragmental Constant', Eds. W.T. Nauta, R.F. Rekker, Elsevier Scientific Publishing Company, Amsterdam, 1977, Vol. 1.

[56] A.K. Ghose, G.M.J. Crippen, *J. Comp. Chem.* **1986**, 7, 565-577.

[57] R. Mannhold, K.P. Dross, R.F. Rekker, *Quant. Struct.-Act. Relat.* **1990**, 9, 21-28.

[58] R. Mannhold, R.F. Rekker, C. Sonntag, A.M. ter Laak, K. Dross, E.E. Polymeropoulos, *J. Pharm. Sci.* **1995**, 84, 1410-1419.

[59] E. Audry, J.P. Dubost, J.C. Colleter, P. Dallet, *Eur. J. Med. Chem.* **1986**, 21, 71-72.

[60] Q. Du, G.A. Arteca, *J. Comput. Aided Mol. Design* **1996**, 10, 133-144.

[61] B. Testa, P.-A. Carrupt, P. Gaillard, F. Billois, P. Weber, *Pharm. Res.* **1996**, 13, 335-343.

[62] H. van de Waterbeemd, M. Kansy, *Chimia* **1992**, 46, 299-303.

[63] O.A. Raevsky, V.Y. Grigor'ev, D.B. Kireev, N.S. Zefirov, *Quant. Struct.-Act. Relat.* **1992**, 11, 49-63.

[64] O.A. Raevsky, K.-J. Schaper, J.K. Seydel, *Quant. Struct.-Act. Relat.* **1995**, 14, 433-436.

[65] K.B. Lipkowitz, B. Baker, R. Larter, *J. Am. Chem. Soc.* **1989**, 111, 7750-7753.

[66] P.N. Craig in 'Cumulative Subject Index & Drug Compendium', Eds. C.J. Drayton, Pergamon Press, Oxford, 1990, Vol. 6, pp. 237-991.

[67] M. Brunner-Guenat, P.A. Carrupt, G. Lisa, B. Testa, S. Rose, K. Thomas, P. Jenner, P. Ventura, *J. Pharm. Pharmacol.* **1995**, 47, 861-869.

[68] D. Barlow, T. Satoh, *J. Contr. Rel.* **1994**, 29, 283-291.

[69] W.J. Dunn III, M.G. Koehler, S. Grigoras, *J. Med. Chem.* **1987**, 30, 1121-1126.

[70] G.L. Amidon, S.H. Yalkowsky, S.T. Anik, S.C. Valvani, *J. Phys. Chem.* **1975**, 79, 2239-2246.

[71] T. Ooi, M. Oobatake, G. Némethy, H.A. Scheraga, *Proc. Natl. Acad. Sci. U.S.A.* **1987**, 84, 3086-3090.

[72] H. van de Waterbeemd, G. Camenisch, G. Folkers, O.A. Raevsky, *Quant. Struct.-Act. Relat.* **1996**, submitted.

[73] M.D. Ribadeneira, B.J. Aungst, C.J. Eyermann, S.-M. Huang, *Pharm. Res.* **1996**, 13, 227-233.

PART III. COMPUTATIONAL APPROACHES IN DRUG DISPOSITION

Chapter 19

QSAR Approaches in Mutagenicity and Carcinogenicity Estimation

by **Romualdo Benigni**[a]) and **Alessandro Giuliani**[b])

[a]) Instituto Superiore di Sanità, Laboratory of Comparative Toxicology and Ecotoxicology, Viale Regina Elena 299, I-00161 Rome, Italy
[b]) Sigma-Tau, Via Pontina Km 30.400, I-00040 Pomezia (Roma), Italy

QSAR Approaches in Mutagenicity and Carcinogenicity Estimation

by **Romualdo Benigni**

Instituto Superiore di Sanità, Laboratory of Comparative Toxicology and Ecotoxicology, Viale Regina Elena 299, I-00161 Rome, Italy

and **Alessandro Giuliani**

Sigma-Tau, Via Pontina Km 30.400, I-00040 Pomezia (Roma), Italy

Although QSAR is primarily a modern technique pioneered in the drug industry, the concept of the relationship between chemical structure and biological activity was established in the nineteenth century in the area of toxicology. Moreover, after the modern formulation of these ideas in the 1960's in the field of medicinal chemistry, it soon became apparent that the QSAR models could be used to formalize and investigate any type of biological activity, including toxicity. The frequency and the systematicness with which the QSAR approaches are used in medicinal chemistry are unparalled in toxicology. There are also great differences in the use of QSAR in the various areas of toxicology. This work presents the QSAR studies produced for the mutagenicity and carcinogenicity endpoint. As in classical QSAR applications to medicinal chemistry, various studies relative to individual classes of mutagens and carcinogens describe how the potency of the active chemicals in each class varies according to the variation of chemical structure/properties. However, in mutagenicity and carcinogenicity QSAR studies, the need for investigating the difference between active and nonactive chemicals has a primary importance, since risk assessment is concerned first with this issue, and second with the potency of the active compounds. A number of examples show how fundamentally distinct types of QSAR models may exist for the same chemical class: those that model the conditions for distinguishing between activity classes, e.g., “actives” and “inactives”, and those that model the conditions for modulating potencies among the “actives”. Moreover, the needs relative to the practice of risk assessment have motivated attempts to construct general QSAR models (e.g., for predicting chemical carcinogenicity), not tailored to congeneric series of chemicals, with the ambitious hope that these models would be valid for all kind of chemicals. The performance of these attempts in validation studies has usually been limited, and the results of the predictions for the individual chemicals have shown that the various SAR and QSAR approaches essentially acted as gross "class-identifiers": they pointed to the presence or absence of alerting chemical functionalities, but were not able to make gradations within each potentially harmful class. The possible reasons of these limitations, together with future perspectives are discussed.

Introduction. - Although QSAR is primarily a modern technique pioneered in the drug industry, the concept of the relationship between chemical structure and biological activity was established in the nineteenth century in the area of toxicology. Moreover, after the modern formulation of these ideas in the 1960's in the field of medicinal chemistry, it soon became apparent that the QSAR models could be used to formalize and investigate any type of biological activity, including toxicity. However, the frequency and the systematicness with which the QSAR approaches are used in medicinal chemistry are unparalleled in toxicology. There are also great differences in the use of QSAR in the various areas of toxicology. The toxicological area which has seen the highest diffusion of QSAR methods is ecotoxicology. Many of these QSARs fall into the category of non-polar narcosis, which is considered as based on a non-specific physiological effect. These models contain only a log P term, and describe what is termed "baseline toxicity", that is the minimal toxic effect that a chemical will exert. However, other chemicals exert their ecotoxic effects by polar narcosis, or by reaction with specific biological targets, and in these cases their activity is more difficult to model and predict [1,2]. Also acute toxicity is sometimes modeled by lipophilicity alone, but in other cases more reactive compounds with other modes of toxic action require more terms (often electronic) in their QSAR equations [3,4]. Chronic toxicity (mutagenicity and carcinogenicity) most often involves more subtle problems than only aspecific interaction, and the damage that chemicals can do to DNA presents a more complex situation. This review presents the QSAR studies produced for the mutagenicity and carcinogenicity endpoint; previous review papers were [5,6,7,8].

QSARs of Chemical Mutagens and Carcinogens. - As in classical QSAR applications to medicinal chemistry, various studies relative to individual classes of mutagens and carcinogens describe how the potency of the active chemicals in each class varies according to the variation of chemical structure/properties (*Table 1*). The interest of the investigators was concentrated mainly on a number of classes with environmental relevance, such as aromatic amines, nitroarenes and N-nitroso-compounds for mutagenicity, and aromatic amines and polycyclic aromatic hydrocarbons for carcinogenicity. Many of the mutagenicity studies relied on the use of the bacterial *Salmonella typhimurium* assay system, since it identifies a large spectrum of point mutations, and it resulted to be the short-term assay with the highest predictive power with respect to rodent carcinogenicity [9,19].

Table 1. *Classes of Mutagens and Carcinogens Studied with QSAR Approaches*

Class	System
Mutagenicity studies	
Benzidines	TA98, TA100
Halogenated aliphatic hydrocarbons	*Aspergillus nidulans*
Aromatic and heteroaromatic amines (with acetamides and azides)	TA98, TA100
Hydroxyfuranones	TA100
Nitroarenes	TA98, TA100, *E. Coli*
Nitrofuranes	*E. Coli*
Nitrostilbene	TA98, TA100
N-Nitroso amines	TA1535
Paracetamol analogues	V79
Platinum	TA92
Quinolines	TA100
Triazenes	TA98
Carcinogenicity studies	
Aromatic amines	
Chloroethanes, chloroethylenes	
N-Nitroso amines	
N-Nitroso compounds	
Polychlorobiphenyls, dioxins	
Polycyclic aromatic hydrocarbons	

Lipophilicity plays a central role in many cases, for example in the bacterial mutagenicity of triazenes [11], aromatic amines [12], nitroarenes [13], and quinolines [14], and in the rodent carcinogenicity of aromatic amines [15] and polycyclic aromatic hydrocarbons [16] (for the role of lipophilicity, see also [7,17]. In a few cases, lipophilicity apparently gave no contribution (e.g. Salmonella mutagenicity of Platinum amines [18] and induction of aneuploidy by halogenated aliphatic hydrocarbons in *Aspergillus nidulans* [19,20]. However, even when lipophilicity is the most important determinant, usually the QSAR model contains other terms, generally electronic [7]. This is rationalized in terms of electronic interaction necessary to the molecules for establishing either weak or covalent bonds with the biological macromolecules.

In recent years, Hansch has applied his comparative QSAR criteria to the analysis of existing models for bacterial mutagenicity. The presence of the log P parameter in the majority of existing equations allowed a determination that its coefficient falls in the range 0.6-1.1. This is in agreement with what has been found on a larger database of about 3000 equations relative to different biological activities, where it normally falls between 0.4-1.1 [21]. When the coefficient is near to 0.5, a substituent contacts a surface, whereas it is near 1.0 when it is engulfed in a hydrophobic pocket [22,23].

Another important source of information is provided by the QSAR studies performed on anticancer drugs. Several classes of antitumor drugs are also carcinogens, and exert their action by binding to the DNA with the same mechanism by which they induce cancer. An excellent review on this topic is [24]. Among the chemical classes of carcinogens which are also used as antitumor drugs, there are the alkylating agents, which bind covalently with nucleophilic centers in vivo. Their QSARs have shown a prevalence of electronic effects, with lipophilicity being of secondary importance. For the nitrosoureas, lipophilicity is the main factor governing the activity, whereas quinones which act against tumors are thought to function as bioreductive alkylating agents, since electronic factors are of primary importance in their QSAR equations [24].

The above examples suffice to indicate how chronic toxicity is characterized by a scenario composed of very different QSARs, contrary to what happens in general toxicity where, very often, only one parameter (lipophilicity) rationalizes the effects of different types of chemicals, in different systems. The complexity of the QSARs for mutagens and carcinogens clearly points to the existence of mechanisms of action, which are specific for the individual classes, whereas general toxicity is often exerted by aspecific mechanisms.

Activity *versus* Potency: Different QSAR Models. - The QSAR applications presented in the above Section are typical of the way the classical QSAR approaches are used in medicinal chemistry, where only the active compounds belonging to one individual chemical class are considered, and a QSAR model describes how the potency of the active compounds is modulated by small changes of the molecular features. Only very rarely is the medicinal chemist also interested in what distinguishes the active compounds from those that show no activity. On the contrary in mutagenicity and carcinogenicity QSAR studies, the need for investigating the difference between active (positive) and nonactive (negative) chemicals has a primary importance, since risk assessment is concerned first with this issue, and secondly with the potency of the active compounds. The following example shows how fundamentally distinct types of QSAR models may exist for the same chemicals class: those that model the conditions for distinguishing between activity classes, e.g. "actives" and "inactives", and those that model the conditions for modulating potencies among a group of chemicals belonging to a common activity class, i.e. "actives".

The QSARs for mutagenic potency of the two classes of aromatic and heteroaromatic amines, and nitroarenes in the *Salmonella* TA98 and TA100 strains had been previously elucidated in the laboratory of Hansch [12,13]. Log P played a major role in modeling potency variations in both classes, in both assay systems. HOMO energy was the second determinant for the amines, and LUMO energy was the second determinant for the nitroarenes. A similar pattern had been found for the mutagenic potency of nitroarenes in the assay for the SOS repair in *Escherichia coli* PQ37 [25]. The data relative to the 3 studies were reanalyzed in our laboratory, in order to identify the determinants of the difference between positives and negatives. In all cases, the F-statistics of log P was very low, and not statistically significant, whereas electronic factors played major roles for the discrimination. For the aromatic amines, the nonmutagenic chemicals tended to have lower HOMO values (poor oxidability), higher MR-ortho values (sterically hindered -NH_2 groups), and lower MR values. On the contrary, LUMO was the major source of difference between mutagenic and nonmutagenic nitroarenes. The accuracy of the individual discriminant models was mediocre, but all together they provided a consistent picture, with the determinants for the potency being largely different from the determinants for the yes/no activity [26]. Debnath et al. [12] proposed that the limitation of the log P-based QSAR in identifying the nonmutagenic amines may be caused by the depression of activity due to concurrent toxicity, and that a number of chemicals would actually be slightly positive without such an interference. Benigni *et al.* [26] proposed the alternative hypothesis that the rate limiting steps for determining yes/no activity and potency, respectively, may actually be different, and may be ruled by different structural determinants. Whereas the operational paradigms of medicinal chemistry assume that all the chemicals in a homogeneous series should be active, even though some may be active at very low levels (asymptotic character of the "no-activity" concept), the evidence provided by [26] points to the view that the lack of activity is manifested as an abrupt discontinuity in activity. This view would imply that the QSAR models for mutagenicity and carcinogenicity should consider the possibility of nonlinear relationships between the potency QSAR and the yes/no activity QSAR.

Presently, further work is being performed in our laboratory to refine the QSAR model for distinguishing between mutagenic and nonmutagenic aromatic amines. This work generated the following results (our unpublished results). The importance of HOMO both in the potency QSAR (combined with log P) and in the yes/no activity QSAR is consistent with the fact that these amino compounds are mutagenic only in the presence of the S9 microsomal preparation, because they are oxidized to the mutagenic hydroxylamine by cytochrome P-450 residing in S9. We have further explored this point by calculating the partial HOMO contributions on the ring Carbons, in order to describe a possible relationship between ease of ring hydroxylation and mutagenicity. We also considered several 3-D parameters (such as inertia moments, lengths of the principal inertia axes, etc..) calculated with the TSAR program (Oxford Molecular, Ltd.). A further attempt was performed by calculating 3-D similarity indices with the ASP program (Oxford Molecular, Ltd). The similarity calculations were performed after having separated the chemicals into homogeneous subsets (single ring, biphenyls, anthracenes, etc...). The indices were calculated between the most potent amine in each subset, and all the other chemicals in the subset. Whereas the first two approaches (partial HOMO, 3D-parameters) were unsuccessful, the ASP-based 3-D similarity index separated fairly well the mutagens from the nonmutagens for the subgroup of diphenyl methanes (about 90% accuracy). This result was confirmed by the fact that, for this specific subclass, an equivalent QSAR model was obtained by using MR-ortho values, thus pointing to a strong steric effect. This result also stimulated us to reconsider the entire matter, by using again the classical descriptors; instead of treating all the amines together, we performed separate analyses for the different subgroups. We found that: a) the chemicals with two or more fused rings were all mutagenic; b) the diphenyl methanes were modeled by MR-ortho (steric effect); c) the biphenyls also displayed influence from steric (ASP shape similarity) effects; d) the single-ring amines were modeled by electronic parameters (*Table 2*).

Table 2. *Salmonella (TA100) Mutagenicity of Aromatic and Heteroaromatic Amines*

Subclass	n	Variable	F	Accuracy
Single ring	61	εHOMO	12.7	70.5%
		εLUMO-εHOMO	8.8	
Biphenyls	15	ASP Shape Similarity	6.2	86.7(80.0)%
Diphenyl-methanes	15	MR-o	48.9	93.3%
Fused rings	35	All positive compounds		

The table reports the QSAR models relative to the yes/no activity of the individual subsets of amines. For each subclass, the number (n), the variables entered into the model, the F-values of the variables, and the resulting discrimination accuracy (jackknifed cross-validated) is given.

This overall evidence indicates that the formulation of a QSAR for yes/no activity may be more difficult than that for the potency of the active compounds. Normally, the active chemicals of a certain chemical class are highly homogeneous, since they belong to the same "nominal" chemical class, and they are active for the same biological endpoint; thus they are likely to act by the same mechanism, and can be modeled by one QSAR. On the contrary, chemicals which look similar to the chemist's eye, but are separated into actives and inactives, may actually be deeply different in terms of interaction with the biological systems, thus making it impossible to use the same QSAR model. This leads to the second lesson learned from the above example, i.e. the need to investigate the degree of homogeneity (congenericity) of apparently similar chemicals.

From Congeneric to Noncongeneric Data Sets: a New QSAR Issue. - The previous study shows that the toxicological QSAR investigations, whereas they use the concepts and tools elaborated for drug design and pharmacology, may face situations which are, in different ways, different from those typical of the latter fields. It should be remarked that the restriction of traditional QSAR approaches to congeneric chemicals, and the scientific rational for such restriction, represents both the key to the success of QSAR approaches and an inherent limitation on their more general application. Nowhere is this more evident than in the area of toxicology, where diverse chemical structures are studied for reasons that rarely take into consideration QSAR modeling concerns (e.g. selection of balanced samples of chemicals in terms of chemical characteristics and activities), where the biological interactions, particularly in vivo, are complex and little understood, and where diverse chemical structures acting by different biological mechanisms often manifest the same toxicity. Another problem that confounds efforts to develop successful QSAR models for toxicity endpoints is the composite, or "impure" nature of many reported endpoints. Examples include the rodent carcinogenicity bioassay, where the results from both sexes and two species of rodents are combined, and the Ames mutagenicity assay, where test results from several different strains of *Salmonella*, with or without metabolic activation, are often combined. The following example is a demonstration of the above concern. The work was performed in the laboratory of Richard, and is reported in [27].

A set of 25 noncongeneric chemicals, considered as hazardous or pollutants under the Clean Air Act [28] were selected. Their genetic activity results were retrieved from the Genetic Activity Profiles (GAP) data base [29]. No single genetic toxicity assay had reported results for all 25 of these chemicals; thus the genetic activity was parametrized by the Mean LED (Lowest Effective Dose), i.e. the geometric average of the LEDs expressed as logarithm for all mutagenicity assays with positive results. A number of physical chemical parameters, likely to relate to general transport and reactivity properties relevant to electrophilic interactions, were screened: log P, LUMO and HOMO energies, and the difference LUMO-HOMO. A single parameter QSAR (based on LUMO-HOMO, with $r = 0.962$) was obtained, with no points identified as outlier: this result was considered remarkable in light of the composite nature of the biological endpoint and the diversity of chemical structures considered. This general QSAR was scrutinized along chemical class lines to probe the contributions of subgroups of chemicals, and to examine the local performance of the general QSAR within such classes. Moreover, the data set was decomposed according to the mutagenicity assays. The chemicals were assigned to three classes: polycyclic aromatic hydrocarbons (PAHs), nitroaromatics (NO2A), and aromatic amines (AA). The NO2A chemicals covered the entire range of potency and LUMO-HOMO values, while the PAHs and AAs formed two fairly compact clusters with respect to these two parameters. Moreover, the regression analysis showed that the overall QSAR reflected primarily the excellent fit of LUMO-HOMO to the Mean LEDs of the NO2As, whereas there was a poor fit to either the PAH or AA Mean LEDs. The implication was that the general QSAR should not be relied upon to prioritize chemicals according to mutagenic potency within either the PAH or AA classes. The general QSAR was also broken down by constituent mutagenicity assay. This analysis indicated that the general QSAR reflected primarily the influence of the *Salmonella* TA100 and TA98 strains data for the NO2A class.

The lessons learned by the above example are: a) a "successful" general QSAR equation for a noncongeneric data set is likely a chimera, with much less general applicability to noncongeneric data outside the training set than overall statistical indicators would imply; b) the local performance of a general QSAR will likely vary considerably within such subclasses of the data base; c) the local performance of a general QSAR must be probed within subclasses of the total data base in order to understand the limits on its general application [27].

As shown above, the difference between QSAR applications to congeneric and noncongeneric sets of chemicals is crucial. Congeneric chemicals are assumed to produce a certain biological action through the same mechanism; different substituents attached to a common molecular skeleton simply modulate the quantitative level of activity. Consequently, the QSAR analysis is aimed at describing in a mathematical model how these small variations of chemical structure -parametrized by physical chemical descriptors or topological indices- affect the biological activity levels [30]. The situation is drastically different when dealing with agents of different chemical classes which induce the same type of activity. Here, the chemicals are likely to induce the same endpoint (e.g. point mutation) through different mechanisms, and it is extremely difficult -if not impossible- to summarize into one model many different mechanisms, or to find some underlying factor which is common to all of them.

Predicting Mutagenicity and Carcinogenicity for Noncongeneric Chemicals Sets. - In spite of the challenges, there is strong motivation to develop QSAR models for toxicity prediction for use in screening, for setting testing priorities, and for reducing reliance on animal testing [2,31,32]. Decisions must be made daily by toxicologists in government and industry to direct limited testing resources to the most urgent public health problems, and to direct the types of chemical synthesis and product development efforts undertaken. This need has motivated attempts to construct general QSAR models, not tailored to congeneric series of chemicals, with the ambitious hope that these models would be valid for all kind of chemicals.

Stouch and Jurs [33] studied a set of 105 compounds, belonging to several different chemical classes and tested for mutagenicity with *E. coli* WP2uvrA. The computerized SAR system ADAPT (Automatic Data Analysis using Pattern Recognition Techniques) was used to generate topological, electronic, geometrical and physical chemical descriptors. A QSAR based on 10 selected descriptors showed an excellent fit to the data, with 90% retrospective accuracy for the training set of 105 compounds. However, when the rule was applied to the prediction of activity of 40 external compounds (test set), the correct prediction was only 50%.

Benigni *et al.*, [34] studied a set of 137 carcinogens and noncarcinogens, belonging to a variety of chemical classes. Five physical chemical descriptors were used within a pattern recognition method (REPAD, RElationships PAttern Discrimination). The model produced 97% correct retrospective classification of the training set. The model was also tested with 11 human carcinogens, not included in the training set: 10 out 11 were correctly identified. However, a more extensive validation of the model with a further 100 chemicals tested for carcinogenicity showed only 60% correct prediction, in spite of several attempts to modify the REPAD procedure (unpublished results). No better results were obtained by considering a wider data base of about 300 carcinogens and noncarcinogens, and by improving the description of the electronic characteristics of the chemicals [35].

Klopman *et al.*, [36] studied a Gene-Tox derived data base of 808 chemicals, tested for mutagenicity in *Salmonella*, with the CASE (Computer Automated Structure Evaluation) approach. CASE is a development of the fragmentation approach by Hodes *et al.*, [37] and is aimed at identifying in the molecules substructural motifs related to the presence or absence of biological activity. It is characterized by the completely automated and unbiased manner in which the substructural descriptors are generated and included in the CASE predictive model. The CASE program decomposes each molecular structure into all possible fragments of a length of 2-10 contiguous non-hydrogen atoms. The analysis of the distribution of each unique fragment among active and inactive molecules in the training set identifies the substructures related with the activity and to be included in the CASE model. Based on the analysis of the Gene-Tox mutagenicity data base, CASE identified 29 activating and 3 inactivating fragments, which correctly predicted the probability of carcinogenicity of 94% of the known mutagens and nonmutagens in the data base. Using the structural determinants identified in this data base, CASE gave excellent predictions of the mutagenicity of chemicals not included in the data base.

Rosenkranz and Klopman [38] applied CASE to a portion of the U.S. National Toxicology Program Salmonella mutagenicity data base, and identified 13 structural determinants which predicted the likelihood of mutagenicity of the 243 chemicals in the data base (about 95% accuracy), as well of chemicals not included in the training set.

A fragmentation approach, similar to that implemented with CASE, was implemented in a new computer system based on graph theory [39]. The structural formulas of 826 chemicals tested for rodent carcinogenicity were fragmented into all possible contiguous-atom subunits with sizes between 2 and 8 (nohydrogenous) atoms. By random sorting, the data base was divided into a training set (80% of chemicals) and a test set (20%). This operation was repeated, with different sortings, eight times. Each time the fragments associated with carcinogenicity, or lack of carcinogenicity, were identified in the training set; then, the model obtained was applied to predict the carcinogenicity of the training set (retrospective classification) and of the test set. The average level of accuracy for the training set was 83.6%, whereas the average level of accuracy for the test set was 67.5%.

In the preceding examples, we particularly remarked on the relationship between descriptive power and prediction ability of the various approaches as the crucial clue. In fact, the SAR or QSAR model description of the noncongeneric training set is often scarcely informative: it is an empirical relationship, which does not provide clues about the mechanisms of action; thus, the prediction ability remains the main goal of these types of studies. Within this context, a recent exercise which was aimed at comparing different prediction approaches for carcinogenicity by challenging them on a common set of chemicals, whose carcinogenic potential was unknown at the time the predictions were formulated, represented a unique occasion for evaluating the state of the art in this field. The exercise considered 44 chemicals, which were in the process of being tested by the U.S. National Toxicology Program (NTP). At the end of the exercise, only 40 chemicals were actually experimentally tested, and the experimental and estimated carcinogenicity were compared. Among the participants to the challenge, some were (Q)SAR approaches; other approaches looked for relationships between carcinogenesis and shorter-term biological events (Activity-Activity Relationships, AAR).

The Tennant AAR method [40] used different types of information: general and target organ toxicity in rodents, *Salmonella* mutagenicity, presence of Structural Alerts according to the list compiled by Ashby [41]. This information was combined into a final carcinogenicity prediction through subjective expert judgement. This approach also used previous carcinogenicity data, when available in the literature, to modulate the predictions. A second AAR approach, called RASH (RApid Screening of Hazard) [42], essentially consisted of a modification of the Tennant approach through alternative means for ranking toxic potencies.

Among the (Q)SAR approaches there was CASE. On this occasion [43], CASE was used in conjunction with MULTICASE. Whereas CASE considers all the fragments in the same way, MULTICASE introduces a hierarchy between the fragments, by trying to discriminate between the functionalities that are responsible for the basic activity, and the functionalities that modulate this activity. This distinction is based on statistical criteria [44].

Another QSAR approach was TOPKAT (TOxicity Prediction by Komputer Assisted Technology) [45]. The quantification of structures was provided by: a) electronic descriptors (charge, electron density, residual electronegativity, effective polarizability); b) connectivity descriptors quantifying topological features; c) shape descriptors (kappa shape indices (46); d) substructure descriptors, selected form a library of about 3000 molecular fragments.

The COMPACT (Computer-Optimized Molecular Parametric Analysis of Chemical Toxicity) system [46] tried to evaluate the ability of chemicals either to be substrate for the cytochrome P450 I family, or to be able to interact with the Ah receptor, based on the analysis of spatial and electronic properties.

Other predictions were made with the computer-based expert system DEREK (Deductive Estimate of Risk from Existing Knowledge), derived from the LHASA (Logic and Heuristic Applied to Synthetic Analysis) [47] the system consisted of sets of rules, which identified fragments, or combinations of fragments putatively associated with carcinogenicity.

Other predictions were based on the experimental measurement of electrophilicity (Ke) [48]; chemicals below a certain Ke threshold were predicted to be noncarcinogenic, and *vice versa.*

Benigni presented predictions [49] based on the combination of two types of information: a) the theoretically estimated electrophilicity (Ke) of a chemical, according to [50]; b) the presence of potentially alerting substructures (Structural Alert (SA), according to [41] in the molecules.

Weisburger and Lijinsky provided two sets of predictions based on expert intuition, and used as their initial information the chemical structure of the NTP chemicals [51].

The different natures of the approaches should be particularly stressed. The Tennant and RASH methods were AAR approaches. Weisburger also incorporated some biological information. Some of the (Q)SAR approaches had a more quantitative character (Benigni, CASE, COMPACT, TOPKAT); other approaches were more qualitative (Weisburger, Lijinsky, DEREK). Three approaches (Tennant, RASH and Weisburger) were not "pure" prediction systems, but were mixed prediction-evaluation systems, because they incorporated, as initial information, previous experimental carcinogenicity results, when available. The Ke approach was not theoretical, but was based on the experimental measurement of a physical chemical property.

The results of the exercise were discussed in several papers (for example, [52,53]). Benigni [54] compared the different prediction sets with multivariate data analysis methods. The analysis pointed to two main results. First, the AAR approaches had similar profiles of predictions for the individual chemicals, whereas most of the SAR and QSAR systems clustered together (including Ke, which was based on a physical chemical experimental measure). Thus, the main difference in the prediction profiles was between the systems that relied on chemical structure as input information, and the systems based on biological information. This result was particularly surprising, in light of the very different rationales and technologies employed by the various SAR and QSAR approaches.

The second result concerned the predictive ability of the various approaches (*Table 3*). When the rodent carcinogenicity results, obtained in two-sexes, two-species experiments, were summarized into a final dichotomous carcinogenicity classification, the Tennant AAR approach had a 75% accuracy in predicting carcinogenicity, and the best (Q)SAR methods ranged between 60% and 65% accuracy. When a carcinogenicity classification maintaining some of the original gradation was considered, the (Q)SAR methods improved their performance to about 70%, whereas the performance of the Tennant approach declined. The lack of stability of these results stresses the importance of the type of formalization of the carcinogenicity results.

When breaking the 40 NTP chemicals into the classes of: a) most powerful carcinogens (homogeneously positive in the four rodent systems); b) chemicals with mixed result profiles; and c) noncarcinogens; most of the prediction systems were concordant in the identification of class a) chemicals. The chemicals with mixed carcinogenicity profiles were often predicted to be positive; in this case, the rate of positive predictions was lower than with the most powerful carcinogens. The real problems were found with the noncarcinogens, since many of them were predicted to be positive by different systems. Thus the clear limitation of almost all the prediction systems (both AAR and SAR) was their excessive sensitivity [8].

The inspection of the results for the individual chemicals also showed that the various SAR and QSAR approaches essentially acted as gross "class-identifiers": they pointed to the presence or absence of alerting chemical functionalities, but were not able to make gradations within each potentially harmful class [8]. The latter observation explains the other results provided by the analysis of the systems performance. In fact it appears that all the predictive systems were strongly influenced by the potentially alerting chemical classes. This large-scale source of variation obscured the other small-scale sources of variation (i.e. the differences among chemicals belonging to the same potentially alerting class). As a consequence, all the systems showed excess sensitivity, and, in practice, acted as expert systems, with little sensitivity to the quantitative differences. Moreover, in this way the potential differences among predictive systems were mostly canceled, and their prediction profiles were similar to a high degree.

Table 3. *Ability of Predicting Rodent Carcinogenicity: Results of a Comparative Exercise on 40 Chemicals Bioassayed by the U.S. National Toxicology Program*

System	Concordance
Tennant *et al.*	0.75
RASH	0.73
Weisburger	0.67
Ke	0.64
DEREK	0.59
TOPKAT	0.58
Benigni	0.57
Lijinsky	0.55
COMPACT	0.54
CASE	0.49

Problems in Establishing QSAR Models for Noncongeneric Chemicals. -The QSAR analyses of well designed congeneric series can provide accurate models of the relationships within the series, and also reliable predictions of the activity of other chemicals belonging to the same chemical context. This essentially holds true when the object of the study is a group of chemicals, which all share, to different degrees, the same type of activity, and act through the same mechanism. Things become more complicated when, together with potency variations, one also wants to know what makes certain chemicals active, and makes other, apparently similar chemicals inactive. Here it is necessary to define, beside the potency modulators, also the determinants of activity, and the boundary conditions (applicability limits) for the potency QSAR. Things become even more complicated when different chemical classes, acting trough different mechanisms, are considered together. Here the goal is the construction of a super-model, able to incorporate various local models. These models can be extremely different from each other, and specific for each chemical class, thus making it extremely difficult, if not practically impossible, to derive a good general mathematical model. Together with this technical difficulty in modeling simultaneously different mechanisms of action, there is another problem, which is more subtle and is often overlooked: the difference between relations, which are apparent on the large scale (e.g. noncongeneric sets of chemicals), and relations, which are proper to the small scale (individual congeneric chemical classes). The two types of relations may be in contrast with each other [8]. For example, a set of noncongeneric chemicals may span a large range of descriptor values; within this range, individual chemical classes may each span a portion of the range. It may happen that no relationships exist within the individual classes, but when the entire range of descriptor values is considered, an overall relationship may appear. In such a case, the QSAR may be purely empirical, and devoid of any relationship with the mechanisms of action, and the QSAR does not permit any accurate prediction for the individual chemicals, but only average estimates relative to the subclasses of chemicals. *Figure 1* gives a pictorial representation of such situations through a hypothetical example. The figure displays: a) two chemical classes for which there is no relationship between the potency and the descriptor Q; and b) two chemical classes with an inverse relationship between potency and Q. However, a good direct, overall relationship (r^2 = 0.8) is apparent when all the chemicals are considered together.

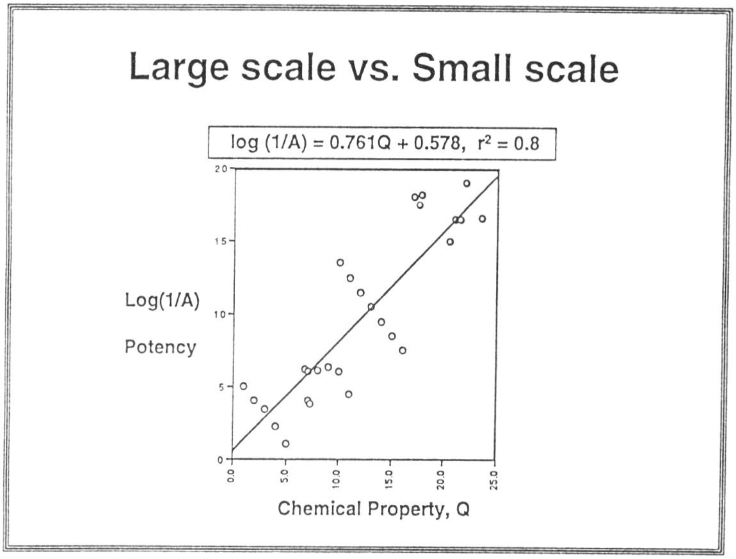

Fig. 1. *Correlations at different scale levels.*

Another serious bias, which accounts for the limited performance of the QSAR applications to noncongeneric chemicals, is the type of classification of biological activity. In fact, the outcome of the rodent carcinogenicity experiments is a complicated pattern, which includes four different experimental groups (rat and mice, male and female), different types of tumors, and different rates of induction. This is usually summarized into a final classification in four categories: No Evidence (NE), Equivocal Evidence (EE), Some Evidence (SE) and Clear Evidence (CE), for each of the four groups. In a further step, any chemical with CE or SE in any of the four rodent systems is classified as a potential carcinogen for man. This dichotomous -/+ carcinogenicity classification is meaningful within a conservative risk assessment perspective; however, it is also clear that the rodent results have an intrinsic gradation and qualitative differences, which are lost when they are summarized in such a way. The lack of a correct parameterization of the biological activity may seriously hamper the derivation of a QSAR model. Together with this, the adequacy of the existing carcinogenicity data base should be taken into account. This is a key point, since good QSARs can only be built from adequate initial information. For example, the data base established by [55] includes about one thousand chemicals tested in carcinogenicity experiments. This number seems large, but the individual character of the mechanisms of each chemical class splits this population into many small subsets. Moreover, the carcinogenicity experiments are not usually planned to fit the requirements of SAR studies. Even the systematic investigation still in progress under the aegis of the NTP [56] follows different criteria: the chemicals to be tested were chosen to represent a sample of those existing in the market. Very often the commercial chemicals have complicated structures, characterized by the simultaneous presence of different functional groups. This further decreases the possibility of studying the effects of the individual chemical functionalities.

Building General QSAR Models: Is it Feasible ?- If and how the difficulties experienced by the various attempts to define general QSARs for noncongeneric chemicals can be surmounted is an open question. In recent years, the research in this field has produced interesting and ingenious attempts (e.g. CASE, or TOPKAT) using as a strategy the analysis of the entire noncongeneric data set. In the view of the authors of these attempts, one advantage of looking at the entire data set was the lack of preconceived notions about chemical classes unnecessarily limiting the outcome of the QSAR analysis. However, the limitations generally shown both in the individual studies and in the comparative exercise sponsored by the NTP seem to confirm the notion that the QSARs are context-specific: in other terms, there is a need for initially identifying -as a subject for the QSAR analysis- a class of chemicals all acting through the same mechanism. So, the central question becomes how to define individual chemical classes of compounds acting through the same mechanism. Presently, there is no general theory to respond to this question. Only experimental mechanistic studies can provide this indication. The experience of the chemist, which selects very "similar" chemicals, is a surrogate to this evidence. Thus, the ability to correctly tailor the congeneric classes will probably depend on the general advancement of our understanding of the chemical biological interactions. Within the QSAR community, there is growing realization of such limits of QSAR theory applied to noncongeneric data. This is demonstrated, for example, in the hierarchical approach taken by MULTICASE [44]. In the first step, the structural descriptors with the highest statistical discriminant power are identified and taken as indicators of class membership. In the second step, less important substructures are identified, and considered as modulators of the activity of the first-step fragments. In addition, recent TOPKAT (Ver.3.0, Oxford Molecular, Ltd.) models have been developed for smaller, more well defined chemical classes and toxicity endpoints than in the past. Another approach along these lines is to combine automated fragment search and traditional QSAR capabilities with codified expert rules that incorporate the current state of knowledge with respect to known mechanisms of action or determinants of toxicity. Examples of these types of hybrid expert systems include DEREK (Lhasa U.K.), HazardExpert (ComGenex Ltd.) and OncoLogic (LogiChem, Inc.).

Expert Systems for Carcinogens Identification. - DEREK (Deductive Extimation of Risk from Existing Knowledge) [47], already introduced in this paper, was designed to initially maintain knowledge of experts and then allow "non-experts" access to this knowledge in order to make decisions concerning risk assessment. DEREK identifies toxicophores (i.e. molecular fragments associated with a specific activity); as such, it is qualitative. The recent addition of an interface to log P calculations should allow the inclusion of this property into rules and their refinement. The rule-base of DEREK is being developed by a collaborative effort, including information from commercial sources.

HazardExpert (ComGenex, Ltd.) also identifies toxic segments in the molecules. It provides species specific information across a range of trophic levels with different dosing regimes. The rules in HazardExpert contain only published information. It also calculates log P and pKa for each molecule, and uses this information to modulate the rule output.

OncoLogic (LogiChem Inc.) is a rule-based expert system that can be applied to a wide range of noncongeneric chemicals, but that relies on a chemical class, mechanism-based approach to cancer prediction [57]. It incorporates thousands of discrete rules for characterizing each of a variety of chemical classes based on the cancer expertise of the Structure-Activity Team within the Office of Pollution Prevention and Toxics at U.S. Environmental Protection Agency. OncoLogic is currently operational only for a limited number of chemical classes. It has a hierarchical structure. In the top-most levels, structural considerations are used to exclude molecules from concern on the basis of factors such as molecular weight, solubility and bioavailability. A chemical is then classified according to properties and structure until sufficient characterization allows application of the SAR rules for a specific chemical class.

A different type of expert system is provided by the Rules Induction Method for

Predicting Chemical Carcinogenesis [58]. This is an automated, decision-tree approach where rules for use in prediction are mathematically induced from available data, rather than obtained by human experts. In contrast to the above expert systems, the rules are derived from both chemical and biological attributes (*Salmonella* mutagenicity, route of administration, maximum tolerated dose, subchronic organ pathology, miscellaneous in vitro mutagenicity short-term test results). As a predictive approach, it obviously depends on the availability of experimental biological data.

A Second Challenging Round Sponsored by the US NTP: Prediction of Rodent Carcinogenicity of further 30 Chemicals. - NTP has recently sponsored a second comparative exercise, inviting participants to present predictions relative to the rodent carcinogenicity of a further 30 chemicals, which are currently being tested. Since the various papers presenting the predictions are still unpublished, we will present here only the rationale for our approach [59]. In our new attempt, we were led by the concerns and reflections exposed above about the general QSAR models for noncongeneric chemicals, and we devised an approach different from that used in the first exercise. We first allocated each chemical to an activity mechanism class, and then, with semiquantitative considerations, we assigned a probability carcinogenicity score. We surrogated the lack of data on mechanisms with the "expert guess" of the chemist's eye. In the next step, for the chemicals with presumed alerting substructures we considered log P and estimated electrophilicity (Ke) values. The consideration of these two parameters led to the assignment of low, medium and high carcinogenicity probability to the chemicals. We were able to rationalize our predictions in terms of supposed mechanism of action only for a portion of the chemicals; for other chemicals, we based our predictions on the sole structural analogy with known carcinogens and noncarcinogens.

Perspectives. - The evidence about the QSAR analyses of mutagens and carcinogens shows how the classical QSAR approaches, created for pharmacology and drug design, apply equally well to this field. On the other hand, practical requirements for assessing chemical risk have stimulated different applications, notably those aimed at generating general models for noncongeneric chemicals. The problems found when facing this "difficult" problem are focusing the attention of the investigators on the need of taking into account, either implicitly or explicitly, the chemical class/action mechanism context of the QSAR models. As shown above, some approaches define the chemical class context with expert systems, which formalize the knowledge of the human experts. Other attempts (e.g. MULTICASE) try to define the chemical classes in a automated way. It is interesting to note that, in this issue, toxicological QSAR research has become very close to pharmacological QSAR research, which is presently facing the issue of chemical similarity. The need to define chemical similarity has recently become important for pharmacological QSAR in order to retrieve, in large databases, compounds similar to other chemicals of interest. As in toxicological QSAR, the investigation on chemical similarity is facing difficult conceptual problems, namely the definition of the "true" or most useful similarity index. For example, Johnson (personal communication) compared two types of similarity indices, and found that the indices: a) were statistically correlated at the large scale level (large sets of heterogeneous chemicals); b) were correlated at the small scale level (i.e. pointed to the same nearest neighbors); c) were (surprisingly) uncorrelated at the intermediate scale (i.e. gave different local ordinations of the chemicals). It can be anticipated that an even more complicated situation will be faced when, after having defined an optimal chemical similarity index, biology is taken into account, because of the possible distortions caused by nonlinear biological effects to the "chemical" space metrics.

The above reflection indicates that toxicological QSAR and pharmacological QSAR, in addition to sharing similar goals (i.e. designing safer chemicals), also face similar conceptual problems. Closer and more frequent contacts between the two communities can be helpful for the advancement of science in both fields.

REFERENCES

[1] M.T.D. Cronin, J.C. Dearden, *Quant. Struct. -Act. Relat.* **1995**. *14*, 1.

[2] M. Zeeman, C.M. Auer, R.G. Clements, J.V. Nabholz, R.S. Boethling, *SAR and QSAR in Environmental Research* **1995**. *3*, 179.

[3] M.T.D. Cronin, J.C. Dearden, *Quant. Struct. -Act. Relat.* **1995**. *14*, 117.

[4] C. Hansch, D. Kim, A.J. Leo, E. Novellino,C. Silipo, A. Vittoria, *CRC Critical Reviews in Toxicology* **1989**. *19*, 185.

[5] M.R. Frierson, G. Klopman, H.S. Rosenkranz, *Environmental Mutagenesis* **1986**. *8*, 283.

[6] R. Benigni, C. Andreoli, A. Giuliani, *Mutat. Res.* **1989**. *221*, 197.

[7] C. Hansch, *Sci. Tot. Environ.* **1991**. *109/110*, 17.

[8] R. Benigni, A. Giuliani, *Medicinal Research Reviews* **1996**. *16*, 267.

[9] R.W. Tennant, B.H. Margolin, M.D. Shelby, E. Zeiger, J.W. Spalding, W. Caspary, M.A. Resnick, B. Anderson, R. Minor, J.K. Haseman, S. Stasiewicz, *Science* **1987**. *236*, 933.

[10] E. Zeiger, J.K. Haseman, M.D. Shelby, B.H. Margolin, R.W. Tennant, *Environ. Mol. Mutagenesis* **1990.** *16*, 1.

[11] A.J. Shusterman, A.K. Debnath, C. Hansch, G.W. Horn, F.R. Fronczek, A.C. Greene, S.F. Watkins, *Mol. Pharmacol.* **1989**. *12*, 939.

[12] A.K. Debnath, G. Debnath, A.J. Shusterman, C. Hansch, *Environ. Mol. Mutagenesis* **1992**. *19*, 37.

[13] A.K. Debnath, R.L. Lopez de Compadre, A.J. Shusterman, C. Hansch, *Environ. Mol.*

Mutag. **1992**. *19*, 53.

[14] A.K. Debnath, R.L.L de Compadre, C. Hansch, *Mutat. Res.* **1992**. *280*, 55.

[15] C. Hansch, T. Fujita, *J. Am. Chem. Soc.* **1964**. *86*, 1616.

[16] L. Zhang, K. Sannes, A.J. Shusterman, C. Hansch, *Chem. Biol. Interact.* **1992**. *81*, 149.

[17] A.K. Debnath, A. Shusterman, R.L. Lopez de Compadre, C. Hansch, *Mutat. Res.* **1994**. *305*, 64.

[18] C. Hansch,, B.H. Venger, A. Panthananickal, *J. Med. Chem.* **1980**. *23*, 459.

[19] R. Benigni, C. Andreoli, L. Conti, P. Tafani, M. Cotta- Ramusino, A. Carere, R. Crebelli, *Mutagenesis* **1993**. *8*, 301.

[20] R. Crebelli, C. Andreoli, A. Carere, L. Conti, B. Crochi, M. Cotta-Ramusino, R. Benigni, *Chem. Biol. Interact* **1995**. *98*, 113.

[21] C. Hansch, *Acc. Chem. Res.* **1993**. *26*, 147.

[22] C. Hansch, T.E. Klein, *Acc. Chem. Res.* **1986**. *19*, 392.

[23] C. Hansch, T.E. Klein, *Methods Enzymol.* **1991**. *202*, 512.

[24] S.P. Gupta, *Chem. Rev.* **1994.** *94*, 1507.

[25] A.K. Debnath, C. Hansch, *Environ. Molec. Mutagenesis.* **1992.** *20*, 140.

[26] R. Benigni, C. Andreoli, A. Giuliani, *Environ. Mol. Mutag.* **1994.** *24*, 208.

[27] R. Benigni, A.M. Richard, *Mutat. Res.* **1996.** in press.

[28] H.F. Stack, M.A. Jackson, M.D. Waters, *Environ. Mol. Mutagenesis* **1993**. *21,* 67.

[29] M.D. Waters, H.F. Stack, A.L. Brady, P.H.M. Lohman, L. Haroun, H.Vainio, *Mutat. Res.* **1988.** *205*, 295.

[30] C. Hansch, in J.A. Keverling Buisman, (Ed.), 'Biological Activity and Chemical Structure', Elsevier, Amsterdam **1977**.

[31] A.M. Richard, *Toxicology Letters* **1995**. *79*, 115.

[32] P.M. Wagner, J.V. Nabholz, R.J. Kent, *Toxicol. Lett.* **1995**. *79*, 67.

[33] T.R. Stouch, P.C. Jurs, *Environ. Health. Perspect.* **1985**. *61*, 329.

[34] R. Benigni, C. Andreoli, A. Giuliani, *Carcinogenesis* **1989**. *10*, 55.

[35] R. Benigni, C. Andreoli, M. Cotta-Ramusino, G. Giorgi, G. Gallo, *Toxicology Modeling* **1995**. *1*, 157.

[36] G. Klopman, M.R. Frierson, H.S. Rosenkranz, *Mutat. Res.* **1990**. *228*, 1.

[37] L. Hodes, G.F. Hazard, R.I. Geran, S. Richman, *J. Med. Chem.* **1977**. *20,* 469.

[38] H.S. Rosenkranz, G. Klopman, *Mutat. Res.* **1990**. *228,* 51.

[39] D. Malacarne, R. Pesenti, M. Paolucci, S. Parodi, *Environ. Health. Perspect.* **1993**. *101*, 332.

[40] R.W. Tennant, J. Spalding, S. Stasiewicz, J. Ashby, *Mutagenesis* **1990**. *5*, 3.

[41] J. Ashby, R.W. Tennant, E. Zeiger, S. Stasiewicz, *Mutat. Res.* **1989**. *223*, 73.

[42] J.D. Jones, C.E. Easterly, *Mutagenesis* **1991**. *6*, 507.

[43] H.S. Rosenkranz, G. Klopman, *Mutagenesis* **1990**. *5*, 425.

[44] G. Klopman, *Quant. Struct. Act. Relat.* **1992**. *11*, 176.

[45] K. Enslein, B.W. Blake, H.H. Borgstedt, *Mutagenesis* **1990**. *5*, 305.

[46] D.F.V. Lewis, C. Ionnides, D.V. Parke, *Mutagenesis* **1990**. *5*, 433.

[47] D.M. Sanderson, C.G. Earnshaw, *Human Experimental toxicology* **1991**. *10*, 261.
[48] G. Bakale, R.D. McCreary, *Mutagenesis* **1992**. *7*, 91.
[49] R. Benigni, *Mutagenesis* **1991.** *19*, 83.
[50] R. Benigni, M. Cotta-Ramusino, C. Andreoli, A. Giuliani, *Carcinogenesis* **1992**. *13*, 547.
[51] Anonymous, Predicting chemical carcinogenesis in rodents. An international workshop. National Institute of Environmental Health Sciences, Research Triangle Park, NC, **1993**.
[52] B. Hileman, *Chem. Eng. News.* **1993**. *june 21*, 35.
[53] J. Ashby, R.W. Tennant, *Mutagenesis* **1994**. *9*, 7.
[54] R. Benigni, *Mutat. Res.* **1995**. *334*, 103.
[55] L.S. Gold, T.H. Slone, G.M. Backman, S. Eisenberg, M. Da Costa, M. Wong, N.B. Manley, L. Rohrbach, B.N. Ames, *Environ. Health. Persp.* **1990**. *84*, 215.
[56] J.K. Selkirk, S.M. Soward, *Envir. Health Perspect.* **1993**. *101*.
[57] Y.T. Woo, D. Lai, M. Argus, J. Arcos, *Toxicol. Lett.* **1995**. *79*, 219.
[58] D. Bahler, D.W. Bristol, in L. Hunter, J. Shavlik, D. Searls, (Eds.), *Intelligent Systems for Molecular Biology* , AAAI/MIT Press, Menlo Park,CA, **1993**, p. 29.
[59] R. Benigni, C. Andreoli, R. Zito, *Environ. Health Persp.* **1996**. in press.

PART III. COMPUTATIONAL APPROACHES IN DRUG DISPOSITION

Chapter 20

Dealing with Metabolic Variability during Drug Development

by **Luc P. Balant, Marianne Gex-Fabry** and **Effie A. Balant-Gorgia**

Clinical Research Unit, Department of Psychiatry,
University of Geneva, 47, Rue du XXXI Décembre,
CH-1207 Geneva, Switzerland

Dealing with Metabolic Variability during Drug Development

by **Luc P. Balant, Marianne Gex-Fabry** and **Effie A. Balant-Gorgia**

Clinical Research Unit, Department of Psychiatry, University of Geneva,
47, Rue du XXXI Décembre, CH-1207 Geneva, Switzerland

This presentation on the impact of metabolic variability on drug behaviour will concentrate on studies performed in humans during the development of new drugs. *Intrinsic variability* is linked to genetic background, age and gender, or concomitant diseases such as renal or hepatic insufficiency. *Extrinsic variability* is caused by environmental factors such as co-administered drugs or smoking. It is important to recognise that the impact of sources of variability on drug metabolism is not constant, but may change with time.

Sources of variability relevant for humans can be explored early in the development of a new drug, using *human microsomes* or *human-DNA-transfected cell systems*. Due to the development of these in vitro techniques in recent years, less emphasis is laid on metabolic findings in animals as predictors of human metabolism.

Studies in healthy subjects (phase I type) provide important information on metabolic variability. Such studies should also explore the influence of gender and age. In the context of world-wide drug development, more emphasis is laid today on interethnic factors as compared to the 1980s.

Studies in patients are usually conducted in order to demonstrate the efficacy and the safety of the new therapeutic agent. The stringent experimental conditions of phase 3 clinical trials lead to very restrictive inclusion and exclusion criteria. As a consequence, most sources of variability encountered in "real life" patients remain unexplored at this stage. One of the possibilities to overcome this serious problem is to use population methods in pharmacokinetics and pharmacodynamics during ad hoc clinical trials and pro-active pharmacovigilance after marketing authorisation has been granted.

Introduction. - More than ten years ago, one of the authors [1] wrote *"During the past 40 years, the tools of pharmacokinetics have been discovered, developed, discarded, and revised. It can thus be stated that, although major theoretical breakthroughs are always possible, the modern pharmacokineticist must use a wide variety of approaches to better understand the behavior of drugs and their metabolites. There is one factor, however, that has progressively emerged from the literature and is far from being under control: the incidence of variability (both intra- and interindividual) on the practice of pharmacokinetics at the experimental, industrial, and clinical levels"*. Since then progress has been made in our understanding of the sources of variability, both in pharmacokinetics and pharmacodynamics, but until recently this knowledge seems to have had little influence on the way drugs are developed. However, new trends in thinking are getting attention and among them one may mention an increased necessity to include elderly patients in large scale clinical trials, the necessity to study kinetics and tolerance early in the development also in healthy women, and the progressive interest in interethnic differences in drug behaviour and action. This presentation of the impact of variability on drug behaviour and action will concentrate on studies performed in humans during the development of new drugs.

Variabilities: Concepts and Definitions. - *Intrinsic Variability:* Each person reacts specifically to the effects of drugs in function of genetic traits in drug metabolism or response, but also gender, age, associated diseases such as renal or hepatic insufficiency. Some of these sources of variability are relatively easy to characterise for patient enrollment in studies; they include age, creatinine clearance or well known genetic polymorphisms for which probe drugs are available. Other sources of variability, such as hepatic function, are less easily defined as far as their precise influence on drug fate and effects are concerned. Subgroups or subpopulations display also specific traits as far as drug effects are concerned. Genetic polymorphisms in drug metabolism may display different frequencies in different racial groups. Some subpopulations may react differently to certain drugs such as Blacks and Caucasians treated with antihypertensive medications. Finally, subgroups may show increased susceptibility to side effects such as patients having some specific types of glucose-6-phosphate dehydrogenase deficiencies. These differences are receiving increased attention in the context of world-wide development of new drugs.

Extrinsic Variability: Environmental factors may also have profound influence on drug behaviour and effects. Some of these factors are of a more individualistic nature such as smoking or alcohol drinking habits, whereas others are linked to cultural background including food or personal appraisal of side effects. In general, with the possible exception of smoking habits, the impact of such factors on study outcome is difficult to quantify. In addition, other environmental factors such as polluants, meteorological conditions, diagnostic approaches or attitude towards medicines certainly play a role, but their precise influence on the outcome of clinical studies and on patterns of medication use in the general population are almost impossible to evaluate.

Inter-individual Variability: Factors responsible for inter-individual variability are both intrinsic and extrinsic. Individuals may differ due to gender, age or genetic causes, but also because of different ways of life.

Intra-individual Variability: Factors responsible for intra-individual variability again may be both intrinsic and extrinsic. The somatic condition of an individual may change with time. Nutritional factors, comedication, exercize may vary from day to day, or even from hour to hour. The influence of such factors is thus closely linked to the time frame of the pharmacokinetic experiment or drug intake, but it has to be distinguished from a possible effect of time per se, e.g. saturation of some processes or tolerance to drug effects.

Time Frame: Living organisms such as humans do not remain "the same" through their life and many factors, including time, may have profound effects on the way they react to drugs. However, in the context of drug development, subjects participating to studies are ideally considered as either invariants within the time frame of the study or displaying variability only as far as clinical target symptoms are concerned. Examples of the first type of assumptions are pharmacokinetic, bioequivalence or tolerance investigations. The second type of assumptions is well illustrated by drug-drug interaction studies and the majority of clinical investigations in patients. Time is usually considered as irrelevant for studies extending for a few months or less, environmental factors are difficult to parametrise and, with a few exceptions such as self-inducing substances, it is assumed that the behaviour of the drug under investigation does not change over time. Accordingly, most designs and data analysis strategies are based on the idea that intra-individual variability may be neglected and concentrate on inter-individual variability.

Sources of Variability. - *Pharmacogenetics:* The pioneering work of Kalow [2] on pseudocholinesterase is widely recognized as the official start of the 'pharmacogenetics era'. As far as xenobiotics are concerned, three major polymorphisms in drug metabolism occupy the front of the scene: polymorphic N-acetylation, debrisoquine/sparteine oxidation and mephenytoin oxidation. In each case the sequence of discoveries was analogous: the initial observation of abnormal response to a drug led to the discovery of a pharmacokinetic anomaly which eventually was shown to be of genetic origin. Clinically, genetic factors in drug metabolism are a source of inter-individual variability and knowledge of these factors may allow for an adaptation of dosage regimen to the individual metabolic capacity of patients. The regulatory consequences arising from such polymorphic drug metabolism are complex and each new chemical entity must be considered as a special case. Clearly, different strategies of clinical development must be implemented depending what is known about the fate of the new chemical entity *in vitro* and in animals. If the major metabolic pathway of a drug is catalyzed by a polymorphically expressed enzyme, the clinical consequences depend on whether the substance is activated or inactivated during the course of biotransformation. If the polymorphic metabolic step leads to inactive metabolites, the active parent compound may accumulate in patients deficient of this pathway and hence toxicity may occur. In contrast, formation of active metabolites via a polymorphic enzyme can lead to lack of drug effect in the subset of patients devoid of metabolic activity. These considerations emphasize that pharmacokinetics should never be dissociated from pharmacodynamic considerations.

Environmental Factors: The pharmacokinetic literature clearly shows that a large number of different environmental and non-genetic host factors can influence the activity of different enzymes participating in the metabolism of drugs. Just a brief perusal of the enormous amount of published data reveals that almost any exogenous influence has been shown to affect at least a certain aspect of drug metabolism in experimental animals. In humans perhaps only practical difficulties and ethical considerations in experimentations have prevented to reach similar conclusions. However, as with genetic polymorphisms, an effect as such does not mean that it is important in terms of physiological, pharmacological or toxicological outcomes. Similarly, a demonstrated effect on a certain enzyme of drug metabolism does not mean that it is discernible in other enzyme systems. Thus, specificity and extent of environmental influences are among the most important considerations of clinical studies. Even more important in the context of new drug development is the assessment of the clinical relevance of the observed effects.

Gender Differences: Male versus female studies have shown, essentially in animals, that gender represents also a potential source of host-induced variability. Differences in structure of microsomal enzymes, substrate binding with P450 enzymes, metabolism of steroid hormones, among other factors, have been reported to alter drug behaviour and action in animals of different genders. It is not intended to provide a comprehensive review of gender differences in clinical pharmacology, this subject having been extensively reviewed [3, 4]. However, some differences that are hormonally influenced and age related may have clinical relevance. If drug studies are conducted primarily on young men information pertinent to the treatment of older men and all women may be missed. Gender differences in humans are not as marked as those in animals but they cannot be ignored and there is, indeed, tendency in some regulatory agencies to promote the idea that this additional source of variability should be taken into consideration in the development plan of a new chemical entity.

Interethnic Differences in Drug Behaviour and Action: In 1982, Kalow [5] published a review on 'Ethnic differences in drug metabolism'. The findings described 15 years ago are basically not different from our present understanding of this topic. However, in the mean time, considerable knowledge has been acquired at the molecular biology level [6]. Understanding these sources of variability and finding strategies to control them have become major issues in the world wide context of drug development and use [7, 8].

The CYP2D6 Story. - *Historical Considerations:* As often, the discovery of the so-called debrisoquine/sparteine oxidation polymorphism was not a result of planned strategy, but rather an incidental observation. Early in the development of bufuralol, a ß-blocking agent, it was suspected that its metabolism was under the control of a polymorphic enzyme [9]. During a phase 1 pharmacokinetic study a volunteer suffered from marked orthostatic hypotension with simultaneous very high blood concentrations of parent drug and very low concentrations of the hydroxylated metabolite. Further work by the same group confirmed this hypothesis [10, 11]. At about the same time, in Great Britain, investigators of the behaviour of debrisoquine, an antihypertensive drug, found that a marked hypotensive response in one volunteer was due to impaired 4-hydroxylation of the drug [12]. Also at about the same time a group of physicians in Bonn observed increased side effects associated with decreased oxidative metabolism of sparteine, an anti-arrhythmic and oxytocic drug [13]. Since these early observations numerous investigations have followed, indicating that the proportion of 'poor' metabolisers is about 7% in Caucasian populations [14]. The discovery of this polymorphism is interesting since today we know the reason for the strange behaviour of some of the volunteers in the early pharmacokinetic experiment. It must, however, be remembered that in 1974, when this early study was performed, polymorphic metabolism was not recognised.

Mechanistic Considerations: Today the molecular bases of the defect of debrisoquine/sparteine oxidation are well known and have been extensively reviewed [15-18]. The defect in poor metabolisers is the result of the lack of synthesis of a cytochrome P450 named CYP2D6, or the synthesis of unstable or inactive forms of the enzyme due to point mutations in the gene [19]. Initially, it was believed that the cause for interindividual variability of CYP2D6 dependent drug metabolism was due to the existence of two phenotypes labelled "extensive" and "poor" metabolisers (EM and PM) of debrisoquine or sparteine. These phenotypes were supposed to be the result of specific DNA changes in the CYP2D6 gene. The situation has become more complicated since more than a dozen variant alleles of CYP2D6 have been identified over the past years [20]. The present knowledge about these mutations may be summarised as shown in *Table 1* [19, 21].

Table 1. *Variant Alleles of CYP2D6.*

ALLELE	EFFECT	ALLELE FREQUENCY	
		Caucasians	Chinese
CYP2D6wt	Wild-type, full enzymatic activity	69.0%	43.0%
CYP2D6A	Frameshift (no activity)	2.3%	0.0%
CYP2D6B	Splicing defect/frameshift (no activity)	21.5%	0.4%
CYP2D6C	Reduced activity		
$CYP2D6Ch_1$	Reduced activity	2.8%	51.0%
$CYP2D6Ch_2$	Unstable enzyme	1.0%	37.0%
CYP2D6D	No enzyme	4.0%	5.7%
CYP2D6E	No activity in vivo		
CYP2D6F	Splicing defect (no activity)		
CYP2D6G	Premature termination (no activity)		
CYP2D6J	Reduced activity		
CYP2D6L	Reduced activity in vivo		
CYP2D6Lx2	Increased activity in vivo	3.0%	
CYP2D6Lx3	Increased activity in vivo	0.1%	
CYP2D6Lx13	Increased activity in vivo	0.1%	
CYP2D6M	Reduced activity		
CYP2D6N	Increased activity in vitro		
CYP2D6T	Premature termination		

In addition to poor metabolisers of debrisoquine, subjects requiring higher doses of drugs which are substrates for CYP2D6 ("ultrarapid" metabolisers) have been described [22, 23]. The molecular basis for this phenomenon became unravelled when genomic DNA from such an individual was subjected to molecular biology investigations. Sequence analysis of this allele ($CYP2D6L_2$) revealed that it contained two mutations causing amino acid exchanges. Evaluation of the enzymatic activity among subjects having only one copy of this gene, showed that the in vivo hydroxylation of debrisoquine was more or less the same as among subjects with the CYP2D6wt gene. The higher capability for hydroxylation of drugs was explained by the presence of two active CYP2D6 genes on the same allele. Recently, alleles carrying two and three extra $CYP2D6L_2$ genes yielding in total three or four active CYP2D6 genes have been found in Caucasian populations [24]. In one family, having the lowest debrisoquine metabolism ratio (i.e. the fastes rate of metabolism) among over 1000 subjects, 12 extra $CYP2D6L_2$ genes were detected. This situation, appearing through gene amplification, had been stably inherited from the father to the two children and represents the first case of a stable inheritance of an amplified active gene.

Pharmacokinetic Consequences: Experience gained with drugs that are under the control of the debrisoquine/sparteine polymorphism show that important differences and high inter-individual variability may be observed [25]. As an example, in one study of bufuralol, the systemic clearance was found to be about 37 L/h in extensive metabolisers and only about 15 L/h in poor metabolisers [26]. It was also shown that polymorphic metabolism had a major influence on the extent of hepatic first-pass elimination as judged by the large variability in systemic availability. Since, for safety reasons, the intravenous doses used in the study were smaller than the oral doses, the calculated apparent absolute availability even exceeded 'one' in a healthy volunteer phenotyped as a 'very poor' metaboliser, reflecting the fact that non-linear kinetics probably occurs at the higher dose levels used orally in the study.

Another pharmacokinetic issue is the search and validation procedures of "probe drugs" for subject phenotyping in order to assess genetic, environmental and ethnic differences in the in vivo metabolism of drugs suspected or known to be under the control of a polymorphic enzymatic system. This is necessary as long as no reliable and easy to perform genotyping procedures are available [27]). The key issues in population phenotyping include separation of risk and exposure, choice of the best experimental index to use, the selection and number of the best experimental index to use, and the use of objective criteria for assessing multimodality in frequency distribution of in vivo data. In order to avoid misinterpretation and overinterpretation of data, it is essential to have a good understanding of the pharmacokinetic basis of indirect measures of in vivo drug metabolising activity, and of appropriate statistical procedures for the analysis of frequency distributions [27]. The pharmacokinetic basis for using various experimental indices for detecting and displaying genetic polymorphism in oxidative drug metabolism, particularly with respect to CYP2D6, has been extensively investigated [28-33]. For example, it was shown that the traditional urinary drug/metabolite ratio for debrisoquine or sparteine (i.e. "metabolic ratio" with low values in extensive metabolisers and high values in poor metabolisers) is the discriminant indirect index of CYP2D6 polymorphism, but that it is in fact "upside-down" in that it is inversely related to intrinsic clearance. The fact that it is directly related to renal clearance of parent drug poses no problem for identifying polymorphism within an otherwise homogeneous population, but systematic differences in renal function between population groups could confound the ability to detect metabolic differences. This example shows that pharmacokinetic considerations are of uttermost importance in the field of pharmacogenetics and that each drug, probe drug and enzymatic system must be thoroughly studied before definitive claims of polymorphisms can be made.

Clinical Relevance: The debrisoquine/sparteine oxidation polymorphism (as it was called until recently) is a important factor of pharmacokinetic variability of several useful drugs [34-37]. The following table gives a listing of known drug substrates of CYP2D6.

The listing is not exhaustive. It must, however, be recognised that not all of these drugs are metabolised to the same extent by this enzyme and that, for some of them, other important routes of elimination are actively involved.

Table 2. *Drug Substrates of Cytochrome P450 2D6 (CYP2D6).*

Anti-arrhythmics: encainide, flecainide, mexiletine, propafenone
Beta-blockers: alprenolol, bufuralol, metoprolol, propranolol, timolol
Tricyclic antidepressants: amitriptyline, clomipramine, desipramine, imipramine, N-desmethyl-clomipramine, nortriptyline, trimipramine
Serotonine reuptake inhibitors: fluoxetine, N-desmethyl-citalopram, norfluoxetine, paroxetine
Neuroleptics: clozapine, haloperidol, perphenazine, risperidone, thioridazine, zuclopenthixol
Opiates: codeine, dextro-methorphan*, ethyl-morphine
Miscellaneous: amiflavine, debrisoquine*, guanoxan, 4-hydroxy-amphetamine,, indoramin, 4-methoxy-amphetamine, methoxyphenamine, perhexiline, phenphormin, N-propyl-ajmaline, sparteine*, tomoxetine
*used as probe drugs for CYP2D6 phenotyping

Despite the ever increasing number of drugs shown to be dependent on this polymorphic enzyme system for their elimination, this monogenetically determined variability becomes clinically significant only in those cases where CYP2D6 is a major metabolic pathway and where pharmacokinetic variability is considered important relative to the efficacy and clinical safety of the drug. In essence, it is applicable to those drugs for which plasma concentration monitoring is usually considered useful, and at present such a 'reduced list' includes tricyclic antidepressants, certain neuroleptics and possibly one or two other non-psychotropic drugs. As an example, excessively high concentrations of imipramine and clomipramine and their demethylated metabolites have been measured in poor metabolisers resulting, paradoxically, in therapeutic failure due to lack of response [38]. Finally, it is interesting to note that persons lacking the enzyme CYP2D6 do not apparently suffer from ill effects, except of course when CYP2D6 deficient individuals are exposed to certain drugs. This contrasts with genetic diseases such as thalassemias, phenylketonuria and others that have dire health consequences. It must, however, be mentioned that extensive research is ongoing to search for possible associations of pharmacogenetic differences and diseases such as cancer [39, 40] or diabetes. The same is true for other known polymorphisms as extensively discussed in the book edited by Kalow in 1992 [41]. It is, however, difficult to predict at the present time the impact of this type of research on the development of new chemical entities.

Interethnic Differences in Drug Metabolism: Numerous studies in Caucasian populations of Europe and North America have shown very similar frequency distribution curves for the debrisoquine and sparteine metabolic ratio measured from urinary data, with a rather consistent incidence of the poor metaboliser phenotype (about 7%) [14]. Oriental populations, however, differ from the Caucasian ones in two respects. Firstly, no clear bimodality of the distribution of the debrisoquine metabolic ratio can be observed, for example in the Chinese population. Using the antimode separating poor from extensive metabolisers as defined in Caucasians, only about 1% of Chinese can be classified as PM [42]. Similar data have been obtained in Japanese and Korean populations [43, 44]. Secondly, the distribution of the debrisoquine metabolic ratio is shifted towards higher values in Chinese extensive metabolisers compared to Caucasian extensive metabolisers, indicating lower average CYP2D6 activity in Chinese [42]. A similar right shift was observed for the metabolic ratio for codeine O-demethylation, a reaction also catalysed by CYP2D6 [45]. The reasons for these differences remained unexplained until the CYP2D6 gene was cloned and the molecular genetic basis of the polymorphism explored. It was subsequently shown that the difference in CYP2D6 activity between Oriental and Caucasian extensive metabolisers is due to a high frequency (about 50%) in the Oriental population of an allele, designated CYP2D6Ch or CYP2D6J (Table 1) giving rise to an unstable gene product with decreased enzymatic activity [46, 47]. On the other hand, the low incidence of poor metabolisers in Oriental populations is inherent to the low frequency of CYP2D6A and especially CYP2D6B, two of the most common defect alleles causing absent CYP2D6 activity among Caucasians (Table 1). Anecdotal reports indicate that for tricyclic antidepressants and some neuroleptics metabolised by CYP2D6, using the same average daily dose in Oriental and in Caucasian patients would result in higher plasma concentrations and increased risk of side effects in Orientals. However, the determination of the clinical relevance of these findings needs systematic studies to elucidate the contribution of interethnic differences to the pharmacokinetics and dosage recommendations of drugs metabolised by CYP2D6 [48].

Population Approaches. - *Pharmacokinetic Screen:* The term "pharmacokinetic screen" was proposed by Temple in 1983 [49] to describe a blood-sampling method and a strategy of data analysis performed during large-scale Phase 3 clinical trials [49, 50]. The rationale behind this approach is that it is improbable (for logistic and ethical reasons) that traditional pharmacokinetic experimentation will be carried out on each and every subpopulation. The best possible hope, therefore, is to obtain one or two blood samples per patient during clinical trials. This proposal recommends the collection of trough levels to detect patient groups at risk of an altered dose-concentration-effect relationship. Interpretation of data collected from pharmacokinetic screens can provide valuable information on the factors susceptible to alter these relationships. However, the one-point-per-patient sampling scheme does not generally lead to full characterization of pharmacokinetic metrics in specific subpopulations. The pharmacokinetic screen is an observational rather than experimental study. Its strengths include simplicity in terms of data collection and statistical analysis, representativeness of the patient population, and early detection of atypical behavior of the drug in some patients or groups of patients. Its weaknesses include limited pharmacokinetic information and the need for a relatively large volume of data to detect significant factors.

Population Pharmacokinetics: This is a more complex approach based on a more intensive data collection scheme [51-54]. Several drug levels (e.g., two to four) are needed from the majority of patients, at different times relative to the last dose. Its main advantage is that it enables the full pharmacokinetic profile to be described and the relevant population parameters to be estimated. The drawback is the need for more complex pharmacokinetic population models and for increasingly sophisticated statistical techniques and software. In the broadest sense, the term "population approach" (as used in pharmacokinetics) applies to any method used to study the variability of drug-metabolite concentrations in blood or plasma and the causes of this variability (i.e. covariables). Population approaches imply that both intrinsic and extrinsic sources of variability are investigated and quantified. It is crucial to understand that despite the name given to these approaches, population methods are basically designed to fulfil the needs of individual patients. Any approach which integrates a kinetic and statistical model can be termed "population kinetics approach", regardless of the study design, the data analysis strategy, or the computer software used. The distinction between integrated kinetic/statistical model specification and software implementation is important because the computer package NONMEM (Department of Laboratory Medicine, University of California, San Francisco), which was developed by Beal and Sheiner in San Francisco in the late 1970s, has often become synonymous with population pharmacokinetics [55]. Since then alternative modelling techniques have been proposed and various software packages have been implemented and become widely available [56-60].

Statistical Issues: Specification of a population model proceeds in two stages. It includes the definition of a pharmacokinetic model, i.e. a function that describes concentrations versus time curves as a function of a number of parameters such as clearance and volume of distribution. Inter-individual variation is then described in terms of a probability distribution and an explicit dependence on cofactors (categorical variables such as gender) and covariates (continuous variables such as age or renal function). The residual variance generally includes intra-individual variance and error. So called non-parametric methods differ in the sense that a particular functional relationship between parameters and covariates does not need to be specified. Successful applications of population approaches to simulated as well as real data have been published. However, as discussed below, a number of questions remain open to investigation. Among them, one may mention the difficulty to anticipate about the potential factors susceptible to influence pharmacokinetic parameters. As a corollary, one may concentrate on a factor that is statistically significant but proves to be clinically irrelevant. In addition, misspecification of the population model at any level may have unpredictable consequences. In a preliminary stage, exploratory analysis of the data may help learning about possible subgroups of patients showing altered kinetics. Dose normalized concentration versus time plots for various subgroups of patients and smoothing techniques have proved useful to detect factors likely to be clinically relevant. Powerful graphical tools are now widely available and graphical exploration of the data is thus believed to be a necessary, although preliminary stage before going on to intensive population modelling. An additional problem encountered with population pharmacokinetic approaches is the likely presence of outliers, i.e. a single or a small number of subjects showing very unusual kinetics. These subjects might be members of a rarely encountered phenotype or receive uncommon comedication. They are generally of high clinical relevance with respect to identification of subgroups of patients at risk. Again, graphical techniques may help identifying them but they may remain unnoticed if many variables are involved. Unfortunately classical statistical methods are very sensitive to their presence, a single outlier being able to mask an otherwise significant regression. Analysis of population data is expected not only to allow identification of outliers on the basis of objective rules but also to provide estimates of parameters which are relatively insensitive to their presence, and thus reflect the behavior of a majority of patients. So-called "robust" techniques are available but their potential usefulness in population pharmacokinetics remains to be shown.

In Vitro and Animal Studies. - Studies performed in healthy volunteers are obviously preceeded by investigations done *in vitro* and in animals. The impact of "variability control" in preclinical studies is of crucial importance, although often unnoticed. For example, toxicological investigations (including toxicokinetics) are performed in strains of animals breeded in order to almost wipe out inter-individual variability. On the other hand, more than one animal species are usually used in order to introduce some well defined variability and improve the predictive power of these safety investigations. There are nevertheless many unresolved issues, and among them one can mention the differences in drug metabolism patterns and/or pharmacodynamic sensitivities often observed between animals and humans. However, in recent years, sources of variability relevant for humans are explored early in the development of a new drug using human microsomes or human-DNA-transfected cell systems. Due to the development of these in vitro techniques, less emphasis is presently laid on metabolic findings in animals as predictors of human metabolism, although animal pharmacokinetic studies remain essential in toxicology to assess exposure to the drug and and check for comparability of metabolic patterns in animals and humans. Sparse sampling methods may also be used with advantage for drug exposure assessment [61].

Studies in Healthy Subjects. - *Dose Escalation, Pharmacokinetics and Tolerance Studies (increasing Variability):* Phase 1 studies should not only aim at obtaining a "Fingerprint" [62] of the pharmacokinetics and effects of the new substance in healthy-young-abstinent-male-Caucasians, but also to explore some aspects of potential variability in the target population. Such variability is thus expected to be allowed by the study design. In this context, it has been advocated that if *in vitro* and animal studies indicate that metabolic polymorphic differences are to be expected in man, it would be prudent to perform, as soon as possible, pharmacokinetic experiments in healthy volunteers, phenotyped as extensive and poor metabolisers for the metabolic pathway under consideration [63]. If these preliminary metabolic studies in man confirm the potential importance of genetically determined pharmacokinetic variability, it will be necessary to investigate the importance of the different disposition pathways, the pharmacokinetic variability of presystemic and systemic clearance, the consequence of such variability for steady-state plasma concentrations and therapeutic margin. This information should then be used to plan the first studies in patients. It would certainly be useful to explore, as stringently as in healthy young volunteers, the behaviour and effects of the new drug also in neonates, children and healthy elderly patients. This is usually not possible using traditional study designs for well understandable ethical reasons. This is why population approaches in patients of different age groups who may benefit from the new drug are particularly advisable in order to explore sources and magnitude of variability. For years it has been advocated not to use female volunteers early in drug development. This attitude is understandable if potential teratogenicity is to be avoided. However, this attitude when implemented too rigidly, can lead to absurd experimental designs such as performing pharmacokinetic or biopharmaceutic investigations of oral-steroidal-contraceptives in healthy males (authors' personal observation). This attitude must therefore be revised and it will probably be necessary in the future to perform reproductive toxicology earlier during drug development in order to be able to include female volunteers in early phase 1 studies, simultaneously minimizing the risks of adverse events in this group of volunteers.

Formal Pharmacokinetic and Bioavailability Studies (increasing Variability): Formal pharmacokinetic studies performed after single dose or multiple dose administration are usually performed in well calibrated volunteers. In the authors' opinion it is questionable whether this is the most adequate procedure. Clearly, curves in different individuals tend to be more similar and confidence intervals tend to decrease with a decreasing variance. As a consequence, graphs and tables of pharmacokinetic parameters tend to "look better" but their information content on variability is poor. This approach is probably the result of the impact of classical statistics on study design and it is possible that the emergence of new "robust" statistical methods may lead to new designs in which the inclusion of a maximum of variability will be the rule rather than the exception. The experience gained during the last year with polymorphisms in drug metabolism certainly points in this direction. However, as already mentioned, variability must be controlled and potential co-variables well documented.

Bioequivalence Studies (reducing Variability): As long as the rules governing bioequivalence studies are those that are presently applied, it is evident that the volunteers of such studies must be considered as "walking test tubes" and that all means to reduce variability should be allowed. In the authors' opinion it is not appropriate to deliberately introduce variability factors in bioequivalence studies, such as to mix young and elderly subjects in the same experiment. If bioequivalence of two drug products is to be judged in elderly subjects, then a homogeneous group of such subjects should be investigated. It must indeed be remembered that bioequivalence should characterize two drug products and not two sets of subjects or patients. Food interaction studies in bioequivalence testing are a special case since the introduced variability addresses questions essentially related to the pharmaceutical form and only indirectly linked to the future patient population. In such a casc, under well defined conditions the inclusion of variability is legitimate in bioequivalence studies.

Early Exploratory Studies in Patients. - The design of the first studies conducted in patients should balance the amount of variability in order to allow clinically relevant observations to be made in a small number of subjects. As a matter of fact, the introduction of a disease represents a major step as compared to healthy subjects so that other sources of variability such as age or co-medication should be kept at a minimum. As phase 2a studies proceed, variability should be introduced stepwise in order to obtain usable information for the preparation of phase 2b and phase 3 clinical trials. In the authors' opinion, concomitant measurement of blood concentrations of the active principle may greatly improve the quality of the information relative to potential kinetic and/or dynamic variability.

Large Scale Confirmatory Studies in Patients. - *Decreasing Variability:* Large scale phase 2b or 3 clinical trials are usually designed to have maximum power for rejecting the null hypothesis of no treatment effect. As such they are usually one-factor (drug), two-level (zero dose, and maximum safe dose) designs which measure a single univariate endpoint per patient (e.g. time to a clinical effect or amount of change after a given time interval). Such studies are usually based on the "randomised double-blind controlled" design and are performed in a homogeneous patient population [64]. In this situation all measures are taken to minimise variability. However, the central issue for modern drug development concerns maximising the benefit/risk ratio for individual patients and minimise the risk of unwanted "surprises" during the post-marketing period. In this context it is questionable whether the classical design represents the best solution to prevent such mishaps. The rigid clinical trials that assign individuals to monotherapy and demand that it is sustained until a statistically significant endpoint is reached, resemble actual clinical practice little if at all [64]. Nevertheless, as long as the pharmaceutical industry will be asked to demonstrate equality or superiority of a new treatment versus a placebo or an accepted therapy, stringent designs will be performed in order to bring unequivocal answers to such questions. Despite these contingencies, the conduct of randomised double-blind controlled clinical trials should not prevent the scientific community to explore the feasability and the potential interest of other types of clinical studies.

Increasing Variability: In theory, information regarding clinical outcome can come from a spectrum of possible data sources. The randomised double-blind controlled trial stands at one extreme, while the informal impressions of a practising clinician stand at the other. The dimension defined by these extremes is that of "stringency of experimental design". The very stringent design demands considerable control for statistical purposes over what is done to whom and when. Physician's impressions represent a very non-stringent design since almost nothing is done for experimental reasons and most of the control is exercised for therapeutic purposes [64]. Between these two extremes there is clearly space for other study designs. If variability is introduced deliberately into the patient population for phase 2b or 3 studies, measures must be taken in order to avoid that the only result of the study be summarised as "There is an important amount of variability in the patient population". One of the possibilities offered by recent developments in the methodology of clinical trials is the measurement of concentrations of active principles in order to separate potential sources of pharmacokinetic variability from those of pharmacodynamic variability. Indeed, for a number of years, it has been advocated that systemic drug concentration data (apart from those related to concentration-response trials, as performed during phase 2) should be obtained routinely in phase 3 clinical trials [49, 50]. These data might explain unusual response, suggest the possible presence or lack of drug-drug and drug-disease interactions, and identify other variables responsible for metabolic heterogeneity. Such data are also useful in the detection of compliance-related problems. The rationale behind blood-concentration measurements during phase 3 trials is that even a well-developed and carefully planned program cannot anticipate all potential problems. Moreover, specific studies of all possible patient subsets and interactions are costly, time-consuming, and probably unnecessary. Blood-concentration measurements during phase 3 could thus lead to substantial gains in information obtained during clinical trials. The pharmacokinetic screen coupled with integrated analyses of effectiveness and safety data, or more formal population models, can be used to identify and quantify important demographic and other subpopulation differences. This would eventually lead to dosage adaptation in population subsets, thus increasing the medication safety.

Open Questions Related to Drug Concentrations Measurement in Phase 2b and 3 Clinical Trials. - Previous sections have attempted to show the potential advantages of measuring blood or plasma concentrations of drug and active metabolites during phase 2b and 3 clinical trials. The conceptual framework for the interpretation of such datasets does exist. As stated by Sheiner and Benet in 1985 [65], it must be determined why this approach is often viewed with some reluctance within the pharmaceutical industry and why it is rarely used in new drug development, even though it has been advocated for nearly 10 years [66]. Possible answers can be found in the practical issues that must be solved before concentration measurements, and consequently population approaches can be fully implemented in traditional new drug development. It is important to note that a drug development program based on clinical trials during which blood levels have been measured in no way implies that the new drug will necessitate the implementation of routine drug monitoring once it reaches the market.

Logistical Issues: Blood-concentration measurements are not easily performed during Phase 3 (multi-center) clinical trials. One common objection is that these kinds of pharmacokinetic data are soft due to poor control of drug administration, sampling time, sample handling and storage, analytical error, or patient compliance. This is certainly a legitimate concern. Furthermore, no population software will ever be able to magically transform the "garbage in, garbage out" problem into "lead in, gold out". In any case, presence of uncontrolled variables will increase the unexplained variability and may result in concealing clinically relevant co-variables. Good clinical practice guidelines are a first step in increasing the quality of such data, and further improvement is certainly possible. Among the possible solutions are concepts developed for the measurement of clinical biology and chemistry data during large-scale clinical trials. For example, it is now an accepted practice to entrust the handling, shipping, and analysis of samples to specialized contract research organizations (CROs). Similar solutions are certainly adaptable to the centralized handling and measurement of drug concentrations. Some CROs and central laboratories are already performing this task with success.

Study Design: A randomized concentration-controlled clinical trial appears to be particularly useful in phase 2 studies for the characterization of the dose-concentration-effect relationships of a new active principle. However, its implementation in large-scale multicenter clinical trials is not easy. On the one hand, most stages necessary to perform randomized concentration-monitored clinical trials are practiced today, though not routinely, and study design is almost identical to the traditional way of performing clinical studies. On the other hand, expertise in population data analysis is not yet widely available, as discussed below.

Sparse Pharmacokinetic Data Analysis: The concept of sparse data or data-poor studies relates to the number of drug concentrations available per patient. Normally, this low number is compensated by the fact that numerous subjects (i.e., more than 100) are included in the analysis. Generally, the plot of blood concentrations as a function of time primarily highlights the lack of structure in the raw data, even after normalization for dose. Different strategies of data analysis have been proposed. Integrated, or one stage approaches include NONMEM and other techniques, as mentioned earlier. However, the complexity of these methods has led a number of people to shy away from them. The way forward is probably through cooperation between traditional pharmacokineticists and invididuals who have the necessary skills to analyze sparse data. This is a crucial issue for the implementation of drug-concentration measurements during phase 2b and 3 clinical trials. This cooperation is also one of the objectives of the European Coordinated Action COST B1 [67, 68]. In the meantime, the use of less sophisticated, or two-stage approaches may prove useful. Individual pharmacokinetic parameters are first estimated using a structural pharmacokinetic model and then related to demographic data using classical statistical methods such as univariate or multivariate analyses and clustering. Such methods have been applied with success both to studies generating data-rich information, such as cyclosporine in 187 patients [69], and studies such as isradipine in 252 patients [70], clomipramine in 150 patients [71] or moclobemide in 89 patients [72].

Pharmacodynamic Data: There is still little experience on the use of population methods for pharmacodynamic data analysis, but examples have shown that the approach can be used in this context [73].

Clinical Data Analysis: Clinicians and statisticians are concerned about the impact of measuring active principle concentrations with respect to the well-established "ended cohort" and "intent to treat" statistical methods. One of the concerns is the ultimate classification in the active group of patients with undetectable drug concentrations or with abnormally high concentrations. It is clear that exclusion of patients based on drug concentrations could modify the overall outcome of the study. This issue must be addressed by scientists in academia and industry and by regulatory authorities if drug measurement during clinical trials is to gain acceptance. In any case, such issues should be settled a priori in the study protocol.

Concluding Remarks. - Dealing with variability during drug development does not mean eliminating variability, but using variability as a means to better characterize drug behaviour in individuals who obviously differ from each other and require treatment adjusted to their specificities. It is also evident that dealing with variability does not mean the introduction of random and uncontrolled variability, but the search for methods allowing ordered and well defined variability sources to be explored during the process of drug development. It is also probable that such strategies will not (at least not immediately) reduce cost and time of drug development, but they may help to implement more adequate dosing strategies once the new drug is on the market and to avoid disagreeable surprises at this later stage. In any case, more work is needed from academia, the pharmaceutical industry and regulatory authorities in order to reach agreement on new strategies for dealing with variability during drug development.

REFERENCES

[1] L.P. Balant, *Clin. Therapeutics,* **1984,** *6,* 112.

[2] W. Kalow , *Lancet* **1956,** *II,* 576.

[3] K. Dawkins, W.Z. Potter, *Psychopharmacol. Bull.* **1991,** *27,* 417.

[4] C.H. Gleiter, U Gundert-Remy, *Eur. J. Drug Metab. Pharmacokin.* **1996,** *21,* 123.

[5] W. Kalow, *Clin. Pharmacokin.* **1982,** *7,* 373.

[6] A. Llerena, J. Cobaleda, C. Martínez, J. Benítez, *Eur. J. Drug Metab. Pharmacokin.* **1996,** *21,* 129.

[7] L.P. Balant, P.G. Welling, in 'Pharmacokinetics of Drugs', Eds. P.G. Welling and L.P. Balant, Handbook of Experimental Phaemacology, Springer-Verlag, Heidelberg, 1994, Vol.110, pp. 233-263.

[8] L.P. Balant, O. Pelkonen, in 'International Pharmaceutical Product Registration', Eds. A.C. Cartwright and B.R. Matthews, Ellis Horwoog, London, 1994, pp. 678-736.

[9] L. Balant, A. Gorgia, J.M. Tschopp, C. Revillard, J. Fabre, *Med. Wochenschrift,* **1976,** *106,* 1403.

[10] P.Dayer, L. Balant, F. Courvoisier, A. Küpfer, A. Kubli, A. Gorgia, J. Fabre, *Eur. J. Drug Met. Pharmacokin.* **1982,** *7,* 73.

[11] P.Dayer, L. Balant, A. Küpfer, F. Courvoisier, J. Fabre, *Eur. J. Clin. Pharmacol.* **1983,** *24,* 797.

[12] A. Mahgoub, J.R. Idle, L.G. Dring, R. Lancester, R.L. Smith, *Lancet,* **1977,** *II,* 584.

[13] M. Eichelbaum, N. Spannbrucker, B. Steincke, H.J. Dengler, *Eur. J. Clin. Pharmacol.* **1979,** *16,* 183.

[14] G. Alván, P. Bechtel, L. Iselius, U. Gundert-Remy, *Eur. J. Clin. Pharmacol.* **1990,** *39,* 533.

[15] F.J. Gonzalez, in 'Pharmacogentics of Drug Metabolism', Ed. W. Kalow, Pergamon Press, New York, 1992, pp. 413-452.

[16] F.J. Gonzalez, U.A. Meyer, *Clin. Pharmacol. Ther.* **1991,** *50,* 233.

[17] U.A. Meyer, U.M. Zanger, D. Grant, M. Blum, in 'Advances in Drug Research', Ed. B.Testa, Academic Press, London, 1990, Vol. 19, pp. 197-241

[18] U.A. Meyer, R.C. Skoda, U.M. Zanger, M. Heim, F. Broly, in 'Pharmacogentics of Drug Metabolism', Ed. W. Kalow, Pergamon Press, New York, 1992, pp. 609-623.

[19] A.K. Daly, M. Eichelbaum, J.R. Idle, M. Ingelman-Sundberg, U.A. Meyer, C.R. Wolf, U.M. Zanger, in 'COST B1 Conference on Variability and Specificity in Drug Metabolism', Eds. G.

Alván, L.P. Balant, P.R. Bechtel, A.R. Boobis, L.F. Gram, G. Paintaud, K. Pithan, European Commission, Luxembourg, 1995, pp. 137-144.
[20] U.M. Zanger, E.U. Griese, B. Evert, M. Eichelbaum, in 'COST B1 Conference on Variability and Specificity in Drug Metabolism', Eds. G. Alván, L.P. Balant, P.R. Bechtel, A.R. Boobis, L.F. Gram, G. Paintaud, K. Pithan, European Commission, Luxembourg, 1995, pp. 125-135.
[21] M. Ingelman-Sundberg, M. Oscarson, I. Persson, C. Masimirembwa, M.L. Dahl, L. Bertilsson, F. Sjöqvist, I. Johansson, in 'COST B1 Conference on Variability and Specificity in Drug Metabolism', Eds. G. Alván, L.P. Balant, P.R. Bechtel, A.R. Boobis, L.F. Gram, G. Paintaud, K. Pithan, European Commission, Luxembourg, 1995, pp. 93-110.
[22] L. Bertilsson, A. Åberg-Wistedt, L.L. Gustafsson, C. Nordin, *Ther. Drug Monit.* **1985,** *7,* 478.
[23] L. Bertilsson, M.L. Dahl, F. Sjöqvist, A. Åberg-Wistedt, M. Humble, I. Johansson, E. Lundqvist, M. Ingelman-Sundberg, *Lancet,* **1993,** *341,* 63.
[24] J.A.G. Agúndez, M.C. Ledesma, J.M. Ladero, J. Benitez, *Clin. Pharmacol. Ther.* **1995,** *57,* 265.
[25] A. Kubli, L. Balant, P. Dayer, A. Balant-Gorgia, J. Fabre, *J. Pharm. Clin.* **1982,** *1,* 301.
[26] P. Dayer, L. Balant., A. Küpfer, R. Striberni, T. Leemann, *Eur. J. Clin. Pharmacol.* **1985,** *28,* 317.
[27] G.T. Tucker, A. Rostami-Hodjegan, S. Nurminen, P.R. Jackson, in 'COST B1 Conference on Variability and Specificity in Drug Metabolism', Eds. G. Alván, L.P. Balant, P.R. Bechtel, A.R. Boobis, L.F. Gram, G. Paintaud, K. Pithan, European Commission, Luxembourg, 1995, pp. 191-203.
[28] P.R. Jackson, G.T. Tucker, M.S. Lennard, H.F. Wood, *Br. J. Clin. Pharmacol.* **1986,** *22,* 541.
[29] P.R. Jackson, G.T. Tucker, H.F. Woods, *Br. J. Clin. Pharmacol.* **1989,** *28,* 647.
[30] P.R. Jackson, G.T. Tucker, H.F. Woods, *Br. J. Clin. Pharmacol.* **1989,** *28,* 655.
[31] P.R. Jackson, G.T. Tucker, M.S. Lennard, H.F. Wood, in 'European Consensus Conference on Pharmacogenetics', Eds. G. Alvan, L.P. Balant, P.R. Bechtel, A.R. Boobis, L.F. Gram, K. Pithan, Commission of the European Communities, Coordinated Action COST B1, Luxembourg, 1990, pp. 43-52.
[32] P.R. Jackson, G.T. Tucker, *Ann. Occup. Hyg.* **1990,** *34,* 653.
[33] G.T. Tucker, P.R. Jackson, M.S. Lennard, H.F. Wood, in 'European Consensus Conference on Pharmacogenetics', Eds. G. Alvan, L.P. Balant, P.R. Bechtel, A.R. Boobis, L.F. Gram, K. Pithan, Commission of the European Communities, Coordinated Action COST B1, Luxembourg, 1990, pp. 59-67.
[34] K. Brøsen, L.F. Gram, *Eur. J. Clin. Pharmacol.* **1989,** *36,* 537.
[35] M. Eichelbaum, A.S. Gross, in 'Pharmacogenetics of Drug Metabolism', Ed. W. Kalow, Pergamon Press, New York, 1992, pp. 625-648.
[36] F. Sjöqvist, in 'Pharmacogenetics of Drug Metabolism', Ed. W. Kalow, Pergamon Press, New York, 1992, pp. 689-700.
[37] D.D. Breimer, in 'COST B1 Conference on Variability and Specificity in Drug Metabolism', Eds. G. Alván, L.P. Balant, P.R. Bechtel, A.R. Boobis, L.F. Gram, G. Paintaud, K. Pithan, European Commission, Luxembourg, 1995, pp. 2-19.
[38] A.E. Balant-Gorgia, L.P. Balant, G. Garrone, *Ther. Drug Monit.* **1989,** *11,* 415.
[39] I. Roots, N. Drakoulis, J. Brockmöller, in 'Pharmacogenetics of Drug Metabolism', Ed. W. Kalow, Pergamon Press, New York, 1992, pp. 815-841.
[40] O. Pelkonen , *Scand. J. Work Environ. Health,* **1992,** *18 (Suppl 1),* 17.
[41] W. Kalow (Ed.) 'Pharmacogenetics of Drug Metabolism', Pergamon Press, New York, 1992, 897 pp.
[42] L. Bertilsson, Y.Q. Lou, Y.L. Du, Y. Liu, X.M. Liao, K.Y. Wang, J. Reviriego, L. Iselius, F. Sjöqvist, *Clin. Pharmacol. Ther.* **1992,** *51,* 388.
[43] K. Nakamura, F. Goto, W.A. Ray, C.B. McAllister, E. Jacqz, G.R. Wilkinson, R.A. Branch, *Clin. Pharmacol. Ther.* **1985,** *38,* 402.
[44] D.R. Sohn, S.G. Shin, C.W. Park, M. Kusak, K. Chiba, T. Ishizaki, *Br. J. Clin. Pharmacol.* **1991,** *32,* 504.
[45] Q.Y. Yue, J.O. Svensson, J.O. Alm, F. Sjöqvist, J. Säwe, *Br. J. Clin. Pharmacol.* **1989,** *28,* 627.

[46] I. Johansson, M. Oscarson, Q.Y. Yue, L. Bertilsson, F. Sjöqvist, M. Ingelman-Sundberg, *Mol. Pharmacol.* **1994,** *46,* 452.
[47] H. Yokota, S. Tamura, H. Furuya, S. Kimura, M. Watanabe, I. Kanazawa, I. Kondo, F.J. Gonzalez, *Pharmacogenetics,* **1993,** *3,* 256.
[48] M.L. Dahl, L. Bertilsson, M. Chang, H.K. Roh, Y.A. Abdi, Q.Y. Yue, J. Säwe, C. Masimirembwa, I. Johansson, M. Ingelman-Sundberg, F. Sjöqvist, in 'COST B1 Conference on Variability and Specificity in Drug Metabolism', Eds. G. Alván, L.P. Balant, P.R. Bechtel, A.R. Boobis, L.F. Gram, G. Paintaud, K. Pithan, European Commission, Luxembourg, 1995, pp. 111-124.
[49] R. Temple, FDA Memorandum, Washington, D.C. 1983.
[50] R. Temple, *Drug Information J.* **1985,** *19,* 483.
[51] L. Aarons, *Brit. J. Clin. Pharmacol.* **1991,** *32,* 669.
[52] S. Vozeh, in 'New Strategies in Drug Development and Clinical Evaluation: The Population Approach', Eds. M. Rowland, L. Aarons, Commission of the European Communities, Coordinated Action COST B1, Luxembourg, 1992, pp. 107-120.
[53] M.E. Ebelin, J.L. Steimer, R. Laplanche, W. Niederberger, in 'New Strategies in Drug Development and Clinical Evaluation: The Population Approach', Eds. M. Rowland, L. Aarons, Commission of the European Communities, Coordinated Action COST B1, Luxembourg, 1992, pp. 131-141.
[54] S. Vozeh, J.-L. Steimer, M. Rowland, P. Morselli, F. Mentré, L.P. Balant, L. Aarons, *Clin. Pharmacokin.* **1996,** *30,* 82.
[55] S.L. Beal, L.B. Sheiner, *Am. Stat.* **1980,** *34,* 118.
[56] E. Chan, T.Y. Ti, H.S. Lee, *Eur. J. Clin. Pharmacol.* **1990,** *39,* 177.
[57] W.F. Dodge, R.W. Jelliffe, J. Richardson, R.A. McCleery, J.A. Hokanson, W.R. Snodgrass, *Clin. Pharmacol. Ther.* **1991,** *50,* 25.
[58] R. Bruno, M.C. Iliadis, B. Lacarelle, V. Cosson, J.W. Mandema, Y. Le Roux, G. Montay, A. Durand, M. Ballereau, M. Alasia, J. Albanese, G. François, A. Iliadis, A. Frydman, *J. Pharmacokin. Biopharm.* **1992,** *20,* 653.
[59] R. Gomeni, G. Pineau, F. Mentré, *Anticancer Res.* **1994,** *14,* 2321.
[60] F. Mentré, R. Gomeni, *J. Piopharm. Stat.* **1995,** *5,* 141.
[61] P. Burtin, F. Mentré, J. Van Bree, J.L. Steimer, *Eur. J. Drug Metab. Pharmacokin.* **1996,** *21,* 105.
[62] L.P. Balant, H. Roseboom, U. Gundert-Remy, in 'Advances in Drug Research', Ed. B. Testa, Academic Press, London, 1990, pp. 1-138.
[63] L.P. Balant, U. Gundert-Remy, A.R. Boobis, Ch. von Bahr, *Eur. J. Clin. Pharmacol.* **1989,** *36,* 551.
[64] L.B. Sheiner, in 'New Strategies in Drug Development and Clinical Evaluation: The Population Approach', Eds. M. Rowland M., L. Aarons, Commission of the European Communities, Coordinated Action COST B1, Luxembourg, 1992, pp. 13-29.
[65] L.B. Sheiner, L.Z. Benet, *Clin. Pharmacol. Ther.* **1985,** *38,* 481.
[66] L.P. Balant, *Méd. & Hyg.* **1985,** *43,* 2470.
[67] L. Aarons, L.P. Balant, F. Mentré, P.L. Morselli, M. Rowland, J.-L. Steimer, S. Vozeh, *Eur. J. Clin. Pharmacol.* **1994,** *46,* 389.
[68] L. Aarons, L.P. Balant, F. Mentré, P.L. Morselli, M. Rowland, J.-L. Steimer, S. Vozeh, *Eur. J. Clin. Pharmacol.* **1996,** *49,* 251.
[69] A. Lindholm, M. Welsch, C. Alton, B.D. Kahan, *Clin. Pharmacol. Ther.* **1992,** *52,* 359.
[70] R. Laplanche, B. Fertil, E. Nüesch, J.P. Jais, W. Niederberger, J.L. Steimer, *Clin. Pharmacol. Ther.* **1991,** *50,* 39.
[71] M. Gex-Fabry, A.E. Balant-Gorgia, L.P. Balant, G. Garrone, *Clin. Pharmacokin.* **1990,** *19,* 241.
[72] M. Gex-Fabry, A.E. Balant-Gorgia, L.P. Balant, *Ther. Drug Monitor.* **1995,** *17,* 39.
[73] M. Jerling, *Eur. J. Drug Metab. Pharmacokin.* **1996,** *21,* 113.

PART III. **COMPUTATIONAL APPROACHES IN DRUG DISPOSITION**

Chapter 21

Molecular Modelling of Cytochromes P450

by **David F.V. Lewis**

School of Biological Sciences, University of Surrey,
Guildford, Surrey, GU2 5XH, UK

Molecular Modelling of Cytochromes P450

by **David F.V. Lewis**

School of Biological Sciences, University of Surrey,
Guildford, Surrey, GU2 5XH, U K

The rationalization of drug metabolism in man is now possible following the construction of human cytochromes P450 from a unique bacterial P450 crystal structure (CYP102). A large number of drugs and other foreign compounds have been modelled within the putative active sites of the particular P450s associated with their metabolism and the results of such studies are consistent with the known metabolic routes of these chemicals. The metabolism of endogenous substrates may also be rationalized from molecular modelling of the relevant P450s. Investigations on the P450s which mediate in the metabolic activation of carcinogens both support the previous findings on substrate structural criteria used to formulate the COMPACT system, and enable further refinement of the technique. The prediction of P450-mediated metabolism should be feasible following validation of the models and development of a suitable methodology for relatively rapid pre-screening of new chemical entities.

Introduction. - There is a need for an indication of the likely routes of metabolism (and their possible consequences) at an early stage of novel compound development. Although the safety evaluation of new chemical entities is conducted on small rodents or other experimental mammalian species, such animal models have been shown to be largely inadequate as human surrogates. One of the reasons for species differences in the metabolism of foreign compounds (and subsequent toxicity) lies in the nature and expression of the enzymes involved in the Phase I oxidative biotransformations, the majority of which have been characterized as cytochrome P450 isozymes.

The cytochromes P450 (CYP) constitute a superfamily of heme-thiolate proteins, present in almost all species, which possess mono-oxygenase catalytic activity for the metabolism of both endogenous and exogenous chemicals [1]. Over 480 CYP genes have been sequenced to date and analyzed in terms of sequence homology across the superfamily [2,3]. Construction of the subsequent phylogenetic tree has led to the view that P450s evolved from an ancestral gene over a period of at least 3,500 million years, and it appears that evolution of the P450 superfamily closely parallels the currently-accepted timescale for biological development. Although involved in the biosynthesis of steroid hormones and in other aspects of endogenous metabolic transformations, P450 plays a pivotal role in the toxicity of foreign compounds. Paradoxically, the particular mammalian P450 isoforms which may have originally evolved as a response to plant toxins are now known to be closely associated with the metabolic activation of pro-carcinogens and other potentially toxic chemicals.

The cytochromes P450 are capable of metabolizing a vast number and variety of organic substrates, estimated to be in the region of 200,000 chemicals or more, and involving around 60 distinct classes of biotransformation reactions which, nevertheless, are thought to proceed via a common catalytic mechanism. This entails the binding and subsequent activation of molecular oxygen by the heme moiety, which leads to a single oxygen atom insertion into the bound organic substrate (RH) with the concomitant formation of a water molecule, according to the following equation:-

$$RH + O_2 \xrightarrow[2H^+,\ 2e^-]{P450} ROH + H_2O$$

The two reducing equivalents (2 protons and 2 electrons) are supplied by NADPH and mediated via a flavoprotein, P450 oxidoreductase, in the case of the microsomal P450 system, or an iron-sulphur redoxin as far as mitochondrial and most bacterial P450s are concerned, with NADH being a common reductant in the latter. It has been recognized that cytochrome b_5 is also able to act as a redox partner in mammalian hepatic microsomal P450, particularly with respect to input of the second reducing equivalent, and these are supplied in two discrete stages during the P450 catalytic cycle. The precise nature of the active oxygen species, which attacks the substrate, is a subject of considerable scientific interest and debate but it is likely that an iron-bound peroxide is formed which rearranges, following the second reduction step, to give the products of the mono-oxygenase reaction, possibly via the mediation of an iron-oxene intermediate. It has been established that there is some degree of substrate specificity towards different structural classes for various P450s, particularly the xenobiotic-metabolizing isoforms [3,4]. Although widely distributed in the organs and tissues of mammalian and other animal species, the major proportion of these P450 isozymes is present in the liver, and *Table 1* shows the average percentages of different P450 subfamilies which make up the human hepatic P450 complement. *Table 1* also indicates the specificities of these P450s towards a number of classes of substrates, and it should be emphasized that P450s of subfamilies CYP1A and CYP2E are most closely linked with the metabolic activation of carcinogens, whereas other P450s are generally associated with detoxifying metabolism. For example, planar polyaromatic hydrocarbons (PAHs) and heterocyclic amines are substrates and inducers of CYP1, and are metabolically activated by the enzyme to form reactive electrophilic intermediates that readily interact covalently with DNA causing miscoding, mutagenicity and, ultimately, carcinogenicity. In the case of benzo(a)pyrene, it is known that CYP1 catalyzes two epoxidation steps in the formation of the highly carcinogenic metabolite, benzo(a)pyrene-7,8-diol-9,10-epoxide, which gives rise to covalent adducts with guanine and other DNA bases.

The regulatory mechanisms which govern the expression of P450s vary, and are also subject to modulation according to tissue type, sex and stage of development. However, it is known for example that induction of CYP1 is mediated via a cytosolic receptor, known as the Ah (aryl hydrocarbon) receptor, which is able to interact with certain regulatory elements on the CYP1 gene, following the binding of potent inducers, such as TCDD and benzo(a)pyrene. Induction of CYP1 is at least as important as substrate activation for toxicity and carcinogenicity because it results in both an increase in cell division and down-regulation of other P450s normally associated with detoxifying metabolism. On the other hand, induction of CYP2E is linked with the generation of oxygen radicals and other reactive oxygen species (ROS) such as $O_2^{\bullet-}$, H_2O_2, $OH^\bullet$, etc. ROS and oxygen radical production have been causally associated with several degenerative disease states and may be related to the process of ageing. Furthermore,

Table 1. Substrate Specificities of Mammalian P450s.

CYP	Substrate structural Characteristics	Typical Substrates	% Human hepatic P450
1A	Planar, medium Mwt. poly(hetero)aromatics and their amino/amido derivatives	Benzo(a)pyrene, caffeine	13
2A	Relatively planar low Mwt. (aromatic) ketones with hydrogen bond potential	Coumarin, aflatoxin B_1	4
2B	Non-planar (V-shaped) lipophilic compounds some with hydrogen bond potential	Phenobarbital, DDT	<1
2C	Non-planar amides and carboxylic acids with hydrogen bond potential	Phenytoin, tolbutamide	20
2D	Non-planar basic compounds, sometimes aromatic and some with hydrogen bond potential	Debrisoquine, metoprolol	2
2E	Low Mwt. compounds of diverse structure, some with hydrogen bond potential	Ethanol, acetone, benzene, p-nitrophenol	7
3A	High Mwt. compounds of diverse structure most with hydrogen bond potential	Cyclosporin, erythromycin	30
4A	Endogenous long chain fatty acids, generally hydrophobic	Lauric acid	~ 20

Steroids such as: testosterone, progesterone, androstenedione, cortisol, estradiol, etc., usually lie on the borderline between specifities for these P450s and are known to be metabolized by different isoforms, sometimes in differing positions on the steroid nucleus, which may also differ from those associated with steroid biosynthesis (eg. CYP11, CYP17, CYP19 and CYP21). The following summarizes steroid metabolism by hepatic microsomal P450s generally associated with xenobiotic transformations.

CYP	**Steroid metabolism catalyzed**
1A	estradiol 2-hydroxylation
2A	testosterone 7α and 15α-hydroxylation
2B	testosterone 16α-, 16β- and 17-hydroxylation
2C	testosterone 6α-, 6β-, 7α-, 16α-, and 17-hydroxylation (other positions can be hydroxylated in progesterone)
2D	none (some evidence of testosterone 15α- and 6α-hydroxylation)
2E	none (acetone and ethanol are endogenous substrates)
3A	testerone 6β-hydroxylation (also progesterone and cortisol)
4A	none (prostaglandins are endogenous substrates of some forms)
	References: Lewis, 1996 [1]; Ortiz de Montellano, 1995 [3].

peroxisome proliferation is a mechanism for carcinogenicity in rodents which could be coordinated with induction of CYP4, and examples of carcinogenic peroxisome proliferators which are also potent inducers of CYP4 in rodents include clofibrate and phthalate esters, such as MEHP. In addition to ROS generation, CYP2E also metabolically activates many small molecular weight halogenated alkenes and alkenes, e.g. dichloromethane, carbon tetrachloride, halothane, and trichloroethene, together with benzene and small-chain aliphatic nitrosamines, such as dimethylnitrosamine (DMN).

Systematic analysis of the structural criteria for P450 specificity has facilitated the development of a computer-based procedure, known as COMPACT, which can be used for the prediction of potential toxicity/carcinogenicity mediated by either CYP1 or CYP2E [1,5]. In this technique, it is found that a combination of three structural parameters: planarity (area/depth^2), activation energy (ΔE) and molecular diameter, is sufficient to discriminate between specificity towards CYP1 or CYP2E as opposed to other P450s. The concordances between COMPACT and rodent carcinogencity vary from 70% to 90%, depending on the nature of the chemicals, and this can be improved by the consideration of structural alerts for the identification of direct-acting agents. As far as CYP1 induction is concerned, the COMPACT radius (a combination of area/depth^2 and ΔE) shows a good correlation ($r = 0.94$) with induction potential for structurally-diverse compounds, and inclusion of the lipophilicity parameter, log P, increases this correlation to over 95%.

Homology Modelling. - Although there are additional structural features which characterize substrate specificity towards other P450 isoforms (eg. CYP2D substrates possess a basic nitrogen at a distance of 5-7Å from the site of metabolism), it is likely that modelling of the relevant enzymes can yield more precise information relating to the features which determine P450 specificity and site of metabolism, together with consolidating the existing knowledge based on substrate overlays and analysis of molecular structural criteria governing P450 substrate specificity. In order to derive three-dimensional models of P450s, the generally accepted procedure involves sequence homology alignment with those isozymes for which crystallographic data are known, followed by amino acid residue replacement, insertion and deletion, as required by the alignment and, finally, energy minimization of the raw structure. To date the molecular structures of four bacterial forms of P450 have been determined by X-ray crystallography [6-9], three of which are compared in an important review published recently [10]. It is apparent that there are considerable structural similarities between these prokaryotic P450s, in spite of relative low primary sequence homology, and most of the regions likely to be in contact with their particular substrates (ie. substrate recognition sites, SRSs) are spatially analogous when the structures are superimposed. The major exceptions to this overall similarity, in what are termed the common core and SRS residues, are variable loop regions between the B' and C helices, and between the F and G helices, which are likely to be involved in substrate access/recognition and are, furthermore, of different peptide lengths in the prokaryotic P450s for which crystal structures are available. It appears that this variation in length of the peptide loops gives rise to altered conformations for the B' and F/G helices which may reflect the character of the relevant substrates. However, of the four bacterial P450 crystal structures, CYP102 ($P450_{bm3}$) is closest in sequence length with the corresponding aligned segments of mammalian P450s in these and other regions, generally [11].

Consequently, although the sequence of all four P450s, which have been resolved crystallographically, are helpful in producing multiple sequence alignments either across the CYP superfamily or within families and subfamilies, it is clear that CYP102 currently represents the best structural template for homology modelling of mammalian and other eukaryotic P450 isozymes. Analysis of such alignments indicates that there are several common features between CYP102 and mammalian P450s which are not shared with other bacterial P450 crystal structures. Moreover, CYP102 exhibits a substrate specificity resembling that of some microsomal P450s, namely hydroxylation of long chain carboxylic acids which are also metabolized by CYP4A, CYP2C and CYP2E subfamilies. CYP102 utilizes an NADPH-dependent oxidoreductase flavoprotein as its redox partner, as do microsomal P450s, whereas mitochondrial and the bacterial P450s (other than CYP102) use iron-sulphur redoxins to supply electrons for catalysis. Furthermore, CYP102 is known to be readily inducible by phenobarbital [12], and other barbiturates, as are P450s of subfamilies CYP2A, CYP2B and CYP2C, pointing to some commonality in regulatory mechanisms.

Table 2. Percentage Sequence Homologies between CYP102, CYP101 and Mammalian P450s†

	CYP102		**CYP101**	
	Identity	**Similarity**	**Identity**	**Similarity**
CYP3A4	27.1	52.1	16.7	45.9
CYP4A4	26.7	54.4	17.4	49.0
CYP4A11	26.0	53.6	11.6	36.2
CYP2E1	24.9	53.6	17.6	47.8
CYP1A1	24.9	54.2	14.5	42.3
CYP1A2	24.3	50.5	16.4	45.7
CYP2A6	22.6	54.0	18.6	46.6
CYP2B1	22.3	52.1	17.6	49.5
CYP2D6	20.8	51.2	19.3	50.2
CYP2C9	20.2	51.6	17.4	48.8

†Based on the alignment presented as *Figure 1.*
% Similarity = % Identity + % Homology
Data kindly supplied by Dr. Luc Koymans (Janssen Research Foundation, Beerse, Belgium)

In terms of sequence comparison, CYP102 shows a greater overall homology with microsomal P450s than do the other prokaryotic P450 crystal structures *(Table 2).* Also, the protein sequences of CYP102 and eukaryotic P450s are of similar length in terms of numbers of residues, apart from an N-terminal membrane 'anchor' of 20-40 amino acids possessed by microsomal P450s; whereas the other bacterial P450s are of shorter sequence. Consequently, alignments based on CYP102 generally show fewer gaps comprising relatively smaller numbers of amino acids as opposed to those produced with either CYP101 or CYP108, for example [10]. There are also several commonalities between CYP102 and eukaryotic P450s, such as the conserved tryptophan which may mediate electron transfer from reductase, together with other motifs which not only aid alignment but reinforce the view that CYP102 belongs to the same class as the mammalian P450s [13].

Therefore, there is generally a greater degree of confidence associated with P450 models generated from the CYP102 crystal structure and, in fact, the modelling

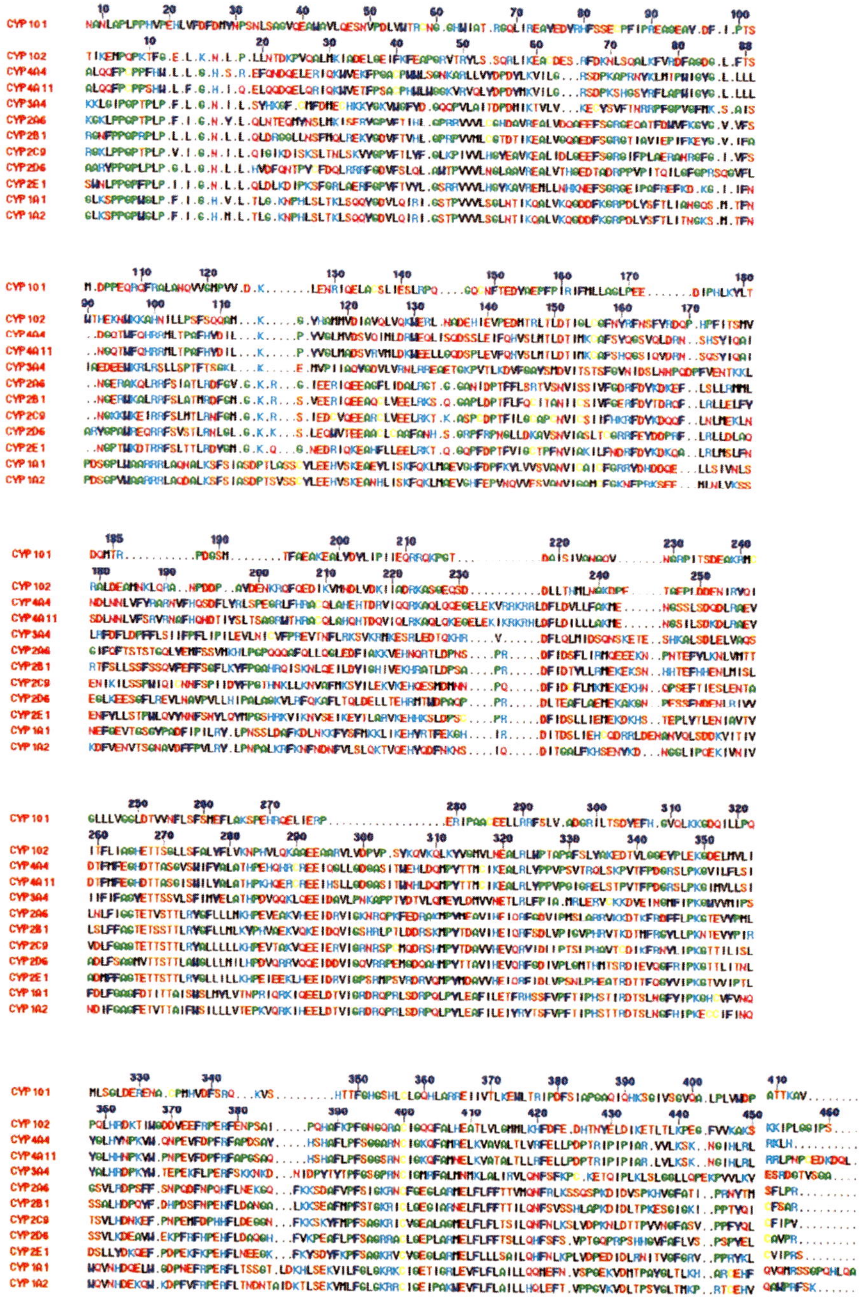

Fig. 1. *A colour-coded multiple sequence alignment* (kindly supplied by Dr. *Luc Koymans*) *between 10 mammalian P450s and the two bacterial forms CYP101 and CYP102* [11]

procedure is facilitated by the close homology between this structural template and the mammalian P450 sequences, especially with relatively few residue insertions to cause any gross distortions of the protein conformation. Moreover, the similarity in redox partners between CYP102 and microsomal P450s enables likely common surface residues to be used in pin-pointing reductase contacts. Consequently, it can be shown that homology alignments, based on site-directed mutagenesis, redox partner interaction points, antibody recognition peptides, allelic variants and retention of secondary structural motifs, give rise to model P450 structures that are consistent with known substrate specificity and positions of metabolism [1,11,14-18].

Results and Discussion. - Using the Sybyl Biopolymer (Tripos Associates, St. Louis, MO) and other modules in the Sybyl molecular modelling software package, a number of microsomal mammalian P450s have been constructed from the CYP102 template [11,14-18]. Although it is found that alignment between the CYP102 sequence and those of P450s within individual families and subfamilies produces more reliable models, a multiple sequence alignment across families makes a useful comparison to highlight similarities and differences. An example of such an alignment is shown as *Figure 1*, which includes the sequences of CYP101 and CYP102, together with those of 10 mammalian microsomal P450s. In fact, a total of 30 eukaryotic P450s have been constructed from CYP102, and *Table 1* indicates the various substrates and inhibitors which have been modelled in the putative active sites of these structures. The findings to date show that the enzyme models are consistent with experimental evidence from protein engineering, and from reported data of specific substrate metabolism, reductase interactions and antibody recognition sites. Furthermore, each P450 structure minimized smoothly over several hundred iterative cycles of geometry optimization to produce minimum energy values in the region of -1000 kcal.mole^{-1}, thus indicating highly stable geometries. The stereochemical quality of the minimized structures has been evaluated using the ProCheck software suite (Oxford Molecular Ltd., Oxford, U.K.) and are within the acceptable criteria for protein geometries. Interactive docking of substrates and inhibitors was facilitated by initial minimization of the substrate lauric acid within the original CYP102 crystal structure, following the removal of water molecules apart from the heme ligand, using information derived from NMR studies [19]. The distance geometry of the substrate-bound enzyme model was thus found to be consistent with the published paramagnetic shift data relating to the interaction of lauric acid with CYP102. The results obtained for some of the mammalian P450s will now be summarized using the interaction with specific substrates or inhibitors as examples.

CYP1A Models. - *Figure 2* shows the putative active site of human CYP1A2 with caffeine, a specific marker substrate, orientated for N-demethylation at the N3-position, which is the characteristic site of caffeine metabolism mediated by human hepatic CYP1A2 [15]. It can be seen from *Figure 2* that a number of key amino acid residues appear to participate in orientation of the caffeine substrate for the preferred position of oxygenation via two hydrogen bonds and π–π stacking interactions with complementary groupings. Two of these amino acid positions (Thr87 and Phe181) correspond with those shown by site-directed mutagenesis to alter substrate selectivity and regiospecificity in CYP2A, CYP2B and CYP2C isozymes [20] and, therefore, may also be associated with enzyme-substrate interactions in the CYP1A subfamily.

It appears that two coplanar aromatic residues (Phe181 and Tyr437) guide the entry of planar polyaromatic substrates to the heme pocket, and a number of other aromatic amino acid side-chains flank the putative active site, thus restricting the overall

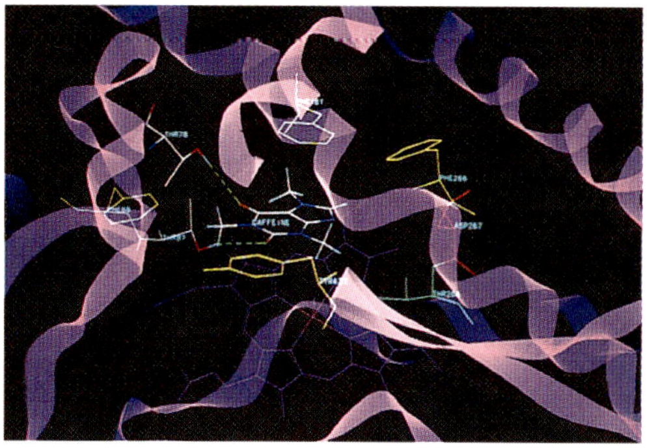

Fig. 2. *Putative active site of human CYP1A2 showing caffeine orientated for N3-demethylation by hydrogen bond interactions with Thr87 and Thr78 side chains* [15]

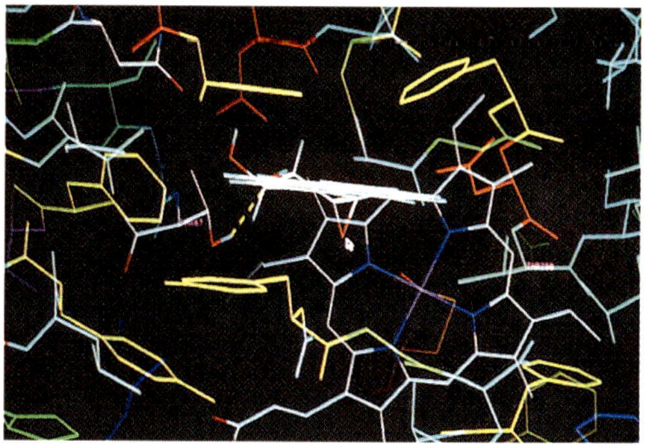

Fig. 3. *Putative active site of rat CYP1A1 showing the diolepoxide of benzo(a)pyrene formed by epoxidation of 9,10-bond of benzo(a)pyrene-7,8-diol* [15]

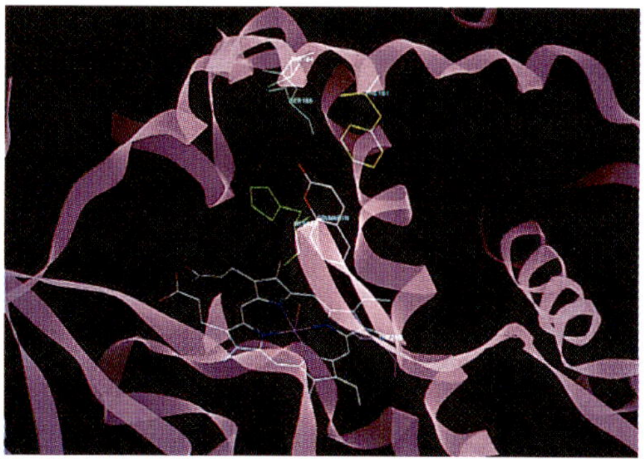

Fig. 4. *Putative active site ot CYP2A6 showing coumarin orientated for 7-hydroxylation by hydrogen bonding with Ser 185* [14]

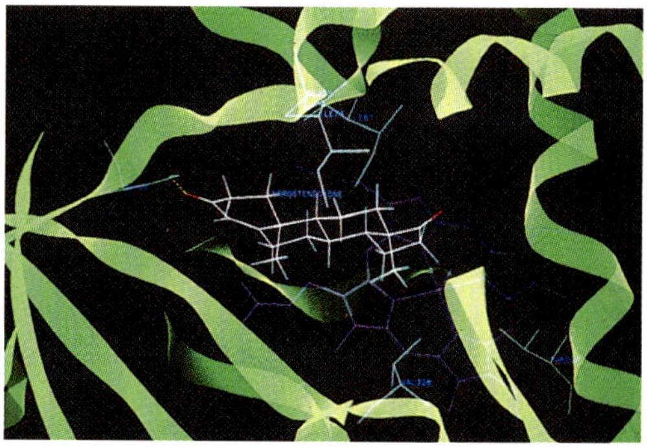

Fig. 5. *Androstenedione is shown orientated in the putative active site of CYP2B1 for 16β-hydroxylation by hydrogen bonding with Ser69* [24]

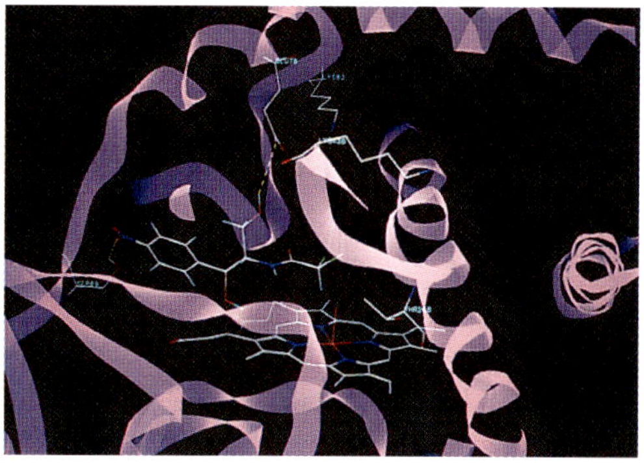

Fig. 6. *Chlorambucil orientated in the putative CYP2B1 active site for dichloromethyl hydroxylation by hydrogen bonding with Ser69* [24]

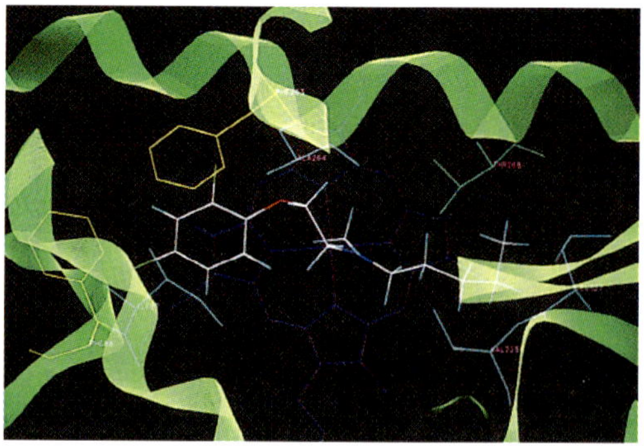

Fig. 7. *N-pentyl isomer of 2,4-dichlorophenoxy N-methylethylamine shown orientated in the putative active site of CYP2B1 for N-demethylation* [24]

dimensions of CYP1A substrates to those possessing relatively small molecular widths in comparison to their molecular lengths; such that the optimal length/width ratio is about 1.5, with an average maximum length of around 15Å.

It is possible that this combination of aromatic amino acids in the vicinity of the substrate binding site is responsible for the specificity of CYP1A enzymes towards planar polyaromatic compounds, whereas the relatively large proportion of amide residues in the active site region of CYP1A2, as opposed to CYP1A1, could explain the former's preference for aromatic amides and aromatic amines as substrates. In addition to the two aromatic residues mentioned previously, the putative active sites of both CYP1A1 and CYP1A2 contain hydrogen bond donor/acceptor amino acid side-chains, and could also orientate the molecule for oxygenation in the experimentally observed positions. One example of such an interaction is given in *Figure 3*, which shows how the 7,8-diol of benzo(a)pyrene may be orientated within the putative active site of CYP1A1 such that epoxidation at the 9,10-position is favoured by hydrogen-bonded contacts with Thr87 and Thr78 side-chains. This would give rise to the diolepoxide metabolite which is, in fact, the major product of the epoxidation specifically-catalyzed by CYP1A1.

CYP2A Models. - In the CYP2A subfamily, site-directed mutagenesis studies carried out by Negishi and coworkers [21-23], involving heterologous expression, have demonstrated the importance of certain amino acid residue positions in the two mouse orthologues, CYP2A4 and CYP2A5, for the different substrate specificities of the isozymes which differ in only 11 amino acids out of 494. Thus, although the two enzymes are 98.3% homologous, CYP2A4 specifically catalyzes the 15α-hydroxylation of testosterone, whereas CYP2A5 hydroxylates coumarin in the 7-position. A single amino acid change from leucine to phenylalanine (L209F) is sufficient to reverse this substrate specificity in the two orthologous proteins and this corresponds to position 181 in CYP102. Molecular modelling of these isozymes shows that the residue at this point lies directly above the heme and is, therefore, likely to possess especial relevance to substrate specificity [14]. Interestingly, the human isoform, CYP2A6, also has a phenylalanine at this position and, in common with the mouse orthologue CYP2A5, specifically catalyzes the 7-hydroxylation of coumarin. *Figure 4* shows how the coumarin substrate may be orientated in the active site of CYP2A6 such that hydroxylation at the 7-position can occur. It can be seen from *Figure 4* that the phenyl ring of Phe181 is able to enter into a π–π stacking interaction with the aromatic nucleus of coumarin, whereas Thr185 can donate a hydrogen bond to the carbonyl oxygen of the substrate, thus orientating the substrate molecule so that the hydrogen atom substituting the 7-position lies directly over the heme iron. Protein engineering of the mouse orthologues, CYP2A4 and CYP2A5, has also shown that other residues are of relevance to substrate binding interactions within the enzymes' active sites [21-23]. In addition to the L209F mutant mentioned above, which maps on to position 181 in the alignment with CYP102 [14], a further two residues, corresponding to CYP102 positions 87 and 330 respectively, appear to play a lesser role in substrate orientations within the CYP2A isozymes, and other investigators have reported similar findings for analogous regions in the CYP2B, CYP2C and CYP2D subfamilies. Consequently, a picture is emerging from site-directed mutagenesis studies of the active site topography of the CYP2 family which is consistent with the current models generated from the CYP102 template [11, 14]. Furthermore, protein sequence analysis of inactive variants in human orthologues of CYP2A, CYP2C and CYP2D isozymes, or those exhibiting altered metabolic profiles, can identify substrate, heme or redox partner interaction sites within the relevant P450 isoform. For example, CYP2A7 is an inactive human P450 (94% identical with

CYP2A6) which is unable to incorporate heme, apparently due to a single residue change R100L, thus suggesting that this position (normally involving a basic amino acid side-chain) is associated with ionic interactions with one of the heme propionates [14], and may also participate in ion-pairing with one of the P450 redox partners. However, a number of basic residues that are likely to be involved in reductase binding have been identified in the CYP2B subfamily, together with various antibody recognition peptides which probably represent surface regions of the protein.

CYP2B Models. - Although constituting only a very small percentage (< 1%) of human hepatic P450, the CYP2B subfamily represents important and highly-inducible forms in rodents and other mammalian species. Molecular models of rat (CYP2B1), rabbit (CYP2B4) and human (CYP2B6) orthologues in the CYP2B subfamily show that these highly homologous isoforms all possess similar active site regions, which is consistent with their known overlapping substrate specificities [24]. For example, the rodent carcinogen, cyclophosphamide, is metabolically activated by both rat and human CYP2B isozymes. Typical substrates of the CYP2B subfamily include: phenobarbital (also an inducer) and other barbiturates, DDT and structurally-related organochlorine pesticides, p-substituted toluenes, phenylbutazone, benzphetamine, nicotine, pentoxyresorufin, feprazone and certain endogenous steroids, such as testosterone and androstenedione, which are characteristically oxygenated in the 16 and 17 positions on the steroid nucleus. Inhibitors of CYP2B isozymes include secobarbital, SKF525-A (proadifen), metyrapone, orphenadrine and chloramphenicol, which binds covalently with an active site lysine residue following activation by the enzyme. It is also known that primary aliphatic alcohols and amines are able to act as inhibitors of CYP2, possibly via heme ligation.

Amino acid sequence alignment and, hence, homology modelling of CYP2B isoforms has been facilitated by experimental evidence from protein engineering [25-28], antibody recognition sites [29,30], chemical modification of surface lysine residues [31], and from regio- and stereo-selectivity of substrate metabolism [24]. For example, the relatively conservative change, I87V, has been shown to alter the regioselectivity of androstenedione metabolism in CYP2B1 from 16β-hydroxylation to 15α-hydroxylation [25], whereas the I87F mutant alters the stereoselectivity of testosterone hydroxylation from 16β to 16α [26]. Moreover, a mutation corresponding to the same alignment position as that mentioned in the previous section on CYP2A, namely, F181L has been shown to confer 6α and 7α-steroid metabolism [27]; whereas positions 328, 332 and 439 are also involved in modifying substrate specificity in CYP2B1 [27,28]. Consequently, site-directed mutagenesis has demonstrated that several active site amino acid residues from different regions of the primary structure are likely to be involved in substrate orientation for CYP2B-mediated metabolism, and these enable identification of important active site residue positions. In addition to protein engineering, CYP2B1 has been investigated using site-specific antibodies [29] such that the probable location of surface regions can be determined, and similar antibody recognition sites have been examined in the rabbit orthologue, CYP2B4 [30]. This enzyme has also been studied by chemical modification of lysine residues which indicate likely points of electrostatic binding interactions with reductase [31]. It can be shown that basic residues corresponding to alignment positions 113, 118, 222 and 349, all lie on the same cytoplasmic face of the enzyme and thus would be able to form ion-pairs with complementary acidic amino acid side-chains on the oxidoreductase redox partner [11]. Furthermore, the serine phosphorylation site (Ser103) in CYP2B isozymes is also present on the surface [11] and is positioned close to one of the heme-binding arginine residues

(R100), such that phosphorylation of serine-103 is likely to lead to a change of orientation in the arginine-100 sidechain, thus bringing about the loss of the heme co-factor and subsequent breakdown of the enzyme as is observed experimentally.

The identification of amino acid residues likely to be involved in the binding and orientation of endogenous steroids (mentioned previously) has facilitated the modelling of these and other substrates within the putative active site region of CYP2B1 and orthologous proteins [11,24]. For example, *Figure 5* shows how the endogenous substrate, androstenedione, may bind to CYP2B1 in the vicinity of the heme pocket where Ile87 and other amino side-chains may orientate the steroid molecule for hydroxylation at the 16β position. The mechanism-based inhibitor, chloramphenicol, could form a Schiff's base (the known mechanism of action) with an active site lysine residue, following metabolic activation by CYP2B, and a possible binding site interaction of this inhibitor is presented as *Figure 6*; moreover, all of the typical substrates of this subfamily fit the CYP2B model. In an homologous series of 2,4-dichlorophenoxy N-alkyl N-methylethylamines, it has been reported that the route of metabolism involving N-demethylation is largely mediated via CYP2B1 in the rat as the homologous series is ascended to a maximum at N-butyl (see *Figure 7*), whereupon there is a dramatic lowering of CYP2B1-catalyzed N-dealkylation and a concomitant switch to CYP3A1 N-dealkylase activity [32]. Molecular modelling shows that the N-butyl congener can readily occupy the putative active site of CYP2B1, whereas there is a significant degree of steric hindrance associated with the binding interaction of the N-pentyl homologue.

The specific inhibitor, secobarbital, is also able to fit the CYP2B1 active site, as can other inhibitors such as SKF-525A and metyrapone, where hydrogen bonding and hydrophobic interactions cooperate in orientating these inhibitors for their respective modes of action. In fact, it is possible to demonstrate that a superimposed template of both substrates and inhibitors can fit the CYP2B1 active site model [24].

CYP2C Models. - There is considerable overlap between substrate specificities for the CYP2C and CYP2B subfamilies although, in man, enzymes of the CYP2C subfamily constitute a major proportion of the hepatic P450 complement, whereas proteins of the CYP2B subfamily represent less than 1% of the total P450 present in human liver. As with the CYP2A and CYP2B subfamilies, site-directed mutagenesis experiments have assisted in the identification of amino acid residues thought to be associated with substrate binding and orientation for metabolism in the known positions [11]. In particular, the mutations V87A and S331T have been shown to affect regiospecificity of progesterone metabolism in the rabbit orthologue CYP2C3 [33], and modelling of enzyme-substrate interactions within the putative active site is able to rationalize the observed findings [11]. *Figure 8* demonstrates how the progesterone molecule can be orientated in the CYP2C3 model for 16α hydroxylation in the S331T variant CYP2C3v, and there are other examples of protein engineering studies which reinforce the findings for the CYP2A and 2B subfamilies. Particular interest in the CYP2C subfamily stems from the known genetic polymorphism in the human orthologue, CYP2C19, which results in impaired metabolic function in some individuals with respect to the clearance of a number of drugs currently in clinical use, such as: mephenytoin, proguanil, warfarin and omeprazole, for example. Substrate template models indicate that hydrogen bond potential is an important feature of CYP2C specificity, and most substrates for this subfamily are amides, carboxylic acids or ketones, where the site of metabolism lies at about 8Å from the hydrogen bond acceptor/donor atom [4,34]. It appears that enzyme modelling of CYP2C subfamily isozymes is consistent with the current substrate

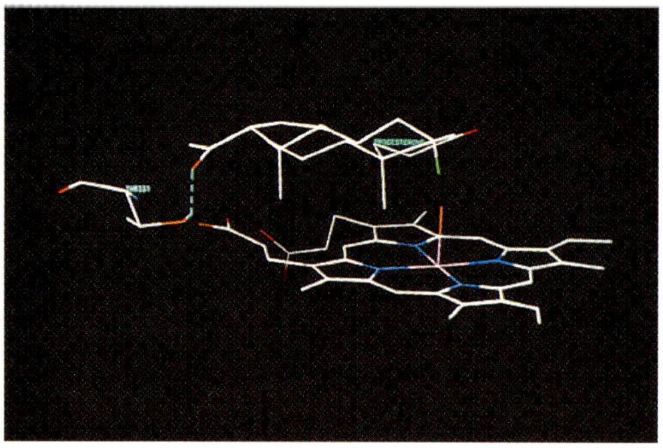

Fig. 8. *Progesterone orientated in the CYP2C3v active site for 6β-hydroxylation by hydrogen bonding with Thr331* [11]

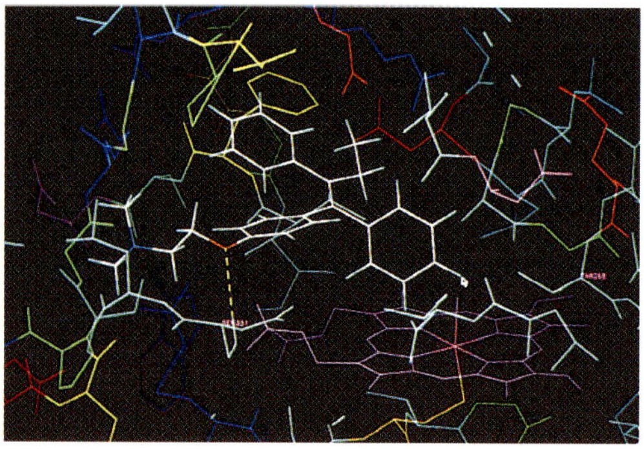

Fig. 9. *Tamoxifen orientated in the CYP2C9 active site for 4′-hydroxylation by hydrogen bonding with Ser331* [18]

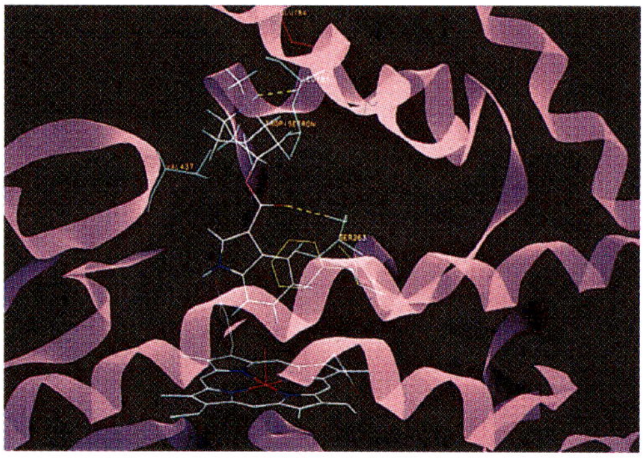

Fig. 10. *Tropisetron is shown orientated in the CYP2D6 active site for aromatic ring hydroxylation by hydrogen bonding with Ser263*

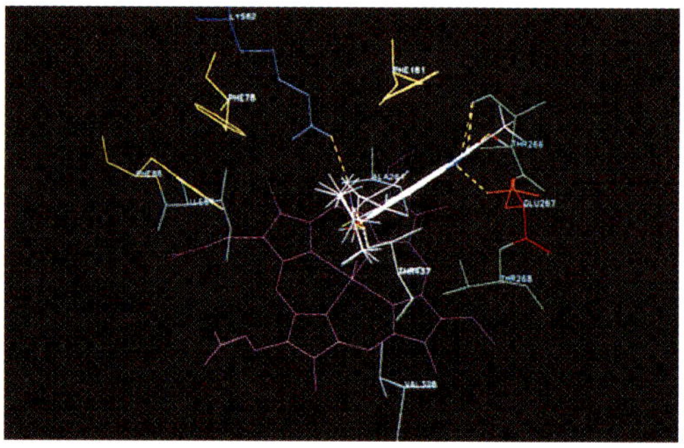

Fig. 11. *A template of CYP2E1 substrates superimposed in the putative active site where hydrogen bonded contacts orientate various substrates for oxygenation at the observed positions* [40]

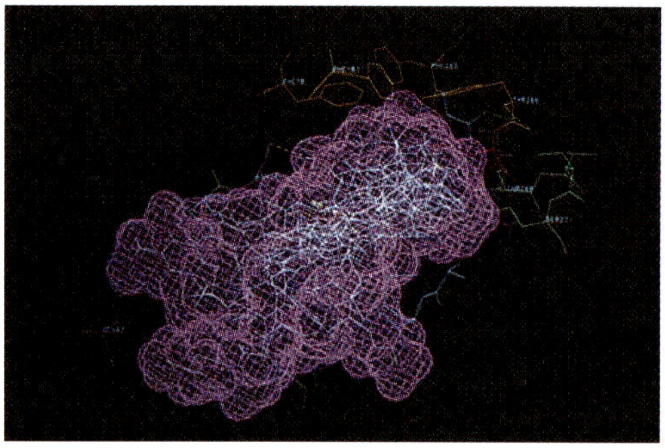

Fig. 12. *A template of CYP3A4 substrates superimposed in the putative active site for oxygenation at the observed positions* via *hydrogen bond interactions with either Asn74 or Thr73* [16]

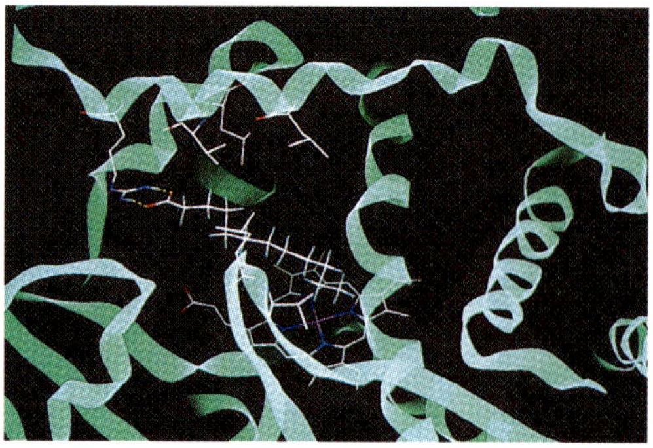

Fig. 13. *Lauric acid is shown orientated in the putative active site of CYP4A11 for ω-hydroxylation by electrostatic interactions with Arg188* [17]

template models and it is likely that Ser331 represents a possible active site hydrogen bond donor/acceptor residue which, in conjunction with other amino acid residues, is able to orientate substrates for oxygenation by the enzyme. *Figure 9* shows how the anti-estrogen tamoxifen may bind to the putative active site of CYP2C9, the human orthologue thought to be involved in its metabolic activation, via p-hydroxylation of one of the phenyl rings, where the side-chain of Ser331 cooperates in substrate orientation with other residues to position the appropriate phenyl ring for oxygenation [18]. Although it is possible to fit all of the known substrates of these isozymes within the 2C models, further studies will be required to rationalize the differences between CYP2C9 and CYP2C19 substrate specificity in terms of the relatively small number of amino acid changes which distinguish these important human isoforms.

CYP2D Models. - The major interest in the CYP2D subfamily lies in the fact that the constitutively expressed human orthologue, CYP2D6, which is present at low levels in human liver, exhibits genetic polymorphism while also being associated with the metabolism of a large number of basic drugs in current clinical use [4]. Several allelic variants of CYP2D6 have been identified which result in impaired or altered metabolism of marker substrates of this enzyme. Site-directed mutagenesis of an aspartate residue (D301) close to the conserved threonine residue (T309) has established that the carboxylate side-chain of this amino acid is involved in ion-paired interactions with the protonated nitrogen moiety of debrisoquine and other typical CYP2D6 substrates [35]. Interestingly, this key residue position (D260 in the CYP102 alignment) was previously identified by molecular modelling of CYP2D6 from CYP101 as representing the likely contact point for the binding of specific substrates [36]. Current substrate template models for CYP2D6 specificity suggest that the positions of metabolism are situated at between 5Å and 7Å from the protonable nitrogen. However, there are some compounds (albeit few) known to undergo CYP2D6 - mediated metabolism at positions which lie outside the 5-7Å limits and, therefore, particularly in the example of tropisetron, require some modification of the generally accepted view of CYP2D6 substrate structural criteria. A possible explanation for CYP2D6 metabolism of chemicals like tropisetron could involve alternative modes of substrate binding, where other acidic active site residues may participate in orientation for metabolism in the experimentally observed positions. Modelling of CYP2D6 from the CYP102 template suggests that a glutamate residue, Glu184, in the F helix which lies above the heme pocket, could represent an ionic contact for the somewhat larger CYP2D6 substrates such as tropisetron. This residue is close to two positions (181 and 182 in the alignment) which have been shown via protein engineering of CYP2A and CYP101 respectively to be involved in modifying substrate specificity, whereas an allelic variant of CYP2D6 (CYP2D6-F) that exhibits lowered catalytic activity contains a change from Gly to Glu at alignment position 180. *Figure 10* shows how the tropisetron molecule may occupy the CYP2D6 active site for metabolism at the known position by interacting with the Glu184 side-chain. Another allelic variant of CYP2D6 has a relatively conservative change from Met to Val at alignment position 332 (374 in CYP2D6), and site-directed mutagenesis experiments in both CYP2D6 and CYP2D1 have demonstrated that substrate regioselectivity can be altered by relatively conservative modifications at this position [37,38]. Consequently, information is developing which enables rationalization of CYP2D - mediated metabolism in terms of substrate interactions with complementary active site residues [11,18].

CYP2E Models. - In contrast with other members of the CYP2 family, there have been very few examples of site-directed mutagenesis on CYP2E reported in the literature, although Fukuda and coworkers have shown that modification of the conserved threonine (T303 in CYP2E) distal to the heme (corresponding to T268 in the alignment) affects the enzymic activity towards dimethylnitrosamine (DMN) and aniline oxygenation, together with altering fatty acid hydroxylase regioselectivity in rabbit CYP2E1 [39]. However, on the basis of mutations in other isozymes of the CYP2 family described previously, certain residue positions can be expected to reside within the active site region, and the CYP2E1 models in rodent, rabbit and human forms are all able to rationalize known substrate specificity for these mammalian P450s. For example, Phe181 is likely to be involved in π-stacking interactions with the benzene ring of the specific marker substrate, p-nitrophenol, and with other aromatic substrates of CYP2E. It is also probable that a threonine residue Thr266 in the distal I helix, two positions upstream from the conserved threonine Thr268, is able to form hydrogen bond interactions with the complementary polar groupings of many CYP2E substrates, such as: p-nitrophenol, aniline, paracetamol, chlorzoxazone and the specific inhibitor, disulphiram. In addition to monoaromatic compounds, a number of small molecular weight alcohols and ketones (eg. ethanol and acetone), together with certain dialkylnitrosamines, such as DMN, are also CYP2E substrates, and modelling shows that these can be orientated for oxygenation in the known positions by a combination of hydrophobic and hydrogen bond interactions within the putative CYP2E active site [40]. The majority of these points of contact have been demonstrated to be of relevance for substrate orientations in other CYP2 family isozymes, and *Figure 11* shows how a template of superimposed CYP2E substrates may fit the putative active site of rat CYP2E1.

CYP3A Models. - Although there have been no published results of site-specific mutations in CYP3A isoforms, there is a relatively high homology (27%) between CYP102 and proteins of this family which suggests that there should be a reasonable conservation of tertiary structure, and this facilitates the generation of CYP3A models. It is generally accepted that CYP3A4 constitutes the major P450 isozyme present in human liver, which is responsible for the metabolism of a large number of drugs and other chemicals ranging in molecular size from endogenous steroids to macrolide antibiotics, such as cyclosporin.

The CYP3A4 model constructed from CYP102 is able to explain how substrates of such structural diversity can occupy the enzyme active site and become oxygenated in the known position [16]. In order to accommodate molecules of the size of cyclosporin, the active site of CYP3A4 is likely to be correspondingly large and open, in contrast with the relatively restricted substrate binding site of the CYP2E enzymes described previously. *Figure 12* shows how a superimposed template if 12 CYP3A4 substrates, including cyclosporin, can readily fit the putative active site of the enzyme, where each known site of metabolism is positioned above the heme iron by interactions with a number of complementary amino acid side-chains, notably: Phe72, Thr73 and Asn74. Together with several hydrophobic residues, these three amino acids appear to be involved in active site contacts with the majority of CYP3A4 substrates. Although no site-directed mutagenesis experiments have been reported for CYP3 family proteins, the multiple sequence alignment (*Figure 1*) shows that Thr73 corresponds to Phe87 in CYP101, which has been found to be important for substrate regiospecificity [41]. The molecular volume of the CYP3A4 substrate template, shown in *Figure 12*, is significantly greater (at 1866.3Å^3) than those of CYP2E (213.7Å^3) and CYP1A

(332.6Å^3), and probably represents the upper spatial limit on CYP3A substrate specificity. The enzyme model is consistent with known CYP3A4-mediated metabolism and, moreover, the antibody recognition site in the rat orthologue, CYP3A1, should correspond to a surface region of the protein [16]. Futhermore, the openness of the CYP3A4 active site may be due to the presence of two consecutive glycine residues mapping onto the 437 alignment position. Consequently, the relatively high sequence homology between CYP3A4 and CYP102, together with the close agreement with known CYP3A4 substrate metabolism, has led to a fair degree of confidence in the authenticity of this structure.

CYP4A models. - After that of CYP3A4, the sequences of CYP4A4 and CYP4A11 (*Figure 1*) exhibit the highest homology with CYP102, and these rabbit lung and human liver orthologues in the CYP4A subfamily are able to metabolize structurally similar substrates to that of the bacterial template. Enzyme models of CYP4A11 and CYP4A4 are in agreement with the preferred substrates specificity of each isoform [17], namely, lauric acid ω–hydroxylation and prostaglandin E_2 (PGE_2) ω–hydroxylation, respectively. *Figures 13 and 14* show a possible mode of substrate binding for each CYP4A isozyme which is consistent with the known position of metabolism and amino acid residue positions identified by site-directed mutagenesis in other P450s mentioned previously. The active site residue, Phe87, which sequesters the end-chain of fatty acid substrates in CYP102, such that ω-2 hydroxylation primarily occurs, is altered to leucine in both CYP4A11 and CYP4A4, thus enabling end-of-chain hydroxylation of lauric acid and PGE_2, respectively. There are also key differences between the two orthologous proteins which rationalize their altered substrate preferences [17]. These include the same three residue positions 72-74 which may be involved in CYP3A substrate binding that become Pro-Arg-Asn in CYP4A4, where the central arginine could form an ion-pair with the carboxylate group of PGE_2, thus representing a key contact for orientation of metabolism. There is also a difference at position 330 where a serine residue in CYP4A4 may represent a hydrogen bond contact with PGE_2 to cooperate with Arg73 and several complementary hydrophobic active site residues (eg. Leu87, Leu181 and Leu437) in positioning the substrate for ω-hydroxylation. It is also possible that a glutamate residue at position 264 (in the I helix distal to the heme) may ensure that the carboxylate head group of fatty acid substrates are orientated away from the heme moiety and thus tend to recognize basic arginine side-chains at the mouth of the active site, such that the end of the hydrocarbon chain of the carboxylic acid substrate will be positioned above the heme moiety for oxygenation at the experimentally observed sites. Although the CYP4A subfamily is primarily associated with endogenous fatty acid and prostanoid metabolism there is also evidence to suggest that phthalate esters, such as MEHP, are metabolized by some of these isozymes [42].

CYP11, CYP17, CYP19 and CYP21 Models. - Although these enzymes are not involved in exogenous metabolism, there is current interest in some of the steroidogenic P450s for the design of specific inhibitors which can exhibit therapeutic activity for the treatment of cancer and other disease states and conditions. Homology modelling of these biosynthetic forms of P450 represent a considerable test for the CYP102 alignment as their substrates all contain the same steroidal nucleus but are metabolized at different positions. A further potential difficulty is that CYP11 isozymes are mitochondrial and utilize adrenodoxin as a redox partner, instead of the common reductase electron source of the microsomal forms which is similar to that of CYP102. However, based on

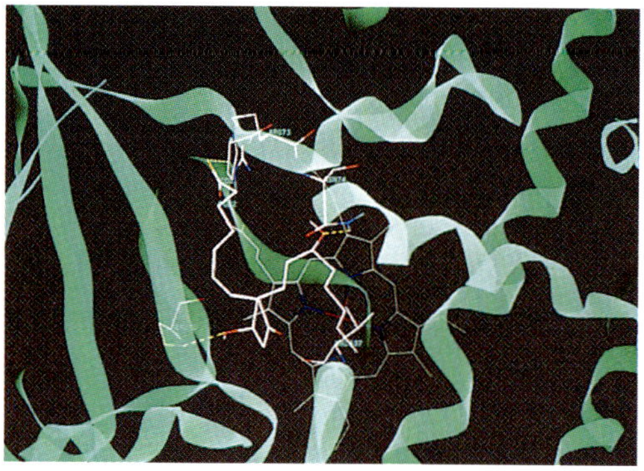

Fig. 14. *Prostaglandin E_2 is shown orientated in the putative active site of CYP4A4 for ω-hydroxylation by hydrogen bonding with Ser330 and ion-pairing with Arg73* [17]

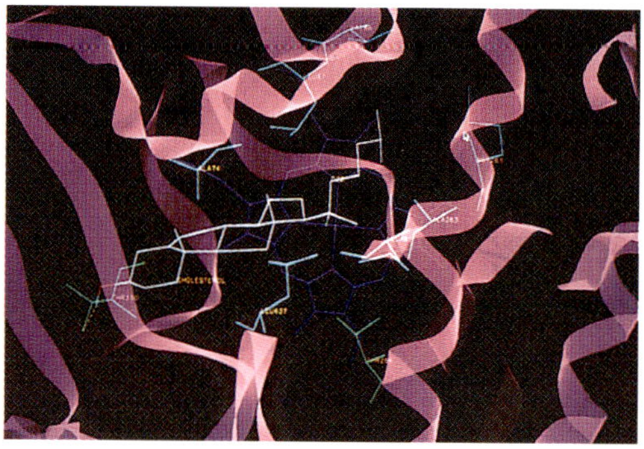

Fig. 15. *Cholesterol is shown orientated in the putative active site of CYP11A1 for C20-oxygenation by hydrogen bonding with Thr330* [1]

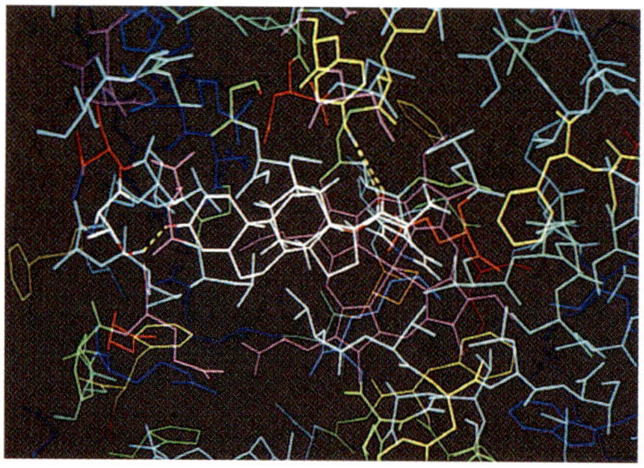

Fig. 16. *Progesterone and an inhibitor are shown superimposed in the CYP17A active site and orientated by hydrogen bonding* [1]

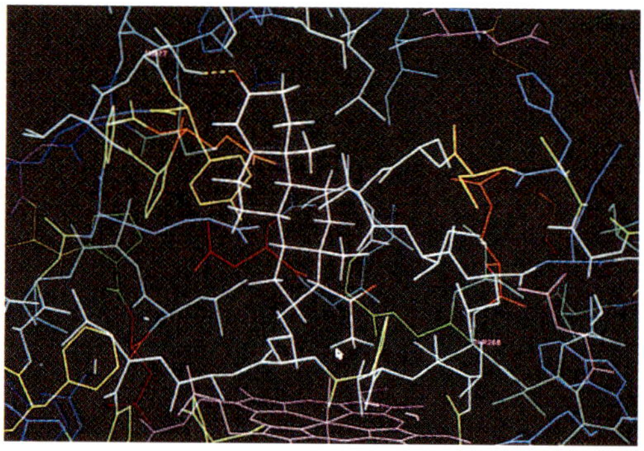

Fig. 17. *Progesterone is shown orientated in the putative active site of CYP21A by hydrogen bonding with the Thr77* [1]

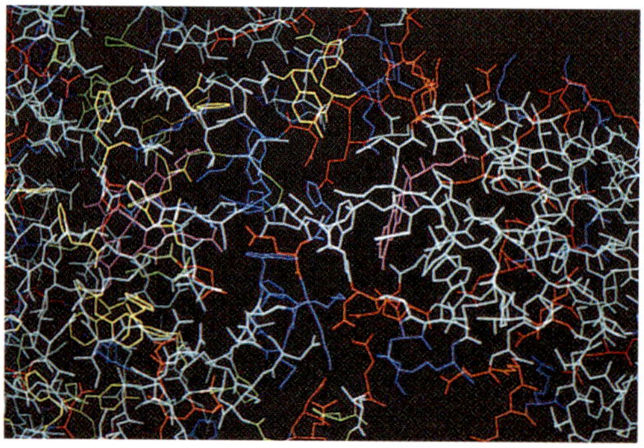

Fig. 18. *The proposed interaction between reductase and P450 in the vicinity of the FMN-Trp96 contact* [1]

previous experience with the xenobiotic-metabolizing P450s, homology alignment between all of the steroidogenic P450s and the prokaryotic template CYP102 is relatively straightforward, leading to enzyme models which both rationalize the known regiospecificity of metabolism for steroidal substrates and are consistent with the interactions of specific inhibitors. For example, *Figure 15* shows how the substrate, cholesterol, is able to fit the putative active site of CYP11A1 for oxygenation at the 22 position on the D-ring side-chain which will give rise to the known metabolite, pregnenolone, by forming complementary interactions with active site residues: Thr330 (which hydrogen bonds to the A ring hydroxyl of cholesterol), Leu87, Leu181 and Leu 437, together with Ala263 and Val 329. These positions correspond closely with those mentioned previously as representing likely substrate contact points in other P450 isozymes, and it is also found that a known inhibitor, (20R)-20-phenyl-5-pregnene-3β,20-diol, can occupy the putative active site of CYP11A.

The model building for CYP17A is also consistent with steroidal substrate metabolism and the likely mode of inhibitor binding. *Figure 16* shows how it is possible to superimpose the structures of the substrate, progesterone, and an inhibitor, cyclohexyl-2-(4-pyridyl) propanoate, within the active site region such that similar interactions with complementary amino acid residues occur. Both molecules can hydrogen bond with Tyr177 and also make a number of hydrophobic contacts with analogous side-chains which orientate them for heme ligation, in the case of the inhibitor, and 17α-hydroxylation of the substrate. It can be appreciated from *Figure 16* that the cyclohexyl moiety on the inhibitor overlays with the C-ring of progesterone, thus indicating that their modes of binding interaction are likely to possess some degree of similarity.

The CYP19A model generated from CYP102 bears certain commonalities with that reported by Koymans, *et al.* based on the CYP101 active site region[43], in that both substrates and inhibitors are able to make favourable contacts with Ser438, Ala264 and leucines at positions 263 and 437; whereas, in addition Ser72 may form a hydrogen bond with the D ring of the steroidal substrate, androstenedione. Interactions between CYP19-specific chemicals and the conserved I helix are also consistent with the results of site-directed mutagenesis experiments.

Moreover, modelling of a further steroid-metabolizing P450, CYP21A, also agrees with the known experimental findings from both protein engineering studies and substrate regiospecificity. In this case, a hydrogen bond from Thr77 to the A-ring carbonyl group and several hydrophobic contacts with Leu181, Ile87 and Ile438, together with Val328 and Val437, orientate the substrate progesterone for hydroxylation at the C21 position, as shown in *Figure 17*. The inhibitory analogue, difluoroprogesterone, is able to fit the putative active site of CYP21 in a similar manner, with one of the fluorines able to ligate the heme iron; whereas the other fluorine could form a hydrogen bond with the conserved distal threonine-268. Site-directed mutagenesis studies on CYP21 have focused on regions close to the enzyme surface and may indicate the position of either an access channel for the substrate or possible binding interactions with the reductase redox partner. However, one of the mutant sites appears to lie in close proximity to a region which may correspond with a proposed steroid-binding motif, GDRL [44], and it may be significant that the analogous GERL tetrapeptide maps onto the F helix in CYP11. Interestingly, another mutation position reported in CYP21 which appears to correspond with the start of the distal I helix is only five residues downstream from a proposed adrenodoxin binding site in CYP11; and homology modelling of this iron-sulphur redoxin from the ferredoxin crystal structure provides a means of investigating redox partner interactions with P450s [1]. Furthermore, based on an alignment with ferredoxin reductase (for the FAD and NADPH binding domains) and flavodoxin (for the FMN domain), for which the relevant crystal structures are available, it is possible to construct a model of P450 oxidoreductase that is consistent with known redox partner-P450 interactions, where the FMN moiety is positioned (see *Figure 18*) close to the heme of P450 for electron transfer via the formation of ion-pairs between the two components.

In conclusion, it is apparent that the models of mammalian P450s generated from the CYP102 template are in close agreement with their known substrate specificities, and are also consistent with available information from mutagenesis studies [20,45-48], antibody binding data [29,30] and chemical probes of various identified redox component interactions [31]. Consequently, it is anticipated that such models could be used effectively to predict the likely routes of metabolism of novel compounds at the developmental stage [49,50].

Acknowledgement. The financial support of GlaxoWellcome Research and Development Limited is gratefully acknowledged.

REFERENCES

[1.] D.F.V. Lewis, *'Cytochromes P450: Structure, Function and Mechanism'*, Taylor and Francis, London, **1996**.
[2.] D.R. Nelson, L. Koymans, T. Kamataki, J.J. Stegeman, R. Feyereisen, D.J. Waxman, M.R. Waterman, O. Gotoh, M.J. Coon, R.W. Estabrook, I.C. Gunsalus, D.W. Nebert, *Pharmacogenetics* 1996, **6,** 1.
[3.] P.R. Ortiz de Montellano (Ed.) *'Cytochrome P450'*, Plenum, New York, **1995**.
[4a.] D.A. Smith, S.M. Abel, R. Hyland, B.C. Jones, *Xenobiotica* 1996, **26,** in press.
b. K.R. Korzekwa, J.P. Jones, *Pharmacogenetics* 1993, **3,** 1.
c. L. Koymans, G.M. Donné-Op den Kelder, J.M. Te Koppele, N.P.E. Vermeulen, *Drug Metab. Rev.* 1993, **25,** 325.
[5.] D.F.V. Lewis, *Toxicol. Mod.* 1995, **1,** 85.
[6.] T.L. Poulos, B.C. Finzel, A.J. Howard, *J. Mol. Biol.* 1987, **195,** 687.
[7.] K.G. Ravichandran, S.S. Boddupalli, C.A. Hasemann, J.A. Peterson, J. Deisenhofer, *Science* 1993, **261,** 731.
[8.] C.A. Hasemann, K.G. Ravichandran, J.A. Peterson, J. Diesenhofer, *J. Mol. Biol.* 1994, **236,** 1169.
[9.] J.R. Cupp-Vickery, T.L. Poulos, *Structural Biology* 1995, **2,** 144.
[10.] C.A. Hasemann, R.G. Kurumbail, S.S. Boddupalli, J.A. Peterson and J. Deisenhofer, *Structure* 1995, **3,** 41.
[11.] D.F.V. Lewis, *Xenobiotica* 1995, **25,** 333.
[12.] A.J. Fulco, *Annu. Rev. Pharmacol. Toxicol.* 1991, **31,** 177.
[13.] K.N. Degtyarenko, A.I. Archakov, *FEBS Lett.* 1993, **332,** 1.
[14.] D.F.V. Lewis, B.G. Lake, *Xenobiotica* 1995, **25,** 585.
[15.] D.F.V. Lewis, B.G. Lake, *Xenobiotica* 1996, **26,** 723.
[16.] D.F.V. Lewis, P.J. Eddershaw, P.S. Goldfarb, M.H. Tarbit, *Xenobiotica* 1996, **26,** in press.
[17.] B.G. Lake, D.F.V. Lewis, in: *'Cytochromes P450-Metabolic and Toxicological Aspects'* Ed. C. Ioannides, CRC Press, Boca Raton, Florida, 1996, Ch. 11. p269.
[18.] H. Wiseman, D.F.V. Lewis, *Carcinogenesis* 1996, **17,** 101.

[19.] S. Modi, W.U. Primrose, J.M.B. Boyle, C.F. Gibson, L.-Y. Lian, G.C.K. Roberts, *Biochemistry* 1995, **34,** 8982.

[20.] E.F. Johnson, *Trends Pharm. Sci.* 1992, **13,** 122.

[21.] R.L.P. Lindberg, M. Negishi, *Nature* 1989, **339,** 632.

[22.] M. Iwasaki, R.L.P. Lindberg, R.O. Juvonen, M. Negishi, *Biochem. J.* 1993, **291,** 569.

[23.] R.O. Juvonen, M. Iwasaki, M. Negishi, *J. Biol. Chem.* 1991, **266,** 16431.

[24.] D.F.V. Lewis, B.G. Lake, *Xenobiotica* 1996, submitted.

[25.] J.R. Halpert, Y. He, *J. Biol. Chem.* 1993, **268,** 4453.

[26.] T. Aoyama, K. Korzekwa, K. Nagata, M. Adesnik, A. Riess, D.P. Lapenson, J. Gillette, H.V. Gelboin, D.J. Waxman, F.J. Gonzalez, *J. Biol. Chem.* 1989, **264,** 21327.

[27.] Y. He, Z. Luo, P.A. Klekotka, V.L. Burnett and J.R. Halpert, *Biochemistry* 1994, **33,** 4419.

[28.] K.M. Kedzie, C.A. Balfour, G.Y. Escobar, S.W. Grimm, Y. He, D.J. Pepper, J.W. Regan, J.C. Stevens, J.R. Halpert, *J. Biol. Chem.* 1991, **266,** 22515.

[29.] C. De Lemos-Chiarandini, A.B. Frey, D.D. Sabatinini, G. Kreibich, *J. Cell. Biol.* 1987, **104,** 209.

[30.] E.F. Kolesanova, S.A. Kozin, A.O. Lemeshko, A.I. Archakov, *Biochemistry and Molecular Biology International* 1994, **32,** 465.

[31.] R. Bernhardt, R.Kraft, K. Ruckpaul, *'Cytochrome P-450: Biochemistry, and and Biophysics'* Ed. I. Schuster, Taylor and Francis, London, 320.

[32.] S.J. Roffey, Ph.D. Thesis, University of Surrey, 1993.

[33.] T.H. Richardson, E.F. Johnson, *J. Biol. Chem.* 1994, **269,** 23937.

[34.] A. Mancy, P Broto, S.Djols, P.M. Dansette, D. Mansuy, *Biochemistry* 1995, **34,** 10365.

[35.] W. Ellis, A.Simula, J. Harlow, G. Hayhurst, G. Smith, M. Lennard, G. Tucker, C.R. Wolf, 1994, *Proceedings of 10th International Symposium on Microsomes and Drug Oxidations*, Toronto, Canada, p.427.

[36.] L.J.H. Koymans, N.P.E. Vermeulen, A. Baarslag, G.M. Donné-op den Kelder, *J. Comp. Aided Mol. Des.* 1993, **7,** 281.

[37] C.L. Crespi, D.T. Steimel, B.W. Penman, K.R. Korzekwa, P. Fernandez-Salguero, J.T.M. Buters, H.V. Gelboin, F.J. Gonzalez, J.R. Idle, A.K. Daly, Proceedings of the 4th International ISSX Meeting, Seattle, Washington, August 27-31, 1995, p.54.

[38.] E. Matsunaga, T. Zeugin, V.M. Zanger, T. Aoyama, U.A. Meyer, F.J Gonzalez, *J. Biol. Chem.* 1990, **265,** 17197.

[39.] T. Fukuda, Y. Imai, M. Komari, M. Nakamura, E. Kusunose, K. Satouchi, M. Kusunose, *J. Biochem.* (Tokyo) 1993, **113,** 7.

[40.] D.F.V. Lewis, M.G. Bird, D.V. Parke, *Toxicology* 1996, submitted.

[41.] E.J. Mueller, P.J. Loida, S.G. Sligar, in: *'Cytochrome P450'*, 2nd Edition (Ed. P.R. Ortiz de Montellano) Plenum, New York, 1995, 83.

[42.] F.P. Guengerich (Ed.) *'Mammalian Cytochromes P-450'*, Volume 2, CRC Press, Boca Raton, Florida, 1987.

[43.] L.M.H. Koymans, H. Moereels, H. Vanden Bossche, *J. Steroid Biochem. Molec. Biol.* 1995, **53,**191.

[44.] O. Gotoh, Y. Tagashira, K. Morohashi, Y. Fujii-Kuriyama, *FEBS Lett.* 1985, **188,** 8.

[45]. A.W. Munro, K. Malarkey, J. McKnight, A.J. Thomson, S.M.Kelly, N.C. Price, J.G. Lindsay, J.R. Coggins, J.S. Miles, *Biochem. J.* 1994, **303,** 423.

[46.] T. Yasukochi, O. Okada, T. Hara, Y. Sagara, K. Sekimizu and T. Horiuchi, *Biochim. Biophys. Acta* 1994, **1204,** 84.
[47.] N.C. Gerber, S.G. Sligar, *J. Amer. Chem. Soc.* 1992, **114,** 8742.
[48.] N.C. Gerber, S.G. Sligar, *J. Biol. Chem.* 1994, **269,** 4260.
[49] J. Wibley, *Pharmaceutical Journal* 1996, **256,** 445.
[50.] A. Coghlan, *New Scientist* 1996, **150,** No. 2027, 22.

Abbreviations: COMPACT - computer-optimized molecular parametric analysis of chemical toxicity; DDT - p,p-dichlorodiphenyltrichloroethane; FAD - flavin adenine dinucleotide; FMN - flavin mononucleotide; MEHP - mono(2-ethyl hexyl) phthalate; NAD(P)H - nicotinamide adenine dinucleotide (phosphate); TCDD - 2,3,7,8-tetrachlorodibenzo-p-dioxin.

PART III. **COMPUTATIONAL APPROACHES IN DRUG DISPOSITION**

Chapter 22

Relative Stabilities of Nitrenium Intermediates in Carbazoles Mutagenic Series

by **Stéphane Chemtob**[a]), **Christiane Mercier***[a]), **Véronique André**[b]) and **Pascal Gauduchon**[b])

[a])Institut de Topologie et de Dynamique des Systèmes, associé au CNRS, URA 34, Université Paris 7 -Denis Diderot, 1 rue Guy de la Brosse, F-75005 Paris, France
[b])Groupe de Recherche Biologie Cellulaire Oncologique et Toxicologie Génétique, Université de Caen, et Laboratoire de Cancérologie Expérimentale, Centre F. Baclesse, route de Lion sur Mer, F-14021 Caen, France

Relative Stabilities of Nitrenium Intermediates in Carbazoles Mutagenic Series

by **Stéphane Chemtob, Christiane Mercier***
Institut de Topologie et de Dynamique des Systèmes, associé au CNRS, URA 34, Université Paris 7 -Denis Diderot, 1 rue Guy de la Brosse, F-75005 Paris, France

and **Véronique André** and **Pascal Gauduchon**
Groupe de Recherche Biologie Cellulaire Oncologique et Toxicologie Génétique, Université de Caen, et Laboratoire de Cancérologie Expérimentale, Centre F. Baclesse, route de Lion sur Mer, F-14021 Caen, France

Thirty-two amino-, nitro- and hydroxycarbazoles were synthesized and evaluated for their mutagenic activity by means of the Ames test in different *Salmonella typhimurium* test strains, with or without the addition of S9, the microsomal fraction containing cytochrome P450. Previous topological QSAR models suggested the importance of factors affecting the stability of their metabolic intermediates, in particular the ultimate electrophilic nitrenium ion, Ar-NH_2^+, in controlling mutagenicity. Full geometry optimizations and heats of formation of the carbazole derivatives and their nitrenium ion intermediates were determined by the semi-empirical quantum mechanical method AM1. Nitrenium ions derived from carbazole are mostly *syn* conformers in the ground-state singlet. In contrast with other studies on aminoaromatic series, the mutagenicity is increased by a lower stability of the nitrenium ion for aminocarbazoles (r^2 = 0.86, n = 8). For nitrocarbazoles it is, as expected, increased by a higher stability of the nitrenium ion but for two separate groups : dinitro derivatives (r^2 = 0.85, n = 8) and 1,4-dimethyl derivatives (r^2 = 0.80, n = 10). Analysis of these two latter groups suggests the importance of hydrophobicity in controlling mutagenicity. Compounds with a 4-nitro group, for which, the mutagenicity is dramatically decreased by an out-of-plane orientation, are not included in these relationships.

Introduction. - Amino- and nitroaromatic compounds, widely used as intermediates in laboratory and industrial synthesis, are major pollutants. Because of their mutagenic and genotoxic properties, they represent a potential risk for human health. This has led to systematic studies on the mutagenicity of the most common derivatives by means of the Ames test [1] and to the development of predictive structure-activity models. Hydrophobicity and frontier orbital energies were found to control the mutagenicity of aromatic and heteroaromatic amines and nitro compounds [2-5]. With a view to predicting the toxicity of carbazoles, used in the dye industry and present in our environment, and to better understand the mechanism by which they act, we tested a large series of carbazole derivatives for mutagenicity in various strains of *Salmonella typhimurium* [6-8] by means of the Ames test. In parallel, we determined physicochemical and electronic properties, such as lipophilicity measured by RP-HPLC method, electrochemical potentials, LUMO and HOMO energies [7-8] and derived, by using our general QSAR search procedure [9], fragmentary predictive models for mutagenic, lipophilic and quantum mechanical data of twenty-two substituted carbazoles [8]. The results of the mutagenicity assays are in agreement with the generally accepted metabolic scheme. Amino and nitro compounds are converted to a common hydroxylamine intermediate, mainly esterified in the case of carbazole derivatives, which is then converted to an electrophilic nitrenium ion which reacts with DNA, as shown in *Figure* 1. Amines are believed to be oxidized before they penetrate the bacterial cell, as they show mutagenic activity only in the presence of the S9 microsomal fraction containing cytochrome P450. Nitro compounds, which do not require microsomal activation, are assumed to penetrate the cell before being reduced by cytosolic reductases.

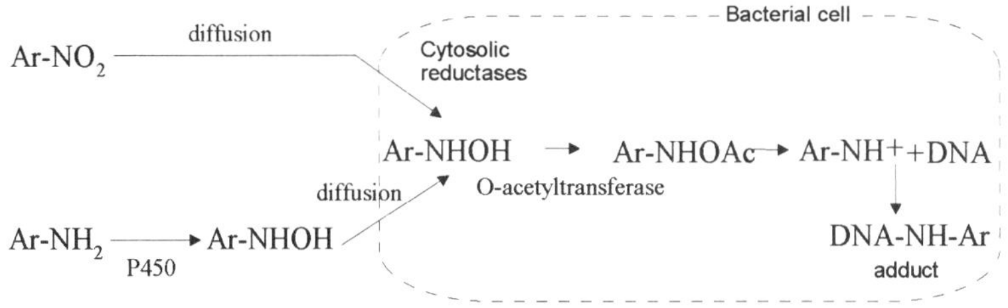

Fig. 1. *Hypothetical mechanism for activation of amino- and nitroarenes.*

The structure-activity models show that mutagenicity is increased by higher lipophilicity, more significantly for nitrocarbazoles than for amines which are metabolized outside the cell. Surprisingly, the absence of correlation with LUMO energy for nitro compounds and an unexpected inverse relationship with HOMO energy show that mutagenicity does not depend on the ease of reduction or oxidation of the parent nitro and amino molecules. Structural analysis of the mutagenicity models suggests the importance of factors affecting the reactivity of the ultimate electrophilic aryl nitrenium ion. It has previously been found that mutagenicity increases with the relative stability of the corresponding nitrenium ion, calculated by using the semiempirical AM1 molecular orbital procedure, for a small group of polycyclic aromatic amines [10] and for small groups of complex heterocyclic amines [11]. In the present paper, we report the mutagenicities of 13 original methyl-substituted nitro- and amino-9H-carbazoles and use the same semiempirical method to calculate the relative stabilities of the carbazole nitrenium ions derived from 11 amino and 25 nitro derivatives. The correlations of mutagenicity with this parameter are discussed.

Mutagenicity. - All the compounds were assayed for mutagenicity in different *Salmonella typhimurium* strains with or without the addition of rat liver microsomes (S9), according to the procedure described for the standard plate-incorporation test, as detailed previously [6]. The mutagenic activities observed with TA98 tester strain are listed in *Table* 1. The results concern nineteen amino, nitro and hydroxycarbazoles [6-8] and thirteen additional methyl-substituted derivatives. The significant reduction of activity observed in a nitroreductase-deficient strain, for nitro derivatives and the strong increase in amine activity upon addition of S9 are in agreement with the metabolic scheme proposed.

Quantum mechanical results. - Semiempirical calculations for the thirty-two carbazole derivatives and their metabolites were carried out with the AM1 approximation [12] using Version 7 of the MOPAC package on a J90 Cray computer. Trial geometries for amino and nitro compounds were generated from standard bonds and angles and in a few cases exploratory calculations were carried out on compounds which present steric interactions, such as the acetamido group, or intramolecular hydrogen bonds, such as $NO_2 \cdots H\text{-}O$. Unconstrained geometries were optimized using the Fletcher Reeves algorithm optimizer with a gradient $\leq$ 0.01 as convergence criterion. Except for the 4- and 5-nitro derivatives, all the substituted molecules have planar structures. For 4-nitrocarbazole [13], the heat of formation is minimal when the nitro group is perpendicular to the phenyl ring and maximal for a planar conformation, due to steric interaction with the proton at the 5 position. The calculated heats of formation of the thirty-two carbazoles and of their corresponding nitrenium ions are presented in *Table* 1. This Table also includes the relative energetics of nitrenium ion formation, $\Delta\Delta H$, defined by the enthalpies of the process described by *Eqn. 1* for amino aromatic compounds [10]:

$$ArNH_2 + PhNH^+ \rightarrow ArNH^+ + PhNH_2$$

$$\Delta\Delta H = (\Delta H_f\, ArNH^+ + \Delta H_f\, PhNH_2) - (\Delta H_f\, ArNH_2 + \Delta H_f\, PhNH^+) \qquad (1)$$

The $\Delta\Delta H$ values quantify the thermodynamic stabilities of the arylnitrenium ions, $ArNH^+$, relative to the simple monocyclic nitrenium ion derived from aniline. A negative value of $\Delta\Delta H$ therefore indicates the amount by which $ArNH^+$ is more stable than $PhNH^+$. The calculated difference in enthalpy of formation between two minima is related to an enthalpy of reaction rather than to an energy of activation. Thus, thermodynamic rather than kinetic criteria are used to measure the relative extent of transformation of parent compounds. Such thermodynamic parameters are appropriate if each reaction is presumed to reach equilibrum. For nitrenium ions derived from nitrocarbazoles, we use a similar equation where aniline is replaced by nitrobenzene:

$$ArNO_2 + PhNH^+ \rightarrow ArNH^+ + PhNO_2$$

$$\Delta\Delta H = (\Delta H_f\, ArNH^+ + \Delta H_f\, PhNO_2) - (\Delta H_f\, ArNO_2 + \Delta H_f\, PhNH^+) \qquad (2)$$

For nitrenium ions derived from aminonitrocarbazoles, we consider the two alternative conformations, according to Ford's notation [10]. In the *syn* conformer the N-H bond is directed towards and in the *anti* conformer away from the ring atom of highest priority, on the basis of the Kahn-Ingold-Prelog sequence rules.

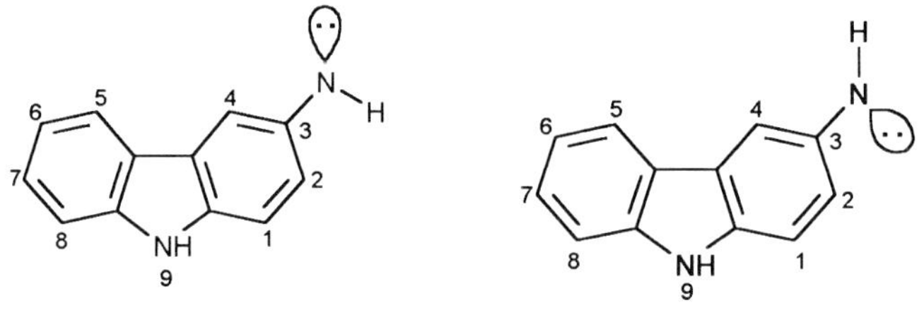

syn *anti*

Fig. 2. *Syn and anti conformations.*

The geometries of the carbazole nitrenium ions were completely optimized with respect to all geometric variables for both singlet and triplet ground states. In similar complete optimizations, using the semi-empirical MNDO technique, Ford *et al.* [14] established that the aryl nitrenium ions exist as singlet ground states resulting from the preferential stabilization of this state relative to the triplet, the ground state of the aliphatic nitrenium ions. Both singlet and triplet species are stabilized by delocalization of the positive charge but electron transfer from the Π orbitals to the electron-deficient nitrogen atom is greater for the singlet than for the triplet where the p orbital is partially filled (*Fig.* 3).

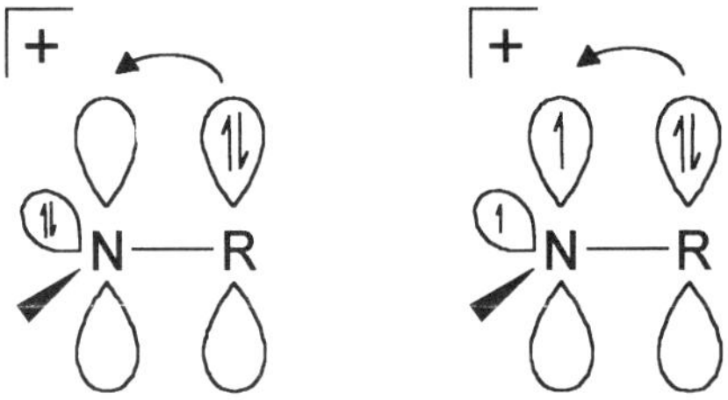

singlet triplet

Fig. 3. *Mesomeric stabilization of singlet and triplet nitrenium ions.*

Singlet states were calculated in the Restricted Hartree-Fock (RHF) formalism, while triplet states were calculated by using the spin-unrestricted Hartree-Fock (UHF) method.

In all cases the nitrenium ions were calculated to be essentially planar in the two alternative conformations. The calculated energies confirm the singlet ground state for carbazole nitrenium ions, except for two species with a barrier between 2 and 4 kcal mol^{-1}: *ortho*-nitro-2-nitrenium ions, which are destabilized by the electron-attracting effect of the nitro group. Nitrenium ions derived from carbazole are mostly *syn* conformers in the singlet ground state, but the barrier between the two conformations is often less than 1 kcal mol^{-1} in the absence of major steric interactions. These energetic differences between *syn* and *anti* conformers are related to electrostatic interaction between the NH^+ substituent and charge distribution within the carbazole ring. However, six nitrenium ions are stabilized in the *anti* conformation with a barrier between 2 and 7 kcal mol^{-1}, five in the singlet ground state and one in the triplet ground state. For three species, 1-nitrenium, 2-amino-3-nitrenium and 3-acetamido-4-nitrenium ions, the syn conformation is hindered due to steric interaction with the heterocyclic nitrogen, an *ortho*-amino substituent or a bulky *ortho*-acetamido group. *Ortho*-nitro 2- and 3-nitrenium ions are stabilized in the *anti* configuration by an intramolecular hydrogen bond.

For the 3, 6- or 3, 5- dinitro or 3, 6- diamino derivatives, nitrenium ions derived from the third position are invariably the most stable because of the simultaneous existence in these compounds of either a 2-hydroxyl or 1,4-dimethyl groups, which stabilize the 3-nitrenium ion by their electron-donation.

The relative energetics of nitrenium ion formation, $\Delta\Delta H$, vary over 40 kcal mol^{-1} and reflect the very different abilities of various substituents to stabilize the ion. The electron-donating effect of the heterocyclic nitrogen tends to stabilize the nitrenium ion when this group is in a conjugated position: 1, 3, 6 or 8. This stabilization can be easily detected and quantitatively measured by a comparison of the 2- and 3-nitreniumcarbazole energies with those of the 2- and 3-nitreniumfluorenes, respectively. The stabilizing effect of the indole heterocyclic nitrogen is only 3 kcal $mole^{-1}$ for the 2-nitrenium ion whereas it rises to 18.6 kcal mol^{-1} for the 3-nitrenium ion.

Structure-mutagenicity relationships. - In order to assess quantitatively the role of nitrenium ion stability on the mutagenicities of the parent compounds, we searched for linear correlations between mutagenicity in TA98, expressed in logarithmic units, and the relative stability, $\Delta\Delta H$, of the possible nitrenium ion intermediate.

For the 11 amino compounds, we find a correlation between mutagenicity with microsomal activation and $\Delta\Delta H$ (R^2 = 0.39, n = 11). Exclusion of three compounds, the 1,4-diMe-6-OMe-3-NH_2, the 9-Me-3-NH_2 and the 2-OH-3,6-diNH_2 carbazoles, detected as outliers, leads to a very significant correlation (r^2 = 0.86) :

$$\log TA98 = 0.10(\pm 0.02)\ \Delta\Delta H + 3(\pm 0.33) \qquad (3)$$

$$r^2 = 0.86,\ F = 36,\ s = 0.50,\ n = 8$$

This relationship (*Fig.* 4) indicates that aminocarbazole mutagenicity is an inverse function of the corresponding nitrenium ion stability. This contrasts with the direct function determined by Ford *et al.* [10-11] for other aromatic amines, but is in agreement with our previous analysis.

Table 1. *Calculated Heats of Formation, Relative Stabilities and Mutagenicity in TA98 of Amino- and Nitrocarbazoles and Related Nitrenium Ions*

Substituents [a]	ΔH_f (kcal mol^{-1})					$\Delta\Delta H$ [c]	TA98 (rev $nmol^{-1}$)	
	Parent	Ar-NH^+ [b] So *syn*	Ar-NH^+ So *anti*	Ar-NH^+ T_1 *syn*	Ar-NH^+ T_1 *anti*		-S9	+S9
2-NH$_2$	66.2	275.5	274.9	273.8	**273.0**	-19.6	-	34 ± 1
3-NH$_2$	67.3	**266.0**	266.2	275.2	275.6	-27.7	-	6.5 ± 0.5
2-OH-**3**,6-NH$_2$	22.9	**211.0**	212.8	211.7	212.6	-38.3	-	18.7 ± 0.4
9-Me-**3-NH$_2$**	72.2	**268.7**	268.9	276.6	276.9	-29.9	-	(115)
1,4-diMe-**3-NH$_2$**	54.5	**249.0**	249.2	259.1	259.4	-31.9	-	(0.35)
1,4-diMe-6-OH-**3-NH$_2$**	11.1	**204.6**	204.8	212.0	212.4	-32.9	-	(0.77)
1,4-diMe-6-OMe-**3-NH$_2$**	17.7	**209.9**	210.0	216.4	216.8	-33.6	-	(55)
1-NO$_2$-**2-NH$_2$**	63.4	290.6	293.9	**285.9**	288.1	- 4.0	-	658 ± 31
1-NO$_2$-2-NH$_2$	"	**251.3**	252.7	262.0	262.1	-33.9	228 ± 5.5	-
2-NO$_2$-**3-NH$_2$**	68.8	**284.6**	288.1	290.5	293.7	-10.5	-	14 ± 0.54
2-NO$_2$-3-NH$_2$	"	**254.1**	260.3	259.6	262.4	-36.4	8.2 ± 0.4	-
3-NO$_2$-**2-NH$_2$**	66.2	296.9	291.7	291.9	**288.1**	- 4.6	-	858 ± 6
3-NO$_2$-2-NH$_2$	"	253.0	**248.3**	266.1	262.8	-39.7	(2)	-
4-NO$_2$-**3-NH$_2$**	71.6	287.8	**286.1**	292.3	291.9	-11.8	-	43 ± 4.1
4-NO$_2$-3-NH$_2$	"	259.3	**252.9**	261.2	259.4	-40.4	0.54 ± 0.03	-
1-NO$_2$	67.3	268.6	**263.3**	281.0	272.5	-25.8	7.0 ± 0.4	-
2-NO$_2$	71.1	275.5	274.9	273.8	**273.0**	-20.0	47 ± 0.6	-
3-NO$_2$	69.6	**266.0**	266.2	275.2	275.6	-25.4	45 ± 0.8	-
4-NO$_2$	74.3	273.8	275.5	**272.1**	275.1	-24.0	(0.47)	-
2-NO$_2$-3-NHCOCH$_3$	34.5	**225.1**	231.1	232.5	234.3	-31.2	(0.86)	-
4-NO$_2$-3-NHCOCH$_3$	40.8	231.0	**224.2**	231.3	228.1	-38.5	0.3 ± 0.04	-
2-OH-**3-NO$_2$**	28.6	**216.5**	217.2	228.2	227.0	-33.9	24 ± 0.7	-
2-OH-**6-NO$_2$**	25.2	**222.3**	222.7	227.4	228.1	-24.7	14 ± 0.2	-
2-OH-**3**,5-diNO$_2$	37.8	**230.4**	230.4	249.4	244.8	-29.3	(0.3)	-
2-OH-**3**,6-diNO$_2$	33.3	**229.6**	230.7	247.9	244.2	-25.5	(49.5)	-
2-OMe-**3**,6-diNO$_2$	39.6	**233.5**	236.6	251.5	252.1	-27.9	294 ± 9.3	-
9-Me-2-OMe-**3**,6-diNO$_2$	45.3	**236.8**	240.7	251.6	253.1	-30.2	(656)	-
9-Me-**2-NO$_2$**	76.2	278.6	279.3	274.4	**273.4**	-24.5	199 ± 11	-
9-Me-**3-NO$_2$**	74.7	**268.7**	268.9	276.6	276.9	-27.8	(279)	-
1,4-diMe-**3-NO$_2$**	58.1	**249.0**	249.2	259.1	259.4	-30.9	6.5 ± 0.7	-
1,4-diMe-**3-NO$_2$**-6-OH	14.9	**204.6**	204.8	212.0	212.4	-32.1	(1.9)	-
1,4-diMe-**3**,6-diNO$_2$	62.0	**262.3**	262.5	276.3	276.5	-21.5	(2.4)	-
1,4-diMe-**3-NO$_2$**-6-NHCOCH$_3$	21.6	**215.8**	216.0	226.0	226.5	-27.5	< 0.2	-
1,4,6-triMe-**3-NO$_2$**	50.8	**240.5**	240.7	248.6	248.7	-32.1	37 ± 1	-
1,4,9-triMe-**3-NO$_2$**	64.7	**253.4**	253.8	261.9	262.0	-33.1	(170)	-
5,8-diMe-**3-NO$_2$**	54.9	**249.7**	249.9	256.8	257.3	-27.0	< 0.09	-

[a] Substituents in bold represent those leading to the nitrenium ion under consideration.

[b] Bold values are associated with the calculated most stable conformer.

[c] Calculated according to *Eqn. 1* for nitrenium ions derived from aminocarbazoles and to *Eqn. 2* for those derived from nitro compounds, with ΔH_f $PhNH^+$ = 246.9 kcal mol^{-1}; ΔH_f $PhNO_2$ = 25.12 kcal mol^{-1}, ΔH_f $PhNH_2$ = 20.5 kcal mol^{-1}.

< Not mutagenic at the highest dose tested; the limiting value is calculated from the dose and the minimum detectable response.

( ) Non-linear response; mutagenicity is calculated for the dose giving the highest response.

This latter suggested that the smaller activity increase due to 3-amino as compared to 2-amino could be related to a decrease of the nitrenium ion reactivity, due to the presence of the heterocyclic nitrogen, which is likely to conjugatively reduce the positive charge. This is consistent with the greater stability of the 3-nitrenium as compared to the 2-nitrenium ion (*Table* 1). It also suggested that the profound effects on mutagenicity of *ortho*-substituents were probably due to their influence on the electrophilic reactivity. An electron-attracting nitro group will enhance the reactivity by increasing the positive charge density at the reactive site, while an electron-donating substituent will reduce it. Again, this is confirmed by the calculated stabilities: 1- and 3-nitro substituents lower the stability of the 2-nitrenium ion by about 16 kcal mol^{-1}, leading to the most unstable species related to the two most mutagenic compounds of this series. Electron-donating substituents, such as 6-hydroxy and 1,4-dimethyl stabilize the 3-nitrenium ion less by about 7.6 kcal mol^{-1}, given their smaller donating effect, but lead to the most stable species associated with the two least active compounds.

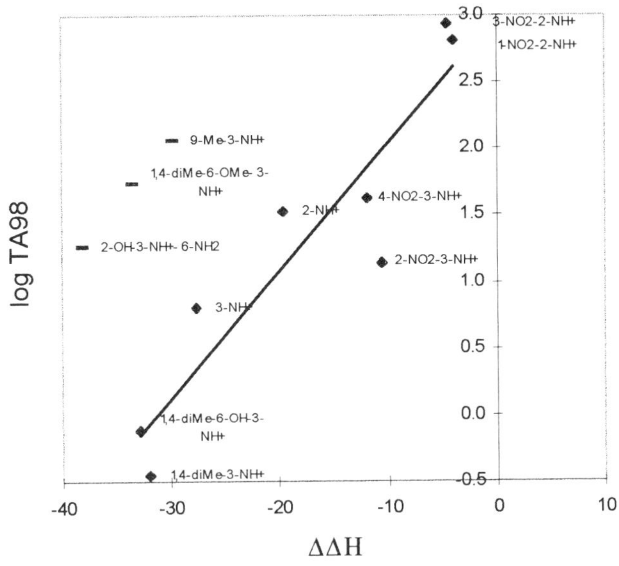

Fig. 4. *Relationship between the relative nitrenium ion stability, ΔΔH, and mutagenicity in TA98 with S9 of aminocarbazoles*

Even if the reactivity of the ultimate electrophilic nitrenium ion is important, the three compounds excluded show that this factor alone cannot explain the expression of mutagenicity, even in this small series. The relative mutagenicities of these three outliers agree their nitrenium ions stabilities but not with the other 2- and 3-aminocarbazoles. Although their nitrenium ions are about equaly stable, 1,4-diMe-6-OMe-3-NO_2 carbazole exhibits mutagenicity, whereas its 6-OH analogue is a very weak mutagen. Introducing an electron-donating methyl on heterocyclic nitrogen slightly stabilizes the 3-nitrenium ion, but its parent compound 9-Me-3-NH_2 is much more mutagenic than 3-NH_2 carbazole. The increased lipophilicity related to the methyl substitution on the 6-OH or on the N9, which favours cellular penetration, could explain the higher mutagenicity of these two outliers. For the 2-OH-3,6-diNH_2 carbazole, there are two possible nitrenium ions and it is not impossible that the 6-NH^+ which is less stable than the 3-NH^+ considered here would fit the relationship.

For all the 25 nitro-substituted compounds we find no correlation between direct mutagenicity and the relative nitrenium ion stability, $\Delta\Delta H$ ($r^2 = 0.03$, $n = 25$). The results, shown in *Fig.* 5, suggest no evident relationship. Given our previous analysis [9], we first excluded two aminonitrocarbazoles, the 2-amino-3-nitro and 3-amino-4-nitro derivatives, whose mutagenicity can be attributed to the amino metabolism, as their mutagenicity is much higher with microsomial activation (*Table* 1). This is in good agreement with the good fit of these compounds to the relationship detected for aminocarbazoles. We also eliminated three carbazoles bearing an out-of-plane 4-nitro substituent which is supposed to disturb access to the nitroreductase and consequently the formation of a 4-nitrenium ion. The remaining points suggest two linear relationships leading, after exclusion of two additional compounds, 2-nitrocarbazole and its 9-methyl derivative, to two significant correlations explaining 80 % and 85% of the variance, respectively :

$$\log \text{TA } 98 = -0.45 (\pm 0.08) \Delta\Delta H - 13.27 (\pm 2.43) \qquad (4)$$

$$r^2 = 0.80 , F = 33 , s = 0.57, n = 10$$

$$\log \text{TA } 98 = -0.31 (\pm 0.05) \Delta\Delta H - 6.32 (\pm 1.37) \qquad (5)$$

$$r^2 = 0.85 , F = 35 , s = 0.36, n = 8$$

These relationships, displayed in *Fig.* 5, indicate that mutagenicity of nitrocarbazoles is, as expected, a direct function of the corresponding nitrenium ion stability, but for two separate groups.

The first group, to which *Eqn. 4* applies, includes di- or trimethylated 3-nitrocarbazoles and mononitrocarbazoles whose nitro group is involved in a hydrogen bond with an amino or hydroxyl substituent. The second group (*Eqn. 5*) includes unsubstituted mononitrocarbazoles and 3,6-dinitro derivatives. The difference between these two groups could be due to the increase in lipophilicity related to the methyl substituents and hydrogen bonds for the first group.

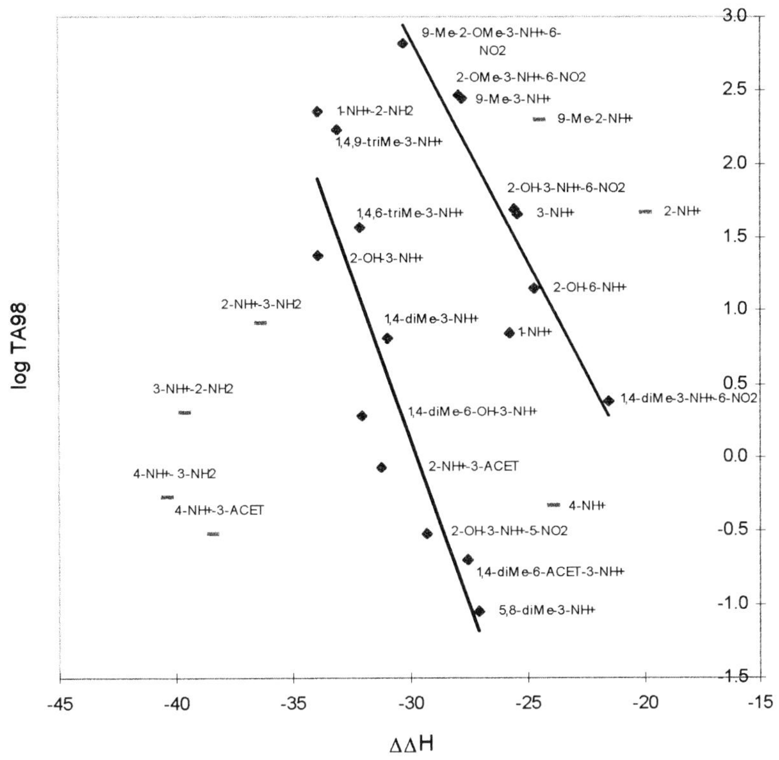

Fig. 5. *Relationship between the relative nitrenium ion stability, ΔΔH, and mutagenicity in TA98 without S9*

Within the first series, when the 1,4-dimethyl-3-nitro derivatives are considered, substituents have the same effects on the nitrenium ions stabilization and the lipophilicity of the parent compound: 6-acetamido decreases these two factors, leading to a weakly mutagenic carbazole, whereas 6- and 9-methyl increase stabilization of the 3-nitrenium ion and lipophilicity of the parent compound, which is the most mutagenic of this small series. Moreover, replacing 6-hydroxy by 6-methyl, which has little effect on 3-nitrenium ion stability, greatly enhances mutagenicity, by about one logarithmic unit, likely due to the increase in lipophilicity. When the unsubstituted 3-nitrocarbazole and the 3,6-dinitro derivatives in the second group are considered, stabilization of the nitrenium ions seems to be a more important factor than lipophilicity. Introducing a 2-hydroxyl on the 3,6-dinitro just compensates destabilization by the electron-attracting 6-nitro group, leading to a nitrenium ion as stable as the unsubstituted 3-nitrenium ion, and the parent compounds exhibit the same mutagenicity despite their different lipophilicity. Replacing this 2-hydroxyl by two methyls increases lipophilicity but lowers the 3-nitrenium ion stability and mutagenicity in parallel.

Conclusion. - This study indicates that for amino-substituted carbazoles, mutagenicity is an inverse function of nitrenium ion stability, i.e. an increasing function of their reactivity, whereas for nitrocarbazoles, mutagenicity is a direct function of this stability, modulated or not by the lipophilicity of the parent compound. It constitutes a second step in the identification of the parameters which will have to be combined in a QSAR equation modelling mutagenicity in carbazoles series.

REFERENCES

[1] D.M. Maron, B.N. Ames, in "*Handbook of Mutagenicity Test Procedures*", Eds .B.J. Kilbey, M.S. Legator, W. Nichols, and C. Ramel, Elsevier, Amsterdam, **1984,** pp. 93-140.

[2] R.L. Lopez de Compadre, A.K. Debnath, A.J. Shusterman, C. Hansch, *Environ. Mol. Mutagen.* **1990,** *15*, 44-55.

[3] A.K. Debnath, R.L. Lopez de Compadre, G. Debnath, A.J. Shusterman, C. Hansch, *J. Med. Chem.*, **1991,** *34*, 786-797.

[4] A.K. Debnath, G. Debnath, A.J. Shusterman, C. Hansch, *Environ. Mol. Mutagen.* **1992a,** *19*, 37-52.

[5] A.K. Debnath, R.L. Lopez de Compadre, A.J. Shusterman, C. Hansch, *Environ. Mol. Mutagen.* **1992b,** *19*, 53-70.

[6] V. André, C. Boissart, M. Lechevrel, P. Gauduchon, J.Y. Le Talaer, J.C. Lancelot, B. Letois, C. Saturnino, S. Rault, M. Robba, *Mutation Res.* **1993,** *299*, 63-73.

[7] V. André, C. Boissart, F. Sichel, P. Gauduchon, J.Y. Le Talaer, J.C. Lancelot, C. Mercier, S. Chemtob, E. Raoult, A. Tallec, *Mutation Res.* **1995,** *345*, 11-25.

[8] C. Mercier, S. Chemtob, P. Vizet, V. André, P. Gauduchon, *Toxicol. Modeling*, **1995,** *1*, 191-206.

[9] C. Mercier, V. Fabart, Y. Sobel, J.E. Dubois, *J. Med. Chem.* **1991,** *34*, 934-942.

[10] G. P. Ford, P. S. Herman, *Chem. Biol. Inter.* **1992,** *80*, 1-18.

[11] G. P. Ford, G. R. Griffin, *Chem. Biol. Inter.* **1992,** *81*, 19-33.

[12] M.J.S. Dewar, E.G. Zoebisch, E.F. Healy, J.J.P. Stewart, *J.Am.Chem.Soc.* **1985,** *107*, 3902-3909.

[13] C. Mercier, S. Chemtob, in, "*Trends in QSAR and Molecular Modelling*", Eds. C.G. Wermuth, ESCOM, Leiden, **1993,** pp. 556-557.

[14] G. P. Ford, J. D. Scribner, *J. Am. Chem. Soc.* **1981,** *103*, 4281-4291.

PART III. COMPUTATIONAL APPROACHES IN DRUG DISPOSITION

Chapter 23

Hydrogen Bond Strength Estimation by Means of the HYBOT Program Package

by **Oleg A. Raevsky**

Laboratory of Comuter-Aided Molecular Design, Institute of Physiologically Active Compounds of Russian Academy of Sciences, 142432, Chernogolovka, Moscow Region, Russia

Hydrogen Bond Strength Estimation by Means of the HYBOT Program Package

by **Oleg A. Raevsky**

Laboratory of Computer-Aided Molecular Design, Institute of Physiologically Active Compounds of Russian Academy of Sciences, 142432, Chernogolovka, Moscow Region, Russia

HYBOT (Hydrogen Bond Thermodynamics), a commercially available program intended for quantitative estimation of hydrogen bond strengths of organic compounds in framework of QSAR and Molecular Modelling, is described. The program **HYBOT.Data Base** is a database of experimental free energy and enthalpy values. It contains data of about 12,000 H-bond complexes between 650 H-bond donors and 2250 H-bond acceptors in 64 solvents and the gas phase. The program **HYBOT.Factor** is a data base of free energy, enthalpy and constant H-bonding factor values for 414 H-bond donors and 1298 H-bond acceptors that have been calculated by a multiplicative approach. It also has a module for the prediction of factor values for new compounds. The program **HYBOT.Complete** contains all modules HYBOT.Data Base and HYBOT.Factor in framework of a common shell.

Two and three dimensional H-bond descriptors are proposed. Successful applications for the quantitative description of lipophilicity, permeability and several types of biological activity are demonstrated.

Introduction. - Interactions between H-bond donor and H-bond acceptor molecules are known to result in the formation of many kinds of molecular and ionic hydrogen bond complexes that are of great importance in chemical and biochemical processes including enzymatic catalysis. Moreover, H-bond complexes are especially important in biological systems where they play a crucial role in macromolecular structures (proteins, DNA and RNA) and in molecular recognition. Nevertheless, up to now only indicator variables (1 = presence and 0 = absence of H-bond donor or acceptor groups [1-3]) have been used as H-bond descriptors in QSAR. These do not reflect the influence of substituents on H-bond donor or acceptor strengths. So at present these qualitative descriptors of H-bond strength in QSAR are not comparable to the quantitative descriptors of two other important interactions: electronic and steric.

The reliable way to describe H-bonding quantitatively is to consider the thermodynamic functions of H-bond complex formation. This kind of functions may be extremely useful for investigations of structure-properties relationships. Knowledge about both enthalpy (ΔH) and free energy (ΔG) or binding constant (K) which are connected with each other by the well known equation:

$$\Delta G = - RT \ln K = \Delta H - T\Delta S \qquad (1)$$

(ΔS is entropy of complexation) is need for complete thermodynamic description of hydrogen bond formation.

Development of empiric scales for H-bond donors and acceptors on the base of thermodynamic parameters began in the middle of the 60 ths. As to the enthalpy H-bond scales it is worth to mention multiplicative-additive approach developed by Drago [4 -6] and multiplicative approach developed by Sherry and Purcel [7] and also by Ioganson [8]. Three pairs of parameters must be used to estimate H-bond enthalpy in the case of additive-multiplicative approach, but only two in the case of multiplicative approach: one for the H-bond donor and one for the H-bond acceptor. Further, in the multiplicative approach it is possible to compare quantitatively the H-bond strengths among donor or acceptor sets. This is why the H-bond scales based on multiplicative approach are much more useful for QSAR and Drug Design compared to scales based on additive-multiplicative approach. In the 80ies Raevsky *et al.* [9,10] developed the multiplicative approach of H-bond enthalpy description and proposed the following equation:

$$\Delta H = k\, E_d\, E_a \qquad (2)$$

where E_d and E_a are enthalpy H-bond donor and H-bond acceptor factors which characterize the relative H-bond donor and acceptor enthalpy ability of compounds. In the framework of *Equation (2)* E_a values have to be positive and E_d values have to be negative because a positive sign was attributed to the standard H-bond acceptor, hexamethylphosphoramide (E_a = 2.50), and a negative sign was attributed to standard H-bond donor, phenol (E_d = -2.50) to demonstrate role of H-bond acceptor and donor compounds in H-bonding.

The similar equation was proposed by Raevsky *et al.* also for free energy [11,12]:

$$\Delta G = k'\, C_d C_a + k_o \qquad (3)$$

where C_d and C_a are free energy H-bond donor and H-bond acceptor factors which characterize relative H-bond donor and acceptor binding ability of compounds. The value C_a = 4.00 was selected for the standard H-bond acceptor, hexamethylphosphoramide and the value Cd = -2.50 was selected for the standard H-bond donor, phenol.

Taft *et al.* were the first authors who proposed a quantitative scale of hydrogen bond strength of H-bond acceptors on the base of logK values of its complexation with 4-fluorophenol in tetrachloromethane [13]. This approach was developed by Abraham *et al.* who finally proposed the equation [14,15]:

$$\log K = k''\, \alpha_2^H\, \beta_2^H + k' \qquad (4)$$

where α_2^H and β_2^H characterize accordingly H-bond donor and acceptor binding ability.

Development and successful application of the above mentioned empiric approaches are possible only on the base of vast experimental data and the ability to calculate factor values for new compounds.. First part of this report contains the description of the commercially available program package HYBOT which is intended for quantitative estimation of hydrogen bond strength of any organic compounds in the framework of QSAR and Molecular Modelling. And the second part of the report contains examples of successful application of H-bond descriptors based on above mentioned factors for investigation of structure-properties relationships.

The program HYBOT.DataBase. - The program Hybot.DataBase is an interactive, menu-driven program, which runs under the Microsoft Windows 3.1 operation system. Its architecture is shown schematically in *Figure 1*. The database consists of a number of data structures, which it is possible to name as files. A user can create a new file or delete an existing one by selection of the appropriate commands from the menu. Each file can contain 5,000 records about structure of H-bond partners, thermodynamic properties and so on. Records may be copied from one file to another or duplicated within a file *via* clipboard.

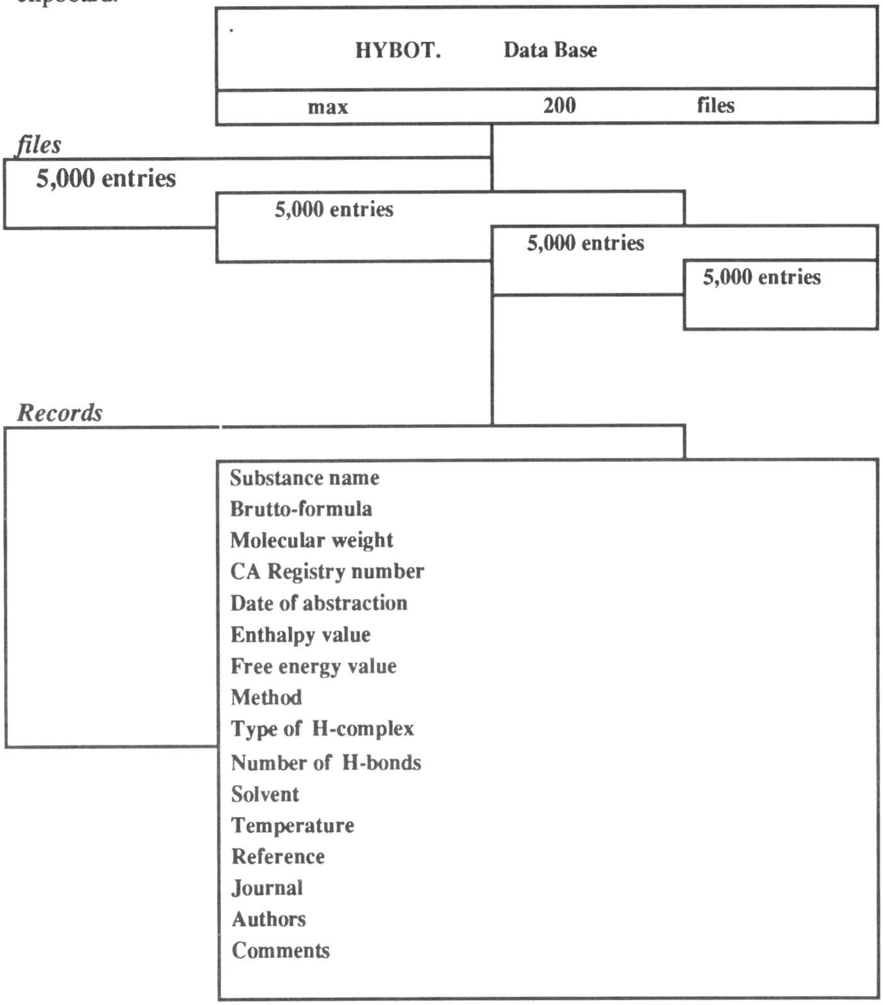

Fig.1. *The architecture of HYBOT.DataBase*

The user can create, delete and modify records. Input of numeric and text data for a record is performed in the usual way for Windows application.
At present there are 12 004 entries in the database which contains information about enthalpy or (and) free energy of H-bonding between 650 H-bond donors and 2250 H-bond acceptors in 64 solvents and gas phase.

The program HYBOT.Factor. - The program has the database of free energy, enthalpy and constant H-binding factor values for 414 H-bond donors and 1298 H-bond acceptors that have been calculated on the base of the multiplicative approach and also the module for prediction of factor values for new compounds.

The majority of the measurements (7793 ones in current version of HYBOT.Data Base) were carried out in tetrachloromethane (CCl_4). This solvent is non-polar, and is thus suitable for the estimation of enthalpy, free energy and binding constant factor values. 414 H-bond donors and 1298 H-bond acceptors that had data from the interaction of more than one partner were selected for estimation of C_a and C_d values. And matrix consisting 414 columns and 1298 rows have been prepared. The computation algorithm includes the following steps:

1. location H-bond acceptors complexed with phenol,
2. locating H-bond donors complexed with these acceptors,
3. constructing and solving equation systems for each pair of complexes, formed by phenol and a given H-bond donor with the same acceptor.

For the majority of H-bond donors one could find more than one such system in the database and solve them independently. In this case the result was presented as the mean value of all independent calculations. In the cases where was impossible to combine such sets of equations for some compounds (they had no complexes with proper acceptor which had reaction with phenol) the above described algorithm was applied again using the factors calculated during the first calculation procedure in the same way as phenol factors were used. The acceptor factors were calculated on the base of the same algorithm using hexamethylphosphoramide as the standard H-bond acceptor.

Factor values for 414 H-bond donors and 1298 H-bond acceptors were estimated as result of above-mentioned calculations. Comparison of experimental values of enthalpy and free energy with calculated ones on the base of equations (2) and (3) demonstrates the adequacy of the multiplicative approach for quantitative description of H-bond thermodynamics:

$$\Delta H_{calc} = -0.49(\pm 0.29) + 0.99(\pm 0.08)\,\Delta H_{exp} \quad (5)$$

$n = 2787, r = 0.970, s = 2.40, F = 44350$

$$\Delta G_{calc} = -0.07(\pm 0.12) + 1.04(\pm 0.05)\,\Delta G_{exp} \quad (6)$$

$n = 3301, r = 0.991, s = 1.12, F = 175000$

By knowing Cd values for 414 H-bond donors, and Ca values for 1298 H-bond acceptors, one can now estimate ΔG values for 537,372 H-bond complexes in tetrachlormethane (in [16] α_2^H values for 150 H-bond donors are given, as well as β_2^H values for 500 H-bond acceptors, and so the possibility to estimate log K values for

75,000 H-bond complexes) .

HYBOT.Factor has also the special module for factor values calculation for new compounds by means of procedure of the search of the nearest neighbour in the abovementioned database of factor values. The algorithm for such calculations is described in [17]. The first step in the calculations is the collation of statistical information (average, standard deviation, minimum, maximum, and number of observations) for each structural fragment. The program uses an atom-centred substructure code like HOSE [18] to describe structural fragments in terms of concentric „spheres“ of neighbouring bonds and atoms, but unlike HOSE the descriptor for each sphere is 4 bytes in length. For the first sphere this is generated by coding information on the atom type (5 bits), charge (2 bits), valency (2 bits), cycle size and type (3 bits) and bond type (2 bits) into a 14 bits variable for each neihbouring atom and then bit shifting and XORing these values to generate a descriptor in the form of a pseudo-random number, derived from the physical environment. Descriptors for subsequent spheres are generated by bit shifting and XORing the first-sphere descriptors centered on the atoms on the outer spheres. Thus, the descriptor for the second sphere is 8 bytes in length, the first 4 bytes being equal to the first sphere descriptor, etc.

The panel containing the results of the calculation of factor values for all H-bond donor and acceptor atom for chlorambucil (as an example) is presented in *Figure 2*

Factor values table

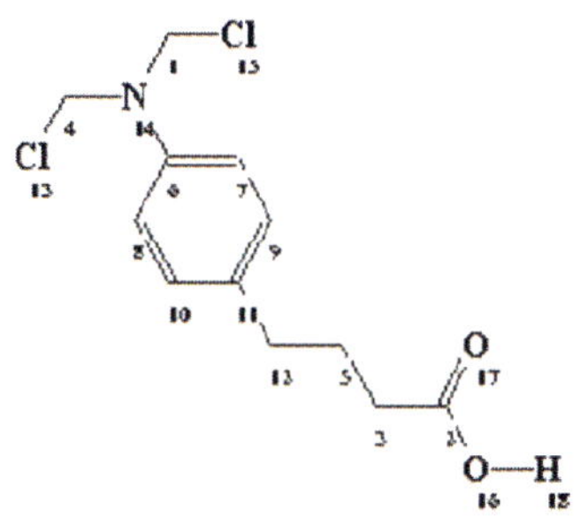

Name: CHLORAMBUCIL

Brutto-formula: C12H15NO2Cl2 MW: 276.16 CAS: 305-03-3

Number	Atom	C-Factor				E-Factor				Alfa(Beta)-Factor			
		Value	SD	N	Sphere	Value	SD	N	Sphere	Value	SD	N	Sphere
13	Cl	0.46	0.15	19	1	0.62	0.07	16	1	0.17	0.01	14	1
14	N	1.37	0.44	7	1	1.58	0.58	2	1	0.41	0.13	4	1
15	Cl	0.46	0.15	19	1	0.62	0.07	16	1	0.17	0.01	14	1
16	O	0.64	0.00	15	2	0.51	0.00	1	3	0.17	0.00	15	2
17	O	1.57	0.18	15	2	1.24	0.00	1	3	0.42	0.04	15	2
18	H	-2.50	0.00	1	5	-2.93	0.52	13	3	0.60	0.00	1	5

Fig. 2. *Calculation of factor values for chlorambucil.*

H-Bond Descriptors Applications. - Twelve H-bond descriptors based on enthalpy and free energy factors were proposed by us for the use in QSAR [19]: free energy and enthalpy factor values for the strongest H-bond donor atom in molecule (C_d^{max} and E_d^{max}) , free energy and enthalpy factor values for the strongest H-bond acceptor in molecule (C_a^{max} and E_a^{max}), the sum of C_d and E_d values for all H-bond donor atoms in molecule (ΣC_d and ΣE_d), the sum of C_a and E_a values for all H-bond acceptor atoms in molecule (ΣC_a and ΣE_a) and also ΣC_d /MW (MW = molecular weight), ΣE_d /MW, ΣC_a /MW and ΣE_a /MW. Successful application of those H-bond descriptors already has been demonstrated in [20-23]. New examples of its applications are presented below.

A Quantitative H-Bond Ability-Lipophilicity Relationship Study. - It has been proposed [24,25], that partition coefficients encode two major structural contributions: a volume-related term (describing steric bulk effects) and a term which reflects such interactions as dipole-dipole and hydrogen bonding. Results of our first investigation of the contribution of H-bond ability in octanol-water partition coefficient (log P) of carbonyl and hydroxyl compounds were published in [26]. Molecular volume (MV) was used in that work to account of the volume-related term. Further work with other classes of compounds showed that in the case of the use of MV and ΣC_a there are non-zero values for the intercepts in equations correlating log P For this reason it was decided to test different volume-related terms together with H-bond descriptors for the training set containing different types of compounds (nitriles, amines, carbonyl compounds, ethers, esters, alcohols, phenols, compounds with phenyl, nitro and halogen groups, acids). Six volume-related terms [molecular weight (MW), integral of spectra of van der Waals interactions (VWI), surface area (SA), molar volume MV, molecular refractivity (MR) and polarizability (Pol) were used together with C_a^{max}, ΣC_a, ΣC_a /MW, C_d^{max}, ΣC_d and ΣC_d /MW. Only one volume-related term among above-mentioned ones could be used because of intercorrelations among the others. The best result was obtained using polarizability and H-bond acceptor strength:

$$\log P = -0.01(\pm 0.08) + 0.249(\pm 0.005)\,\text{Pol} - 0.68(\pm 0.02)\,\Sigma C_a \qquad (7)$$

$n = 234, r = 0.961, s = 0.34, F = 1383, \text{SDEP} = 0.35$

The intercept is non-significant. A graphical comparison of experimental and calculated log P values is presented in *Figure 3*.

Here it is also needed to note that including a H-bond donor descriptor in *Equation (7)* did not improve the relationship in spite of the presence of eighty of H-bond donors in the training set.

It is possible to conclude on the base of *Eq. (7)* that H-bond acceptor ability plays an essential role in the lipophilicity of organic compounds. The values of contributions of polarizability and H-bond acceptor ability in log P for some polyfunctional compounds and drugs (joint research with Drs. H. van de Waterbeemd and G. Camenisch) calculated on the base of *Eq. (7)* are presented in *Table 1*.

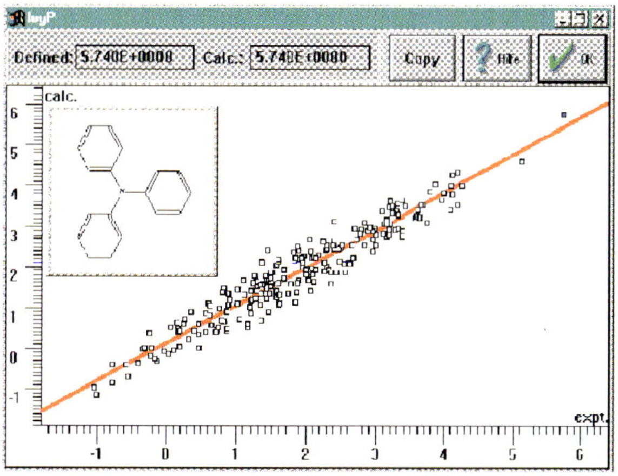

Fig. 3. *Comparison of calculated vs experimental log P values using Eq. 7.*

Table 1. *Estimation of log P from Polarizabilty and H-Bonding Factor.*

NN	Name	ΣC_a	Pol	Contribution in log P		log P	
				ΣC_a	Pol	calc	exp
1	Chloracetamide	3.80	8.00	-2.57	1.99	-0.59	-0.53
2	Acetaldoxime	2.64	5.59	-1.79	1.39	-0.41	-0.12
3	Ethanolamine	4.16	6.43	-2.82	1.60	-1.23	-1.30
4	Nicotinamide	5.88	13.19	-3.98	3.28	-0.71	-0.37
5	Hydroquinone	3.00	11.32	-2.04	2.82	0.77	0.59
6	Dimethyldisulfide	1.50	10.44	-1.02	2.60	1.57	1.77
7	Acyclovir	10.90	21.44	-7.39	5.34	-2.06	-1.52
8	Cefoperazone	24.10	60.97	-16.34	15.18	-1.17	-0.84
9	Fluorouracil	5.78	10.32	-3.92	2.57	-1.36	-0.85
10	Hydroflumethiazide	9.31	25.16	-6.31	6.26	- 0.06	0.04
11	Paracetomol	5.49	16.01	-3.72	3.99	0.26	0.64
12	Warfarin	7.41	32.83	-5.02	8.17	3.14	2.47

A Quantitative H-Bond Ability-Permeability Relationship Study. - A number of physicochemical descriptors, such as log P, the difference between octanol-water partition coeffecients and cyclohexane-water partition coefficients (Δlog P), MW, water-accessible surface area (ASA) are commonly used for estimation permeability properties [27]. And it is proposed that Δlog P values assess hydrogen bond capacity. We decided to use direct H-bond descriptors for QSAR of permiability. *Table 2* contains experimental human skin permeation data (log Kp) taken from [27], polarizability and H-bond descriptors and also calculated log Kp values (joint reseach with Dr. K.-J.Schaper) for 14 steroid hormones.

variables are not scrambled in the 3D space but grouped in chemically meaningful regions (see *Figure 7*).

The same dataset was used to compute the prediction ability of the models obtained using the Norinder and Tropsha methods (see Methods section for details). *Table 1* reports a comparison of the results of the calculations.

The highest Q^2 value in *Table 1* (the smallest prediction error SDEP) is obtained with the FFD/RD procedure. The standard Tropsha method improves the predictivity but the final prediction error (SDEP) is about 40% larger than that obtained with FFD/RD procedure.

Moreover, a graphical analysis reveals that the Tropsha procedure removes all the variables in the α-pocket of the active site (see *Figure 8*). Therefore the final model obtained by this procedure appears to be unreliable. The Norinder approach gives results numerically similar to those of the Tropsha approach. Besides this, many more variables are retained in the final model and the importance of the α-pocket is recognized by the method. The small improvement in the prediction error is probably due to the method used to build the regions.

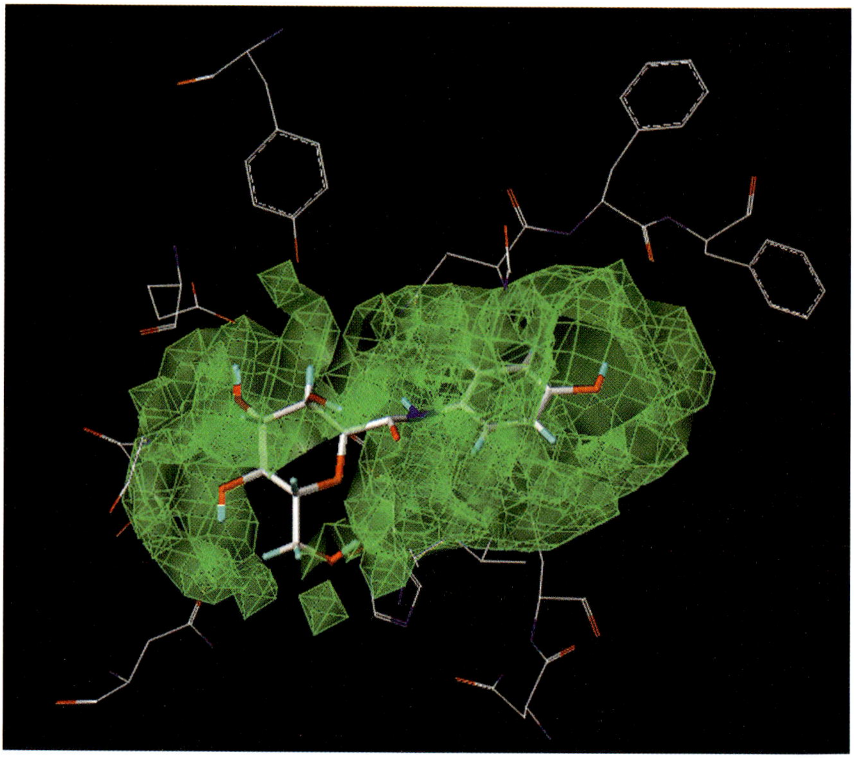

Fig. 7. *The global retained regions after the RD/GOLPE procedure. Wire frame represents the active site residues*

Table 1. *Results of the PLS Modeling with Different Variable/Region Selection Procedures.*

vars. sel.	vars.[a]	dimens.[b]	$R^{2\,c}$	$Q^{2\,d}$	SDEP[e]
none	2087	4	0.92	0.50	0.69
FFD	379	4	0.93	0.73	0.51
RD/FFD	457	4	0.95	0.79	0.45
Tropsha	336	4	0.82	0.59	0.63
Norinder	1542	4	0.94	0.60	0.63

[a] Number of variables used in the PLS model. [b] Dimensionality of the model. [c] Squared correlation coefficient. [d] Cross-validated squared correlation coefficient. [e] Standard Deviation of Error of Predictions.

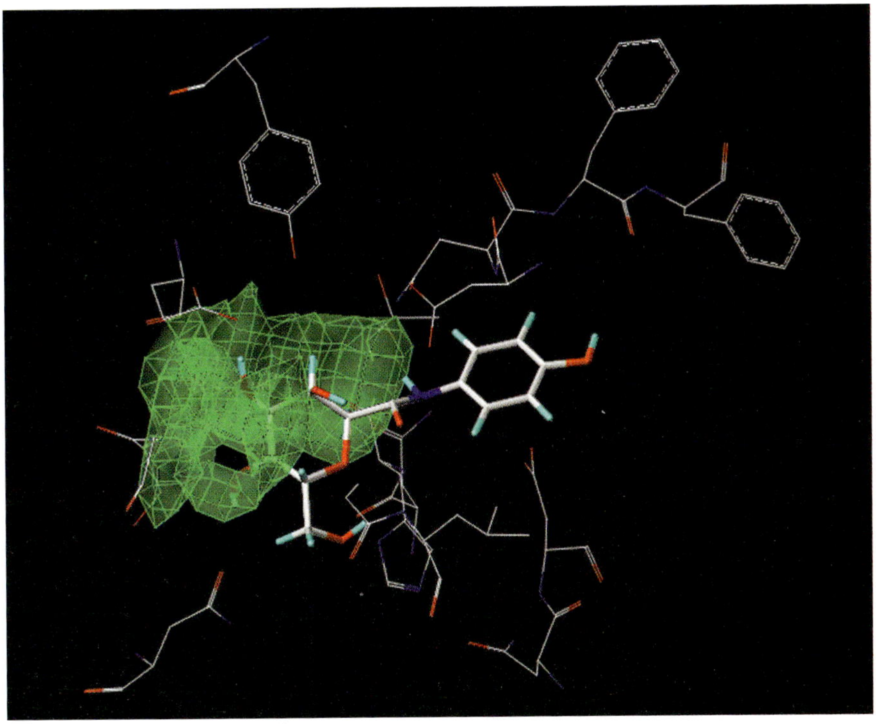

Fig. 8. *All the retained variables after the Tropsha guided region selection. It should be noted that all the variables in the α-pocket are removed*

External Predictions. - So far we have quantified the predictive quality of the model on the basis of internal validation (crossvalidation) only. Although the crossvalidation method utilised (random groups crossvalidation) is very conservative, there is the risk that it can fail to detect overfitting in the models, thus giving overoptimistic indicators of their predictive performance. In order to compare the methods of variable and region selection, it is of critical importance to confirm that the predictivity parameters (SDEP and Q^2) derived from internal validation actually reflect the real predictive quality of the models. Therefore external validation was carried out using six new recently synthesized Gpb-inhibitor compounds. The results are presented in *Table 2.*

Table 2. *External Predictions Obtained Using Models Reported in Table 1.*

compound	Experim.	none	FFD	RD/FFD	Norinder	Tropsha
bnac2	4.06	4.04	4.09	4.10	4.25	3.54
cmpd	3.50	3.30	3.32	3.21	2.96	1.97
hyda	3.26	1.39	1.57	1.83	1.11	2.04
hydo	4.40	3.38	3.31	3.24	3.06	2.62
old13	1.70	2.44	2.32	2.26	2.73	1.86
old29	2.00	1.16	1.55	1.75	0.92	2.67
SDEP[a]	- -	0.98	0.88	0.80	1.22	1.15

[a] External Standard Deviation of Error of Predictions.

It should be noted that the models obtained using the FFD procedures produce better external predictions (smaller SDEP). The classical FFD procedure on single isolated variables decreases the external errors of prediction, but the best result is obtained with the FFD procedure applied on regions. On the contrary, both the Tropsha and Norinder methods fail to improve the external prediction if compared with the plain model.

A detailed analysis on the single prediction experimental values reveals that the models behave in an excellent way, expecially for compounds with high activity values (good inhibitors). The apparently modest results measured by the total SDEP is mainly due to a **hyda** compound which is really peculiar because it contains substituents in positions not explored by any other compounds in the series (see *Figure 8*). In fact, without this compound, the total SDEP for FFD/RD region selection is 0.59, not far away from the predicted internal SDEP of 0.45.

Discussion. - The numerical results, listed in *Tables 1* and *2*, indicate that PLS models obtained with region selection RD/FFD as implemented in GOLPE are better than the simple PLS model both in internal and external validation. Moreover the RD/FFD method in this dataset produces models which are more stable and simpler to interpret. In our opinion the power of the procedure is in the chemically and statistically homogeneity of the regions selected by the RD algorithm together with the design criteria method to select the regions.

On the other hand, it is doubtful that the regular boxes of variables can be successfully used in variable selection because, as mentioned above, they do not contain unique information. This is the reason why these procedures report only marginal improvements in the internal predictivity and no improvement at all in the external predictivity tests.

Conclusions. - We have described a method of Region Definition and Selection which helps to define chemically and statistically homogeneous regions around the molecules under study and to select only those important for the predictivity of the model. However the RD procedure can be further used for other purposes. For example, to identify regions characterized by a poor chemical variability, or to identify highly correlated regions in the series pointing out the lack in the design aspect.

The overall procedure permits to obtain more reliable results than other variable or region selection procedures, especially when external prediction results are compared. Moreover, it allows to extend the application of 3D-QSAR methodologies in all the cases in which is important to take into account indivisible properties of molecules such as water.

Work is in progress to find merits, limits and drawbacks in the application of the RD method in 3D-QSAR problems.

Methods. - *GRID/GOLPE Models.*- The 47 inhibitors are described elsewhere in this book (*Table 1* of M.Pastor and G.Cruciani, The role of water in receptor-ligand interactions. A 3D-QSAR approach). All the inhibitors were considered in the conformation and position found in the crystal, and no further superimposition operation was applied. The energy calculations were performed with the GRID program [1], using the phenolic hydroxyl group OH probe. The size of the box was defined in such a way that it extends about 4Å from the structure of the inhibitors, resulting in a box of 22x20x18Å. GRID calculations were performed using 1Å grid spacing, thus giving 7920 probe-target interactions for each compound, which were unfolded to produce a one-dimensional vector of variables. A cutoff of +5Kcal/mol was applied to produce a more symmetrical distribution of the X matrix. The matrix was imported into GOLPE 3.0 [10] and further pretreated zeroing values having absolute values smaller than 0.1

Kcal/mol, removing variables with standard deviation below 0.1 and deleting variables with skewed distribution (two and three-levels variables, see GOLPE manual [10] for details).

Region Definition. - Variables were grouped using 450 seeds, a cut-off distance of 1Å and a collapsing distance of 2Å. The RD method yields a group 0 containing more than 6000 variables (not used in the analysis) and 424 regions, containing between 2 and 20 variables each.

Variable Selection. - The X matrix was analyzed by PLS and variables were selected using FFD method to improve their predictivity. The standard FFD variables selection required two consecutive FFD selections and a D-optimal variable preselection, resulting in the selection of 379 variables. The RD/FFD region selection procedure requires only the FFD, with no preselection, resulting in a model with 457 variables.

Other Methods. - All the calculations were performed in GOLPE 3.0, using our own implementation of Tropsha and Norinder methods. In theTropsha method 5x5x5 boxes were defined, using the cross-validation method of 5 random groups repeated 20 times. Only the 12 boxes with a Q^2 higher than 0.2 were used in Norinder approach.

Acknowledgments. - We thank our colleagues L. N. Johnson, K. A. Watson, G. W. J. Fleet and N. G. Oikonomakos for sending data regarding some compounds in the training set and compounds in *Table 2* in advance of publication, the EC for providing financial support (project BIO2-CT943025) and grants for one of us (MP). The Italian funding agencies of MURST and CNR are thanked for financial support .

REFERENCES

[1] P.J. Goodford, *J.Med.Chem.* **1985**, *28*, 849.

[2] R. D. Cramer, D. E. Patterson, J. D. Bunce, *J. Am. Chem. Soc.* **1988**, *110*, 5959.

[3] P. Floersheim, J. Nozulak, H.P. Weber, Experience with Comparative Molecular Fiels Analysis. In: '*Trends in QSAR and Molecular Modelling 92*', C.G. Wermuth Ed. ESCOM, Leiden, **1993**, pp.227.

[4] G.E. Kellogg, D.J. Abraham,. *J Mol.Graph.* **1992**, *10*, 212.

[5] G. Cruciani, From variable selection to region selection in 3D-QSAR studies. 4th Scand. Symp. on Chemometrics, Lund, Sweden, **1995**.

[6] S. Wold, Proceedings of the 4th Scand. Symp. on Chemometrics, Lund, Sweden, **1995**.

[7] M.Pastor, G.Cruciani, The role of water in receptor-ligand interactions. A 3D-QSAR approach, this book.
[8] S.J. Cho, A. Tropsha, *J.Med.Chem.* **1995**, *38*, 1060.
[9] U. Norinder, *J. Chemometrics* **1996**, *10*, 95.
[10] GOLPE version 3.0, Multivariate Infometric Analysis, Perugia, Italy, **1996**.
[11] M.Pastor, G.Cruciani, S.Clementi, *J.Med.Chem.* **1996**, submitted
[12] M.Baroni, G.Costantino, G.Cruciani, D.Riganelli, R.Valigi, S.Clementi, *Quant. Struct-Act. Relat.* **1993**, 12, 9-20.

PART IV. 3D-QSAR AND MOLECULAR MODELLING

Chapter 25

The FLEX Approach: An Alternative for Receptor-Ligand Docking and Computing Crystal Conformations

by **Thomas Lengauer**

Institute for Algorithms and Scientific Computing (SCAI),
GMD-German National Research Center for Information
Technology, Schloß Birlinghoven, D-53754 Sankt Augustin

The FLEX Approach: An Alternative for Receptor-Ligand Docking and Computing Crystal Conformations

by **Thomas Lengauer**

Institute for Algorithms and Scientific Computing (SCAI),
GMD-German National Research Center for Information Technology,
Schloß Birlinghoven, D-53754 Sankt Augustin

Computing molecular interactions is one of the basic challenges of computational chemistry. We found that discrete modeling can be a very helpful tool in this process, especially if one aims at screening databases of small organic molecules and rating the compounds with respect to their properties in forming molecular complexes. Such applications arise in biomolecular docking or in the prediction of crystal structures. While discrete models cannot achieve the accuracy of more involved methods, they allow for extremely fast conformational search. If tuned properly, they can discern between attractive and uninteresting compounds in a computing time that is several orders of magnitude smaller than for numerical methods of molecular modeling.

We report on the *Flex* approach, which combines a discrete model of three-dimensional molecular structure with appropriate statistically derived heuristic scoring functions for energy. With this approach, we obtain several interesting results on biomolecular docking and the prediction of crystal structures of small organic molecules. The approach also promises to be helpful in ligand superposition.

Introduction. The computer-aided analysis of molecular interactions is a basic methodical tool for understanding chemical processes and developing bioactive compounds. If this analysis can be based on the knowledge of the three-dimensional structures of the involved molecules, then we can hope to get more accurate predictions of the structures of the molecular complexes as well as of the strength of binding. There are at least two areas, in which structure-based analysis and design is of major advantage.

In drug design the process of generating new lead compounds based on structural knowledge of the receptor belongs to the area of *rational drug design*. This approach is chosen mainly for the design of inhibitors for enzymes. There are two different variants of the approach. In *de novo* design, an inhibitor that complements the receptor pocket sterically and chemically is assembled on the graphics screen from single atoms or larger molecular fragments. This process is supported with tools for selecting the fragments from a molecular library and for ranking their fitness in steric and energetic terms [1]. In *molecular docking* a suitable inhibitor is selected from a data base of complete ligands [2,3]. The ligand is placed inside the receptor pocket and the free energy of binding of the molecular complex is estimated computationally. Both approaches can be configured with varying degrees of interactive and automatic procedures. However, in practice, *de novo* design is the more interactive process, whereas molecular docking usually is an automatic procedure. *De novo* design is the preferred approach if no extensive collection of substances is at hand. On the other hand, docking is preferred if a large collection of potential inhibitors is accessible, for which we know properties such as ease of synthesis, toxicity, bioaccessibility etc.

In the computation of crystal structures the goal is to find small, mostly organic, molecules that aggregate in crystal structures with desired chemical or physical properties, such as, color, conductivity, special nonlinear optical effects, ferromagnetism, etc. Such properties can be derived from the crystal structure by appropriate computational procedures – at least in the case that the properties are known for related compounds. Again, we could envision a *de novo* design procedure that assembles the molecule and the unit cell of the crystal simultaneously, although we do not know of any implemented instance of this method. Alternatively automatic screening procedures can scan through databases of existing molecules, compute the crystals they form, and rank these structures with respect to the degree to which they have the desired property.

In this paper, we focus on molecular docking and largely disregard *de novo* design. In both application scenarios, inhibitor design as well as crystal structure prediction, the goal of the computer-assisted procedures is to guide, focus, and limit laboratory experimentation. For this purpose, we need to be able to clearly discern between desirable and undesirable substances. We need just as much accuracy in our computational procedures as is necessary for achieving this goal. In general, this means that neither do we need to compute molecular complexes with extremely high degrees of accuracy, nor do we have to compute a highly accurate energy estimate. We will be content with an rms-deviation of the computed molecular structures from the measured data up to around 1Å and an energetic accuracy to within about 8 kJ/mol[1]. This is to be contrasted with the higher demands on the accuracy, especially of the energy estimate, for thoroughly analyzing single molecular complexes.

All of today's models used for this purpose are so inaccurate that it does not pay to compute molecular complexes that are globally optimal with respect to (estimated) free energy of binding. Rather, for each molecule that is to serve as an inhibitor or generator of a crystal structure, we need to compute an ensemble of molecular complexes that rates favorably with respect to our energy estimate. If the naturally occurring binding mode is within this ensemble then an inspection of these structures can elucidate important molecular interactions that are at the basis of inhibitor function or crystal formation, and identify a small set of compounds that is to be investigated more thoroughly with classical molecular modeling methods or by laboratory experimentation.

The *Flex* approach is designed for this sort of application scenario. It uses appropriately defined discrete conformational models, discrete definitions of short-range molecular interactions, and specially designed and calibrated scoring functions for energy estimation. The fact that the conformational models are dis-

[1] For crystals we cannot make any statements on the deviation of our energy estimates from the measured energy, at this point.

crete leads to larger structural inaccuracies than would be expected with the classical modeling methods (energy minimization and molecular dynamics in a conformational continuum). On the other hand, discrete models allow for extremely efficient global search procedures through the conformational space. This feature, on the one hand, supports the demand for generating a large variety of possibly structurally quite different molecular complexes that are energetically favorable. On the other hand, discrete search and optimization are fast enough to handle a single problem instance within a matter of a few minutes on a common-day workstation. This enables us to screen reasonably large molecular databases in acceptable time.

In this paper, we will describe the *Flex* approach in general terms. Then, we detail the specific demands of inhibitor design and crystal structure prediction. We describe the correspondingly adapted methods FlexX for biomolecular docking and FlexCryst for crystal structure prediction. Furthermore, we report on experiments carried out with the respective software. Finally, we make remarks on how to extend the *Flex* approach to the problem of ligand superposition. A corresponding tool FlexS is currently under development at GMD-SCAI.

The *Flex* Approach.– The three basic components of the *Flex* approach are (a) a discrete model of molecular conformation, (b) a discrete model for short-range molecular interactions and (c) a statistically derived potential function for scoring the energetic contributions of the individual interactions realized in a molecular complex.

(a). The discrete conformational model provides us with a finite set of conformations defined by internal coordinate values. This model is different for inhibitors and crystals but, in both cases, it has the property that the conformational space can be efficiently searched.

In the case of receptor-ligand docking, the coordinates identify the conformation of the ligand in the receptor pocket which, at present, is kept rigid. (The approach can be extended to making selected parts of the receptor flexible.) The discrete conformations that are allowed by the model are taken from a special database and are derived from molecular structures in the Cambridge Structure Database (CSD) [4] by an analysis of energetically favorable torsion angles [5]. The hypothesis is that, for inhibitor design, some such conformation is close to any conformation that is adopted by the ligand inside the receptor pocket. The argument for this hypothesis is that inhibitors bind strongly to the receptor and therefore can be assumed to bind to it in a conformation in which the torsion angles assume low-energy values. Computing experiments have shown that the inaccuracy incurred by the discretization is about 0.5Å rms. A tree structure is imposed on the conformational space of a flexible ligand, in a natural manner. This tree structure facilitates a variety of optimization techniques for the conformation of the ligand inside the receptor pocket with respect to the scoring function estimating the free energy of binding (see section on receptor-ligand docking below).

In the case of crystal structure prediction, currently, the molecule making up the crystal is considered to be rigid. The objective of a structure prediction is to determine the symmetry elements making up the unit cell of the crystal and the position of the one or several instances of the molecule inside the unit cell. The symmetry elements can be represented by vectors in three-dimensional space. The discretization is performed by assuming that both endpoints of these vectors lie on a cubic grid with mesh size 1Å. This mesh size implies an inaccuracy due to the discretization which is smaller than roughly 0.85Å for relative locations of molecules in adjacent unit cells.

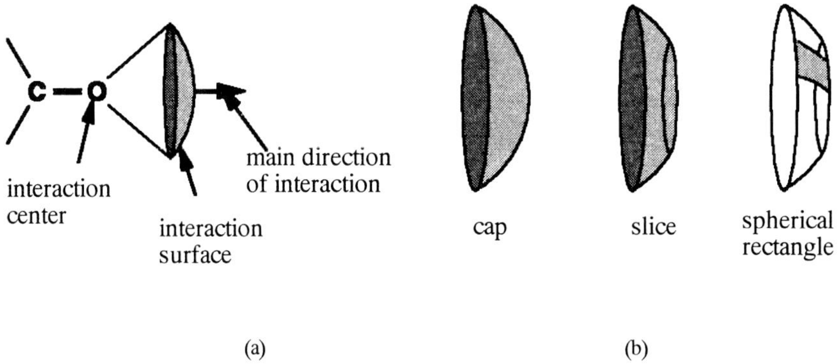

Figure 1. *Modeling partners of molecular interactions. (a) An O-atom as the interaction center of an H-bond acceptor. (b) Three different interaction surfaces.*

(b). Molecular short-range interactions are modeled the same way in both application scenarios. These interaction are used in the Flex approach, in order to select favorable molecular structures during the traversal of the conformational space of the problem.

An interaction is assumed to take place between two *interaction partners*, both of which are single atoms or consist of a few atoms. A certain point inside the atomic group is considered the *interaction center* of this group. Mostly, this point coincides with the place of an atom, but the interaction center can also be the center of a phenyl ring, for example. Besides the interaction center, we assign an *interaction surface* to each interaction partner. This surface is part of a sphere. *Figure 1* illustrates the concept of interaction center and interaction surface.

An interaction is formed, if the interaction centers of two interaction partners of compatible types (e.g. H-bond donor and H-bond acceptor) are located on each others interaction surface (see *Figure 2*). This puts the two interaction centers at the required *interaction distance* and within the required angular ranges. We discretize this requirement by replacing one of the two participating interaction surfaces by a set of approximately equidistant points, the *interaction points*. The distance between two nearest neighbors in this point set is a parameter δ_s, which is being set suitably for each application. In order to account for inaccuracies in the model as well as in the measured data, sometimes this requirement for an interaction to be realized can only be met approximately. The level of tolerance allowed here is another parameter δ_Δ, which is being set appropriately for each area of application.

A large variety of short-range interactions can be modeled in this way. The solid angle that defines the interaction surface determines to what degree an interaction is *directional*. Lipophilic interactions are much less directional than H-bonds, for instance. Note, however, that this model disregards long-range interactions such as mediated by the Coulomb force, for example.

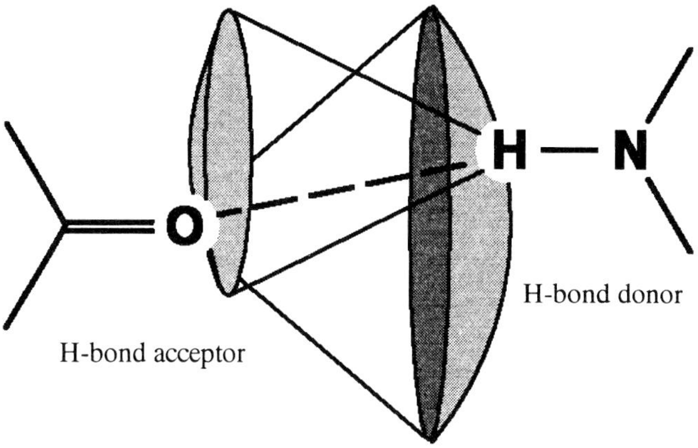

Figure 2. *Model of an H-bond.*

(c). The conformational set defines what three-dimensional structures are admissible for the molecular partners to form the molecular complex. The model of interactions determines, when and where short-range interactions take place between the molecular partners. These interactions are used to select favorable molecular structures. What is left is to find a way of estimating the energy of the resulting complex. For this purpose, we use a statistically derived scoring function that is tailored towards the specific application. This scoring function comprises terms of the interactions modeled in (b) as well as additional enthalpic and entropic terms.

In the case of receptor-ligand docking, the scoring function adds up terms stemming from the realized short range interaction plus additional terms resulting from hydrophobic surfaces contacting in the ligand and the receptor, respectively. Also, additional entropic terms arising from the ligand flexibility are added. Each term is associated with a statistically derived weight [6].

In the case of crystal structure prediction the scoring function is a sum of atom-pair potentials. In contrast with receptor-ligand docking, a larger variety of interactions needs to be covered by the scoring function. The variety of interactions in receptor-ligand docking is limited because one of the docking partners is a protein which is composed of a limited set of different kinds of building blocks, namely the amino acids. Entropic terms are not of major importance in crystal structure prediction.

The three elements described above make up the *Flex* approach. Combined and tailored to an application they can analyze single problem instances within a matter of minutes on a common-day workstation. In the following sections we describe how the *Flex* approach is tailored to receptor-ligand docking and to the prediction of crystal structures.

Receptor-Ligand Docking. In receptor-ligand docking, we have the 3D-coordinates of the atoms in the protein receptor, and we are told where the active site is located. Furthermore, we receive a structure formula of the ligand together with bond lengths and angles. The task is to generate a set of energetically favorable configurations of the receptor-ligand complex. For each configuration, we should identify the molecular interactions stabilizing it and give an estimate of the free energy of binding of this complex [3]. For the present, we assume the receptor structure to be rigid. We will make remarks later on how to weaken this assumption.

Tailoring the Flex Approach to Receptor-Ligand Docking. We tailor the *Flex* approach to this application in the following manner. The molecular complex is completely determined by the conformation of the ligand together with a transformation of the ligand into the active site. We discretize the conformational space of the ligand using a database of molecular patterns for energetically favored torsion angles for acyclic single bonds and a software tool for computing a set of low-energy conformations for each ring system containing elementary rings with up to seven atoms [7,5]. This data base contains a finite set of conformations (typically up to six but may be as large as several hundred) for each ring system and a finite set of torsion angles (up to 12) for each single bond. Observing these patterns yields a finite set of conformations of the ligand, whose size may be in the many millions, however.

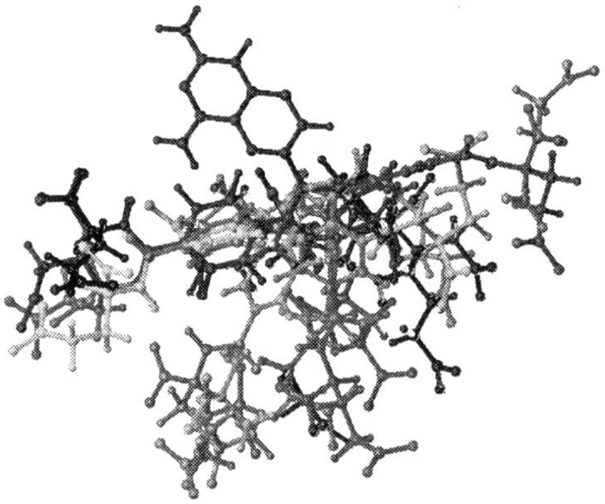

Figure 3. *Several legal conformations of methotrexate.*

Figure 3 shows several of these conformations for the ligand methotrexate which is an inhibitor for dihydrofolate reductase. All conformations in the figure are superimposed at the pteridine-ring system (top middle).

The model of molecular interaction has been described above and is taken over unchanged. The scoring function has the following form. We use an additive model of energy, i.e., the scoring function sums up energetic contributions for each realized interaction and add in additional entropic terms [6]. For each type of interaction, we use a simple heuristic model for determining the contribution of the interaction to the overall score. In general, this contribution is of the form $g(\Delta r,\Delta\alpha)=f(\Delta r)f(\Delta\alpha)$, where Δr is the deviation of the interaction distance from an optimal value r_{opt} and $\Delta\alpha$ is an angular deviation of the interaction from the main direction identified by the center of the interaction surface. *Figure 4* depicts two types of such a function *f(x)*. The form (a) has been used by Böhm [6]. The form (b) is more attractive from an optimization point of view and equally successful, in practice [8].

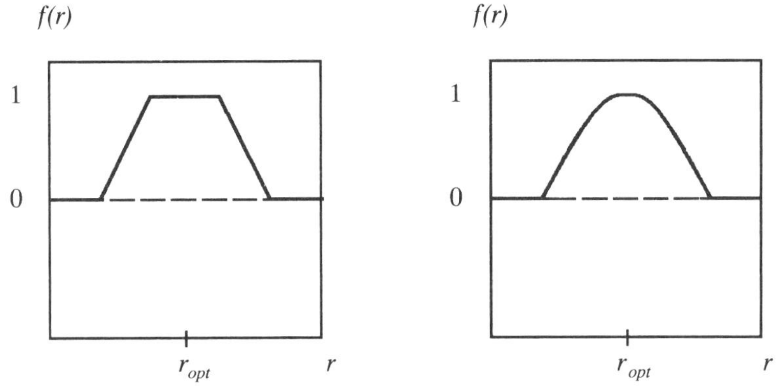

(a) a stepwise linear function (b) a quadratic function

Figure 4. *Potentials for the contribution of a single interaction to the energy score.*

Thus, the scoring function $\phi(p)$ for a configuration p has the following general shape:

$$\phi(\mathrm{p}) = \sum_{\mathrm{t\ interaction\ type}} \mathrm{W_t} \sum_{\substack{\mathrm{i\ interaction}\\ \mathrm{of\ type\ t}}} \mathrm{g}(\Delta \mathrm{r_i}, \Delta\alpha_\mathrm{i}) + \mathrm{W_S F_S(p)} \tag{1}$$

$F_S(p)$ is the term by which we estimate the entropic contribution to the energy of the whole molecular complex. The weight factors W_t, W_S are calibrated such that measured binding constants of a set of several dozen complexes are reproduced.

A scoring function of this type has several desirable properties. First, it can handle more than just atom pair interactions, because the model of interaction that the scoring function is based on allows for considering atomic groups as interaction partners. Second, the scoring function can be calibrated to existing data. By this procedure, it can incorporate phenomena that are not explicitly modeled in the potential or the underlying interaction model, such as interactions with the solvent. Third, since the scoring function is additive, it can be applied to partial molecular assemblies. Thus it supports the incremental construction of the molecular complex, which will be our method for receptor-ligand docking. Finally, it can be computed quickly, since it does not involve complex mathematical operations.

The next step is to organize the conformational space of the ligand in a tree structure. To this end, we fragment the ligand into small and (almost) rigid pieces, by essentially cutting the ligand at each acyclic single bond that joins atomic groups which are modeled by more than just a united atom. This process decomposes the ligand into small atomic groups and (always complete) ring systems.

Figure 5 (a) shows the result for Methotrexate. Since all ring systems are contained inside single fragments, the fragmentation of the ligand always results in an unrooted tree structure. The decomposition tree for methotrexate is shown in *Figure 5 (b)*.

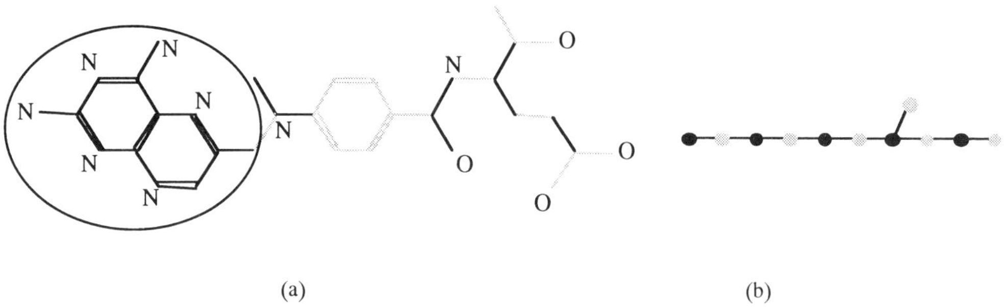

Figure 5. *Fragmentation of Methotrexate. (a) decomposition into 11 fragments (the base fragment is circled). (b) decomposition tree.*

One of the fragments is chosen as the root of the tree. We call this fragment the *base fragment*. The base fragment is placed into the receptor pocket first. For this purpose, it should be large and have many partners for directional molecular interactions. The natural base fragment of methotrexate is circled in *Figure 5 (a)*.

The location of the base fragment inside the receptor pocket determines the transformation of the ligand, given a certain conformation of the ligand. On the other hand, the base fragment effectively roots the decomposition tree of the ligand. Thus we obtain the rooted tree structure depicted in *Figure 6* for the conformational space of this problem. The first level below the root enumerates the placements of the base fragment inside the active site and, therefore, the transformations of the ligand. The succeeding levels specify the conformation of the ligand by determining the torsion angles along single bounds and the conformations of ring systems of the succeeding fragments of the ligand, as the ligand decomposition tree is traversed. The conformational tree depicted in *Figure 6* has as many levels as the ligand has fragments, and as many leaves as there are legal placements of the ligand inside the active site. The number of arcs emanating from an interior node varies even on the same level. This is, because different placements of the same part of the ligand may imply different steric and chemical conditions for extending the next fragment. Some partial placements are not extendible, at all. They represent dead ends during the search for a viable conformation of the molecular complex. After adding a fragment, the transformation of the partial ligand is recomputed in order to optimize molecular interactions and reduce steric overlap.

A complete search through the conformational tree would amount to an exhaustive coverage of all possible conformations. This is not possible, for performance reasons. But the scoring function affords us with energy estimates for complete as well as partial conformations of the ligand. At each node inside the tree, some interactions have been realized, and the potential from Equation (1) can be used to score the respective partial placement. Then any of a large variety of branch&bound procedures can be used to traverse the tree. Many of these procedures potentially facilitate not only a quick generation of conformations with low energy but even the computation of lower bounds on the energy that can be expected from this problem instance.

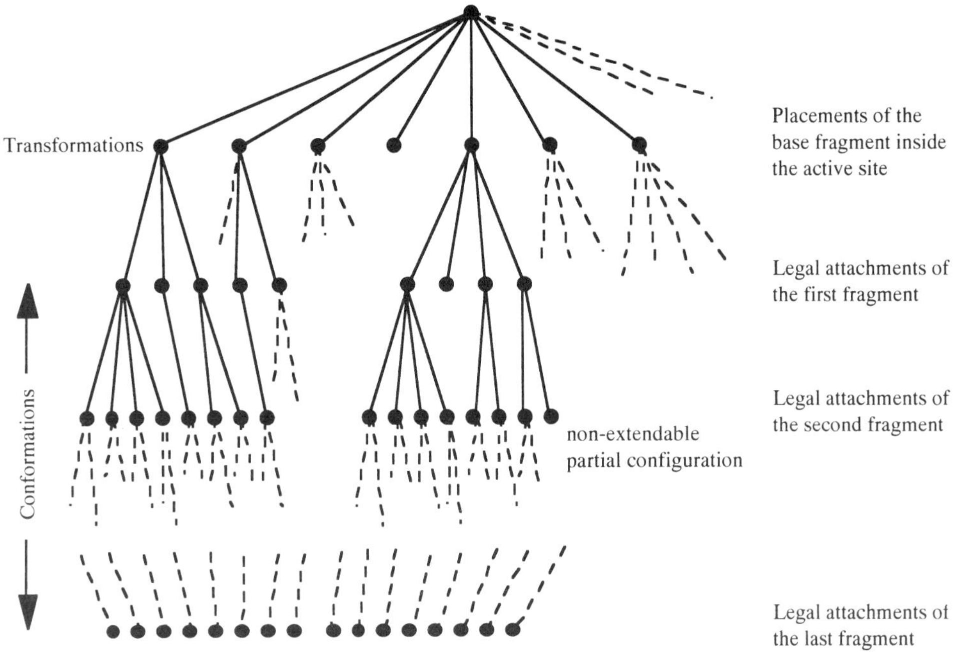

Figure 6. *Tree-structured configuration space of the ligand.*

We have implemented a quite straightforward breadth-first optimization procedure. In each level of the tree, the up to 500 best partial placements are retained and expanded further. This procedure amounts to an incremental construction of the ligand inside the receptor pocket. The base fragment is placed first. All placements that realize at least three interactions, two of which are directional are retained and expanded. Then the other fragments are attached one by one. While this procedure is very fast and operates successfully on a large suite of examples, it has the potential disadvantage that it generates low-energy conformations only if all partial conformations on the way are also low-energy. We have found that, nevertheless, breadth-first optimization can handle problem instances with ligands with up to roughly fifteen fragments.

Several computational features were added, in order to make the *Flex* approach workable on receptor-ligand docking.

First, in order to place the base fragment into the receptor pocket, a geometric hashing technique is adapted from computer vision. The receptor pocket is preprocessed. The result of this procedure is a hash table of pairs of interaction points in the pocket. The base fragment has up to a few dozen conformations, each of which is processed independently. Triples of interaction centers in the ligand are matched with triples of interaction points in the pocket, using the hash table. Each matching yields a transformation of the ligand into the pocket. The number of matchings found in this process goes into the tens of thousands, many of which are quite similar. Testing angular requirements on the ligand side reduces this number to between 100 and 1000. A subsequent clustering process merges almost identical matchings and reduces the number of placement to a few dozen. The process of placing the base fragment takes anywhere from several seconds to a minute. Afterwards, the first level of the tree depicted in Figure 6 has been constructed entirely. Details of the procedure for placing the base fragment can be found in [9].

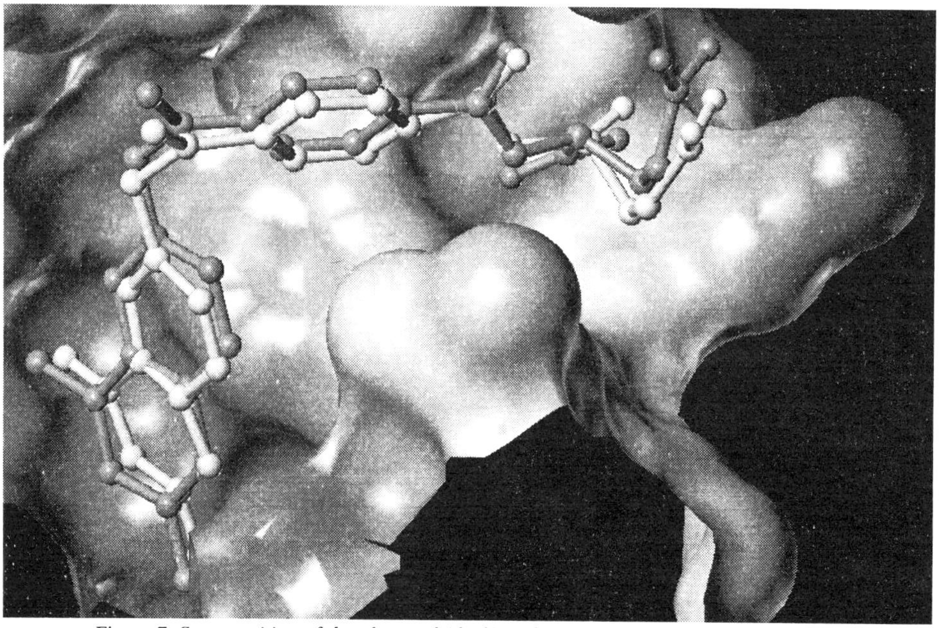

Figure 7. *Superposition of the observed (dark) and computed (light) conformation of methotrexate inside dihydrofolate reductase (DHFR).*

The succeeding construction of the ligand encompasses data structures for incrementally collecting information on partial placements of the ligand in the receptor pocket. We use a pointered tree-like data structure in order to avoid storing information on partial conformations multiply. A special caching technique is used in order to render the frequent clash tests efficient. The whole tree search takes about a minute to compute. Details of this process can be found in [10,11].

All of the above algorithms have been implemented in the FlexX system.

Results on Receptor-Ligand Docking. We have tested the FlexX system in two different ways. First we have analyzed 19 different receptor-ligand complexes from the PDB [12] with FlexX. The details are described in [11]. In summary, we can make the following statements.

Runtimes are invariably from about 30 seconds to 90 seconds per problem instance on a common-day workstation. Depending on the problem instance, placing the base fragment or extending the ligand search are the dominating phases.

Table 1. *Results on 19 complexes from the PDB. Columns from left to right: PDB identifier, number of solutions generated by FlexX, predicted energy and rms-deviation (from the crystal structure) of the solution with minimal score, predicted energy, rms-deviation, and rank of the minimally scoring solution with an rms-deviation of about 1Å or less (if such a solution has been generated), same data for the generated solution with minimal rms-deviation, measured free energy of binding. Energies are in kJ/mol.*

PDB	# sol.	min. score		best prediction			min. deviation			measured	runtime
		ΔG	rms	ΔG	rms	rank	ΔG	rms	rank	ΔG	min:sec
3ptb	30	-29.2	0.49	-29.2	0.49	1	-29.2	0.49	1	-27.2	12
2phh	198	-23.4	0.61	-23.4	0.61	1	-21.5	0.35	4	-26.7	27
1ldm	28	-32.8	0.62	-32.8	0.62	1	-32.8	0.62	1	-30.8	8
5tim	6	-16.3	1.99	-10.3	0.87	3	-10.3	0.87	3	-13.1	2
1ulb	164	-16.8	5.18	-16.7	0.67	3	-14.8	0.50	10	-30.2	30
4dfr	147	-63.9	1.51	-63.5	0.85	2	-54.8	0.61	36	-55.3	1:28
1stp	213	-26.2	0.98	-26.2	0.98	1	-23.6	0.82	4	-76.4	17
2ctc	150	-31.0	0.65	-31.0	0.65	1	-21.2	0.54	46	-22.2	48
1dwd	425	-42.8	1.16	-41.2	0.58	3	-41.2	0.58	3	-48.6	2:11
3tpi	475	-24.0	0.58	-24.0	0.58	1	-24.0	0.58	1	-24.5	1:06
6rsa	188	-35.9	0.85	-35.9	0.85	1	-35.8	0.59	3	-28.5	37
1tmn	550	-35.3	2.33	-31.5	0.94	111	-31.5	0.94	111	-41.6	2:46
3cpa	215	-35.5	2.63	-33.7	0.97	2	-33.7	0.97	2	-22.1	3:11
1rnt	174	-39.8	1.48	-26.9	0.98	122	-26.7	0.82	123	–	1:53
4phv	32	-40.5	1.60	–	–	–	-38.0	1.04	5	-52.2	3:13
4ts1	252	-25.7	0.64	-25.7	0.64	1	-23.9	0.51	2	-32.0	1:35
121p	296	-65.5	2.21	–	–	–	-41.4	1.14	95	–	8:07
4tln	374	-21.5	4.11	-8.7	0.92	349	-8.4	0.85	359	-21.2	40
1dwc	344	-36.9	8.24	–	–	–	-18.5	1.74	317	-44.0	7:08
1nsc	171	-48.8	6.32	–	–	–	-42.2	1.09	45	–	2:45

For half of the investigated problem instances, the placement of the ligand that FlexX generates and ranks highest w.r.t. energy is also the placement whose rms-deviation from the observed crystal structure is lowest. In all these cases, the rms-deviation is around or below 1Å. Details of results which we achieved recently using a variant of the algorithm that optimizes over several automatically chosen alternatives of a base fragment and uses a scoring function as in *Figure 3 (a)* are given in *Table 1*. *Figure 7* shows the superposition of the observed (dark) and computed (light) conformation of methotrexate inside the active site of dihydrofolate reductase (PDB:4dfr). The computed conformation shown is the one with the smallest rms deviation to the observed conformation. Its rank by energy is 2. The corresponding rms deviation is 0.85Å. For a detailed analysis of this example see [11].

The actual scenario that FlexX is designed for is screening databases of ligand for their strength of binding to a given receptor. We have tested FlexX on three sets of ligands,

(I) containing 68 inhibitors of dihydrofolate reductase (DHFR) [13],

(II) containing a congeneric series of 108 inhibitors of thrombin [14],

(III) 357 ligands from the Cambridge Structure Database CSD [4] (all ligands in Release 5.09 of the CSD that contain the words „active“ or „drug“ in their description).

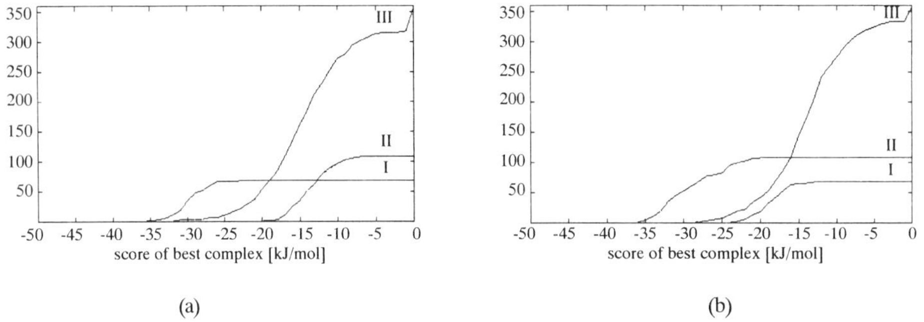

Figure 8. *Two graphs illustrating the ability of FlexX to separate inhibitors from ineffective ligands. The ordinate y(x) of each curve displays the number of ligands that FlexX docked into the respective receptor with an optimal score of at most x. There are curves for each of the three ligand sets (a) Docking different ligands into DHFR. (b) Docking different ligands into thrombin.*

Figure 8 shows the performance of FlexX on docking all of these ligands into DHFR and thrombin, respectively. Apparently, FlexX can separate the inhibitors of each of the two proteins from the other ligands quite clearly. On ligand sets I and II we have used a version of FlexX that is able to dock all ligands in a congeneric ligand set simultaneously. Identical parts of different ligands are docked only once. This is achieved by extending the tree depicted in *Figure 6* to cover a whole congeneric set of ligands or, at least, all those ligands with the same base fragment. With this version of FlexX we achieve a speed-up of a factor between 2 and 3 compared to the original version of FlexX [15].

There are several problem instances for the coordinates computed by FlexX differ from the experimentally observed data. We assume that the following limitations of the model are responsible for these difficulties:

a) Sometimes, critical solvent molecules inside the receptor pocket mediate the binding between the receptor and the ligand. In some examples, e.g. HIV-1 protease/VAC (PDB:4phv), the inclusion of the critical solvent molecules in the receptor pocket leads to dramatically improved docking predictions. We are currently adding a feature to FlexX that computes the locations and orientations of such solvent molecules.

b) The scoring function leaves a lot of room for improvement. Especially, FlexX does not explicitly model the long-range Coulomb force, since the inclusion of these force terms require large computation costs. Also, the estimation of the entropic part of the energy is still quite rough. These limitations preclude FlexX from solving some problem instances. On other instances, they lead to large deviations in our score from the measured free energy of binding. An example of this phenomenon is streptavidin/biotin (PDB:1stp).

c) FlexX needs a suitable base fragment to start. This fragment should exhibit a sufficiently large number of directional molecular interaction, e.g., H-bonds. Ligands that are stabilized mostly by non-directional, e.g., hydrophobic, interactions are much harder to handle by FlexX. In most cases that we investigated, there was a clear choice for the base fragment. We also have an automatic procedure for determining the base fragment, based on fragment size and on a quick analysis of the interaction partners in the fragment. Currently we are working on starting with several base fragments.

Finally, so far, FlexX assumes that the receptor molecule is rigid. This restriction does not weaken our results, because the structure of the receptor is taken from the observed complex. We have also experimented (more realistically) with receptor structures taken from a molecular complex with another ligand, usually with the natural substrate of the protein. In general, the results depend on the problem instance. As long, as the structures of the active site do not differ significantly in complexes with different ligands, the computed interactions and complex geometries do not depend much on the actually used receptor structure. We have developed basic concepts of extending FlexX such that it incorporates restricted flexibility of parts of the active site, e.g., end-standing groups in some critical side chains. These features are not implemented yet. Currently we believe that a more general model of receptor flexibility would require different optimization methods or could even be beyond the *Flex* approach, altogether.

Computing Organic Crystals. The computer-based prediction of crystal structures aims at inferring physical and chemical properties of crystals formed by medium-size organic molecules. Our aims are similar with this application as with receptor-ligand docking. We want to compute a reasonable set of favorable crystal structures for the molecules within minutes on a common-day workstation. It is important that we find representatives of as many low-energy crystals of the molecule as possible, as such an analysis will enable us to make conclusions about the tendency of the crystal to degrade into similar but undesired crystal structures. (Our prime application for crystal structure prediction is the design of pigments. In this case, a necessary condition for the pigment to be color-fast is that the tendency of the crystal to change into other crystal structures be low.) We do not object to certain inaccuracies geometrically as well as w.r.t. our energy score. For this leniency, we gain substantial computational speed.

Tailoring the Flex Approach to Organic Crystals. Currently, we assume that the molecule making up the crystal is rigid. (This is a reasonable assumption, for pigments, as these substances are mostly composed of organic compounds made up of more or less complex ring systems which are practically rigid.) Thus, predicting the crystal structure amounts to determining the space group of the unit cell of the crystal, the shape of that cell, and the placement of the one or several instances of the organic molecule inside the cell.

Space groups are processed independently. We have implemented the space groups P1 and $P\bar{1}$, so far. In principle, the *Flex* approach should be able to handle all space groups. Also, we started with the restriction that there is a single molecule inside the unit cell. Again, this restriction is not fundamental and will be removed, in the future.

We determine the crystal structure by starting with a certain conformation and orientation of the molecule making up the crystal and then selecting a set of three or more vectors spanning the unit cell of the crystal and determining the symmetry elements of the space group. The placement of the molecule inside the unit cell follows from these computations. We now make another critical assumption, namely, that all of the vectors that determine the unit cell are represented by short-range molecular interactions between molecules inside adjacent cells. Essentially, this means that the crystal is stabilized by such interactions. This assumption enables us to use the *Flex* model of molecular interactions to generate a set of interaction vectors (dashed in *Figure 9*), from which we can derive the vectors making up the unit cell of the crystal (solid in *Figure 9*). Such vectors emanate from interaction centers inside one molecule and converge into interaction centers inside an adjacent molecule in the crystal. Thus, these vectors not only determine the shape of the unit cell but simultaneously the placement of the molecule inside that cell.

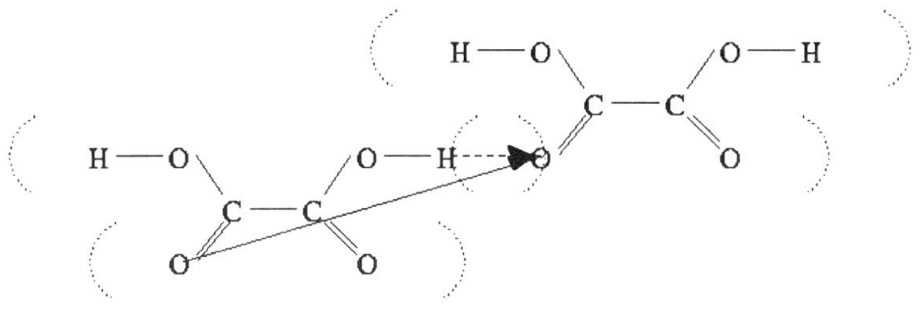

Figure 9. *An interaction between two adjacent oxalic acid molecules. The dashed arrow displays the interaction. The solid arrow displays the vector derived from that interaction.*

Figure 9 also illustrates that we do not need to maintain several molecules for this analysis. Rather, the vectors that are candidates for the spanning vector set of the unit cell can be determined by analyzing just a single molecule w.r.t. interaction centers and interaction points.

Due to the above observations, we use the *Flex* interaction model. The discretization of our vector set is achieved by rounding the vector coordinates to a cubic grid with 1Å mesh size., i.e., vector coordinates are rounded to integer Å-units. We find this discretization to be a reasonable tradeoff between accuracy and computing time. The inaccuracy introduced through the model is at most 0.85Å rms, and the computing time is seconds per crystal in space group P1 on a common-day workstation. For $P\bar{1}$, this figure increases to about 90 minutes per crystal, due to the three additional degrees of freedom resulting from the additional symmetry element.

In order to find a spanning vector set for the unit cell, we need to be able to score vectors by energy. Again, we take an additive model for scoring. Each vector receives a score that is supposed to estimate the energy of the dimer imposed by the vector. The score for a vector set is the sum of the scores for all vectors in the set plus a few linear combinations of these vectors that take care of diagonally adjacent cells in the crystal.

	H	C	N	O	F	P	S	Cl
H	◆	◆	◆	◆	◆	◆	◆	◆
C	◆	◆	◆	◆	◆	◆	◆	◆
N	◆	◆	◆	◆	◆		◆	
O	◆	◆	◆	◆	◆		◆	◆
F	◆	◆	◆	◆	◆			
P	◆	◆						
S	◆	◆	◆	◆			◆	
Cl	◆	◆		◆				◆

Figure 10. *Pair interactions, for which we have derived potential functions (indicated by ◆).*

In this application, we chose to formulate the scoring function for each vector as a sum of atom-pair potentials. The reason is that, as described in the introductory part of this paper, we need to take into account a larger variety of interactions in crystals than in receptor-ligand docking. Modeling all of these interactions as depicted in *Figure 2* is quite complex. Atom-pair potentials were first introduced to the conformational analysis of organic molecules by [16,17] and later applied to crystal structures by [18]. Atom pair potentials for different pairs of atom type together can take into account directional aspects of interaction. For instance, the atom pair potentials of O ... H together with O ... N can approximate an H-bond such as depicted in *Figure 2*. Furthermore, atom-pair potentials take into account medium-range interactions (in our case up to 5Å distance), which are important in crystal structure formation.

We computed atom-pair potentials for the atom type pairs depicted *in Figure 10*. The potentials are derived by inverse-Boltzmann statistics on 1000 randomly selected observed crystal structures from the Cambridge Structure Database (CSD) [4]. Details of the derivations of these functions can be found in [19]. We are adding interactions to the scoring system as we encounter them. Our goal is to also increase the variety of interactions accounted for in the construction of the unit cells.

We now detail the algorithm for space group P1. In a first step, the molecule making up the crystal is analyzed with respect to short-range interactions that have the potential of forming dimers. These interactions are modeled as shown in *Figure 2*. The resulting vectors are rounded to a grid with a mesh size of 1Å and sorted increasingly by their estimated dimer energies. For the energy estimation we use the atom-pair potentials described above. In general, the resulting set of vectors is still enormously large, up to several 10000, depending on the molecule. As a heuristic measure, we consider only the 500 lowest-scoring vectors in the subsequent analysis. This may preclude the system from predicting crystals, for which the unit cell is spanned by a set of vectors, several of which are energetically so favorable that others can afford to be high in energy. Long chain-like molecules, such as depicted in *Figure 11* can be examples of this phenomenon. In general, however, we found that such examples occur quite rarely (see below). In the future, we will extend our selection process to deal with chain-like molecules.

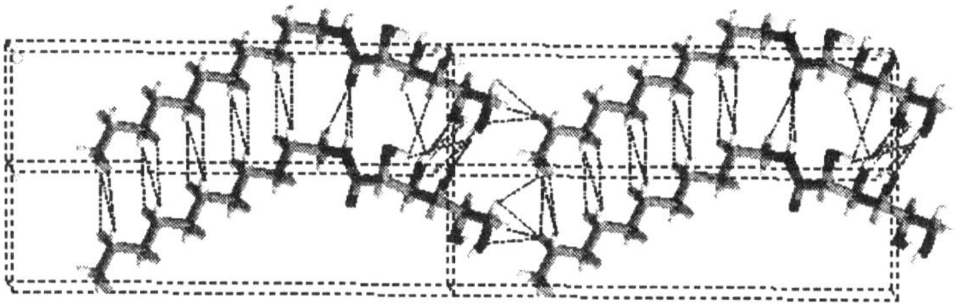

Figure 11. *A chain-like molecule for which the vector selection precludes the algorithm from finding the observed crystal structure.*

From the list of (up to 500) vector candidates, we now form unit cells. For the space group P1, we have to find three linearly independent vectors. An exhaustive search is computationally quite expensive. We limit the search time by imposing the additional restriction that the two vectors a, b spanning the base plane of the unit cell meet a *triangle constraint* $a-b\pm d=0$ with a third vector d from the candidate set. If we assume that the crystal realizes a densest packing of molecules, then this constraint is necessarily fulfilled for strictly convex molecules[2] but may be violated for other molecules (see *Figure 12*).

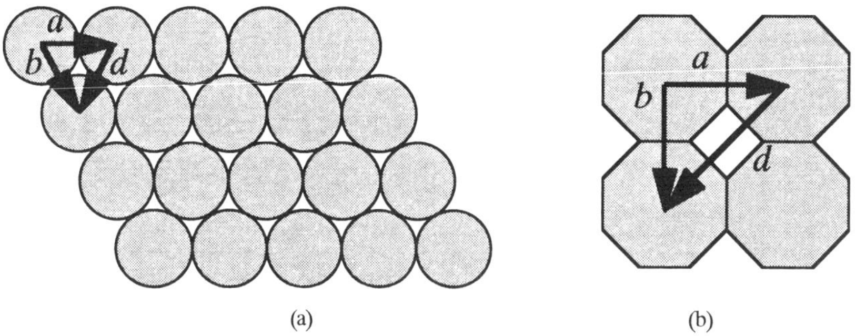

Figure 12. *The triangle constraint. (a) The constraint is met by strictly convex shapes. (b) The constraint may be violated by shapes that are not strictly convex.*

[2] A molecule is *strictly convex* if its surface is positively curved everywhere.

In *Figure 12 (b)*, *d* is not among the set of candidate vectors since the two corresponding shapes do not touch. We found that the triangle constraint holds for the observed crystal structures of all of our test systems. The 200 highest-ranking triangles by energy score are retained for further processing.

We now add a third vector from the set of 500 highest-ranking vectors to each triangle in order to produce a unit cell. Here we limit our search by requiring the density of the crystal to be within the range from 1.1 g/cm^3 to 1.6 g/cm^3. The density is an easily accessible property of crystals.

The result of the search is a large number of crystal structures (up to about 10^6) that can be ranked by score. A subsequent clustering step reduces the number of structures by about a factor of five.

For space group P$\bar{1}$, this algorithm is preceded by a stage that constructs a suitable set of 50 low-energy dimers of the molecule with an inversion of itself. Instead of 1000 clusters we form 10000 clusters at the end of the algorithm.

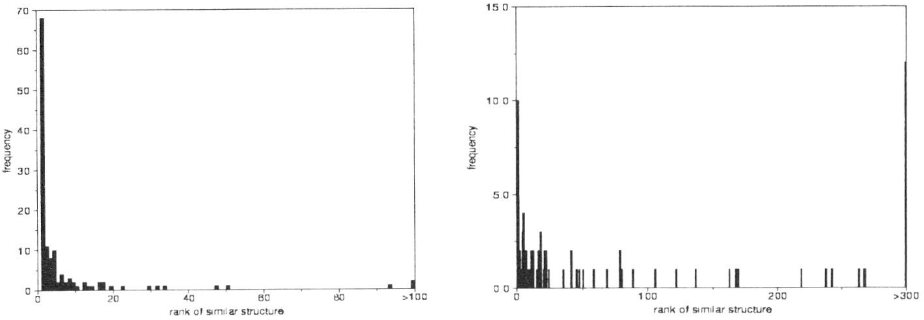

(a) (b)

Figure 13. *Ranks of the lowest-energy structure computed by FlexCryst that is similar to the observed crystal structure. (a) for 131 crystals in space group P1. (b) for 95 crystals in space group P$\bar{1}$.*

Results on Organic Crystals. We have implemented this method of crystal structure prediction in the program FlexCryst [19] and tested it on crystals from the CSD that are in space groups P1 and P$\bar{1}$. We considered the program successful, if it generated a crystal structure that was *similar* to the observed structure in the sense that it was identical to that structure up to rounding to the unit Å-grid. We required this crystal structure to be among the 1000 highest-ranking structures for P1 and the 10000 highest-ranking structures for P$\bar{1}$. In all cases, in which the program was successful, we analyzed the lowest rank of a similar crystal structure computed by the program. The result is depicted in *Figure 13. Figure 14* shows the superposition of the observed (dark) and computed (light) crystal structure in space group P$\bar{1}$ of Bis(dimethylamino)-tetrasulfur-tetranitride (CSD:BASVOG). The cell origins (upper front right corners) are superposed, the *b*-axes and the *b,c*-planes of the cells are aligned.

We looked at all 131 crystals in space group P1 that are contained in the current version of the CSD. For 129 of them, FlexCryst was successful in finding a crystal structure that is similar to the observed one. The ranks of the lowest-energy similar structures found by FlexCryst for each of the crystals are shown in *Figure 13 (a)*. In 68 cases the lowest ranking structure found by FlexCryst was similar to the observed crystal structure.

For our test of crystals in space group P$\bar{1}$, we took (the first) 95 crystals in that space group that are contained in the current version of the CSD. The program succeeded in 81 of the cases. For 77 cases, the similar structure was among the 1000 best ranking computed structures. For 4 cases the rank was between 1000 and 10000. In 14 cases, the program did not find the correct structure. We are currently investigating the reasons for this phenomenon. Partly, we attribute this problem to the lack of important atom-pair potentials. Sometimes , some of the constraints have to be relaxed. And chain-like molecules present a problem (see *Figure 11*). Nevertheless, our approach is quite competitive to other published alternatives with respect to runtime as well as with respect to the quality of the predictions.

While we were testing FlexCryst we found quite a number of errors in the CSD data. This proves that a program like FlexCryst can be useful in validating crystal structure data.

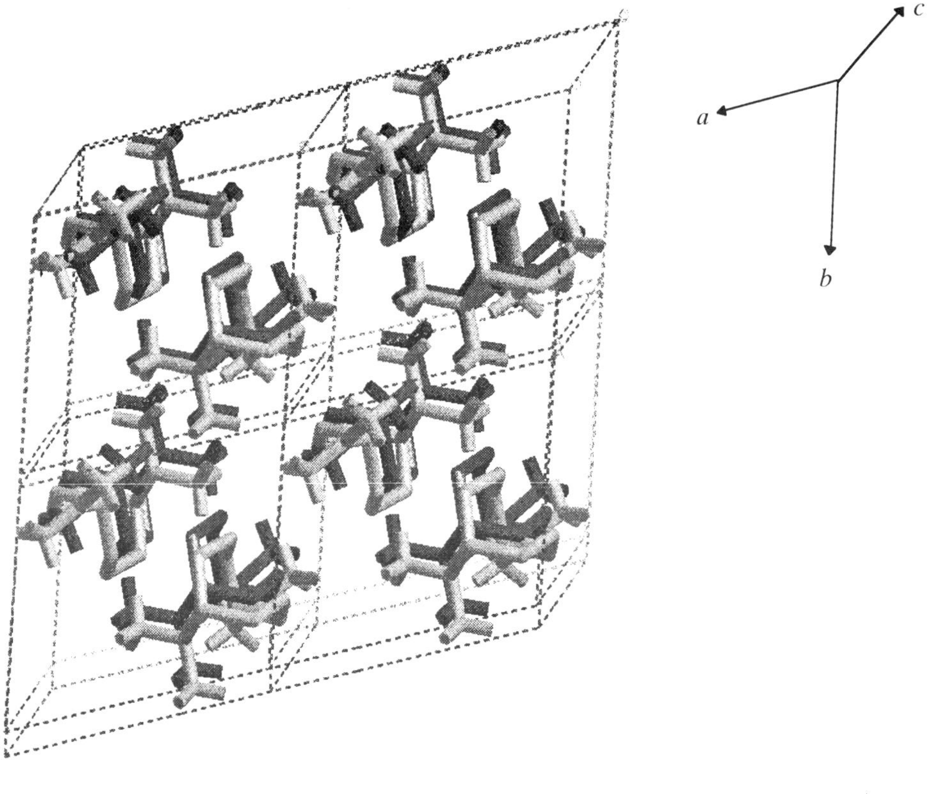

Figure 14. *Observed (dark) and computed (light) crystal structure of Bis(dimethylamino)-tetrasulfur-tetranitride.*

Ligand Superposition. This application addresses problems related to receptor-ligand docking. However, here we assume that the receptor structure is not known. In its place we are given a set of ligands that are known to bind to the receptor. By superposing the ligands structurally, we hope to gain information on the steric and chemical properties of the receptor pocket [20]. In the best case, we would be able to compute a „negative imprint“ of the receptor pocket which can then be used for further studies. Alternatively, we can use the superimposed ligand set itself as a complement of the receptor pocket, align new ligands to it and, in this way, try to estimate their strength of binding to the receptor.

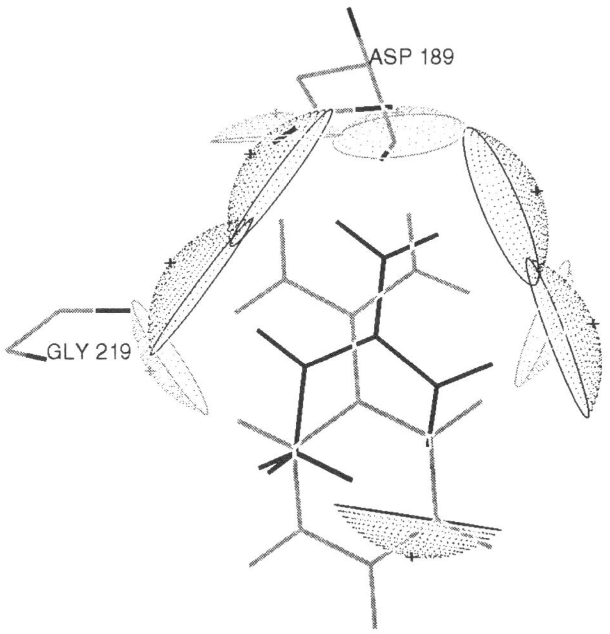

Figure 15. *Two ligands with the same interaction partner in the receptor pocket but with different directions of the interaction.*

An important problem that is to be addressed if we want to adapt the *Flex* approach to ligand superposition is illustrated in *Figure 15*. In general, the interaction center on the receptor side is located inside or close to the intersection of the two corresponding interaction surfaces in each of the ligands. Inversely, we cannot assume that the interaction centers inside the ligands are superposed. Rather, any superposition of the ligands in which the two interaction surfaces of the ligands intersect is potentially able to realize interactions of each ligand with the same partner in the receptor. As an example, in *Figure 15* we see a superposition of (parts of) two inhibitors of thrombin, namely argatroban (dark) taken from PDB:1dwc and NAPAP (light) taken from PDB:1dwd. Both inhibitors form H-bonds with GLY 219 and ASP 189. The directions of the H-bonds differ in the two inhibitors. The corresponding interaction centers on the receptor side are

cover more space groups, and in ligand superposition to allow for the flexibility of both ligands to be superposed.

Acknowledgments. Many people have contributed to the work reported on in this paper. FlexX has been developed by Matthias Rarey, Stephan Wefing and Bernd Kramer. Christian Lemmen is developing FlexS. All of this work has been supported by the German Ministry for Science, Education, Research, and Technology (BMBF) through the project RELIWE, grant no. 413-4001-01 IB 302 A/1. Detlef Hofmann has developed the program FlexCryst. My collaborators also provided many helpful remarks on this manuscript. Furthermore, I gratefully acknowledge many insightful conversations with our project partners from BASF Ludwigshafen, Hans-Joachim Böhm, Peter Erk, Gerhard Klebe, Hugo Kubinyi, and Thomas Mietzner, and from Merck, Darmstadt, Gerhard Barnickel and Friedrich Rippmann. Part of this research was performed within GMD's contribution to the High Performance Computing Center (HLRZ).

REFERENCES

[1] R.A. Lewis A. Leach, *J Comput-Aided Mol Design* **1994,** 8, 467-475.

[2] I. D. Kuntz, E. C. Meng, and B. K. Shoichet, *Acc Chem Res* **1994,** 27, 117-123.

[3] T. Lengauer, M. Rarey, *Curr Opin Struc Biol* **1996,** 6, 402–406.

[4] F.H. Allen, S. Bellard, M.D. Brice, B.A. Cartwright, A. Doubleday, H. Higgs, T. Hummelink-Peters, O. Kennard, W.D.S. Motherwell, J.R. Rodgers, D.G. Watson, *Acta Cryst B* **1979,** 35, 2331–2339.

[5] G. Klebe, T. Mietzner, correlation of crystal data to analyze and predict ligand/receptor interactions, In *Organic Crystal Chemistry,* Ed D.W. Jones, Oxford University Press, Oxford, U.K. **1992,** 135–158.

[6] H.-J. Böhm, *J Comput-Aided Mol Design* **1994,** 6, 243–256.

[7] J. Hoflack and P.J. DeClercq, *Tetrahedron Comput Methodol* **1988,** 44, 6667

[8] B. Kramer, *In preparation* **1996.**

[9] M. Rarey, S. Wefing, T. Lengauer, *J Comput-Aided Mol Design* **1996,** 10, 41–54.

[10] M. Rarey, B. Kramer, T Lengauer, Time-efficient docking of flexible ligands into active sites of proteins, *Proceedings of the Third International Conference on Intelligent Systems for Molecular Biology (ISMB'95),* AAAI Press, Menlo Park, CA **1995,** 300–308.

[11] M. Rarey, B. Kramer, T. Lengauer, G. Klebe, *J Mol Biol* **1996,** 261, 470–489.

[12] F.C. Bernstein, T.F. Koetzle, G.J.B. Williams, E.F. Meyer, Jr., M.D. Brice, J.R. Rodgers, O. Kennard, T. Shimanouchi, M. Tasumi, *J Mol Biol* **1977,** 112, 535–542.

[13] C.D. Selassie et al., *J Med Chem* **1989,** 32, 1895–1905.

[14] H.-J. Böhm, *Private communication* **1995.**

[15] C. Bernd, *Strukturelles hierarchisches Clustern von Liganden,* Diploma thesis, Dep. of Computer Science, University of Bonn, Bonn, Germany **1995.**

[16] T.L. Hill, *J Chem Phys* **1948,** 16, 938–949.

[17] F.H. Westheimer, S.E. Mayer, *J Chem Phys* **1946,** 14, 733–738.

[18] A.I Kitaigorodskii, *Izvest. Akad. Nauk. S.S.S.R., Ser. Fiz.* **1951,** 15, 157–163.

[19] D. Hofmann, T. Lengauer, A discrete algorithm for crystal structure prediction of organic molecules, *submitted for publication* **1996.**

[20] G. Klebe, Structural alignment of molecules, In *3D QSAR in Drug Design. Theory, Methods and Applications,* Ed. H. Kubinyi, ESCOM Science Publishers, Leiden, Netherlands **1993,** 173–199.

[21] S.K. Kearsley, G.M. Smith, *Tetrahedron Comput Methodol* **1990,** 3(6C), 615–633.

[22] G. Klebe, T. Mietzner, F. Weber, *J Comput-Aided Mol Design* **1994,** 8, 751–778.

PART IV. 3D-QSAR AND MOLECULAR MODELLING

Chapter 26

Molecular Partial Similarity Using Surface-Volume Comparisons

by **Tim D.J. Perkins*** and **Philip M. Dean**

Drug Design Group, Department of Pharmacology, University of Cambridge, Tennis Court Road, Cambridge, UK, CB2 1QJ

Molecular Partial Similarity Using Surface-Volume Comparisons

by **Tim D.J. Perkins*** and **Philip M. Dean**

Drug Design Group, Department of Pharmacology, University of Cambridge, Tennis Court Road, Cambridge, UK, CB2 1QJ

A computer program, SNAP, has been developed to determine alignments between a pair of molecules by optimizing their steric molecular partial similarity. It aligns only the most similar parts of the molecules, which are predicted to interact with the receptor, and will therefore allow both molecules to have unmatched portions. The algorithm uses point-based molecular-surface volumes, which are partitioned into a number of overlapping regions; poorly matched regions are ignored. Multiple possible alignments can be filtered in terms of similarity of surface properties, such as hydrogen bonding and electrostatic potential. The program has correctly generated X-ray-crystal alignments for both FKBP and thermolysin ligands that were not identified by other techniques.

Introduction. – Molecular similarity tools can be a useful resource in drug design by suggesting optimal superpositions between apparently dissimilar molecules that bind to a common receptor site. This type of approach is particularly important when detailed structural information about the target receptor or enzyme is absent. Determination of the optimal alignment between any set of molecules is complex, since each molecule will adopt an orientation and conformation at the receptor to give maximum steric, electrostatic, and hydrogen-bonding complementarity and hydrophobic similarity over the binding surface. This is not necessarily identical to optimum similarity for these parameters between entire matched molecules.

A number of methods are available for unbiased and automated superposition of dissimilar molecules, which tend to be based around either atom [1–4] or molecular-surface [5–7] matching. A particularly promising approach is superposition on the basis of a surface layer with thickness, described as a skin or surface volume [8–10]. The use of surfaces is especially important when matched molecules differ significantly in size and when there is no commonality between the atom skeletons of the matched molecules.

The concept of molecular partial similarity has been generally neglected to date. When a drug binds to its receptor, only a fraction of its surface is involved in binding. The size of this fraction will vary between different drugs at a given receptor, and also between classes of receptor site. For example, the extent of surface in contact will generally be lower for a shallow receptor-surface cleft site than a deep site. This implies that when a molecular alignment is performed, it is not necessary to match entire molecules, but merely to align those parts of each molecule that comprise the binding face(s). This can be restated as a search for subsets of molecular structures with highest similarity. Both matched molecules may therefore have unmatched portions (see *Fig. 1*).

There have been a small number of attempts to solve this and related problems. Barakat & Dean [11] devised an atom-matching procedure for two molecules that allows a specified

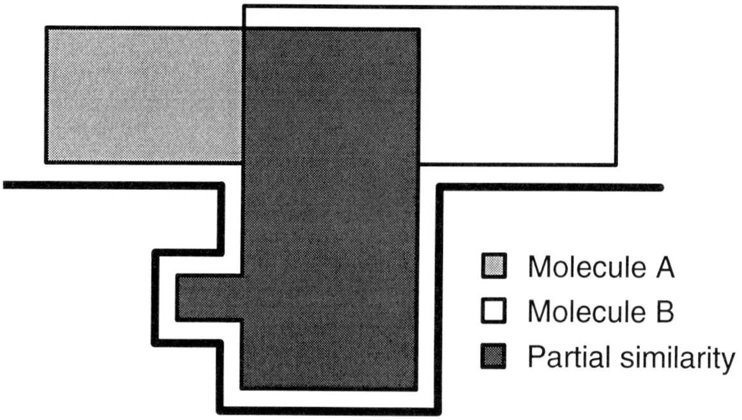

Fig. 1. *An illustration of molecular partial similarity. The optimal overlap shown here reveals only partial similarity between molecules A and B; many algorithms would generate a superposition with molecule A flipped about its vertical axis to maximize gross similarity.*

number of atoms in the smaller molecule to be ignored (termed 'null correspondences'). The number of atoms to be omitted must be fixed in advance by the user, but the selection of atoms is made by the algorithm to optimize its match statistic. This approach comes close to the requirements for molecular partial similarity; however, multiple runs of the same match using different parameters must be performed because of the necessity to predetermine the number of atom pairs to ignore. An alternative is to search for a best-matched hemispherical window on the surface of two molecules [12]. Other methods have analysed existing matches to obtain information on the parts of the molecules that match best [13]. This does not optimize partial similarity, and significantly different results may therefore be obtained.

Overview. In this publication, we present a novel method for determining an alignment between two molecules by maximizing the molecular partial similarity between their surface volumes. The method has been implemented as the program SNAP and has been developed from PLM [10]; the divide-and-conquer approach to molecular matching has again been applied by searching for matches with high steric similarity and then selecting between them on the basis of additional surface properties. A molecular-surface volume (typically between the van der Waals and accessible surfaces), comprising points taken from a rectangular grid, is created around each molecule and then subdivided into a number of overlapping regions. The common matched points are calculated [9, 10], but only those regions that contain a specified fraction of matched points are included in the resulting alignment. This allows badly matched regions to be completely ignored. The process is optimized using simulated annealing.

Methods. – The algorithms presented here for molecular partial matching have been developed and extensively modified from the program PLM [10].

Steric Matching. One of the two molecules to be matched is selected by the user for rotation/translation and is designated molecule A; the other is termed molecule B and remains fixed. For molecule A, points are generated from a rectangular grid (typical spacing 1.0 Å) and lie between the van der Waals surface and an accessible surface (typical probe radius 1.4 Å). For molecule B, a grid of points is placed around the molecule: each is designated one of INTERIOR, EXTERIOR, or SURFACE, using the same criteria as for molecule A.

In order to cope with partial similarity, the surface points of molecule A are divided into a number of overlapping regions (see *Fig. 2*). Without any prior knowledge about the optimal superposition between the molecules, any such procedure must be arbitrary. However, the use of overlapping regions helps to reduce dramatic edge effects at region boundaries. The regions are divided as follows: the two mutually most distant points become sink points 1 and 2; additional sink points are added (1 per region required) with the condition that they maximize the minimum distance (d_{min}) to existing sink points. The smallest value of d_{min} over all regions is then used in the assignment of the remaining points to regions. For each sink point, if any point is within a given multiple, R, (e.g., 0.8–2.0) of d_{min}, then it becomes part of that sink point's region. If a point does not qualify for membership of any region, it is assigned to the region with the closest sink point.

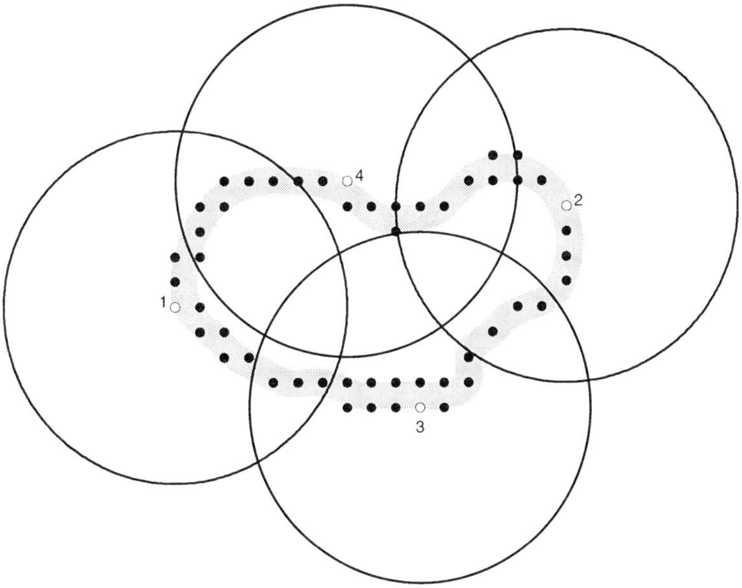

Fig. 2. *Division of surface points of molecule A into 4 overlapping regions. The surface volume is drawn as a thick grey line and ordinary grid points are marked in black. The sink points of each region are white and labelled with the region number 1–4. The large circles indicate to which overlapping regions each grid point belongs. See text for details of the procedure.*

The molecules are placed initially so that their centroids overlap, before molecule A is given a random rotation about each of its euler axes, and a random displacement along a randomly selected vector. At each step in the optimization scheme, molecule A is given a random rotation (default probability 0.9) or a random translation (default probability 0.1). If the translation places the centroid of molecule A outside the van der Waals surface of molecule B, new random translation vectors are applied until this is not the case. The steric score coefficient (ξ) is then calculated (see below), and this value is optimized using simulated annealing with a dynamic cooling schedule [14].

For any given superposition of molecule A and molecule B, the quality of the match is assessed using the following two-stage procedure. Firstly, for each point on molecule A, the status of the closest point (one of INTERIOR, EXTERIOR, or SURFACE) in the molecule B grid is determined (internal scaling ensures that this operation consists merely of an array lookup). If the grid point is marked as SURFACE, the corresponding point in molecule A is marked as a possible common-surface point. Secondly, the fraction of points with each region that are marked as possible common-surface points is calculated; only regions that exceed a user-specified cutoff q (e.g., 0.5) are included in the list of N_{reg} successful regions. Only common-surface points that are members of at least one successful region are included in the final common surface. On completion of the optimization, the number of points in this surface is divided by the number of points in the smaller molecule to give the match quality, termed the steric score coefficient, ξ.

In an attempt to discover a number of discrete local minima for the steric match, the SNAP program includes two major modifications to standard simulated annealing. Firstly, whenever a new global minimum is discovered, a small random local search is made. This search is performed independently of the progress of the simulated annealing but does allow interesting local areas of the configuration space to be examined. Secondly, the configuration of each newly discovered global-minimum energy is stored on a stack. Very similar configurations are immediately overwritten, otherwise the match with the lowest value of ξ is overwritten. This is a particularly useful feature as it allows a number of good steric matches to be obtained simultaneously.

Surface-Feature Discrimination. Once a number of high-quality steric matches have been obtained, they are assessed in terms of properties in the surface volume: measures of hydrogen-bond and electrostatic-potential similarity are based on the methods described by Perkins *et al.* [10]. Only the common-surface points generated by SNAP are used in the discrimination.

For hydrogen-bond similarity, the probability of hydrogen-bond formation at each point for both molecules is calculated using data from an extensive crystal survey [15] of the Cambridge Structural Database [16]. If both molecules predict the same type of hydrogen bond (i.e., donor or acceptor), that point becomes a common hydrogen-bond point; the total number of such points, N_{hbp}, is the hydrogen-bond similarity score.

For electrostatic-potential similarity, atomic-partial charges and electrostatic potentials at each common surface point for the two molecules are calculated as previously described [17]. The difference between potentials at the same point and the root-mean-square (r.m.s.) difference of all common matched points (R_{ev}) are calculated.

Molecules Matched. Two compounds that bind to the FK506 (tacrolimus) binding protein (FKBP), (1*R*)-1,3-diphenyl-1-propyl (2*S*)-1-(3,3-dimethyl-1,2-dioxophentyl)-2-piperidinecarboxylate (**1**) and rapamycin (**2**), were taken from the Brookhaven Protein Databank (PDB) [18] files 1fkg and 1fkb, respectively. Chemical structure diagrams were created with CACTVS [19]. A crystal alignment was determined by superposing the main-chain atoms of the binding residues (6, 26, 46, 55–56, 59, 99; r.m.s. distance 0.16 Å) and using the same matrix transformation to superpose the two ligands.

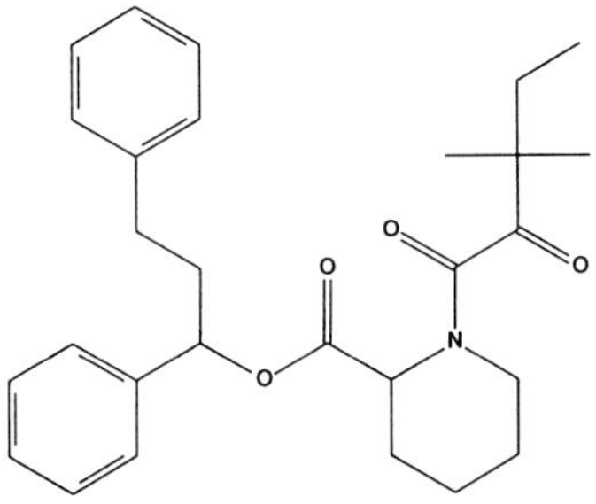

1

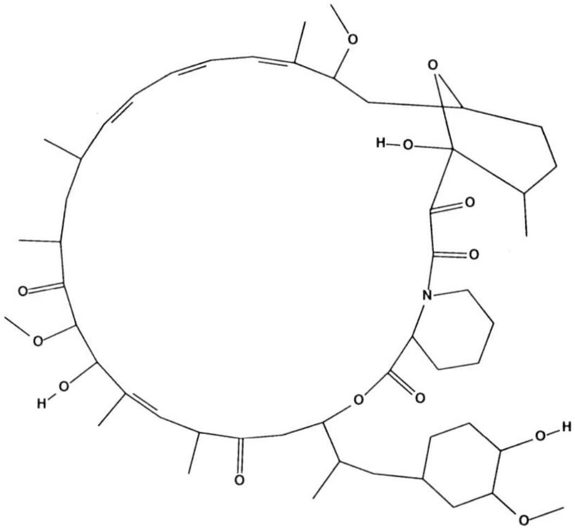

2

The other pair of compounds used are thermolysin inhibitors: benzylsuccinate (**3**) and Cbz-Gly^{P}-(O)-Leu-Leu (**4**) were taken from the PDB files 1hyt and 6tmn, respectively. Both compounds were placed in the fully ionized state for electrostatic calculations. The main-chain atoms of residues 112, 113, 143, 203, 231 (r.m.s. distance 0.12 Å) were used to determine the the crystal orientation of **3**.

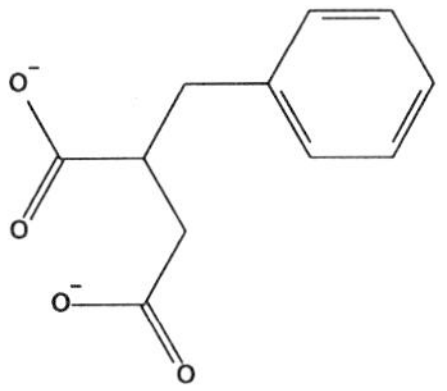

3

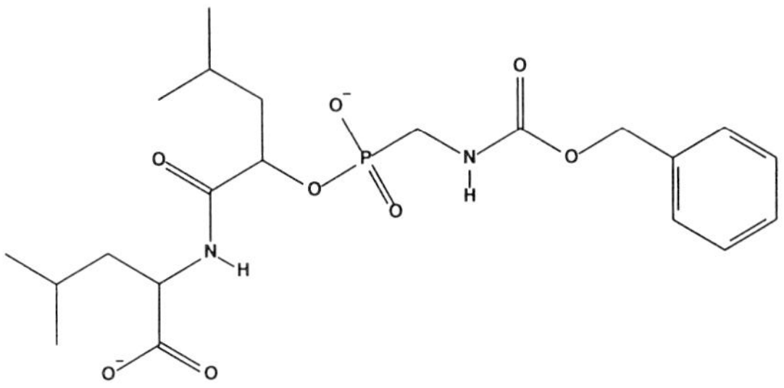

4

Computational Parameters. All matches were performed with a 1.0 Å grid size and probe radius of 1.4 Å. Any CPU timings given are for a single run on a DEC AlphaStation 600 5/266 for code compiled with the O4 flag.

Results. – The approach used and molecules selected for this section are chosen primarily to test and validate the program SNAP.

FKBP Ligands. High-resolution crystal structures of complexes of a number of ligands, including FK506 (tacrolimus), with FKBP are available. These molecules have been used as a validation set for our earlier program PLM [10] (data not shown). The relative orientations indicated by the crystal structures were correctly predicted for all pairs of molecules, apart from the superposition of **1** onto **2**. For this pair, the closest distance from the X-ray-crystal orientation of **1** was greater than 3 Å. These compounds thus provided an ideal test of partial similarity matching.

SNAP was run 10 times from different random starting configurations using 50 regions with $q = 0.5$ and $R = 1.0$. Each run required 45 s CPU time. The 10 resulting alignments are shown in *Fig. 3*. This shows that although a number of matches reproduce the X-ray-crystal alignment for these molecules, some are distinctly different. The numeric descriptors for these alignments, including those for the X-ray-crystal structure alignment, are shown in *Table 1*. This table shows that matches close to the X-ray-crystal alignment have a smaller electrostatic-potential deviation, in terms of R_{ev}, between the matched molecules. In addition, the number of common hydrogen-bonding points is higher for these orientations. The range of values for the steric similarity ξ is small: each of the alignments has a good match in steric terms.

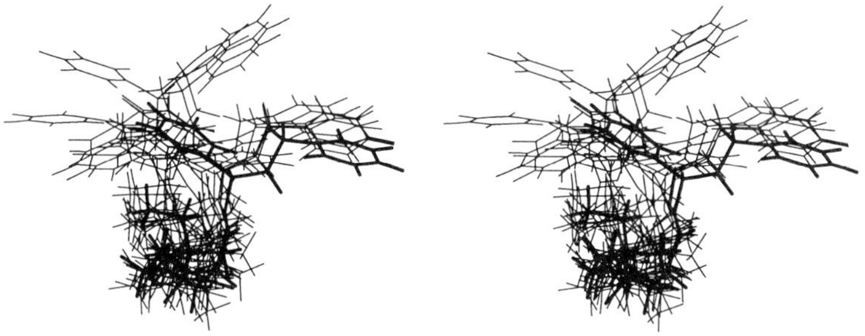

Fig. 3. *Stereoview of the 10 best orientations of* **1** *following alignment onto* **2** *using molecular partial similarity. The equivalent X-ray-crystal alignment is drawn in thick lines for comparison.*

Table 1. *Results of 10 Matches between* **1** *and* **2** *Ranked by r.m.s. Distance from the X-Ray-Crystal Alignment* (R_{xtal})

Run	ξ	N_{reg}	R_{ev} [kJ mol^{-1}]	N_{hbp}	R_{xtal} [Å]
1	0.60	38	21.8	76	0.61
2	0.55	34	24.5	75	1.15
3	0.59	36	28.8	57	1.18
4	0.59	37	28.4	62	1.42
5	0.62	37	28.0	44	1.49
6	0.61	41	29.9	22	3.34
7	0.60	39	34.1	29	3.51
8	0.59	35	34.2	19	3.88
9	0.59	36	44.6	27	7.21
10	0.62	37	46.3	20	7.71
X-ray	0.61	36	20.2	67	0.00

The alignment of **1** that scores best in both electrostatic-potential and hydrogen-bonding similarity is closest to the X-ray-crystal alignment. It is shown superposed on **2**, together with the map of electrostatic-potential difference in *Fig. 4*. The majority of the matched surface exhibits only a small difference in electrostatic potential for the matched points. The corresponding view for common hydrogen-bonding points is shown in *Fig. 5*; the common points appear to cluster around the end of matched molecules that has the largest atomic overlap.

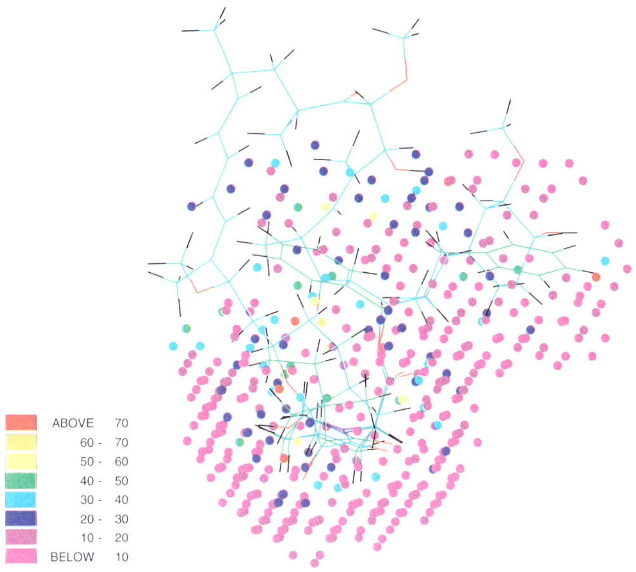

Fig. 4. *Superposition of* **1** *(green carbon atoms) onto* **2** *(cyan carbon atoms) with the partial-similarity matched points colour-coded by electrostatic-potential difference between the molecules (kJ mol*$^{-1}$*).*

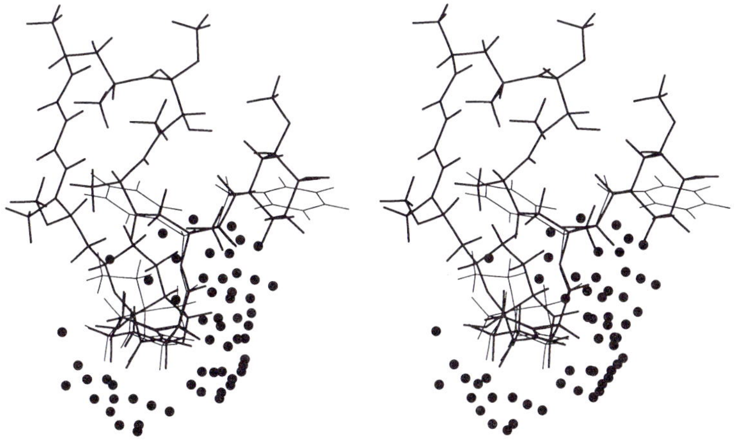

Fig. 5. *Stereoview of the optimum superposition of* **1** *(thin lines) onto* **2** *(thick lines) with common hydrogen-bonding points (all relating to ligand acceptor atoms) shown as dots.*

Thermolysin Inhibitors. The two thermolysin inhibitors included here were matched using PLM, and, like the FKBP ligands, the X-ray-crystal alignment was never obtained.

This pair of molecules was matched 10 times under the same conditions as above, but in addition, 3 local minima were saved during the annealing procedure. Each run required 25 s CPU time. The 40 resulting alignments were then given equal status in the analysis, irrespective of their steric score. The values for ξ ranged from 0.52 to 0.70, although only 1 alignment scored less than 0.6. The values of R_{ev} ranged from 100.9 kJ mol^{-1} to 289.2 kJ mol^{-1}, while N_{hbp} ranged from 12 to 54. The corresponding values for the X-ray-crystal orientation were ξ = 0.63; R_{ev} = 119.0; N_{hbp} = 56. Hydrogen-bonding and electrostatic-potential similarity measures do not agree on the ranking of the matches. However, there are 9 matches in the top 10 (25%) for both property matches. *Fig. 6* shows these comprise two overall orientations of 7 and 2 matches, respectively, with the larger group being close to the X-ray-crystal orientation (r.m.s. deviation 0.89–1.27 Å).

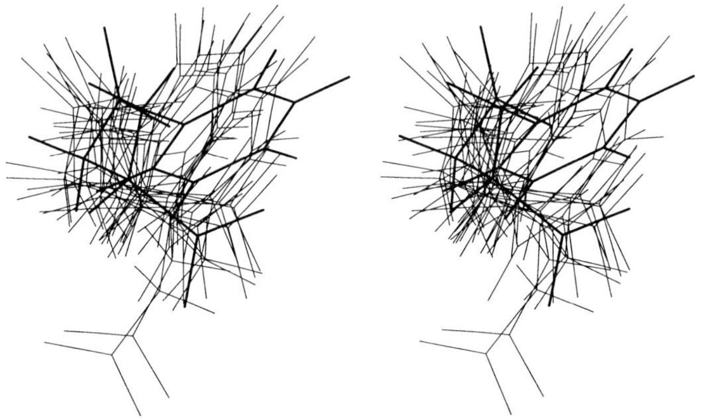

Fig. 6. *Stereoview of 9 orientations of* **3** *that are ranked in the top 10 for both hydrogen-bond and electrostatic-potential similarity following steric partial-similarity matching onto* **4**. *The X-ray-crystal alignment of* **3** *is drawn with thick lines for comparison.*

The highest-ranked electrostatic-potential (ξ = 0.66; R_{ev} = 100.9 kJ mol^{-1}; N_{hbp} = 42; R_{xtal} = 3.7 Å) and hydrogen-bond (ξ = 0.65; R_{ev} = 136.0 kJ mol^{-1}; N_{hbp} = 54; R_{xtal} = 0.97 Å) matches of **3** superposed on **4** are shown in *Fig. 7*, together with their surface properties. These two orientations, and the portions of surfaces that match well, differ considerably between the matches. The measures of surface property similarity are comparable, and also close to those for the X-ray-crystal conformation. Despite the highly charged nature of the ligands, the electrostatic-potential similarity does not seem to be the most important factor here. These ligands provide a case where molecular similarity alone cannot provide a definitive answer to the correct superposition; however, it can suggest two highly plausible superpositions.

Discussion. – We have presented a novel superposition algorithm that attempts to use molecular partial similarity to extract only the most similar parts of two matched molecules. When used in conjunction with evaluation of surface-property similarity, it is able to identify X-ray-crystal alignments where whole-surface approaches previously failed. This suggests that molecular partial similarity is indeed an important problem to solve.

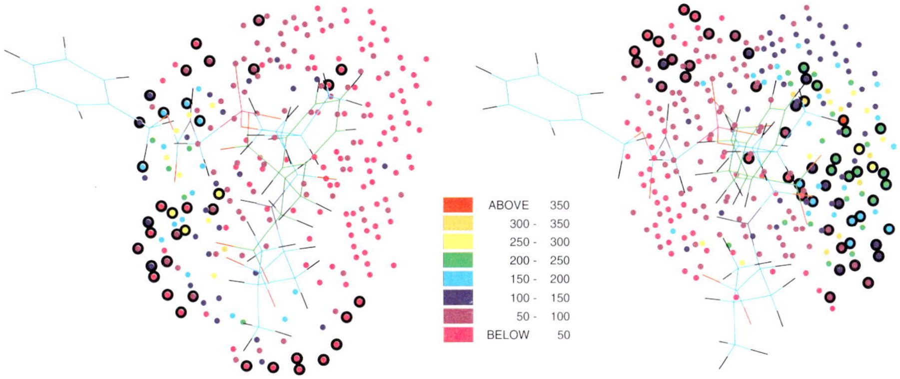

Fig. 7. *Two alternative superpositions of* **3** *(green carbon atoms) onto* **4** *(cyan carbon atoms). The match that scored best in the electrostatic-potential score (left) and the match that scored best according to hydrogen-bond similarity (right) are shown together with their respective common matched surfaces. These are coloured according to electrostatic-potential difference (kJ mol^{-1}) at each point, while common hydrogen-bonding points (ligand acceptor atoms) are coloured black.*

The concept that steric complementarity is a prerequisite to ligand–receptor binding is implicit in the work described here. The alternative approach is to attempt to optimize both steric and surface-property features simultaneously [3, 4]. The major problem with this is that some scaling factor must be defined to relate steric and property scores. It is not possible to correlate empirical scores to free energy differences, so scaling factors tend to be arbitrary and must therefore differ between receptor sites. Since it is relatively easy to collect a number of reasonably good steric matches using simulated annealing (or genetic algorithms) as the optimization scheme, it seems preferable to split the search for optimal alignments in this way. This may be particularly appropriate for partial similarity, where the number of local minima can be far greater than for whole molecule matching. The inclusion of simultaneous property matching may frustrate any optimization scheme.

The conformations of the molecules used in this study have been those of the X-ray-crystal. This was a deliberate choice to focus attention on partial similarity. The extension of molecular partial matching to flexible molecules should be straightforward since this has already been achieved for whole-surface matching in two ways: representative conformers [20] and dynamic conformational matching [21]. These approaches have been discussed elsewhere [22], but dynamic matching is useful if rigid analogues or other conformational information are available for one of the molecules to be matched.

The generation of a number of matched orientations that cannot be distinguished in terms of the match criteria may not as great a problem as it appears on first sight. Since many molecules bind to more than one receptor site, it is possible that a given pair of molecules will adopt alternative binding modes at different sites. It may be possible to obtain selectivity for the desired site by modification of existing compounds or the use of *de novo* design using the information from multiple possible alignments.

One of the drawbacks of the algorithm for molecular partial similarity described here is the number of adjustable parameters required for its operation: notably, the number of regions, region size R, and match quality for points in a region to be included in the final common surface q. While reasonable default values are provided for these parameters, they may have to be modified by the user. The overall effect they have is to control the size of the region of overlap. It is possible to set certain combinations of parameters to unsuitable values. For example, if q is set very high (> 0.6) with a small number of regions (< 10), the algorithm may fail to find any solutions, or may quickly become trapped in a poor local minimum. A certain amount of experimentation may therefore be necessary if the default values do not provide sensible solutions.

In summary, the program SNAP is an effective tool for uncovering molecular partial similarity between apparently dissimilar chemical structures. In combination with measures of surface-property similarity, it can be used to suggest a number of high-quality superpositions to the drug designer, to form the basis of future experiments.

Acknowledgements. – The authors wish to thank Rhône-Poulenc Rorer (TDJP) and the Wellcome trust through the PRF scheme (PMD) for personal financial support. Part of this work was carried out at the Cambridge Centre for Molecular Recognition, funded by the BBSRC.

REFERENCES

[1] Barakat, M.T., Dean, P.M., *J. Comput.-Aided Mol. Design* **1990,** *4,* 295.
[2] Barakat, M.T., Dean, P.M., *J. Comput.-Aided Mol. Design* **1990,** *4,* 317.
[3] Kearsley, S.K., Smith, G.M., *Tetrahedron Comput. Methodol.* **1990,** *3,* 615.
[4] Klebe, G., Mietzner, T., Weber, F.J., *J. Comput.-Aided Mol. Design* **1994,** *8,* 751.
[5] Chau, P.-L., Dean, P.M., *J. Mol. Graph.* **1987,** *5,* 97.
[6] Masek, B.B., Marchant, A., Matthew, J.B., *J. Med. Chem.* **1993,** *36,* 1230.
[7] Blaney, F., Finn, P., Phippen, R., Wyatt, M., *J. Mol. Graph.* **1993,** *11,* 98.
[8] Masek, B.B., Marchant, A., Matthew, J.B., *Proteins* **1993,** *17,* 193.
[9] Perkins, T.D.J., Mills, J.E.J., Dean, P.M., in 'QSAR and Molecular Modelling: Concepts, Computational Tools and Biological Applications', Eds. Sanz, F., Giraldo, J, Manaut, F., Prous, Barcelona, 1995, pp. 374–379.
[10] Perkins, T.D.J., Mills, J.E.J., Dean, P.M., *J. Comput.-Aided Mol. Design* **1995,** *9,* 479.
[11] Barakat, M.T., Dean, P.M., *J. Comput.-Aided Mol. Design* **1991,** *5,* 107.
[12] Dean, P.M., Callow, P., Chau, P.-L., *J. Mol. Graph.* **1988,** *6,* 28.
[13] Borea, P.A., Dean, P.M., Martin, I.L., Perkins, T.D.J., *Mol. Neuropharmacol.* **1992,** *2,* 261.
[14] van Laarhoven, P.J.M., Aarts, E.H.L., Simulated Annealing: Theory and Applications, Reidel, Dordrecht, 1987.
[15] Mills, J.E.J., Dean, P.M., *J. Comput.-Aided Mol. Design* **1996,** in press.
[16] Allen, F.H., Davies, J.E., Galloy, J.J., Johnson, O., Kennard, O., Macrae, C.F., Mitchell, E.M., Mitchell, G.F., Smith, J.M., Watson, D.G., *J. Chem. Inf. Comp. Sci.* **1991,** *31,* 187.
[17] Perkins, T.D.J., Dean, P.M., *J. Comput.-Aided Mol. Design* **1993,** *7,* 155.
[18] Bernstein, F.C., Koetzle, T.F., Williams, G.J.B., Meyer, E.F., Brice, M.D., Rodgers, J.R., Kennard, O., Shimanouchi, T., Tasmui, M., *J. Mol. Biol.* **1977,** *112,* 535.
[19] Ihlenfeldt, W.D., Takahashi, Y., Abe, H., Sasaki, S., *J. Mol. Biol.* **1994,** *34,* 109.
[20] Perkins, T.D.J., Dean, P.M., *J. Comput.-Aided Mol. Design* **1993,** *7,* 155. .
[21] Dean, P.M., Barakat, M.T., Garland, S.L., Mills, J.E.J, Perkins, T.D.J., Todorov, N.P., in 'Bioactive Compound Design: Possibilities for Industrial Use', Eds. Ford, M.G., Greenwood, R., Brooks, G.T., Franke, R., Bios Scientific Publishers, Oxford, 1996, pp. 57–66.
[22] Perkins, T.D.J., in 'Molecular Similarity in Drug Design', Ed. Dean, P.M., Blackie, London, 1995, pp. pp. 89–109.

PART IV. 3D-QSAR AND MOLECULAR MODELLING

Chapter 27

Multiple Field Alignment of Flexible Molecules

by **Michael Lipkin**[a]), **David Salt**[a]), **Watcyn Wynn**[a]) and **Martyn Ford**[b])

[a]) School of Mathematical Studies, University of Portsmouth, Portsmouth PO1 2EG, UK
[b]) Centre for Molecular Design, University of Portsmouth, Portsmouth PO1 2QF, UK

The new algorithm differs from previous ones by automatically adjusting conformation and matching shape and field simultaneously.

Non-Linear Optimisation. - Moving the conformers relative to each other to obtain the optimal match often involves non-linear optimisation. An alignment function is defined such that minimising this function moves the conformers to optimise the match. There is no single best method for non-linear optimisation; it depends on the character of the function to be optimised, as illustrated below.

Type	Shape	Character	Optimisation procedure
A		Smooth and differentiable	Conjugate gradient, BFGS etc.
B		Like a Fractal. Both large and small scale structure	Simulated Annealing, Genetic Algorithms.
C		Depth of minima unrelated to area occupied in space.	Exhaustive search necessary, but the time taken can be improved by ensuring that effort spent is proportional to depth of minima explored.

The algorithm is based on the assumption that the alignment function is type C. As more fields are included, the shape of the alignment function changes - see type C1 and C2 below.

Type	Shape	Description
C1		Small number of fields simultaneously matched. The alignment function has many deep minima
C2	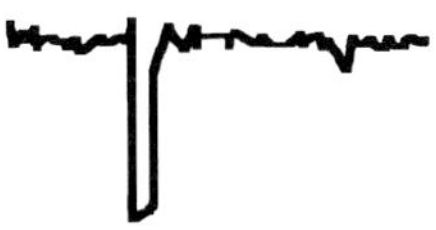	Many fields simultaneously matched. The alignment function has a small number of deep minima corresponding to matches of the molecules.

The current blind search algorithm has an optimisation time that increases as a power of the size of the match. By matching several fields simultaneously the size of unproductive matches should minimised providing a faster algorithm.

Description of the Algorithm. - The atom centres of each molecule are represented by points in Cartesian space. The state of the algorithm at any particular stage is represented by the atom centres plus a set of points which define the region of match. These points re all compounded into one large vector. If there are N molecules each with n_i atoms and M match points the dimension of this vector is

$$3M + 3\sum_{i=1}^{N} n_i$$

An error function is defined on this parameter space so that when this function is minimised a number of things happen simultaneously:

- the molecules stay within reasonable conformational bounds
- the match points tend to stay on the VDW surface of the superimposed molecules
- the field error at the match points is minimised
- the match points maintain a pre-set distance from each other.

The algorithm consists of an iterative process whereby the error or alignment function is first minimised to a value below some pre-set tolerance. This procedure saves computational time since the disproportionate length of time spent finding the minimum is avoided. A new point is then added and the minimisation repeated until the value of the error function cannot be reduced below the pre-set tolerance.

Improving Efficiency. - For minimisation at each stage, the algorithm uses the conjugate gradient method. This procedure is very efficient provided the function being minimised approximates to a quadratic bowl at the minimum.

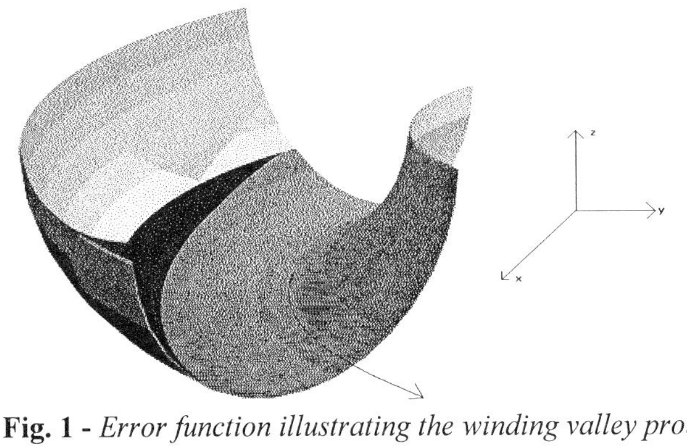

Fig. 1 - *Error function illustrating the winding valley problem*

A particular problem for these types of line search algorithms is the "winding valley problem". In Figure 1, the parameter space is represented by the x-y plane and the value of the error function is represented by the height on the z-axis. The function illustrated has the shape of a curved gutter with the direction of the downhill slope indicated by the arrow. In this situation, conjugate gradients or any straight line search algorithm will be very slow, proceeding in small steps down the valley. In a parameter space with dimensionality > 2, the problem is compounded. This situation occurs when we represent a flexible molecule using Cartesian co-ordinates. The bond lengths are assumed to a first approximation to remain constant. Atoms will therefore follow curved paths as the molecule flexes. To overcome the winding valley problem, the following co-ordinate transformation is used. First a spanning tree is drawn on the molecule. An atom is then chosen as the root atom, and its position represented in Cartesian co-ordinates. Atoms one branch away from the root are represented in polar co-ordinates centred on the root atom. Subsequent atoms in the tree are then represented by polar co-ordinates centred on their parent atoms. The axis of the polar co-ordinates is aligned with the bond.

Match points are also included in the transformation by finding the atom whose VDW surface they lie on, and incorporating in the tree a link from this atom to the point. In this new co-ordinate system the molecular conformation error approximates a quadratic and hence its minimisation is efficient. Another way of improving efficiency is by judicious placement of new matching points in the system. The values of the field error around the boundary of the matching region can be probed to enable a new point to be placed so that its contribution to the field error is minimised. In many cases this results in no minimisation being required at all on the next iteration, the total alignment function error being already less than the pre-set tolerance.

Application of the Algorithm. - The algorithm has been applied to two molecular alignment problems, the first involving superimposition of four water molecules each with a different initial orientation. The following diagram illustrates four stages in the process of alignment which results in superimposition with identical orientation. In these pictures (Figs 2 and 3), atoms are represented by spheres and bonds by cylinders. Lines are drawn between the match points to form a triangulation which gives a three-dimensional effect to the molecules. The top row shows the starting positions for four water molecules which are to be matched simultaneously. Note that the molecules occur in different orientations. The second row shows the situation after a few match points have been added. As the match grows (row three), the molecules orientate themselves to maximise the alignment. The increasing degree of match is indicated by the extent of triangulation. Row four shows the final superimposition for which the entire molecular surfaces of all four molecules have been matched.

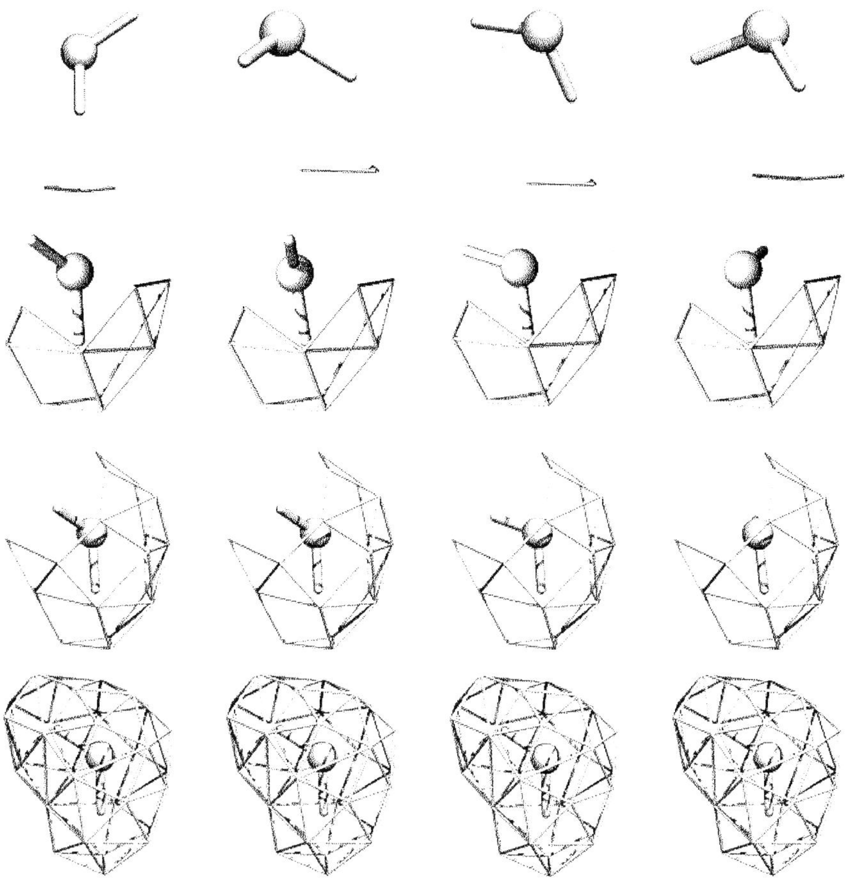

Fig 2. *Four water molecules matched simultaneously*

Although different starting points were adopted for each water molecule, the same final orientation was achieved. The degree of matching, in this case 100%, can be estimated from number of completed (matched) triangles as a proportion of the total number of possible triangles. This example provides evidence that the algorithm produces a perfect alignment for small, identical objects with imperfect orientations.

The second application was investigated to determine whether perfect alignments were possible for large molecules containing approximately 20 or more skeletal atoms. (Many biologically active molecules are of similar size). The example chosen is based on two leukotriene antagonists, LY171883 and LY137617 used by Herman and Herron to evaluate the overlap programs, OVID and SUPER [4]. Figure 3 presents final alignments

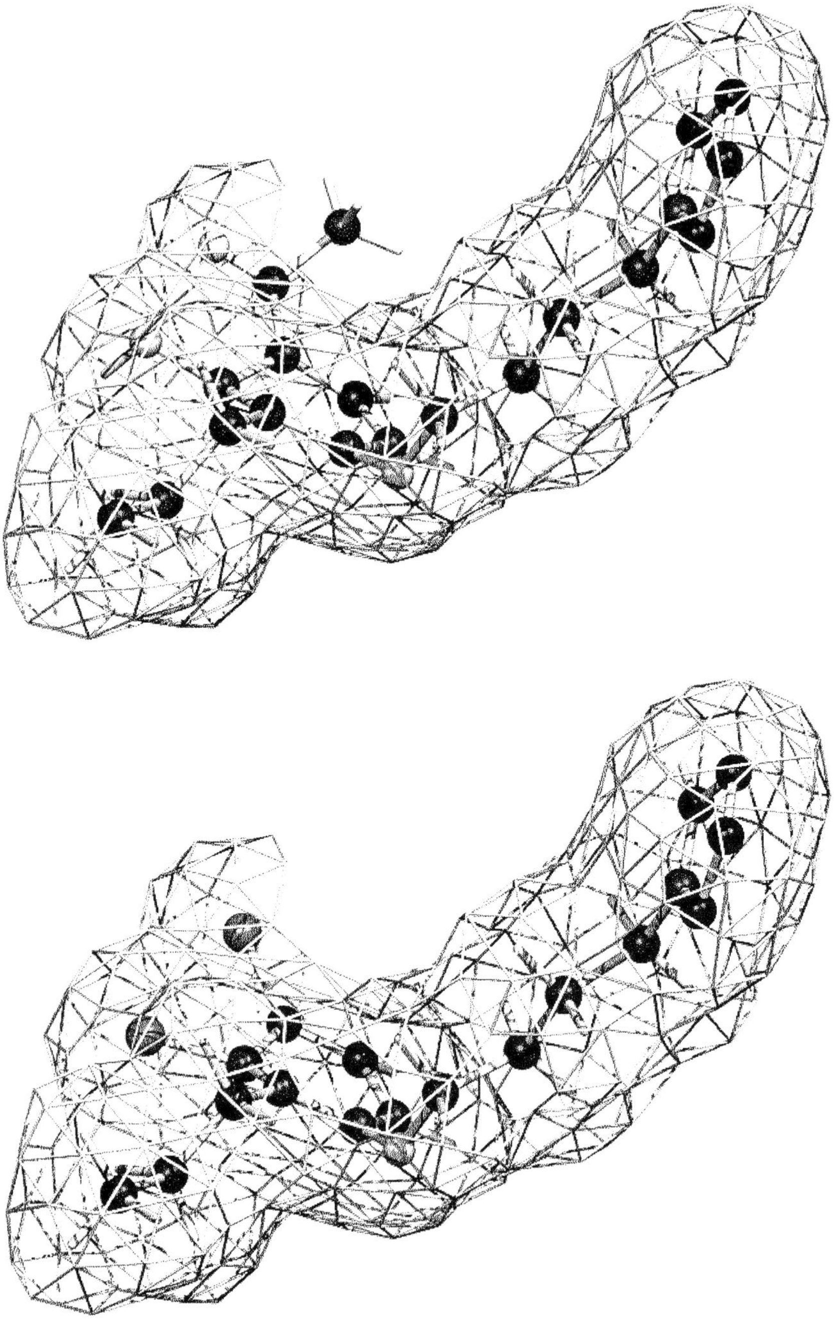

Fig 3 - *Comparison of LY171883 and LY137617*

for these molecules aligned by the algorithm from randomly chosen starting positions. Again the alignment is close, but LY171883 has a small region (without triangulation) which cannot be matched by a corresponding region on LY137617. This feature illustrates the utility of triangulation as a graphic device to identify regions of match or contrast. Further studies are required to determine whether the same match is obtained if the algorithm commences from different, randomly selected conformations.

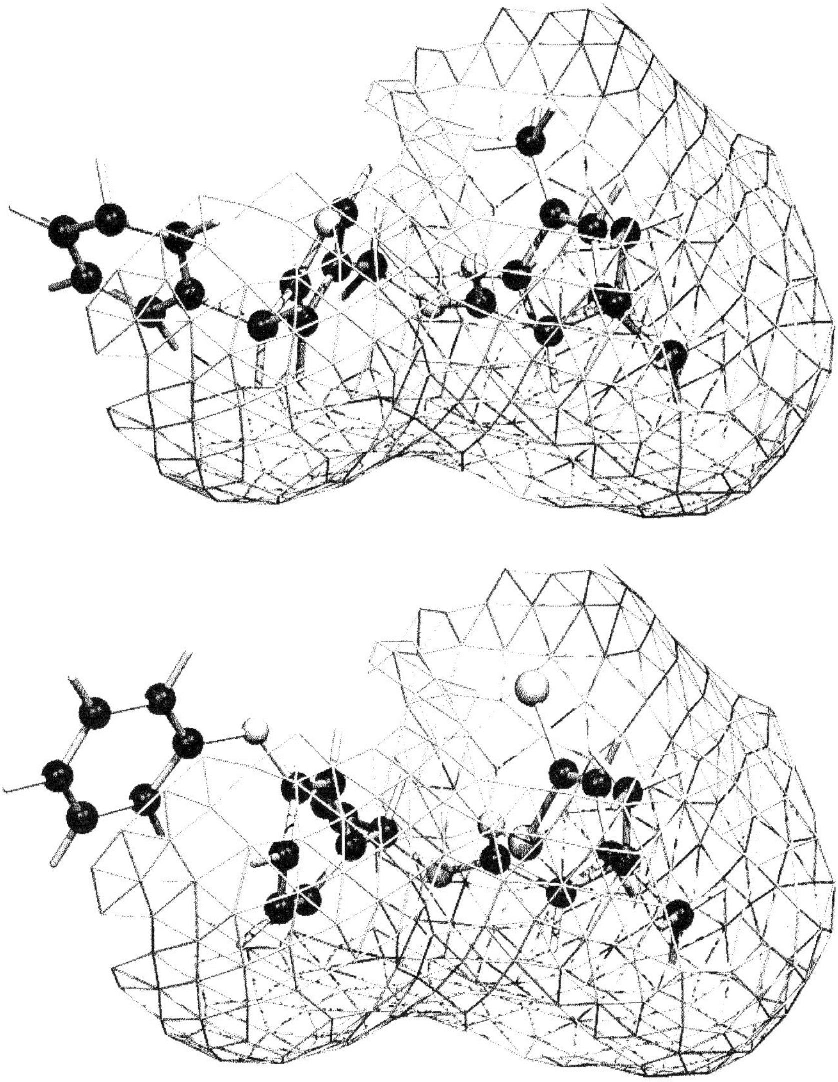

Fig 4 - *Comparison of cismethrin and deltamethrin*

When the method was applied to large, flexible molecules, alignment was often prevented by the ocurrence of atoms on opposite sides of a match point. Such a superimposition represents an invalid match. A procedure has been developed to overcome this difficulty, based on use of a function whose minimum occurs whenever atoms from different molecules are on the same side of the match point. The derivative of the function is identified and used to obtain the minimum function value using a standard procedure such as Congruent Gradients Minimisation. This process ensures that only valid matched points are considered. Figure 4 shows the result of applying the alignment procedure to match two flexible insecticides, deltamethrin and cismethrin. These esters have acid moieties of similar, but alcohol moieties of dissimilar, chemical structure.

Although the current implementation matches only the VDW fields, the algorithm has been designed so that incorporation of additional field properties, *e.g.* electrostatics, interaction energies and hydrogen bonding probabilities, should be relatively straightforward to implement. This work is already in hand.

Acknowledgements. - This investigation has been funded by the University of Portsmouth in collaboration with Pfizer Central Research, Sandwich, Kent. The authors are particularly grateful to Dr Paul Finn (Pfizer Limited) for valuable support, advice and discussion.

REFERENCES

[1] S. K. Kearsley, G. M. Smith. *Tetrahedron Comp Method.* **1990**, *3,* 615-633.

[2] B. B. Masek, A. Merchant, J. B. Matthew. *PROTEINS: Struct, Funct Genet.* **1993**, *17,* 193-202.

[3] P. M. Dean. in *'Concepts and Applications of Molecular Similarity',* Eds. M. A. Johnson, G. M. Maggiora, Wiley, 1990, 211-238.

[4] R. B. Hermann, D. K. Herron, *J. Comput. Aided Mol Des.* **1991**, *5,* 511-524.

[5] *Molecular Similarity in Drug Design,* Ed P. M. Dean, Chapman & Hall, 1995.

PART IV. **3D-QSAR AND MOLECULAR MODELLING**

Chapter 28

A Stereoselective Pharmacophoric Model of the Serotonin Re-uptake Site

by **Klaus Gundertofte**[a]), **Klaus P. Bøgesø**[a]) and **Tommy Liljefors**[b])

[a]) Research and Development, H. Lundbeck A/S, Ottiliavej 9, DK-2500 Valby-Copenhagen, Denmark
[b]) Department of Medicinal Chemistry, Royal Danish School of Pharmacy, Universitetsparken 2, DK-2100 Copenhagen, Denmark

A Stereoselective Pharmacophoric Model of the Serotonin Re-uptake Site

by **Klaus Gundertofte, Klaus P. Bøgesø**

Research and Development, H. Lundbeck A/S, Ottiliavej 9, DK-2500 Valby Copenhagen, Denmark

and **Tommy Liljefors**

Department of Medicinal Chemistry, Royal Danish School of Pharmacy, Universitetsparken 2, DK-2100 Copenhagen, Denmark

Exhaustive conformational analyses on four selective serotonin reuptake inhibitors resulted in a pharmacophoric model explaining observed differences in enantioselectivities. A number of test compounds from a diverse set of chemical structures was included in the evaluation of the model. Furthermore, selectivity towards noradrenaline reuptake is explained.

Introduction. - Many of the successful antidepressant drugs with a clinically documented effect on depression, including citalopram, fluoxetine, paroxetine and sertraline (*Fig.1*), exert their effect through inhibition of the serotonin reuptake mechanism.

Regulation of the neurotransmitter transporters is important in determining the concentration of neurotransmitters in the synaptic cleft. Hence, inhibition of serotonin reuptake causes the neurotransmission in serotonergic synapses to be increased and this effect is believed to be causal in the treatment of depression and related diseases with selective serotonin reuptake inhibitors (SSRIs).

Despite the structural diversity of the compounds in *Fig. 1*, they are all relatively specific inhibitors of the serotonin transport and do not interact significantly with other receptors or transport mechanisms.

The cloning of several transporters from various species has led to increasing knowledge regarding the structure and function of the transporter molecules. In spite of functional complexity it is likely that all transporter molecules conform to a single organizational plan. However, information concerning the structure of these molecules is not available. Certain members of the subfamily of Na^+- and Cl^--dependent transporters, especially transporters for the biogenic amines NA and 5-HT, have a potential as targets for antidepressant drugs. These transporter molecules are functional at the plasma membrane as opposed to the H^+-dependent transporters functional in intracellular vesicles [1]. Like other receptor and transporter molecules these are peptides of a varying number of amino acids, generally 600-650 amino acids for transporter molecules. Based on hydrophobicity studies, showing that they all apparently contain 12 stretches with 20-24 hydrophobic residues, it is believed that they consist of 12 membrane spanning helices with both terminals located intracellularly. All loops in the Na^+- and Cl^--dependent transporter molecules apart from the second extracellular are very short.

The transmembrane domains are generally highly conserved within the Na^+- and Cl^--dependent transporters. Outstanding is the first putative helix with homologies between the biogenic amine transporters as high as 85-90%. The highest homology among the cloned human transporters is found between the dopamine transporter and the noradrenaline transporter.

Site-directed mutagenesis studies of the dopamine transporter strongly suggest a highly conserved aspartic acid residue in the first hydrophobic domain to be involved in the binding of the amine group of the neurotransmitters [2].

Since only indirect structural information about the topology of the transporter molecules is available, a number of studies have been performed in order to generate pharmacophoric models applicable in the attempt to design / develop new ligands. Also in the area of the HT transporter a few studies have been published [3-7]. The present work is going to discuss these models and present a new pharmacophoric model based on *S*-citalopram (**1**), *S*-fluoxetine (**2**), (3*S*, 4*R*)-paroxetine (**3**) and (1*S*, 4*S*)-sertraline (**4**).

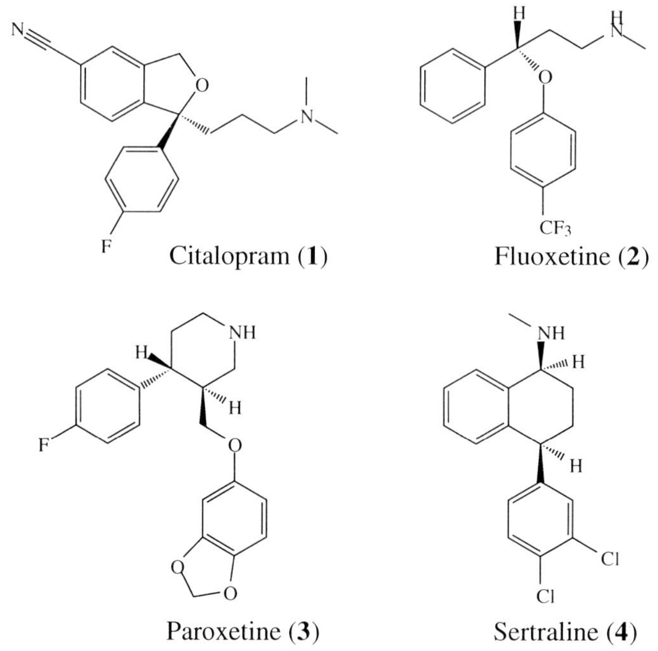

Fig. 1. *Chemical structures of four selective serotonin reuptake inhibitors*

Computational Methodology. - A frequently used approach in pharmacophoric modelling is to identify common motifs contributory to the interaction with the receptor or transporter molecule. In the structures chosen for this modelling study it seemed reasonable to focus on the two aromatic rings and the nitrogen atom. Due to the short distance between the fused aromatic ring in sertraline (**4**) and the nitrogen it is, however, obvious that superimposition of these three pharmacophoric elements could not be achieved.

As we assume that the amino groups of all the compounds studied in this work interact with the same amino acid residue of the transporter, we have instead of the nitrogen atom used a "site point" as a pharmacophore element. This site point is located 2.8 Å from the nitrogen atom in the nitrogen - lone pair direction. The site point represents an amino acid residue with which the amino groups of the molecules, included in the present study, are interacting through a hydrogen bond. Inspection of intermolecular hydrogen bond interactions in structural databases, such as the Cambridge Structural Database and the Brookhaven Protein Data Bank, reveals that although hydrogen bond donor-acceptor distances are fairly constant for each type of hydrogen bond interaction, the relative angular orientation of the donor-acceptor moieties may vary significantly. For ligand-receptor interactions this implies that, for instance, amino groups of different ligands may interact essentially equally well with an amino acid residue, even though the nitrogen atoms of the ligand amino groups are located in different spatial positions relative to the interacting residue. In the development of pharmacophore models this variation of the direction of interaction can not be taken into account by the use of a classical ligand point, that is, by using the nitrogen atom as a pharmacophore element. However, a site point, located at a typical hydrogen bonding donor-acceptor distance, has shown to be very useful in this respect and provides a more realistic model for the elucidation of different possible interaction modes in ligand-receptor interactions [8-10].

Conformational Analyses. - Exhaustive conformational analyses were applied in order to fully explore the conformational space of these rather flexible molecules. Besides rotational degrees of freedom, various conformations in the individual structures (e.g. pseudo-axial and pseudo-equatorial substituents) were examined. Application of the molecular mechanics method in conformational analysis of chiral amines poses an extra complication, since asymmetric secondary amines have two distinct invertomers with different conformational profiles. The difference can be quite substantial. Therefore, the two invertomers were in those cases treated separately.

Conformational energies and optimized geometries were calculated by the molecular mechanics program MM2(91) [11]. All conformational analyses were done on the unprotonated amines with the unshared electron pair represented by a pseudo-atom.

It is not known whether the amino groups in the compounds, studied in this work, are protonated or not, when they are interacting with the transporter. However, as described below, we are using *in vacuo* force-field calculations to calculate the conformational energy difference between a candidate bioactive conformation and the global energy minimum. In order to be able to relate conformational energy differences to observed affinities, the global energy minimum conformation should be that for the molecule in water solution. If a protonated amino group is employed in gas phase calculations, and there is a possibility of strong hydrogen bond interactions between the ammonium group and an electronegative atom, as is the case for citalopram (**1**) and fluoxetine (**2**) in the present study, such calculations may be dominated by the strong hydrogen bonding (electrostatic "collapse") [6]. In water solution, solvation of the ammonium group and the electronegative hydrogen bond acceptor should efficiently compete with the intramolecular hydrogen bond. Thus, if gas phase calculations are employed to simulate conformational properties in water solution of compounds containing ammonium groups, the electrostatic "collapse" - generally observed in gas phase calculations - may be avoided by employing the amine as a model for the ammonium group. This should give a more faithful simulation of the conformational properties of the molecule in solution. The same arguments are valid for candidates for the bioactive conformation with the difference that in this case the strongly polar groups are "solvated" by interactions with amino acid residues or the protein backbone. This makes electrostatically "collapsed" conformations less likely.

Model Building of the Serotonin Reuptake Site. - The four rather different and flexible molecules represent a widespread conformational space. We therefore applied certain criteria in the selection process of candidates for the hypothetically biological active conformations. Distances between pharmacophoric elements and other topological parameters, such as dihedrals and planes, were used. The most important criteria was, however, conformational energies. Since we are dealing with highly active compounds (*Table 1*), the conformational energies of the bioactive conformations should be low.

This follows from a thermodynamic consideration of the equilibrium between the transporter-ligand complex and the separated ligand and transporter molecules. From the relationship between the free energy and the equilibrium constant

$$\Delta G = -RTlnK = -1.4logK \quad (T=310\ K)$$

it follows that a conformational free energy increase of 1.4 kcal/mol reduces the affinity of the ligand to the transporter by a factor of 10. For each additional 1.4 kcal/mol of conformational free energy, the affinity decreases by a further factor of 10.

If two similar compounds with the same pharmacophore elements and equally strong interactions with the receptor sites are compared, the difference in conformational energies of their bioactive conformations should largely determine their relative affinities. (This also requires that the two compounds have closely similar free energies of hydration). In analogy with the discussion above, a conformational free energy difference of 1.4 kcal/mol for the bioactive conformations of the two molecules implies a relative affinity ratio of 10 for the two compounds due to the "conformational effect". The compound with the highest conformational energy, of course, has the lower affinity. The validity of this argument has been unambiguously demonstrated [12].

From the discussion above, one may draw the conclusion that structurally similar compounds, all displaying high affinities to the transporter, should have bioactive conformations with low and similar conformational energies.

The calculated conformational energy is, of course, dependent on the quality of the force field used. A recent comparison of several force fields [13] proved the MM2(91) force field to be amongst the very best. More than 10 distinct conformations with conformational energies of less than 1.5 kcal/mol were found for each of the four compounds. Therefore, a low energy cut-off could initially be applied in order to reduce the large number of candidates for the bioactive conformation.

In the process of model building, three points, each representing the two centres of the aromatic rings and a point 2.8 Å from the basic nitrogen in the direction of the lone pair, have been used. The latter point is used to simulate an interaction with a hypothetical acidic residue, presumed to be the primary recognition site in the transporter molecule. As discussed above, this is a reasonable hypothesis and also a necessary assumption, if the concept of superimposable aromatic rings is correct.

From the analyses of the conformational data on citalopram (**1**) and sertraline (**4**) it is obvious that the distance between the centers of the two aromatic rings in the bioactive conformations of fluoxetine (**2**) and paroxetine (**3**) have to be confined to a small interval.

This imposed distance interval was then used in the process of selecting candidate conformations from the conformational analyses of fluoxetine and paroxetine. In the stepwise selection other criteria were applied, e.g. distances to the site point, in order to further reduce the number of possible candidates. Eventually, least squares superimpositions of the various selected conformations were carried out by means of the molecular modelling program MacMimic [14]. The superimpositions were evaluated on the basis of their RMS values and on visual inspection.

Results and Discussion. - A thorough investigation of the huge amount of conformational data resulted in a very limited number of candidates fulfilling the conditions described above. In the initial phase of the model building process the focus, therefore, was concentrated on the energy criteria and the visual impression of the fits.

In the selected conformations, citalopram (**1**), paroxetine (**3**) and sertraline (**4**) all had conformational energies of less than 0.2 kcal/mol. The applied conformation of *S*-

fluoxetine (**2**) had a somewhat higher energy of 1.2 kcal/mol. The resulting superimposition of citalopram, fluoxetine and paroxetine is shown in *Fig. 2*.

Table 1. *Effect of Antidepressants on the 5-HT Reuptake In Vitro (values in nM) and Selectivity Data [15]*

	CITALOPRAM	FLUOXETINE	PAROXETINE	SERTRALINE
IC_{50}	1.8	6.8	0.29	0.19
5-HT SELECTIVITY	3400	54	280	840
EUDISMIC RATIO	167	1.6[a)]	130[b)]	7.9[c)]

a) See [16]. b) See [17]. c) See [18].

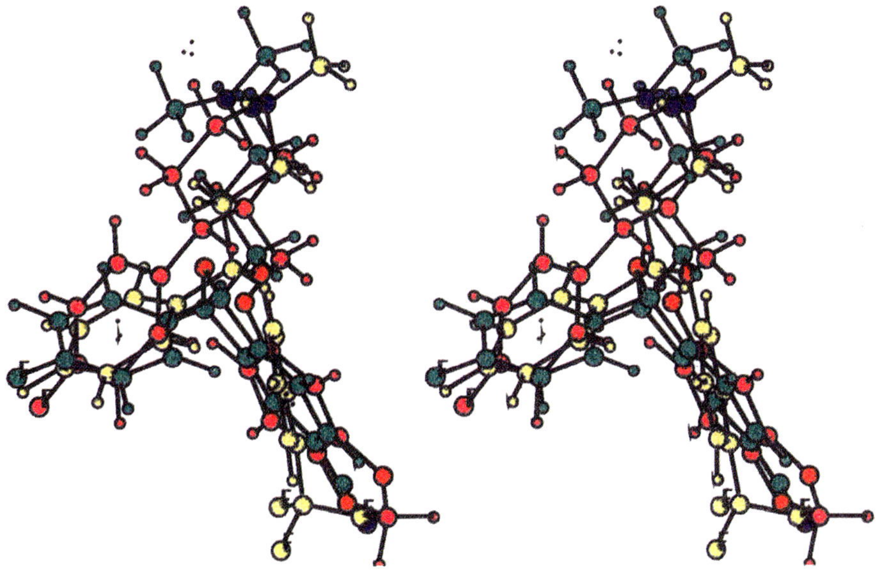

Fig. 2. *Least squares superimpositions of citalopram (**1**, green), fluoxetine (**2**, yellow) and paroxetine (**3**, magenta). Nitrogen atoms are marked in blue, oxygen atoms in red and pseudoatoms in cyan.*

Included in the superimpositions were the two centres of the aromatic rings and the putative site point. RMS values for the superimpositions were both less than 0.25 Å. In the case of these three compounds a superimposition including the nitrogen atom is certainly possible, and this four-point fit is almost as good as the three-point fit with RMS values less than 0.37 Å.

As described above, in sertraline (**4**), the distances from the basic nitrogen to the centres of the aromatic rings deviate from the corresponding distances in the other model compounds. The distances to the site point were, however, comparable. A three-point superimposition of sertraline and the model with an RMS value of 0.31 Å is shown in *Fig. 3*. It is evident that in this case the inclusion of the nitrogen atom would seriously deteriorate the result, and no conformation exists that simultaneously fulfils all criteria.

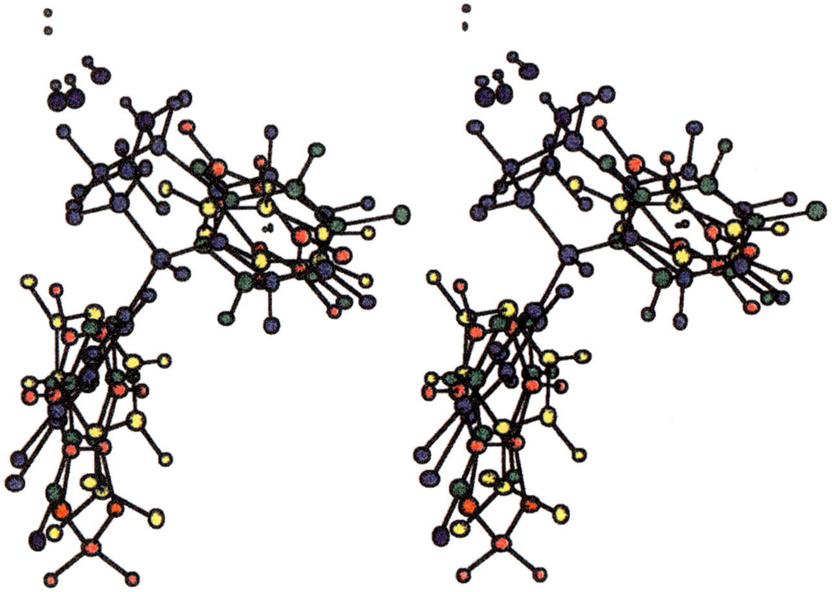

Fig. 3. *Superimposition of sertraline (**4**) on the model.*
*Only the pharmacophoric elements of **1-3** are shown. (Colour coding as in Fig. 2).*

Model Description. - As described above, four model compounds were used in the development of the 3D-pharmacophore. Geometric parameters are listed in *Table 2*. As can be seen, the distance between the two aromatic rings, R1 and R2, in all model compounds is confined to a narrow range from 4.6 Å to 5.2 Å. The distance between R1 ring and site point Sp is in the range of 7.4 Å and 8.3 Å, and the distance from the other aromatic ring R2 to Sp is between 6.2 Å to 6.9 Å.

On the other hand, variation in distances from the center of the aromatic rings to the nitrogen atom is pronounced. This is also illustrated by the variation in the pseudo-dihedral angle R1R2NSp, describing the angle between a plane, defined by the mid-points of each aromatic ring and the basic nitrogen atom, and the site point. As mentioned above, amino groups of various ligands in different spatial positions may interact essentially equally well with an acidic amino acid.

The two pharmacophoric aromatic rings R1 and R2 in the model are perpendicular to each other (*Fig.4*). In the superimpositions, no attempt has been made to force coplanarity between individual R1 and R2 rings, respectively. The R1 rings of the compounds in *Fig. 2* are more or less coplanar. Coplanarity between the R1 ring of sertraline (**4**) and the other R1 rings can be obtained by rotating the dichloro-substituted phenyl ring about 30° at the expense of 0.9 kcal/mol. In the case of the R2 ring, some variation in the relative position of the individual rings is allowed.

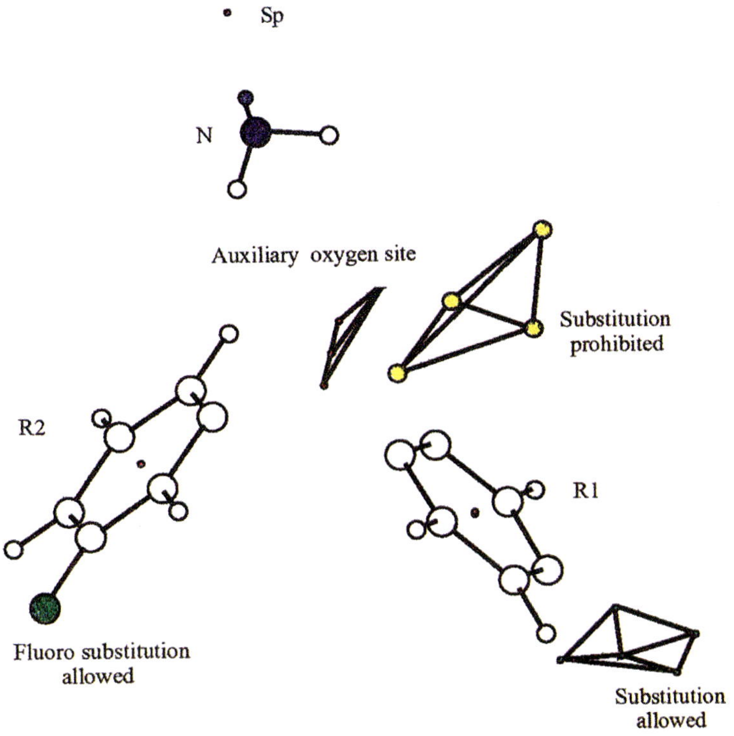

Fig. 4. *The model*

Substituents in the primary ring (R1) of the four compounds (cyano, trifluoromethyl, chloro and dioxolo substituents) occupy the same region in space (*Fig. 4*). Furthermore, fluoro substituents in the secondary ring (R2), as present in **1** and **3**, point to the same region in space.

Comparison with Other Models. - A number of models of the 5-HT uptake site have been described in the literature [3-7].

A common feature of these models is that in all cases only one aromatic moiety is included in the pharmacophore building process. This aromatic moiety is placed at a distance from the basic nitrogen, comparable to the distance found in serotonin itself. The secondary phenyl ring in most of the described models shows no spatial correlation. In the work of Rupp *et al.* [6] there seems to be some correlation, however, but still there is a lot of variation in the actual position of the secondary ring. It is believed that the secondary ring - if present - as well as the primary, takes part in hydrophobic interactions with the transporter molecule; they should therefore be included as an additional pharmacophoric element. Additional support to this theory is the finding that the distances between the two aromatic rings in molecules with conformationally fixed distances, e.g. in citalopram (**1**) and sertraline (**4**) are essentially the same.

Evaluation of the Model. - *Enantioselectivity*. - As can be seen in *Table 1*, the enantiomers of fluoxetine are almost equipotent as 5-HT uptake inhibitors, although somewhat weaker than the other compounds. As a consequence of the equipotency of the two enantiomers, a low-energy conformation of the *R*-enantiomer of fluoxetine should exist, which fit the model. Indeed, a conformation of *R*-fluoxetine with an energy of 1.6 kcal/mol was found. This could, in spite of the fact that enantiomers are, by definition, non-superimposable, be superimposed on the postulated biologically active conformation of *S*-fluoxetine (conformational energy 1.2 kcal/ mol) with an RMS value of 0.26 Å (*Fig. 5*). The lower affinity of this enantiomer is correlated to the somewhat higher conformational energy of the selected conformation.

The distomer of citalopram, i.e. the *R*-enantiomer, has an affinity 167 times lower than its eutomer. It is, therefore, anticipated that a conformation of *R*-citalopram should fit the model. On the other hand, none of the lowest energy conformations should fit. A thorough investigation of a large number of reasonable conformations revealed a single candidate fitting the model. This had a conformational energy of 2.9 kcal/mol, as compared to 0.1 kcal/mol for *S*-citalopram, explaining a difference in affinity between the two ligands with same physico-chemical properties by a factor of about 100 (*Fig. 6*).

As opposed to the fluoxetine case, in citalopram the pharmacophoric rings R1 and R2 are interchanged in the two enantiomers.

Table 2 *Summary of model data*

MODEL COMPOUNDS	R1Sp	R2Sp	R1R2	R1R2NSp	R1N	R2N
CITALOPRAM	8.22	6.91	4.85	84	7.67	6.00
S-FLUOXETINE	7.78	6.56	4.6	102	7.05	6.15
PAROXETINE	8.35	6.24	5.17	103	7.54	5.70
SERTRALINE	7.39	6.38	4.85	85	5.70	3.74
TEST COMPOUNDS						
R-FLUOXETINE	7.48	5.93	4.59	142	5.81	5.76
ZIMELDINE	8.68	6.22	4.89	162	6.01	4.68
DHMMTPI	8.51	6.47	4.73	122	7.03	5.72
McN-5652	6.85	6.09	4.89	161	4.30	3.82
Ru25591	9.58	7.16	4.86	127	7.51	5.60
INDATRALINE	6.67	6.05	4.90	63	5.78	3.78

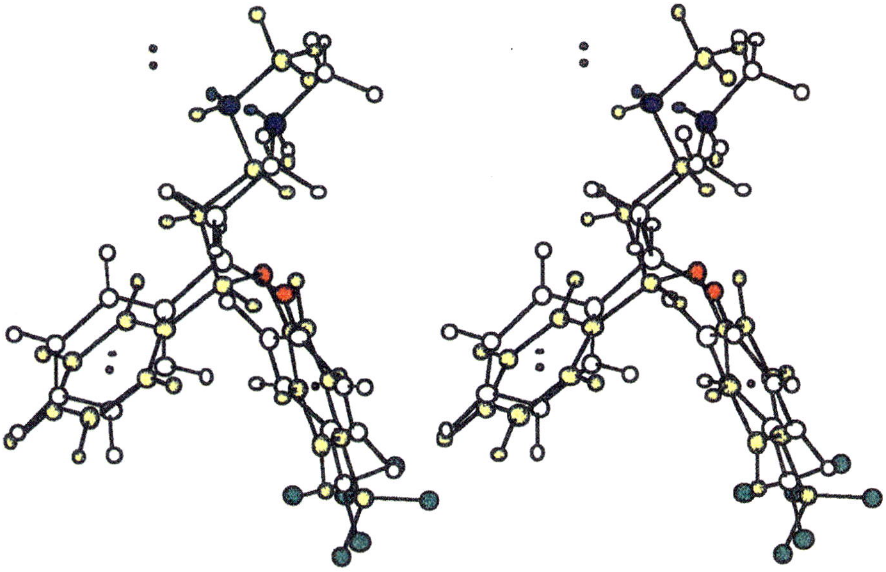

Fig. 5. *Superimposition of S-fluoxetine (yellow) and R-fluoxetine (open circles). Nitrogen atoms are marked in blue, oxygens in red, fluoro in green and pseudo atoms in cyan.*

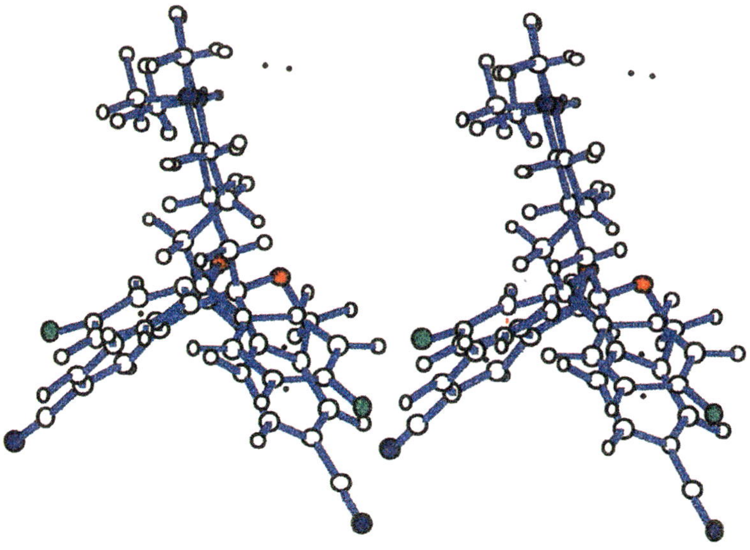

Fig. 6. *Superimposition of S-citalopram and R-citalopram. Heteroatoms are colour coded as in Fig. 5.*

Other Test Compounds.- For further evaluation purposes, a number of other 5-HT reuptake inhibitors, representing a variety of structural classes, were tested in the model (*Fig. 7*). Zimeldine (**6**) is a potent and selective serotonin reuptake inhibitor [19]. It has been registered as an antidepressant drug, but is now withdrawn from the market. The indan derivative **7** can be viewed as a ring-closed analogue of fluoxetine (**2**) and the absolute configuration of the eutomer has been published [20]. McN-5652 (**8**) is an uptake inhibitor on serotonin as well as dopamine and noradrenaline systems [21] and compound **9** is reported to be a selective, long-acting serotonin reuptake inhibitor [22]. Indatraline (**10**) is a potent inhibitor on all three biogenic amine reuptake systems [23] and is like sertraline (**4**) and **8** representatives of compounds having a short distance between the aromatic moiety R2 and the nitrogen atom, as well as a fixed distance between the two aromatic rings. Indatraline fits, as expected, well to sertraline (*Fig. 8*). Interestingly, it fits in the same conformation that was earlier proposed [23] as the biological active conformation.

All of the applied conformations are low energy conformations. The highest conformational energy is found in **8** with an energy of 2.3 kcal/mol above the global energy minimum. *Table 2* lists detailed information and fits of **7** and **9** are shown in *Fig. 9*.

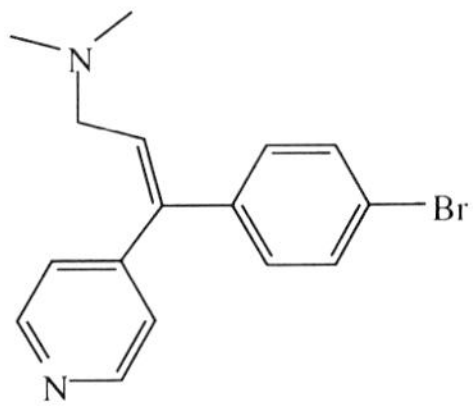

Zimeldine (**6**)

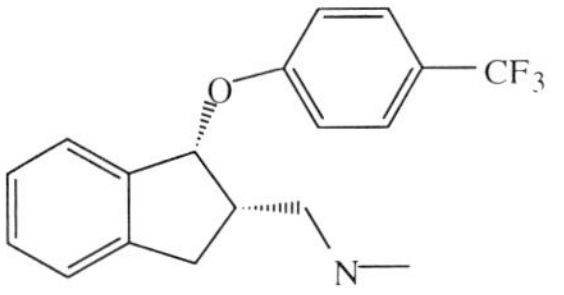

DHMMTPI (**7**)

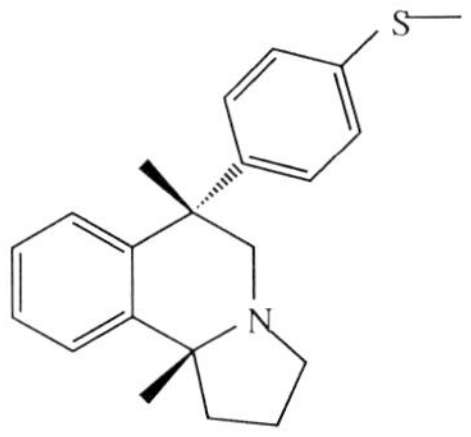

McN-5652 (**8**)

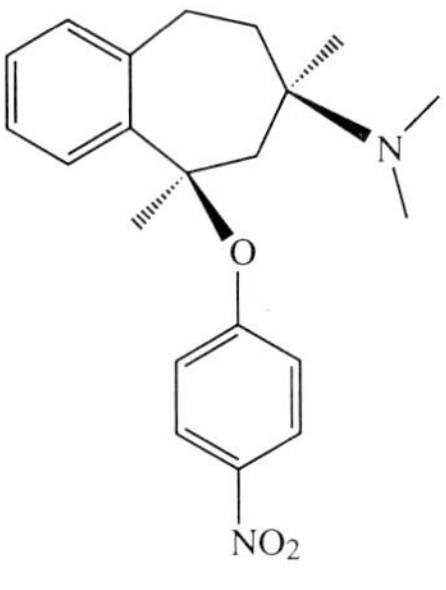

Ru25591 (**9**)

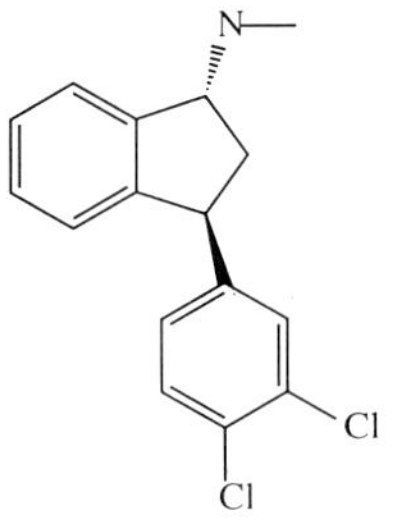

Indatraline (**10**)

Fig. 7. Chemical structures of test compounds

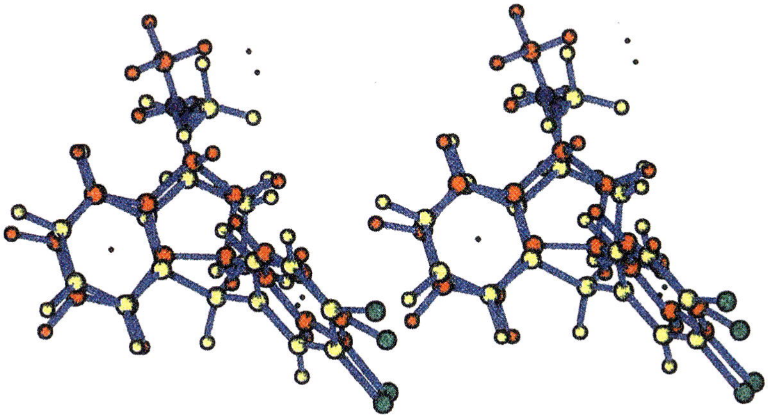

Fig. 8. *Superimposition of sertraline (yellow) and indatraline (red). Nitrogen and chloro atoms are marked in blue and green, respectively.*

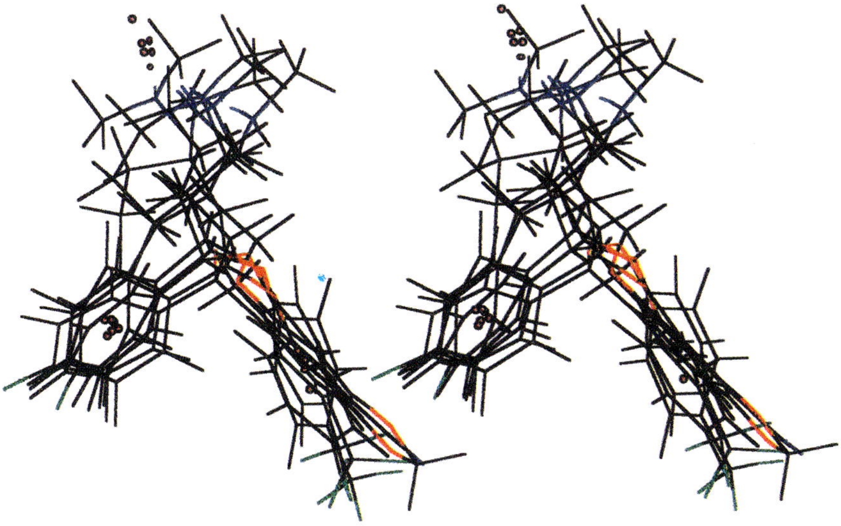

Fig. 9. *Superimposition of S-citalopram, S-fluoxetine, R-fluoxetine, paroxetine,* **7** *and* **9.** *Oxygen atoms fall in the same region of space. Nitrogen, oxygen and fluoro atoms are marked in blue, red and green, respectively.*

Examination of the various superimpositions revealed that an extra pharmacophoric element may exist. In many of the applied structures, there is an oxygen atom. Evidently, this oxygen atom seems to be placed in the same area in all these cases, although these oxygen atoms are not explicitly used in the superimpositions (*Fig. 9*). A similar observation was made in the work of *Rupp et al.* [6], suggesting a favourable interaction between an oxygen in this position and the transporter.

Selectivity towards Noradrenaline. - As mentioned, high homology among the cloned human transporters of biogenic amines is found. This is reflected in the structure-activity relationships of some of the compounds applied in this study. This prompted us to use the developed model in an attempt to explain selectivity towards the various biogenic amine transporter systems. Some very close analogues of citalopram (**1**) (talopram (**11**), a 3,3-dimethyl-substituted analogue) and fluoxetine (**2**) (nisoxetine (**12**), and tomoxetine (**13**)) (*Fig. 10*) were analysed. These are all highly selective and potent NA reuptake inhibitors [24-26], yet they only differ from citalopram and fluoxetine, respectively, in their aromatic substitution patterns. Superimpositions of these structures in the above proposed biological active conformation revealed that the methyl and methoxy substituents in **12**-**14** occupy a common region in space, facilitating NA reuptake inhibition at the expense of serotonin reuptake (*Fig. 11*).

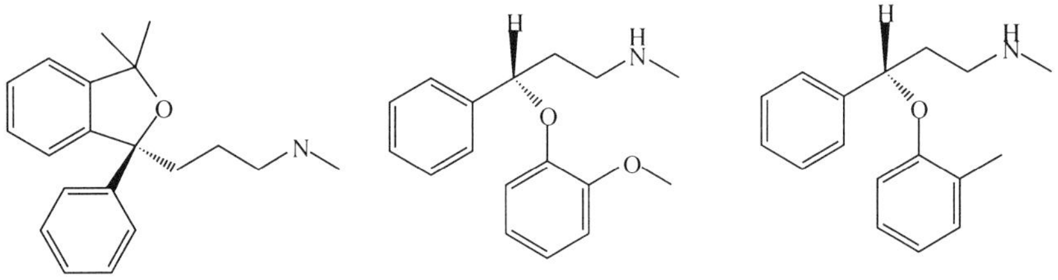

Talopram (**11**) Nisoxetine (**12**) Tomoxetine (**13**)

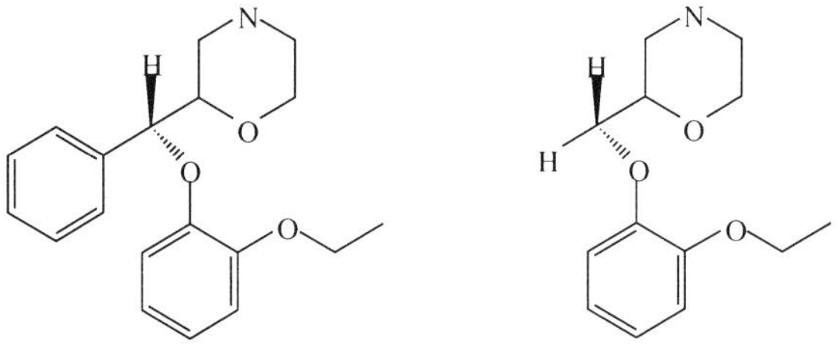

Reboxetine (**14**) Viloxazine (**15**)

Fig. 10. *Chemical structures of noradrenaline reuptake inhibitors.*

Furthermore, a number of other selective NA reuptake inhibitors seems to have substituents in the same area, e.g. reboxetine (**14**) [27] and viloxazine (**15**) [28] (*Fig. 9*). These structures represent another class of compounds with a less flexible amine group. It was therefore of great interest to fit these structures to the model. Conformational studies were initiated in order to test the hypothesis and, indeed a low-energy conformation of reboxetine (conformational energy < 1 kcal/mol) was found that fitted almost perfectly to the model. The superimpositions of nisoxetine, reboxetine and viloxazine on the proposed biologically active conformation of fluoxetine are shown in *Fig. 12*.

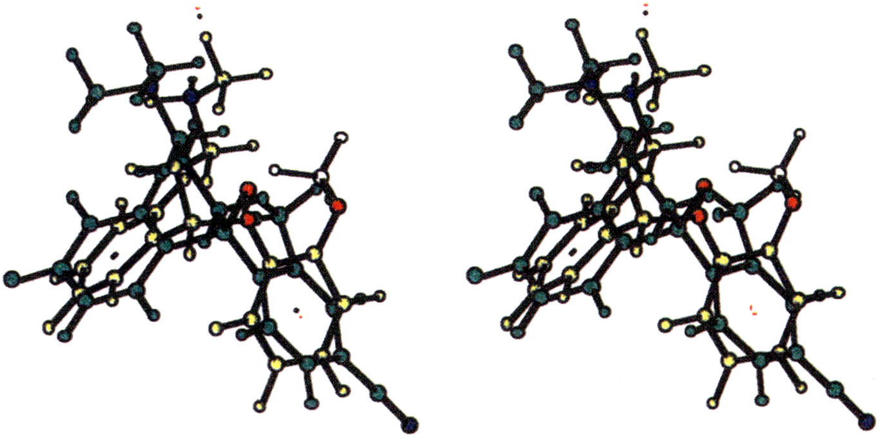

Fig. 11. *Superimposition of citalopram (**1**, green) and nisoxetine (**12**, yellow). Nitrogen and oxygen atoms are marked in blue and red, respectively. The methoxy substituent in nisoxetine occupies the same space as methyl substituents in talopram (**11**) (not shown).*

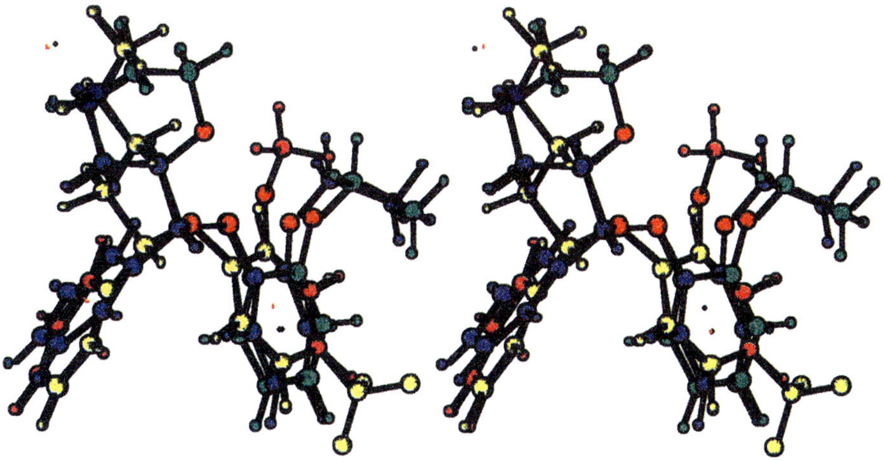

Fig. 12. *Superimpositions of fluoxetine (yellow), nisoxetine (magenta), reboxetine (cyan) and viloxazine (green). Nitrogen and oxygen atoms are marked in blue and red, respectively.*

Conclusions. - Conformational analyses of citalopram, fluoxetine, sertraline and paroxetine suggest a common pharmacophore, including both aromatic rings and a primary site point at the transporter site.

Substituents in the primary ring (R1) of the four compounds (cyano, trifluoromethyl, chloro and dioxolo) occupy the same region in space.

Fluoro substitution in the secondary ring (R2), if present, occupies the same region in space.

The model has the potential of explaining enantioselectivity.

An increase in conformational energy is reflected in reduced affinity. A direct comparison of the conformational energies and the affinities of the compounds, used for evaluation purposes, is not possible, since these data originate from different laboratories.

Low-energy conformations of each of the compounds **5-10**, representing a diverse series of compounds, were found to fit the model.

An explanation of the sudden change in selectivity towards noradrenaline in very near analogues of citalopram and fluoxetine (**11-13**) is presented.

This hypothesis is further supported by the inclusion of the more rigid analogues of fluoxetine (**14-15**).

REFERENCES

[1] B. Borowsky, B.J. Hoffmann, *Int. Rev. Neurobiology* **1995**, *38*, pp. 139-199.

[2] S. Kitayama, S. Shimada, H. Xu, L. Markham, D.M. Donovan, G.R. Uhl, *Proc. Natl. Acad: Sci.* **1992**, *89*, 7782.

[3] U.H. Lindberg, S-O. Thorberg, S. Bengtsson, A.N. Renyi, S.B. Ross, S-O. Ögren, *J.Med.Chem.* **1978**, *21*, pp. 448-456.

[4] D.F. Smith, *Neurosci. Biobehav. Rev.* **1986**, *10*, pp. 37-46.

[5] L.G. Humber, D. Lee, S. Rakhit, A.M. Treasurywala, *J.Mol.Graphics* **1985**,*3*, pp. 84-89.

[6] A. Rupp, K-A. Kovar, G. Beuerle, C. Ruf, G. Folkers, *Pharm. Acta. Helv.* **1994**, *68*, pp. 235-244.

[7] A.S-S. Chang, S.M. Chang, D.M. Starnes. *Eur.J. Pharmacol.* **1993**, *247*, pp. 239-248.

[8] J.E. Lloyd, P.R. Andrews, *J. Med. Chem.* **1986**, *29*, pp. 453-462.

[9] I. Pettersson, T. Liljefors. *J. Computer-Aided Mol. Design* **1987**, *1*, pp. 143-152.

[10] K. Andersen, T. Liljefors, K. Gundertofte, J. Perregaard, K. P. Bøgesø. *J. Med. Chem.* **1994**, *37*, pp. 950-962.

[11]U. Burkert, N.L Allinger. *Molecular Mechanics;* American Chemical Society: Washington, DC, 1982. (The Macintosh version of the MM2(91) program is available from InStar Software AB, Ideon Research Park, S-223 70 Lund, Sweden).

[12]K.P. Bøgesø, T. Liljefors, J. Arnt, J. Hyttel, H. Pedersen, *J. Med. Chem.* **1991**, *34*, pp. 2023-2030.

[13]K. Gundertofte, T. Liljefors, P-O. Norrby, I. Pettersson, *J. Comput. Chem.* **1996**, *17*, pp. 429-449.

[14]MacMimic is available from InStar Software AB, Ideon Research Park, S-223 70 Lund, Sweden).

[15] J. Hyttel, J. Arnt, C. Sánchez, *Rev. Contemp. Pharmacother.* **1995**, *6*, pp. 271-285.

[16]R.W. Fuller, D.T. Wong, D.W. Robertson. *Medicinal Chemistry Reviews* **1991**, *11*, pp. 17-34 .

[17] P. Plenge, E.T. Mellerup, T. Honoré P.L.F. Honoré, *J.Pharm. Pharmacol.* **1987**, *39*, pp. 877-882.

[18]K. Koe, A.Weissman, W.M. Welch, R.G. Browne. *J.Pharm. Exp. Therapeut.* **1983**, *226*, pp. 686-700.

[19]A. Coppen, V.A. Rama Rao, C. Swade, K. Wood. *Psychopharmacology (Berlin)* **1979,** *63*, pp. 125-129.

[20]D.R. Michals, H.E. Smith. *Chirality* **1993,** *5*, pp. 20-23.

[21]B.E. Maryanoff, J.L. Vaught, R.P. Shank. *J.Med.Chem.* **1990**, *33*, pp. 2793-2797.

[22]C. Dumont, J. Laurent, A. Grandadam. J.R Boissier. *Life Sci.* **1981**, *28*, pp. 1939-1945.

[23]K.P. Bøgesø, A.V. Christensen, J. Hyttel, T. Liljefors. *J.Med.Chem.* **1985**, *28*, pp. 1817-1828.

[24]P.V. Petersen, N. Lassen, V. Hansen, T. Huld, J. Hjortkjær, J, Holmblad, I. Møller Nielsen, M. Nymark, V. Pedersen, A. Jørgensen, W. Hougs. *Acta. Pharmacol. Toxicol.* **1966**, *24*, 121.

[25]D.T. Wong, F.P. Bymaster, J.S. Horng, B.B. Molloy. *J. Pharmacol. Exp. Ther.* **1975**, *193,* pp. 804-811.

[26] D.T. Wong, P.G Threlkeld, K.L. Best, F.P. Bymaster. *J. Pharmacol. Exp. Ther.* ***1982***, *222,* pp. 61-65.

[27]G. Cocchiara, R. Battaglia, P. Pevarello, M. Strolin Benedetti. *Eur. J. Drug Metab. Pharmacokinet.* **1991**, *16*, pp. 231-239.

[28]M. Bourin, A. Ortega, C. Larousse. *Neuropsychobiology* **1987,** *17*, pp. 35-40.

PART IV. 3D-QSAR AND MOLECULAR MODELLING

Chapter 29

3D-QSAR Analysis of the Binding of Triazine Herbicides to a Monoclonal Antibody

by **Roger Lahana**[a]), **Peter McWilliams**[a]), **John Holland**[a]) and **W. Graham Richards**[b])

[a]) Oxford Molecular Ltd, Medawar Centre, Oxford Science Park, Oxford, OX4 4GA, UK
[b]) Physical Chemistry Laboratory, Oxford University, South Parks Road, Oxford, UK

3D-QSAR Analysis of the Binding of Triazine Herbicides to a Monoclonal Antibody

by **Roger Lahana, Peter McWilliams, John Holland**

Oxford Molecular Ltd, Medawar Centre, Oxford Science Park, Oxford, OX4 4GA, UK

and **W. Graham Richards**

Physical Chemistry Laboratory, Oxford University, South Parks Road, Oxford, UK

A complete 3D QSAR analysis involves consideration of the conformations, alignment and comparison of many molecules. Whilst traditional grid-based QSAR methods deal with some of these aspects, we suggest an integrated approach which tackles all of these problems. The use of a newly designed architecture has allowed us to apply many disparate algorithms to a diverse range of QSAR applications.
We present the use of this method applied to the QSAR analysis of a set of triazine herbicides. Recent environmental studies have focused on the binding of these species to antibodies, as a means of monitoring the toxicological effects. We have used our integrated algorithm to predict the binding of triazines to a monoclonal antibody, and offer an insight into the means of binding.

Introduction. - Atrazine and its related triazine analogs are amongst the most heavily used class of herbicides worldwide. Their interatomic binding to plant proteins has already been extensively studied in relation to their ability to block photosynthetic electron transport [1]. In this paper we now focus upon their binding to a monoclonal antibody.

Antibodies specific to these triazine analogs are potentially useful for environmental monitoring and detection as well as related toxicological research. With this in mind, Ward et al. [2] cloned a triazine-specific monoclonal antibody and determined the DNA sequence, with the eventual aim of engineering new specifities.

We used the combined algorithm in ***AbM*** [3] to model the 3D structure of the combining site for this monoclonal antibody. Furthermore, we have developed a 3D QSAR, by combining the immunological cross-reactivity data and the physicochemical properties as well as molecular similarity descriptors, that may help to determine how triazines fit into this antibody binding site.

Data Set. - The 3D structures have been generated (as outlined in the next section) for 26 triazine herbicides based upon the following generic structure:

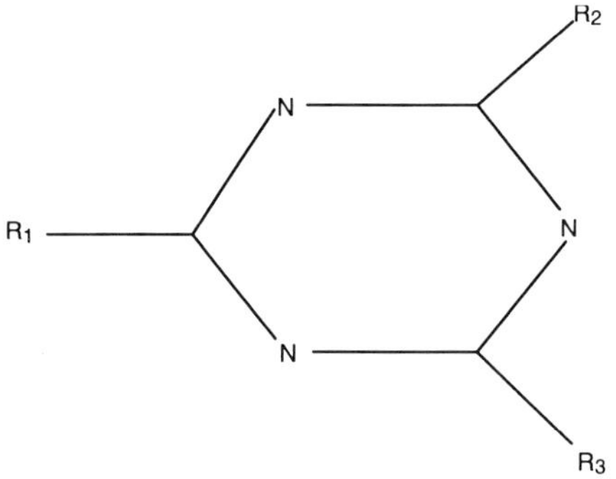

The experimental data was generated as follows [4] a direct enzyme immunoassay was performed for 26 triazine analogs with simazine N(C2)-alkaline phosphatase as the competitor. The concentration of each analog giving half-maximal inhibition (I_{50}) were calculated from multi-point dose-response curves. The « percent cross-reactivity » is then given by the ratio of the I_{50} for the analyte to the I_{50} for the analog.

Molecular Modeling. - *Conformational Analysis.* A 3D structure for each molecule is generated using the ***Cobra*** [3] package. ***Cobra*** uses artificial intelligence techniques [5] to generate a user-specified number of low energy conformations. Beginning with a 2D representation, each molecule is decomposed into a series of unique 3D molecular sub-units stored in a database.

These units are then joined together, one at a time, using a series of energy and distance constraints. In the present study, the lowest energy conformation is selected for each molecule, as this conformation will be optimized at a later stage.

Molecular Superposition. The molecules are superimposed relative to the most active compound using the ***Anaconda*** [3] package. ***Anaconda*** uses a gnomonic projection technique [6] to project either a steric or electrostatic potential onto the surface of a sphere. These spherical projections, rather than the molecular framework itself, are then directly compared. The package thereby provides an automatic technique to locate the best superposition in terms of molecular shape or electrostatic potential.

Optimization of Alignment and Conformation. The precise conformations and superpositions that have been derived are now refined using the ***Asp*** [3] program. ***Asp*** is an automatic similarity package [7] that provides a quantitative measure of molecular similarity in terms of shape as well as electrostatic, lipophilic and refractive field and potential. Similarity between two molecules i and j is expressed in terms of a normalized index S_{ij}.

Using a SIMPLEX algorithm [8], both the translational and rotational degrees of freedom are refined to maximize S_{ij} relative to the most active compound. To avoid highly unfavorable conformations for flexible molecules, a Boltzmann energy penalty is applied to all rotations about torsional bonds.

A Gaussian approximation [9] is used to calculate both the shape and potential similarity indices. This procedure is nearly two orders of magnitude faster than traditional grid-based approaches and shows excellent correlation with biological data.

Calculation of Atomic Charges. The electrostatic potential used by ***Asp*** is evaluated from the atomic charges. These charges have been calculated using ***Vamp*** [3], a semi-empirical molecular orbital program for immediate calculation of optimised geometries and molecular orbitals. The ***Vamp*** NAO-PC model provides an intermediate level between atom-centred charges and electron density estimations [10]. The NAO-PCs give modelling systems a more detailed description of charges, electrostatic potential, polarisability and dispersion effects at the atomic level due to their anisotropic nature. The solvent surrounding a molecule may be simulated to investigate the change produced in the energy and spectroscopic properties of the molecule. In addition, interactions with ordered environments (zeolites, enzymes, DNA) can be explored employing a varied quantum mechanical and force field approach.

Calculation of Similarity. Using the precise alignments and conformations that have now been derived, the similarity indices S_{ij} for all pairwise comparisons within the data set are recalculated using ***Asp***. These indices are conveniently arranged as an N by N matrix. This is a symmetric matrix, where all diagonal elements Sij, that compare a molecule to itself, are equal to 1.

Similarity matrices [11] provide a complete numerical representation of how all the molecules interrelate and, through the precise alignments and conformations that were input, implicitly include a location dependence.

Structure-Activity Relationships. Statistical analysis of the similarity matrix using the correlation tools in ***Tsar*** [3] has been shown to yield strongly predictive SARs. Through its innovative molecular spreadsheet, ***Tsar*** gathers a wide range of molecular and topological descriptors, as well as a comprehensive set of statistical techniques [12]. Principal components analysis (PCA), together with stepwise regression and partial least squares (PLS) regressions are used to explore the structure of the matrix and provide quantitative predictive relationships. Depending on the nature and quality of activity data, discriminant analysis (DA), neural networks (NN) and cluster significance analysis (CSA) may also be used.

An Integrated Approach. The overall integrated approach may be summarised as follows:

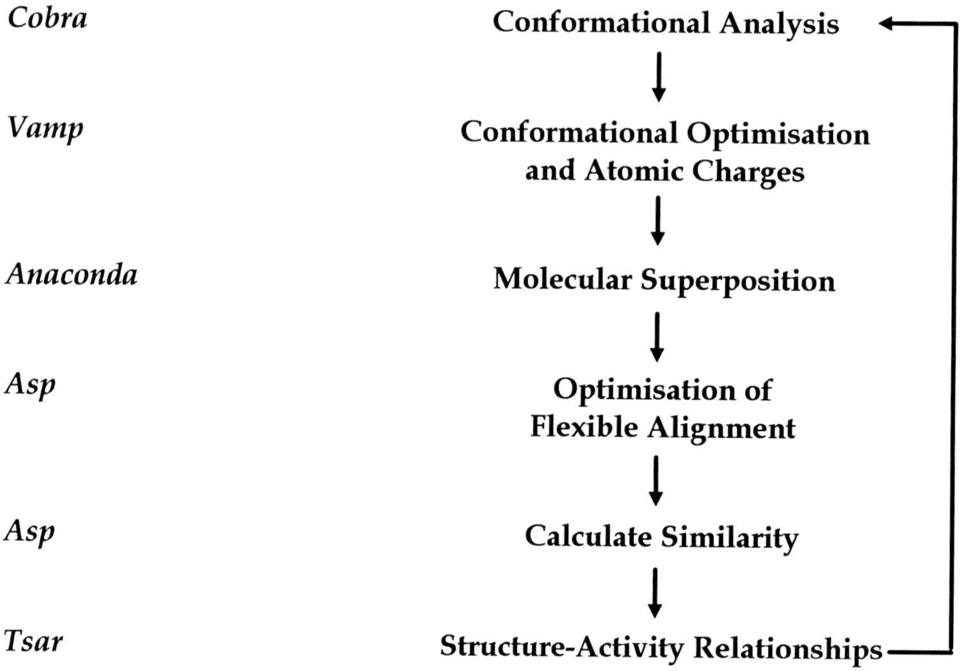

In order to make all these products communicate in a seamless way, we have used a new technology developed by Oxford Molecular, the Comms Manager (Communication Manager). This module allows a seamless linkage of software and data. Its technology is described in a more detailed way in an annex at the end of this paper.

Tsar-Asp 3D-QSAR Analysis. - *Lipophilic Similarity.* Previous work [11] has illustrated the utility of molecular similarity matrices for investigating electrostatic and steric interactions. The present study extends this approach to examine the lipophilicity.

A lipophilic potential function is generated using the Ghose and Crippen parameters [13] for atomic log P values. We assume this heuristic potential function to follow a 1/r decay and calculate the lipophilic similarity indices in an analogous manner to the corresponding electrostatic indices. The Hodgkin [7] similarity index was used, and for two molecules A and B is defined as:

$$S_{AB} = \frac{\int \rho_A \rho_B dv}{\int \rho_A^2 dv + \int \rho_B^2 dv}$$

where ρ_A is the lipophilic potential of molecule A and ρ_B is the lipophilic potential of molecule B.

Details of the Calculation. The lowest energy conformations from ***Cobra*** are first aligned relative to the most active compound using a steric superposition from ***Anaconda***. The steric superposition, based on molecular shape, is chosen to ensure that a steep repulsive potential is included in the model. Using ***Asp***, the relative molecular superpositions, together with the individual conformations, are then refined by optimizing the lipophilic, shape or electrostatic similarity indices respectively. It is also feasible to use a combined index, based on a weighted sum of any combination of the three indices. A similarity matrix is then generated using these orientations and conformations.

Results. - *Tsar: Principal Components Analysis (PCA).* PCA, as a non-parametric method, is ideal for analyzing the structure of the similarity matrices. A striking result is obtained from analysis of the shape similarity matrix. A plot of the first two principal components show the active and inactive compounds tightly clustered into two distinct groups. As shown in the diagram, however, the most active compound, propazine, is grouped with the inactive compounds. This is strongly suggestive that more than one binding mode is in operation and this is currently the subject of further investigation.

Tsar: Stepwise Regression. The linear regression of log(cross-reactivity) with log P produces a model for which the correlation coefficient, $r^2 = 0.8$, and the standard error, $s = 0.41$. This very basic representation of lipophilicity, simply in terms of log P, is a reasonable first approximation.

However, the regression can be significantly improved by using the lipophilic similarity matrix. This alternative representation of lipophilicity yields a three variable regression model with $r^2 = 0.87$ and $s = 0.34$. We can therefore describe up to 87% of the variance in the data by using the lipophilic similarity matrix.

An improved model was constructed by simultaneously considering both the lipophilic and the shape similarity matrices: in this case, a five variable model with $r^2 = 0.92$ and $s = 0.28$ is obtained (significant variables are chosen to be included in the model on the basis of a partial F-test with $F = 4$).

Table 1. Statistics of the MLR Analysis

	log P	Lipophilic Similarity	Lipophilic & Shape Similarity
r^2	0.80	0.87	0.92
s	0.40	0.34	0.28

Tsar: Partial Least Squares (PLS). PLS analysis of the lipophilic similarity matrix produces a highly predictive model as indicated by the cross-validated correlation coefficient, $r(cv)^2$, calculated by the leave one out method. With only two PLS vectors, selected from the original 26 columns of similarity indices, we find that $r(cv)^2 = 0.75$.

The steric or shape similarity matrix yields a value of $r(cv)^2 = 0.36$ and the electrostatic matrix a value of $r(cv)^2 = 0.07$. By itself, therefore, the electrostatic potential similarity matrix contains essentially no predictive power whatsoever.

By combining two similarity matrices a strongly predictive model can often be obtained and PLS is particularly well suited to analysis of these wide matrices. The best model we obtained from this combined approach was a 2 variable model of lipophilic potential and shape, where $r(cv)^2 = 0.73$.

Table 2. Statistics of the PLS Analsysis

	Lipophilic Similarity	Shape Similarity	Electrostatic Similarity	Lipophilic & Shape Similarity
$r(cv)^2$	0.75	0.36	0.07	0.73

Discussion. - We have been able to derive a strongly predictive structure-activity relationship by modeling hydrophobic and steric interactions in terms of molecular similarity matrices. The antibody that binds to the triazine derivatives in this particular system has been modeled separately [14] and was shown to have a major effect in determining how the triazines will react.

We find that, unless specific lipophilic and steric requirements are fulfilled, the reactivity will be severely limited. Furthermore, we have been able to model and predict these requirements and to highlight the importance of considering the spatial distribution of the lipophilicity. Strong evidence for the existence of more than one binding mode was uncovered from analysis of the shape similarity matrix.

Further investigations are now underway including, in particular, elucidation of different binding modes, determination of the precise role of electrostatic interactions and extension of our approach to different data sets.

Annex: The Comms Manager Technology. - In August 1993, following consultation with major customers, Oxford Molecular embarked on the development of a Communication Manager (« Comms Manager ») together with an associated Molecular Function Library. The purpose of this development is to provide a seamless linkage of software and data spanning the entire drug discovery process, from gene discovery through molecular biology to small molecule chemistry and databases.

The overriding requirement from customers is to have a single software framework that will allow the assembly and analysis of different types of molecular data from a wide variety of sources. This new software system will encompass all common hardware platforms. The system is modular and object oriented, providing a highly flexible environment in which developers and customers can link their existing and future proprietary and third-party applications. A schematic view of a product using Comms Manager is shown below:

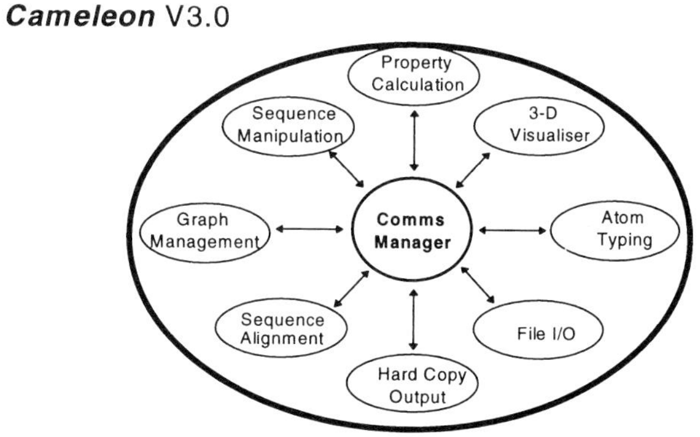

Fig. 1. *A schematic view of Cameleon, an OMG product based on the Comms Manager Architecture*

PART IV. **3D-QSAR AND MOLECULAR MODELLING**

Chapter 30

The Role of Water in Receptor-Ligand Interactions. A 3D-QSAR Approach

by **Manuel Pastor** and **Gabriele Cruciani***

Laboratory of Chemometrics, University of Perugia, Via Elce di Sotto 8, 06123, Perugia, Italy correlation coefficient (q^2) for different model dimensionalities. Four components were found to be significant ($r^2 = 0.95$ and $q^2 = 0.79$).

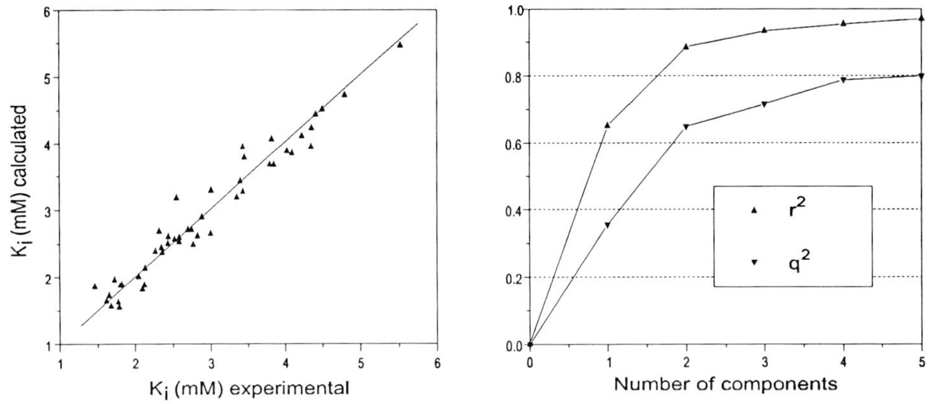

Fig. 1. *Standard GRID/GOLPE model. Calculated vs experimental activity (left). Squared correlation coefficients (r^2) and cross-validated squared correlation coefficients (q^2) (right).*

Even if the model was "numerically" satisfactory both in fitting and in prediction, their interpretability was not good. The coefficients of the PLS model were plotted in their positions in the space around the ligands (*Figure 2*) and they clearly identify some regions important for activity, as the carbonylic oxygen of Hys377 and the side chains of Asp283 and Asn284 [12]. However, the regions related to the C1 substituents (which shows the highest variability in the series) contain confusing coefficients. Moreover, no coefficients account for the effects of long substituents as, for instance, the effect of the phenyl ring of **42**, represented in the figure.

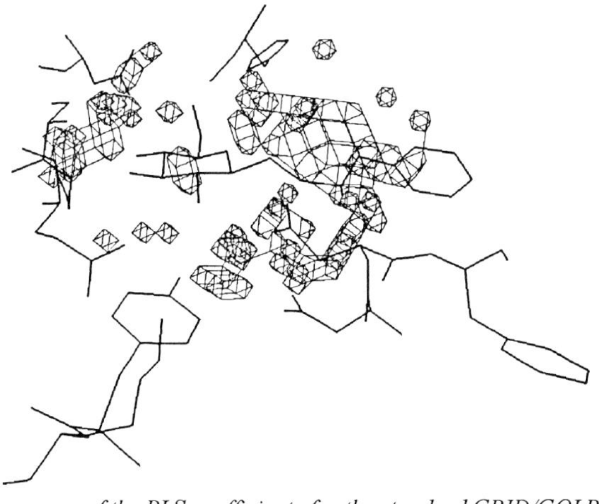

Fig. 2. *Contour map of the PLS coefficients for the standard GRID/GOLPE model, using four components.*

In fact the series contains C1 substituents of very different size, for example, *Figure 3* represents compound **1** (glucose) superimposed to compound **42**. If the ligands are considered on their own, they occupy different parts of the receptor and can be expected to interact in a different way, but if the water present in the glucose complex is considered they are not so different.

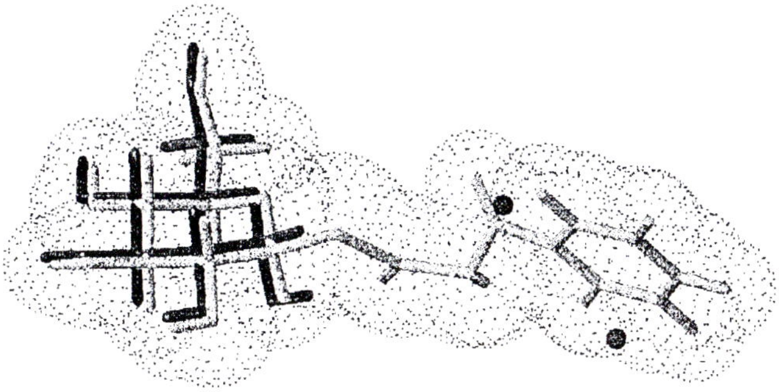

Fig. 3. *Superimposition of compound **1** (dark gray) and **42** (light gray). The Van der Waals volume of compound **42** and two water molecules present in the complex **1** are also shown.*

It becomes evident that depending upon their size and the number of water molecules displaced, the receptor-ligands interactions are different and the 3D-QSAR fails to produce a general model representing the receptor. The problems that can arise when the water is not considered are further illustrated in *Figure 4*. Compounds **23** and **31** are represented superimposed together with the residue Asn284 of the receptor. The distances between **31** and the residue in *Figure 4a* show that the ligand makes an hydrogen bond with the peptidic N of Asn284, while **23** is simply too far away for making an hydrogen bond to the receptor. However, when the water molecules present in the complex of compound **23** are considered, as in *Figure 4b*, it comes out that this compound actually makes an hydrogen bond with the receptor through two bridging water molecules.

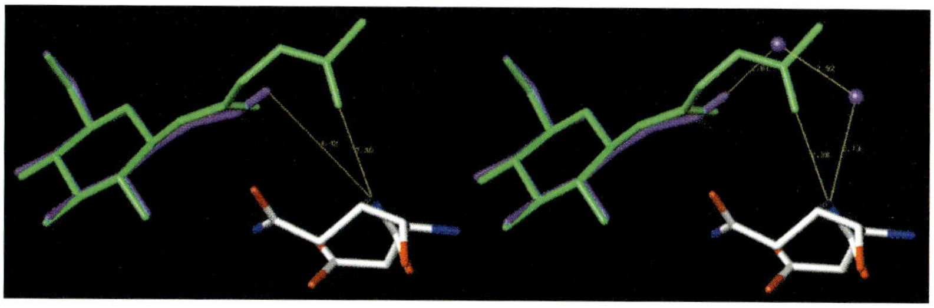

Fig. 4. *The figure represents compound **23** (violet) and **31** (green) plus the residue Asn284. On the left (4a) are represented the distances between the nearest heteroatoms of the compounds and the peptidic N of the Asn284, showing that **23** can make a hydrogen bond while **31** does not. On the right (4b) is presented a similar picture, but now including two water molecules present in the **23** complex, through which the compound can make an hydrogen bond with the receptor.*

Conclusions. - We have described a method which helps to understand the role of the water molecules in the receptor-ligand interactions in a quantitative, general and objective way. It permits to extend the 3D-QSAR methodologies in order to make more general, predictive and easy to interpret models. The method, as it is described here, requires to know the position of the water in the active site.

Work is in progress to extend this ideas to situations where the coordinates of the water molecules are unknown. It is, however, clear that the models will depend to a great extent on the positions chosen for the water molecules and therefore, some experimental data will be always required.

Acknowledgments. - We thank our colleagues L. N. Johnson, K. A. Watson, G. W. J. Fleet and N. G. Oikonomakos for sending data regarding the compounds **35** and **46** in advance of publication, the EC for providing financial support (project BIO2-CT943025) and grants for one of us (MP). The Italian funding agencies of MURST and CNR are thanked for financial support to GC.

REFERENCES

[1] P. J. Goodford, *Pharmacochem. Lib.* **1991**, *16*, 49.
[2] C. S. Poornima, P. M. Dean, *J. Comput.-Aided Mol. Des.* **1995**, *9*, 500.
[3] J. T. Bolin, D. J. Filman, D. A. Matthews, R. C. Hamlin, J. Kraut, *J. Biol. Chem.* **1982**, *257*, 13650.
[4] R. D. Cramer, D. E. Patterson, J. D. Bunce *J. Am. Chem. Soc.* **1988**, *110*, 5959.
[5] G. Cruciani, K. A. Watson, *J. Med. Chem.* **1994**, *37*, 2589.
[6] GOLPE 3.0, Multivariate Infometric Analysis, Perugia, Italy, 1996.
[7] GRID v.14, Molecular Discovery Ltd., West Way House, Elms Parade, Oxford, 1996.
[8] J. L. Martin, L. N. Johnson, S. G. Withers, *Biochem.* **1990**, *29*, 10745.
[9] K. A. Watson, E. P. Mitchell, L. N. Johnson, J. C. Son, C. J. F. Bichard, M. G. Orchard, G. W. J. Fleet, N. G. Oikonomakos, D. D. Leonidas, M. Kontou, A. Papageorgioui, *Biochem.* **1994**, *33*, 5745.
[10] C. J. F. Bichard, E. P. Mitchell, M. R. Wormald, K. A. Watson, L. N. Johnson, S. E. Zographos, D. D. Koutra, N. G. Oikonomakos, G. W. J. Fleet, *Tetrahedron Lett.* **1995**, *36*, 2145.
[11] T. M. Krülle, K. A. Watson, M. Gregoriou, L. N. Johnson, S. Crook, D. J. Watkin, R. C. Griffiths, R. J. Nash, K. E. Tsitsanou, S. E. Zographos, N. G. Oikonomakos, G. W. J. Fleet, *Tetrahedron Lett.* **1995**, *36*, 8281.
[12] K. A. Watson, E. P. Mitchell, L. N. Johnson, G. Cruciani, J. C. Son, C. J. F. Bichard, G. W. J. Fleet, N. G. Oikonomakos, M. Kontou, S. E. Zographos, *Acta Cryst.* **1995**, *D51*, 458.
[13] M. Pastor, G. Cruciani, S. Clementi. *J. Med. Chem.* (submitted).

PART IV. **3D-QSAR AND MOLECULAR MODELLING**

Chapter 31

Modelling of the Interactions of Some Inhibitors with the PGHS-1 by BIODOCK - a Stochastic Approach to the Automated Docking of Ligands to Biomacromolecules

by **Alessandro Pedretti, Anna Maria Villa***, **Luigi Villa** and **Giulio Vistoli**

Instituto de Chimica Farmaceutica, Università di Milano, Viale Abruzzi 42, I-20131 Milano bring to the fore necessary common features between all studied compounds. But more fundamentally, as far as the benzodiazepine ligands and their analogs are concerned, the pharmacophore elucidation still remains the major challenge for drug design [15,16].

Very often, the comparison of different ligands is based on rigid or flexible pairwise superimposition of the molecular skeletons using the three-dimensional (3D) coordinates of the constituting atoms [17-19]. In the following study, we choose, on the one hand, to compare several benzodiazepine like compounds on the basis of more complex and significant data, *i.e.*, their 3D electron density and, on the other hand, on the use of a genetic algorithm technique, an optimization method which mimics natural evolution by selecting, recombinating, and mutating potential solutions until the best are found. To test the application of such an original approach, we choose to consider a relatively small number of ligands, but properly selected to avoid fortuitous results due to excessive diversity. Four molecules were thus selected owing to their ability to bind with comparable affinities to the CBR and MBR: **diazepam** (7-chloro-1,3-dihydro-1-methyl-5-(phenyl)-2H-1,4-benzodiazepine-2-one), **Ro 5-4864** (7-chloro-1,3-dihydro-1-methyl-5-(p-chlorophenyl)-2H-1,4-benzodiazepine-2-one), **QUIZ** (6-chloro-N1-methyl-4-phenylquinazolin-2-one), and **TZQ** ([1,2,4]triazolo[1,5-c]quinazoline-5(6H)-one) (*Fig. 1*). The procedure adapted for their comparison can be decomposed in three consecutive steps.

• First, three-dimensional electron density maps are calculated at medium resolution for each ligand from calculated structure factors obtained by the simulation of an X-ray diffraction experiment using the crystal resolution programs XTAL [20].

• As the resulting three-dimensional maps contain too many data points to allow a direct comparison of the electron densities of the four chosen ligands, these maps are then simplified by topological analysis [21]. Critical points of the electron density function, *i.e.*, points where the density gradient is equal to zero, allow to gather the information of the whole map on a limited number of points. This kind of analysis is performed by the program ORCRIT [22], which also connects the critical points in order to represent the ligands as "connected graphs".

• A genetic algorithm method [23,24] is then especially implemented to compare these graphs and superimpose the ligands. We will show that the design of such an artificial intelligence type algorithm implies the use of appropriate and performant coding and evaluation of the candidate solutions to compare simultaneously the representations of the four studied ligands and finally generate graph superimpositions.

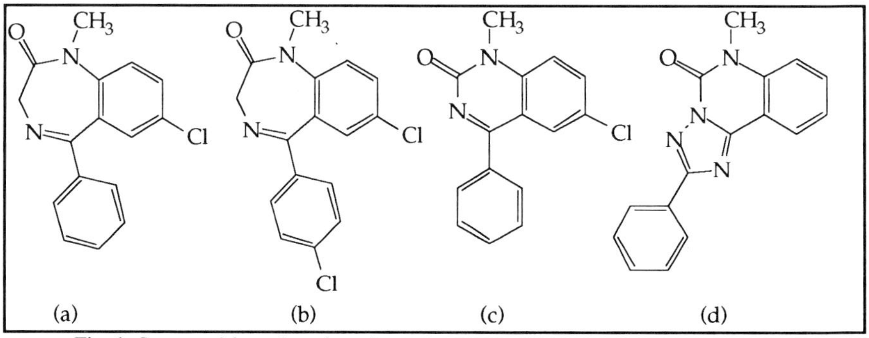

Fig. 1. *Structural formulae of (a) diazepam, (b) Ro 5-4864, (c) QUIZ, and (d) TZQ.*

In this first paper on the subject, we will briefly develop the methods elaborated to carry out the study, *i.e.*, the topological analysis of electron density maps and the genetic algorithm applied to graph similarity search. Then we will describe the obtained results in terms of connected graphs of critical points as representations of the four selected ligands, the optimization of the parameters characterizing the genetic algorithm code, to finally obtain a superimposition of the four ligands depicting the arrangement of the discovered common features between these ligands, *i.e.*, the potential pharmacophore elements.

Methods. - *Topological Analysis of Electron Density Maps.* Crystallographic data relative to the compounds diazepam (CSD code: DIZPAM10) and Ro 5-4864 (CSD code: FULWUE) were retrieved from the Cambridge Structural Database (CSD) [25] while the crystalline structures of the ligands QUIZ and TZQ were resolved in collaboration with the laboratory of "Chimie Moléculaire Structurale" of our University [26].

On the basis of the crystallographic data - unit cell parameters and atomic coordinates, space group, isotropic temperature factors - the programs XTAL [20] allow to create a three-dimensional ideal electron density map ρ(x,y,z) by simulation of an X-Ray diffraction experiment: the reflections for each vector (h,k,l) from the reciprocal space are generated at a given resolution, and the structure factors are computed. The electron density is then obtained by Fourier Transform.

The major role of the topological analysis method is the representation of ρ(x,y,z) as a set of critical points. The topological approach requires the calculation of the gradients of the electron density, $\nabla\rho(\mathbf{r})$, Hessian matrices, $H(\mathbf{r})$, Laplacians, $\nabla^2\rho(\mathbf{r})$, and rank of $H'(\mathbf{r})$, $H'(\mathbf{r})$ being the diagonalized form of the real and symmetric matrix $H(\mathbf{r})$.

$$H(\mathbf{r})=\begin{bmatrix} \frac{\partial^2}{\partial x^2}\rho & \frac{\partial^2}{\partial x\partial y}\rho & \frac{\partial^2}{\partial x\partial z}\rho \\ \frac{\partial^2}{\partial y\partial x}\rho & \frac{\partial^2}{\partial y^2}\rho & \frac{\partial^2}{\partial y\partial z}\rho \\ \frac{\partial^2}{\partial z\partial x}\rho & \frac{\partial^2}{\partial z\partial y}\rho & \frac{\partial^2}{\partial z^2}\rho \end{bmatrix} \qquad H'(\mathbf{r})=\begin{bmatrix} \frac{\partial^2}{\partial x'^2}\rho & 0 & 0 \\ 0 & \frac{\partial^2}{\partial y'^2}\rho & 0 \\ 0 & 0 & \frac{\partial^2}{\partial z'^2}\rho \end{bmatrix}$$

The rank of $H'(\mathbf{r})$ is given by the number of non-zero eigenvalues, and its signature s is the algebraic sum of their sign. If the rank of $H'(\mathbf{r})$ is 3, four cases can be met: s = 3, which corresponds to a local maximum, called **Peak**; s = -1, which corresponds to a saddle point, called **Pass**; s = +1, which is another kind of saddle point, called **Pale**, and s = +3, which is a local minimum, or **Pit**. These particular points, defined by $\nabla\rho(\mathbf{r}) = 0$, are known as "critical points".

If the electron density is computed at atomic resolution, the peaks and passes are located on atoms and bonds, respectively, while pales and pits correspond to rings and cages, respectively. Such a procedure has been implemented in the program ORCRIT [22]. If the resolution is lowered, the critical points are shifted away from the atomic positions and give a more global representation of the ligand.

Once they are found, the critical points are connected to form "forests" of acyclic minimal graphs, called "minimal spanning trees". All the connections between two critical points i,j are characterized by an undirected "connection weight", computed according to the following equation:

$$w_{ij} = \Delta r\ R\ F\ \Delta s,\quad R=\frac{1+\frac{|\rho_i-\rho_j|}{(\rho_i+\rho_j)}}{\sqrt{\rho_i+\rho_j}}$$

where Δr is the distance between i and j; Δs is 1.1, 1.0, or 5.0 if the number of negative eigenvalues difference between i and j is 0, 1, or 2, respectively; $F=(3.0 - F_i - F_j)$, F_i being the normalized projection of the eigenvector of i along the vector Δr between the two critical points.

procedure, which, if activated, ensures the survival of the fittest chromosome of the population in the next generation without undergoing any genetic operation. The performance of the algorithm is evaluated by plotting the average performance and the cumulative average of the average performance *vs* the generation number. *Table 2* shows the performed optimization tests and results for a total number of 5000 generations. All tests and calculations were performed on an IBM RS 6000 model 340 workstation.

Table 2: *Parameters and Best Set of Parameters, Evaluated on the Basis of the Cumulative Average of the Average Performance of the GA Code.*

Parameter	Starting Value	Final Value	Step	Best performance
Mutation rate (%)	0.1 0.1	5 0.5	0.5 0.1	0.1 (1)(2)
Crossover rate (%)	45	75	5	60 (1)(2)
Population size	100	500	100	100 (1) 300 (2)

(1) Elitism not activated; (2) elitism activated.

The most performant algorithm is thus obtained, considering 5000 generations, with 0.1% mutations and 60% crossover, which is supported by the information found in the literature [28]. The best population sizes are 100 and 300 chromosomes, if the elitist option is inactive or active, respectively. These two best sets lead to the following performance values: the final cumulative averages are 369.398 (CPU time = 2407 sec) and 309.503 (CPU time = 7131 sec) for the first and second sets of parameters, respectively. The performance is thus better when the calculation is performed under elitist conditions. However, as it is known that the elitist procedure also increases the probability of premature convergence towards a local minimum, we choose to examine the results of both GA runs.

GAGS: Results of the Graph Comparison. The final solutions taken into account are the best fitted chromosomes which appeared during all 5000 generations, and the solutions kept for the analysis are statistically determined according to their occurrence frequency. The study of the first results obtained with an elitist population of 300 chomosomes suggests that premature convergence has occurred, leading to a chemically senseless graph superimposition. The results obtained by the experiment carried out with a population of 100 non elitist chromosomes lead to two interesting remarks. On the one hand, a pertinent mapping of principal features of the two benzodiazepine ligands (Ro 5-4864 and diazepam) as well as the quinazolinone (QUIZ) is observed, *i.e.*, a good superimposition of the chlorine atoms, of the free rotating aromatic regions, and more globally of the two nearly planar bicyclic heterocycles. On the other hand, the mapping of meaningless critical points suggests that the fitness function tries to match significant critical points to points rising from the background noise of the electron density map. In order to check the validity of our fitness function, in a next step we removed these "noisy points" from the four graphs, and performed two new experiments under the same conditions.

As far as the 100 non elitist experiment is concerned, we obtained a critical point mapping result without any chemical sense; for example, the chlorine atom of diazepam is superimposed to the π-aromatic region of Ro 5-4864, and *vice versa*, which makes the global superimposition of the two ligands totally impossible. However, the results from the optimization of the population composed of 300 elitist chromosomes suggest that the superimposition of the four benzodiazepine like ligands is possible, on the basis of five hypotheses.

The first hypothesis concerns the superimposition of the free rotating aromatic region (FRA) of the benzodiazepine ligands diazepam and Ro 5-4864 to the benzo ring (π-aromatic region, PAR) of the benzodiazepine like molecules QUIZ and TZQ, which implies an inversion in the relative orientation of the two types of ligands. The second one only shows the probable superimposition of the chlorine atoms of the two benzodiazepine ligands diazepam and Ro 5-4864. The third hypothesis matches the imine groups of diazepam and Ro 5-4864 to the triazole group of TZQ. As far as the fourth hypothesis is concerned, the critical points located at the junction between the two heterocycles of the benzodiazepines are superimposed to the junction point between the FRA region and the bicyclic moiety of the ligand QUIZ. The last hypothesis suggested by this GA experiment involves only diazepam, QUIZ, and TZQ, and matches their PAR, FRA, and FRA regions, respectively.

These five propositions allow to perform three-points superimpositions of the four ligands using IFMFIT [17]. The hypotheses 1, 2, and 4 lead to a good mapping between diazepam and Ro 5-4864 (*Fig. 6a*). Hypotheses 1, 4, and 5 lead to three superimposition points for the ligands diazepam and QUIZ (*Fig. 6b*), and hypotheses 1, 3, and 5 lead to the mapping of diazepam and TZQ (*Fig. 6c*).

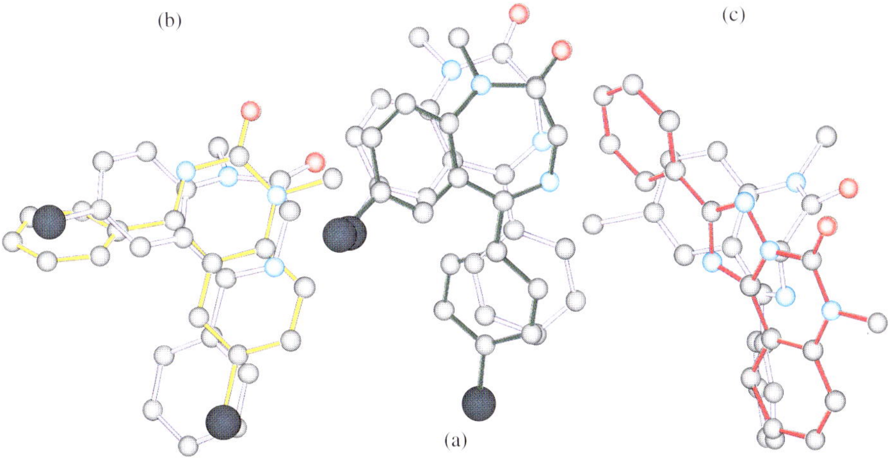

Fig. 6. *Pairwise superimposition of the four studied benzodiazepine like ligands: (a) diazepam (blue skeleton) and Ro 5-4864 (green skeleton), (b) diazepam and QUIZ (yellow skeleton), and (c) diazepam and TZQ (pink skeleton). The identification of the atoms is similar to Figure 5.*

Conclusions. - In this very first preliminary paper dedicated to the use of GA methods applied to pharmacophore elucidation, we mostly concentrate on a brief survey of the techniques used, *i.e.*, topological analysis of 3D electron density maps and genetic algorithms.

Most of our present efforts have been put on the development, elaboration, and writing of a totally original GA code, especially designed for the problem of graph comparison, and which is applied to the elucidation of the benzodiazepine receptor pharmacophore.

At this early stage of the development of the code, we however obtained encouraging preliminary results which in a sense justify the use of such an optimization technique for the problem of the comparison of molecular graphs and hence for the development of pharmacophore models.

PART IV. 3D-QSAR AND MOLECULAR MODELLING

Chapter 33

Signal Transduction *via* G Protein-Coupled Receptors: Ternary Complexes As Secondary Proton Pumps and GTP Synthases

by **Paul H.J. Nederkoorn** and **Henk Timmerman**

Leiden/Amsterdam Center for Drug Research (LADCR),
Division of Medicinal Chemistry, Department of Pharmacochemistry, Faculty of Chemistry, Vrije Universiteit,
De Boelelaan 1083, NL-1081 HV Amsterdam,
The Netherlands

Signal Transduction *via* G Protein-Coupled Receptors: Ternary Complexes As Secondary Proton Pumps and GTP Synthases

by **Paul H.J. Nederkoorn** and **Henk Timmerman**

Leiden/Amsterdam Center for Drug Research (LADCR), Division of Medicinal Chemistry, Department of Pharmacochemistry, Faculty of Chemistry, Vrije Universiteit, De Boelelaan 1083, NL-1081 HV Amsterdam, The Netherlands

A G protein-coupled receptor (R), its agonist (H) and an appropriate heterotrimeric $G_{\alpha\beta\gamma}$ protein (G) form together a so-called ternary complex (HR*G). G protein activation is currently thought to be mediated via conformational changes of the receptor upon agonist binding. Subsequently, the G_α* subunit dissociates from the ternary complex accompanied by an exchange of pre-bound GDP for GTP. However, experimental data - regarding among others guanine nucleotide levels measured on light activation of rod outer segments (where G proteins are abundant) and phosphorylation reactions occurring at the G protein - are available challenging this widely accepted G protein activation mechanism. In addition, theoretical calculations show that a ligand-mediated proton transfer is possible as a signal transduction mechanism for G protein-coupled receptors.

Here, a new model for ternary complex activation is presented which aims at explaining all experimental findings. Within this model, the ternary complex is compared with the secondary proton pump $F_1 \cdot F_0$-ATP synthase: the activated receptor (R*) transports protons from the extracellular site across the membrane toward the cytosol and hence strongly resembles the aforementioned membrane-bound F_0 unit in a functional manner. Subsequently, the G protein, being part of the ternary complex, uses the translocated proton(s) to synthesize GTP from pre-bound GDP and P_i and thus functions analogous to the F_1 unit of the ATP synthase. Within this concept HR*G acts as a GTP synthase. Upon GTP synthesis, the G_α*·GTP complex dissociates from the ternary complex, which enables the liberated G_α subunit to participate in subsequent GDP/GTP exchange reactions. In our scheme, the re-association of G_α with $G_{\beta\gamma}$ induces the deactivation of a G protein. We claim that within this new model the characteristic exchange patterns of GDP and GTP can be explained. In addition, the guanine nucleotide levels monitored on the activation of rod outer segments, phosphorylation reactions in which the G protein participates and proton transfer mechanisms can now be addressed accordingly.

Introduction. - G proteins (G), also called guanine nucleotide-binding regulatory proteins, consist of three subunits: α, β, and γ. Recently, the crystal structures of trimeric $G_{\alpha\beta\gamma}$ (complexed with GDP), dimeric $G_{\beta\gamma}$, the G_α (complexed with GDP) and the activated G_α* subunit (complexed either with the non-hydrolyzable GTP analogue GTPγS or with the transition state analogue GDP·AlF_4^-) have been elucidated [1-6]. G proteins have been reported to display two characteristic reactions: (**i**) dissociation of pre-bound GDP with subsequent binding of GTP followed by the dissociation of the G_α subunit from the G protein and yielding the activated G_α*·GTP complex plus the $G_{\beta\gamma}$ dimer; and (**ii**) *in situ* hydrolysis of GTP at G_α* which eventually leads to reassociation of $G_{\beta\gamma}$ with G_α [7,8]. Although the GDP/GTP exchange reaction (Eq.1) is exergonic ($\Delta G > 0$), the barrier for this exchange mechanism and subsequent G_α*·GTP dissociation is probably high, which prevents the G protein from being activated spontaneously. The GTP hydrolysis is regarded as the inactivation step [7,8].

$$G_{\alpha\beta\gamma}{\cdot}GDP + GTP \rightleftharpoons G_{\alpha}{*}{\cdot}GTP + G_{\beta\gamma} + GDP \qquad (1)$$

GDP/GTP exchange mechanism for G protein activation; from experiments with GTPγS it has been estimated that the K_{eq} is approx. 100M ($\Delta G >> 0$) [7].

G protein-coupled receptors (GPCRs) belong to a large class of membrane-spanning receptors which all probably share a seven α-helix motif [9]. GPCRs function as antennae for external signals consisting of chemical ligands or light (in case of opsins). Once the external signal is detected by the GPCR, the GPCR is activated and subsequently the signal is transmitted to a cytosolic G protein, whereupon the GDP/GTP exchange mechanism (Eq.1) is initiated. The role of an activated receptor could therefore be regarded as to lower the barrier of G protein activation [7].

Receptors are considered to exist in two states, denoted by R and R* [10,11]. R and R* represent the resting and activated state of a GPCR, respectively. Within this so-called two-state model (R vs. R*), the different molecular species are: H, R, R*, G, HR, HR*, and HR*G (the ligand, or hormone, is denoted by H; HR*G is called the ternary complex) [10]. It is thus assumed that only the activated GPCR is able to couple the G protein. At physiological GTP concentrations, both the ternary complex and the trimeric G protein are ephemeral, since GTP is thought to bind instantaneously to the ternary complex at the G_{α} subunit leading to dissociation of $G_{\alpha}{*}{\cdot}GTP$ and subsequently to the transfer of (H)R* into (H)R. After dissociation of the ternary complex, the receptor plus its agonist are available to catalyze the activation of another G protein. One GPCR can therefore activate many G proteins (amplification factors of up to 500 have been reported) [7,12]. Both $G_{\alpha}{*}{\cdot}GTP$ and $G_{\beta\gamma}$ can interact with effector molecules, thereby regulating the second messenger production in the cell. After GTP hydrolysis, the G_{α} reassociates with $G_{\beta\gamma}$, a process which terminates the signal transduction. Agonist-independent modulation of effector molecules (so-called basal activity) is explained by the presence of the R*G combination.

The proportion of the two states of the receptor and the relative affinity for agonists of these two states, the so-called high and low affinity, vary with the concentration and the intrinsic activity of the agonist on the one hand and with the presence of G proteins and guanine nucleotides on the other hand. The high affinity state corresponds with HR* and HR*G, whereas the low affinity state corresponds with the HR state, respectively. Constitutively active (mutant) receptors give rise to an increase in basal activity and show an increased affinity for (partial) agonists and an improved (partial) agonist-induced signal transduction, resulting in a more efficient regulation of the effector system.

Evidence exists that some antagonists display a high and low affinity pattern as well. This pattern is, however, exactly opposite to the one displayed by agonists; in other words, some antagonists bind more tightly to R and more loosely to R* [13]. This type of antagonists are called inverse agonists (or negative antagonists) and have reverse intrinsic activity, which means that the basal activity decreases in their presence. Other antagonists, so-called neutral antagonists, have null intrinsic activity. This latter antagonistic class does not reveal any high and low affinity pattern and does not alter the basal activity.

In the absence of agonists, basal activity of receptors can be observed. This implies that a part of the receptor population must be present in the R* state to form R*G. From the Boltzmann distribution rule the energy difference between R and R* can be calculated once the [R]:[R*] ratio is known. G protein coupling to a GPCR will stabilize the R* state with respect to R. Also coupling of H to R* or R*G will cause a stabilization of the R* state at the expense of the amount of R. Constitutively active receptors have a preference for the R* state with respect to the wild type GPCR. Inverse agonists lower basal activity by stabilizing R (at the expense of R*) and therefore induce nett uncoupling of G proteins since R does not couple to G. Hence, neutral antagonists are not expected to alter the [R]:[R*] ratio.

Theories Regarding GPCR Activation. - At present there are roughly three different viewpoints on how GPCRs may activate the coupled G protein: (**i**) via the conformational change; (**ii**) by binding the missing entity (so-called deletion model for the origin of receptors); and (**iii**) by proton translocation. We will summarize these viewpoints briefly.

ad i) Coupling of the G protein to a GPCR is mediated via the intracellular loops of the receptor that join the α-helices I-II, III-IV, and V-VI [14 and refs. *loc. cit.*]. Samama *et al.* [10] and Jung *et al.* [15] suggest that within the inactive (R) conformation of the receptor these intracellular regions are shielded from interaction with the G protein. According to their concept, a conformational change of the GPCR (R=>R*) is responsible for its activation. Only after GPCR activation (yielding R*), an interaction with the G protein is established as a result of structural changes. Also Weinstein [16] points to conformational changes as the origin for the activation of the G protein by a GPCR. Theoretical studies were performed on a seven α-helix model of the 5-HT_2 receptor in order to study ligand-induced displacements of the C_α carbons of the protein backbone; intra- and extracellular loops were not included in this model. From these data an activation model was inferred in which agonists distort the protein from the resting state with the extent of distortion related to the known pharmacological efficacies; antagonists did not distort the protein. The agonist-induced conformational changes were most pronounced in helices V and VI at the cytosolic side, strongly suggesting that the third intracellular loop will be affected and will lead to G protein activation. However, it should be noted that the above calculations were restricted to the transmembrane helices solely and that besides the neglect of the intra- and extracellular loops also any role for the membrane was omitted. The helices were allowed to move freely in a medium with a dielectric permittivity of 4. It is important to mention that the above model does not address inverse agonism and neutral antagonism explicitly since all agonists are assumed to distort the protein, whereas all antagonists are assumed not to alter the protein.

Finally, we give an example from Oliveira *et al.* [17] who assign a switch function to a highly conserved arginine residue in the third membrane-spanning helix. This arginine also plays a crucial role in Timms's proton pumping model [18] (*vide infra*). The arginine is thought to be located close to both the polar receptor pocket and the cytosol. Therefore, according to Oliveira *et al.* [17] this residue could be shielded in the R state, whereas it becomes exposed in the R* state. Within these conceptions of a conformational change, the amount of R* is directly related to the amount of G protein activation.

ad ii) Topiol [19] postulated a so-called deletion model for the origin of receptors: receptors, together with their agonists (complementary ligands), are considered to be derived from a common parent system. The "parent" system is a collection of entities necessary for the functioning of a complete biochemical system. Deletion of one entity (α_i subunit) generates an inactive subsystem. Only when the α_i subunit and the inactive subsystem are recombined (to constitute a ligand-receptor pair) does activity return. The relationship between endogenous ligands and biological building blocks (such as histamine and histidine, respectively, or serotonine and tryptophan, respectively) follows directly from this deletion model. For a successful application of this deletion model to the histamine H_2 receptor the reader is referred to Nederkoorn *et al.* [20]. Combining the idea of the ligand-mediated proton pump (sub iii) with this deletion model, an amine receptor may be seen as a proton pump that lacks one entity in its proton shuttle for pumping continuously; agonist binding restores the pumping function.

ad iii) Timms *et al.* [18] suggested that the majority of G protein-coupled receptors are capable of catalyzing proton translocations as an integral part of G protein activation, that is, a receptor acts as a proton pump during excitation. Proton conduction ends with the transfer of at least one proton to an intracellular location per agonist molecule bound. This concept may further be supported by data presented in a mini-review [21] in which the proton pumping abilities of a phototaxis receptor sensory rhodopsin (SR) from *Halobacterium salinarium* and of bacteriorhodopsin (BR) are compared. Both BR and SR release a proton from their Schiff base on photo-activation and both share a similar seven transmembrane α-helical motif. This same motif is found for GPCRs [9]. Like the G protein-coupled rhodopsin, SR signals to a transducer molecule (i.e., the G protein in case of rhodopsin and the halobacterial transducer (HtrI) in case of SR). BR is a so-called primary proton pump, which means that on (light-) activation it starts to pump protons across the membrane thereby generating a proton gradient (*vide infra*, cf. *Fig.2*). On photo-activation SR functions as primary proton pump as well but only in the absence of HtrI. In the presence of HtrI, SR does not pump protons against a gradient and therefore, the role of the transducer is likely to absorb the proton that is released from the Schiff base of SR on photo-activation. The released proton can thus be used in the signal transduction from SR to its transducer. Also the (G protein-coupled) rhodopsin activation is associated with proton transfers for which experimental evidence is reported by Hofmann and co-workers [22-24]. A key function of the seven transmembrane α-helical motif in both *Archea* and *Eucarya* may be to create the means for proton transfer reactions.

Experiments that Challenge the Two-State Model. - In literature the concept of the conformational change is often related to the two-state model [e.g. 10,15,16], thereby explaining G protein activation by the presence of the R* state. In this respect we would like to make the following observations: it is probably not correct to consider the agonistic high affinity state as the pre-coupled (i.e., to the appropriate G protein) state of the receptor, because Leeb-Lundberg *et al.* [25] claim (from the fact that high affinity binding of an agonist was still present at the uncoupled state of the B_2 bradykinin receptor) to provide the experimental evidence for the existence of both the HR* and the HR*G state: when a non-hydrolyzable GTP analogue was added, preventing coupling of the G protein to the receptor, a considerable amount of high affinity binding (corresponding to HR*) could still be observed. Furthermore, Strader *et al.* [26] even claim to have mutated a GPCR in such a way that it has become unable to couple to the G protein, but yet it displayed enhanced agonist affinity. This raises the question whether the R* form in the two-state model combinations R*, HR*, R*G and HR*G really is one and the same form. In addition, by showing that a mutation of a G protein-coupled (β_2-adrenergic) receptor impairs agonist activation of the effector system (*in casu* adenylyl cyclase) without affecting high affinity binding, Hausdorff *et al.* [27] have proved that the rate of G protein activation is not solely dependent on the level of high affinity binding of the agonist . In other words, the formation of high affinity ternary complexes is not sufficient to activate the intracellular G protein pool adequately. It has therefore been concluded [27] that the molecular determinants of ternary complex formation are not identical to those that transmit the agonist-induced stimulatory signal to the G protein. This shows that the concept of one particular conformational change (R=>R*) for explaining G protein activation is questionable. The existence of more than one agonistic binding site at a GPCR (for an example of the histamine H_1 receptor, see Leurs *et al.* [28]) does not comply with the alleged one unique conformational change of the recepor either. Furthermore, for the G protein-coupled rhodopsin mutants have been reported that are able to bind the G protein in a light-dependent manner but lack the ability to produce the activated G_α*·GTP complex [29]. In conclusion, explaining the amount of G protein activation by only a straightforward Boltzmannian relationship between high affinity binding of agonists and the amount of (H)R*G present might be an oversimplification.

Experiments that Challenge the Present-Day Mechanism for G Protein Activation. - There are three literature findings in particular that do not comply with the current model for G protein activation as described in the introduction. (**A**) The first finding stems from experiments reported by Robinson and Hagins [12], who call it the usual large drop in GDP observed within 4s after photolysis of rhodopsin present in intact frog rod outer segments (ROSs; cf. *Fig.1*).

The data depicted in *Fig.1* clearly contradict the widely accepted exchange mechanism of GDP for GTP in which GDP levels are likely to rise and certainly not to drop in the period directly after activation. In the period between 4 and 64s after the flash, total GTP levels dropped, while GDP rose by similar amounts (indicating GTP hydrolysis) [12].

In intact frog ROSs, at least 100 GTP molecules are hydrolyzed for each photon absorbed by rhodopsin during the 20s following a flash [12]. The concentration of rhodopsin in ROS is about 3mM [7], while approximately 1% of the total photopigment is bleached in the photolysis experiments reported by Robinson and Hagins [12]. Since an amplification factor of more than 100 for R* activating G is claimed [7,12], a huge GDP release is to be expected in case of an exchange mechanism (mM range) instead of the GDP drop shown in *Fig.1*.

ROSs contain guanyl cyclases catalyzing the conversion of GTP into cGMP and cyclic nucleotide phosphodiesterases catalyzing the conversion of cGMP into GMP. The catalytic activity of these enzymes affect guanine nucleotide levels as well and might therefore interfere with conclusions drawn from the aforementioned experiments in which GTP and GDP concentrations are directly related to G protein activation alone [12]. Although the catalysis by phosphodiesterase is highly efficient (turnover ~ $800s^{-1}$ [7]), total acid-extractable GMP levels are much lower than extractable GDP and GTP levels in ROSs before and after activation [12]. Moreover, GDP appears to be the main product of GTP metabolism in a time period of 64s after illumination [12]. Hence, GTP hydrolysis to GDP is the main metabolic pathway, unless GMP (the end product of the combined guanyl cyclase and cyclic nucleotide phosphodiesterase activity) is very rapidly phosphorylated to GDP. For the latter phosphorylation no evidence, however, has been reported. It is also shown that cGMP directly opens sodium channels at μM concentrations [7]. Hence, the concentration effects regarding guanine nucleotides due to the G protein activation (substrate at mM concentration) is more pronounced than the effect of phosphodiesterase (product at μM concentration). Furthermore, if the total concentration of phosphodiesterase and/or guanyl cyclase is much lower than that of the G protein (*in casu* transducin), the effect on total guanosine levels of the last mentioned protein will, despite its lower turnover number, be much larger again. Robinson and Hagins [12] measured total acid-extractable nucleotide levels in ROSs, and we have here referred to these levels throughout. Since the mass balance of guanine nucleotides in these experiments is not complete (cf. *Fig.1*), a significant part of the guanine nucleotides must become acid-inextractable upon activation. This phenomenon is explained by the fact that the X-ray structures of G_α and G_α* reveal that the nucleotide binding site in G_α* is much more shielded from the environment [1-3]: GDP can be extracted from G_α, whereas the closed binding site of G_α* prevents nucleotide extraction. In conclusion, the guanosine levels depicted in *Fig.1* (in particular the large drop in GDP receptor activation) do not comply with the GDP/GTP exchange mechanism for G protein activation.

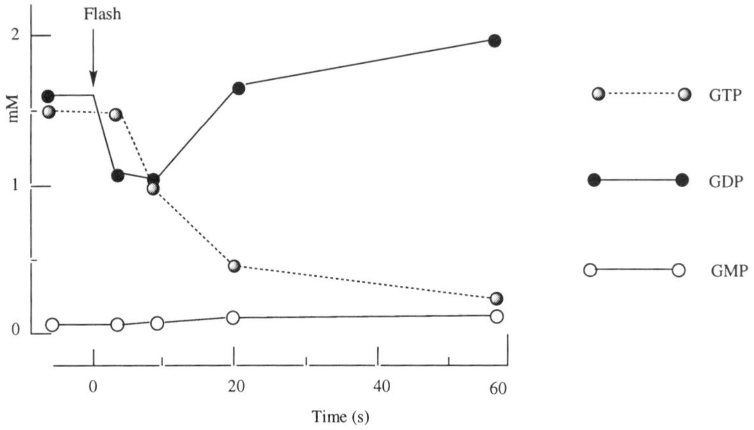

Fig.1. *Draft of (acid-) extractable GMP, GDP and GTP levels measured on intact isolated frog ROSs [12]. Time-scale starts at the beginning of a 2-ms flash of green light that bleached 1% of the rhodopsin chromophores. GTP has a half-time of approx. 10s.*

(**B**) The second finding is reported by Fung [30] and relates to influences of the G_α:$G_{\beta\gamma}$ ratio on overall GTP hydrolyis. A hydrolysis optimum of 10 is found for such a ratio [30]. In the GDP/GTP exchange scheme the influence on GTP hydrolysis of this ratio is not obvious, because here deactivation is explained by one sole GTP hydolysis at the G_α*. Because the amount of trimers designates GTP binding within the exchange scheme, this implies that the more trimeric G proteins are present, the higher GTP binding upon receptor activation and, in the end, the higher the GTP hydrolysis. Within the exchange theory it is thus expected that a G_α:$G_{\beta\gamma}$ ratio yielding as many G_α subunits in the trimeric form as possible (i.e., a ratio of 1 or even less) will lead to a GTP binding as high as possible and hence to an overall GTP hydrolysis as high as possible.

(**C**) The third finding stems from phosphorylation reactions reported for several types of G proteins [e.g. 31-33]. A phosphate transfer has been described which proceeds via the G_β unit of G proteins: a phosphorylated histidine residue on the G_β subunit becomes dephosphorylated upon G protein activation, resulting in an overall phosphorylation of GDP to GTP. This GDP phosphorylation reaction is regulated by agonist-activated receptors. Wieland and co-workers [33] suggest that these phosphorylation reactions accelerate and amplify the receptor-mediated signal. Also thiophosphorylation reactions with GTPγS, which are able to thiophosphorylate the G_β subunit of the G protein coupling to the G protein-coupled fMet-Leu-Phe receptor, have been reported [32]. The thiophosphate group can then be transferred to GDP yielding GTPγS (*in casu*, G·GTPγS) [32]. Thus, a direct phosphate transfer reaction to the GDP bound to G_α in the trimeric G protein seems to be part of the G protein activation mechanism (*vide infra*) [34]. Within the GDP/GTP exchange mechanism for G protein activation there is no place for these phosphotransfer reactions.

These phosphotransfer reactions reported for G proteins have stimulated the scientific debate. Hohenegger *et al.* [35] question the phosphotransfer from G_β toward the GDP bound at G_α within the trimeric G protein. The latter authors attached the nucleotide bound at G_α in a covalent manner by first oxidizing all hydroxyl groups (among others the 2'- and 3'-ribosyl hydroxyl group of the nucleotide) to aldehydes by adding $NaIO_4$. Subsequently, by reducing with $NaCNBH_3$ and $NaBH_4$ the nucleotide becomes covalently bound to the ε-amino group of the lysine residues (via Schiff base formation) [35], present at the nucleotide binding site at G_α [cf. 1-6]. It can, however, not be ruled out that such a harsh treatment of the proteins interferes with the phosphotransfer from G_β toward G_α (N.B. Adenylyl cyclase activity, for instance, is disturbed by these subsequent oxidation/reduction reactions) [35]. Furthermore, by applying immunoprecipitation techniques, Hohenegger *et al.* failed to detect any phosphotransfer from G_β toward the GDP bound at G_α and these authors therefore stick to the GDP/GTP exchange mechanism for protein activation [35]. Kowluru *et al.* [36], on the other hand, report experiments (in some cases even rather similar to those reported by Hohenegger *et al.* [35]) which, to their opinion, indicate that the GDP bound at the G_α subunit does act as the (thio)phosphate acceptor and claim that G protein activation indeed starts with a GTP synthesis. Kowluru *et al.* [36] also show that the immunoprecipitation procedure itself can affect the phosphorylation state of the G protein (the G_β looses its phosphate group during the immunoprecipitation), which of course prohibits any GTP synthesis at the G protein. For a detailed review of these experiments *pro* and *contra* phosphotransfers at the G protein, we refer to Nederkoorn *et al.* [37]. We conclude that current evidence is in favour of a GTP synthesis mechanism at the trimeric $G_{\alpha\beta\gamma}$ protein, rather than a GDP/GTP exchange mechanism.

Bioenergetics. - The proton pumping theory of Timms and co-workers [18] has stimulated us to compare the activation of ternary receptor complexes and the field of bioenergetics. At the centre of the bioenergetical theory, two proteins play a crucial role by forming a closed proton circuit: a primary pump pumps protons against a gradient, and a secondary pump consumes this gradient. The consumption of the proton gradient delivers the energy to synthesize ATP (*Fig. 2*). The circuit is closed [38], which means that the protons pumped by a primary pump do not diffuse into the extracellular bulk, but rather stick to the membrane and are transported along the membrane directly to a sink (i.e., a secondary pump). The proton circuit can be short-circuited by so-called uncouplers such as carbonyl cyanide *m*-chlorophenylhydrazone (CCCP), which function as highly selective proton carriers nullifying any Δp.

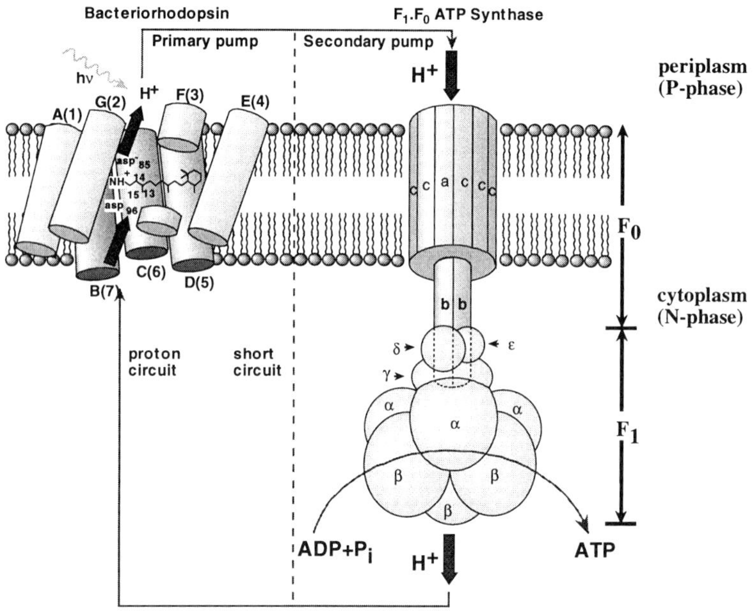

Fig. 2. *A simple (closed) proton circuit in an energy-transducing membrane. Light-activation results in the primary pump bacteriorhodopsin pumping protons "uphill" from the inside to the outside. A proton gradient, Δp, is established resulting in a negative charge for the inside and a positive charge for the outside. The secondary pump, F_1·F_0-ATP synthase, uses Δp by translocating the pumped protons (via the F_0 unit) and synthesizing ATP from ADP and P_i (in the F_1 unit). a, b, c and α, β, γ, δ, ε indicate the different subunits of the F_0 and F_1 unit, respectively. Short-circuiting can be induced by adding CCCP.*

A New Mechanism for Ternary Complex Activation, Proton-Transporting Receptors and GTP Synthesis. - Our new mechanism, presented in *Fig.3*, starts with G protein binding to an activated receptor (induced by a hormone, neurotransmitter, odorant or light). In accordance with the two-state model only the activated state of R* is presumed to bind the appropriate G protein. We do, however, not restrict ourselves to one unique conformational change (R=>R*), but rather interpret R* as the state of the GPCR able to couple to a G protein (N.B. High affinity agonist binding does not guarantee G protein binding to a GPCR; *vide supra* [26]). The next step in our scheme is a GTP synthesis reaction at the G protein. At this GTP synthesis level (1) (*Fig.3*) we predict amplification. This step explains why GDP concentrations drop dramatically on rhodopsin activation monitored in ROSs: because GDP (and a phosphate, P_i) is consumed in the synthesis process. Simultaneously, G_α changes into G_α*, the latter having a closed nucleotide binding site [1-6,43] (preventing acid-extraction of the

GTP molecule and explaining why the mass balance of guanosine levels depicted in *Fig.1* seems to be incomplete). In strong analogy with ATP synthases we propose that both the ternary complex HR*G and R*G (basal activity) convert GDP plus P_i into GTP with the aid of the translocation of protons (by consuming Δp the necessary energy is delivered into the G protein; *Fig.4*). Hence, the ternary complex acts as both a secondary proton pump and a GTP synthase.

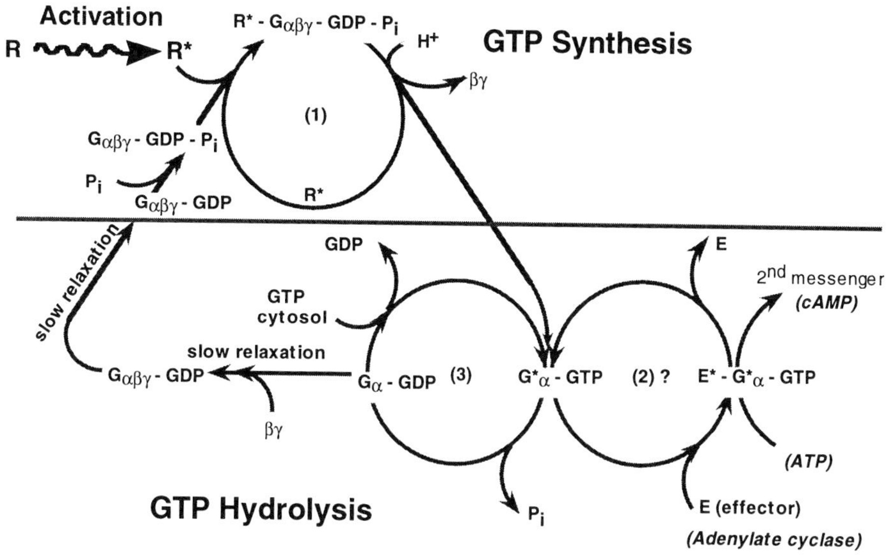

Fig.3 *A new scheme for G protein activation, see text for further details.*

Applying Timms and co-workers' proton pump mechanism [18], protons are introduced into the G protein. The P_i which is needed for the GTP synthesis originates from the phosphorylated histidine at the G_β subunit. Phosphoramidates are acid labile [39] and the introduction of a proton to the appropriate (phosphorylated) histidine at G_β liberates the P_i. The introduction of P_i at G_α helps in the conversion of G_α into the activated G_α* (precisely as AlF_4^- can induce the activated state of the G_α subunit). It is important to notice that the molecular mechanism of G protein coupling and the actual G protein activation mechanism (via proton translocation) are clearly distinct in our scheme. Whether or not the GTP synthesis step is a *conditio sine qua non* for G protein activation may be tested experimentally with the GDP analogue [NH]ppG. Since [NH]ppG cannot participate in a synthesis reaction, loading the G protein with [NH]ppG in the absence of GDP should inactivate the G protein. It is noted that GDPβS should not be considered as a negative control, since Kowluru *et al.* [36] have demonstrated that this GDP analogue can indeed be converted into GTPβS, leading to activated G_α*·GTPβS complexes, which strongly supports our GTP synthesis hypothesis [34].

Whether one G_α*·GTP can activate more than one effector molecule is not known at present [10], which explains the questionmark at level (2) in *Fig.3*. Depending on the local [GTP]:[GDP] and $[G_\alpha]$:$[G_{\beta\gamma}]$ ratios, it can be thermodynamically favourable for a dissociated G_α subunit to hydrolyse several GTP molecules (amplification level (3)), thereby flipping between the G_α and $G_\alpha{}^*$ state. In other words, the dissociated G_α subunits are considered to be involved in the "classical" GDP/GTP exchange mechanism. During the dissociated state of the G protein, both $G_\alpha{}^*\cdot$GTP and $G_{\beta\gamma}$ can couple to effector molecules. Effectors thus stabilize the dissociated state of the G protein, thereby prolonging this dissociated state of the G protein. Since in *Fig.3* only dissociated G_α subunits are seen to participate in the GDP/GTP exchange (each exchange is followed by GTP hydrolysis), it can now be explained why (at least some) effector molecules have been reported to make the GTPase activity of the G protein more efficient (effector molecules can speed up GTP hydrolysis from 5 min^{-1} up to 50-250 min^{-1}) [41-43, and refs. *loc. cit.*]. Given a half-time of 10s (*Fig.1*) in which 0.5 mM acid-extractable GTP is converted into GDP (i.e., 3 mM/min [12]) and a k_{cat} of 50-250 min^{-1}, the amount of G protein (i.e., transducin) that is (at least) present in ROSs can be estimated to range from 0.6 up to 3 mM which is comparable to the known concentration of rhodopsin being approx. 3 mM [7].

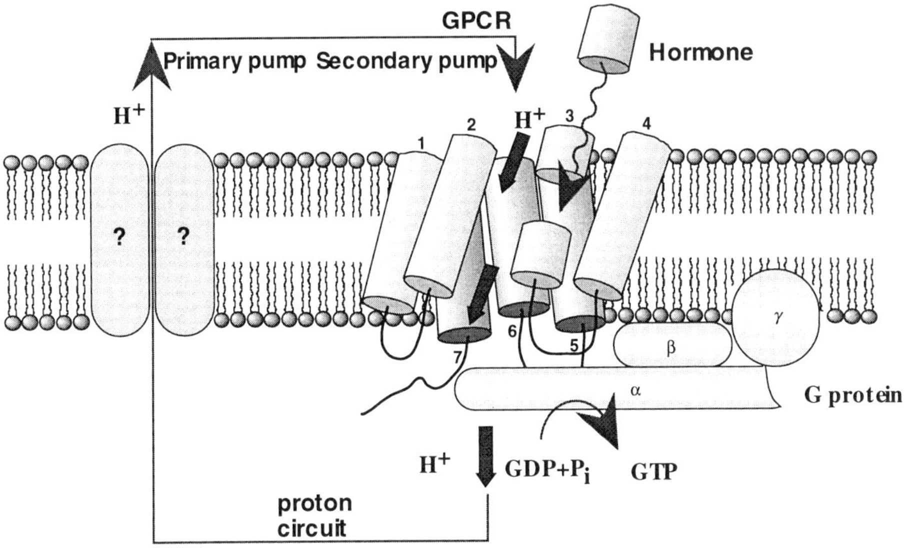

Fig.4. *A (Not yet identified) primary pump generates a protonmotive force, which is consumed by a secondary pump: the ternary complex. This enables GTP synthesis at the G_α subunit, which in turn triggers $G_\alpha{}^*\cdot$GTP dissociation. The presence of primary pumps is confirmed by experiments with the Cytosensor Microphysiometer System [40], which monitors primary pump activities upon GPCR activation.*

GTPase activity of $G_\alpha{}^*\cdot$GTP results in the formation of G_α. Subsequently, G_α can either bind another GTP molecule from the cytosol accompanied by the release of GDP and P_i (amplification level (3) for which the presence of effector molecules and a high enough local concentration of GTP are important) leading to $G_\alpha{}^*$ with effector coupling and hydrolysis activity again, or G_α can reassociate with $G_{\beta\gamma}$. This latter reassociation results in the $G_{\alpha\beta\gamma}$ trimer formation and inactivation. To our opinion, not the intrinsic

GTP hydrolysis but the recombination to the trimer represents the real inactivation of G proteins. This is in perfect accordance with the observation that $G_{\beta\gamma}$ suppresses the "noise" generated by ligand-unoccupied receptors [44], i.e., $G_{\beta\gamma}$ is known to lower basal activity. Furthermore, since the trimer reassociation is considered to be the inactivation step of the G protein, it can now also be explained why the G_{α}:$G_{\beta\gamma}$ ratio can influence the overall GTP hydrolysis directly [30]. *Obiter dictum*, it is noted that in ROSs an inhibitory factor for the effector is available that is independent of GTP hydrolysis [45]. This inhibitory factor, which remains to be identified, accounts for the necessary quick deactivation process rather than the G protein's intrinsic GTP hydrolysis [45].

Summarizing, in our model G protein activation starts with GTP synthesis, yielding G_{α}*·GTP and $G_{\beta\gamma}$. As in the classical model, we assume that two states of the dissociated G_{α} subunit exist: one with a closed nucleotide binding site, G_{α}*, and one with a more open site G_{α}. The only difference between level (3) activity of the dissociated G_{α} subunit (*Fig.3*) compared to the classical model is the possibility that a dissociated G_{α} participates in the classical GDP/GTP exchange mechanism without trimer interference; effector molecules can speed up this exchange (N.B. In the absence of $G_{\beta\gamma}$, activated GPCRs are known to catalyze exchange reactions at G_{α} subunits, hence R* may control amplification level (3) as well). In other words, the trimeric G protein is unable to be involved in GDP/GTP exchange reactions, whereas a once dissociated G_{α} subunit is considered to have a sufficiently low barrier of activation for this exchange reaction. If enough GTP is present in the direct surroundings of the dissociated G_{α}, it will cycle between the G_{α} and G_{α}* state (level (3), *Fig.3*). In contrast to GTP and GTPγS, p[NH]ppG is unable to participate in transphosphorylation reactions and will thus bind to and activate G proteins via the classical nucleotide exchange mechanism (level (3) in our scheme). Since p[NH]ppG is expected to be involved at level (3) (exchange) only, whereas GTPγS is indicated to interact both at levels (1) (synthesis) and (3) (exchange), this may explain why GTPγS activates the G protein pool up to a two-fold higher level than p[NH]ppG does [32].

The Role of the Agonist in Signal Transduction and a Three-Dimensional Model of a Ternary Complex. - In our model, protons are transported via so-called proton wires. These wires exist of a hydrogen-bonded chain of polar groups. The polar character of the receptor's central cavity enables the formation of extended hydrogen-bonded networks via which protons can be transported. In the resting state of the receptor (R) the proton wire within the cavity is considered to be interrupted. In other words, the receptor in its R state is unable to translocate any protons. This blockade of the proton wire inside the receptor is caused by either the presence of a so-called hydrophobic gate (i.e., the hydrogen-bonded chain is interrupted by the side chain(s) of hydrophobic amino acid(s); for an example of the presence and function of such a hydrophobic gate in the proton pump mechanism of BR we refer to Dencher *et al.* [46]) or by the absence of one essential entity in the chain. The function of an agonist is either to open the hydrophobic gate (cf. the activation mechanism of BR [46]) or to bind to that position where an entity is missing in the proton wire. In the latter case, Topiol's deletion theory is applicable: after binding of the agonist, an active proton wire is constituted via which protons can be transported (*Figs.4 and 5*). For the G protein-coupled opsins, the "ligand" is light, isomerizing retinal.

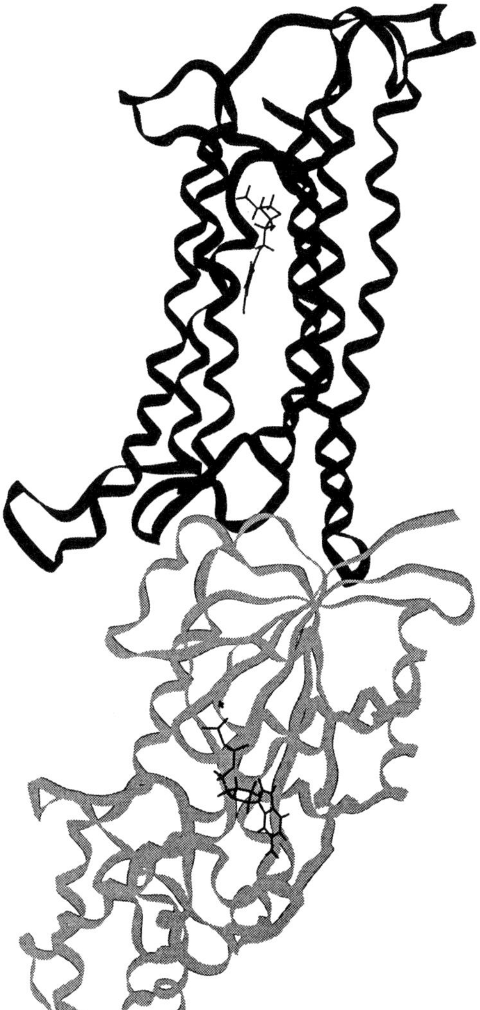

Fig.5. *A three-dimensional model of an agonist-β_1-adrenoceptor-$G_{S\alpha}$ complex. The α-carbons of the receptor's seven α-helical proton pump model and noradrenaline are depicted in dark gray. The $G_{S\alpha}$ subunit is coloured light-gray with the guanine nucleotide in black. The figure is rendered with the programme Setor [47]. For further details the reader is referred to Nederkoorn et al. [48].*

Basal receptor activity may be explained by the presence of transient water molecules inside the receptor's cavity. These water molecules either temporarily fill the empty position in the proton wire or open the hydrophobic gate (cf. ref. [46]). In both ways these transient water molecules contribute to the formation of a weakly active proton wire. The proton conductance of an active translocating proton wire due to the presence of an agonist is, however, considered to be higher than a wire in which (a-specific) water molecules participate. A proton wire with a partial agonist is suggested to have a lower proton conductance than a comparable hydrogen-bonded chain to which a full agonist binds. It is noted that different agonists can contribute in distinct ways to the constitution of an active proton-transporting hydrogen-bonded chain. In this way, the phenomenon of different agonistic binding sites for different agonists (e.g. ref. [28]) can now be addressed as well.

Raising the temperature results in an impaired GPCR-mediated physiological response. Since the proton conductance of an ordered hydrogen-bonded chain is higher than that of a distorted chain (N.B. The proton conductance of ice is higher than that of water), we explain this temperature effect by regarding an ordered proton wire to have more efficient proton transport abilities. Inverse agonists are considered to distort the proton wire (e.g. by expelling the aforementioned a-specific transient water molecules from the wire or by acting as a hydrophobic gate itself).

Timms and co-workers' proton pump mechanism has been combined with the GTP synthase concept to construct a three-dimensional model for a ternary complex consisting of a heptahelical model for the β_1-adrendergic receptor (based on the structure of BR), the endogenous ligand noradrenaline, and the $G_{s\alpha}$ subunit [48] (*Fig.5*). The presence of a G protein is found to be essential for proton pumping, which prevents an uncoupled receptor from short-circuiting a protonmotive force. As a next step in developing the new model, we currently construct a ternary complex model using X-ray data on trimeric G proteins [1-6]. These three-dimensional ternary complex models enable a close inspection of G protein coupling to the appropriate GPCR, provide new insights into the actual signal transduction process mediated by the GPCR, and offer the means of examining the possible influences of mutations on these molecular phenomena.

For further details of the molecular mechanisms underlying transmembrane signalling, the thermodynamics of receptor and G protein activation, and in particular proton transfer processes, we refer to Nederkoorn *et al.* [35,37,48]. We end by quoting the late Thomas Samuel Kuhn [49]: "(..) since no paradigm ever solves all the problems it defines and since no two paradigms leave all the same problems unsolved, paradigm debates always involve the question: Which problems is it more significant to have solved?" (*sic*)

REFERENCES

[1] J.P. Noel, H.E. Hamm, P.B. Sigler, *Nature* **1993,** *366,* 654.

[2] D.G. Lambright, J.P. Noel, H.E. Hamm, P.B. Sigler, *Nature* **1994,** *369,* 621.

[3] D.E. Coleman, A.M. Berghuis, E. Lee, M.E. Linder, A.G. Gilman, S.R. Sprang, *Science* **1994,** *265,* 1405.

[4] M.A. Wall, D.E. Coleman, E. Lee, J.A. Iñiguez-Lluhi, B.A. Posner, A.G. Gilman, S.R. Sprang, *Cell* **1995,** *83,* 1047.

[5] D.G. Lambright, J. Sondek, A. Bohm, N.P. Skiba, H.E. Hamm, P.B. Sigler, *Nature* **1996,** *379,* 311.

[6] J. Sondek, A. Bohm, D.G. Lambright, H.E. Hamm, P.B. Sigler, *Nature* **1996,** *379,* 369.

[7] L. Stryer, *Ann. Rev. Neurosci.* **1986,** *9,* 87.

[8] A.G. Gilman, *Angew. Chem.* **1995,** *107,* 1533.

[9] G.F.X. Schertler, C. Villa, R. Henderson, *Nature* **1993,** *362,* 770.

[10] P. Samama, S. Cotecchia, T. Costa, R.J. Lefkowitz, *J. Biol. Chem.* **1993,** *286,* 4625.

[11] P. Leff, *Trends Pharmacol. Sci.* **1995,** *16,* 89.

[12] W.E. Robinson, W.A. Hagins, *Nature* **1979,** *280,* 398.

[13] W. Schütz, M. Freissmuth, *Trends Pharmacol. Sci.* **1992,** *13,* 376.

[14] T.P. Iismaa, T.J. Biden, J. Shine, in: 'G Protein-Coupled Receptors', R.G. Landes Company, Bioscience Publishers, Austin, USA, 1996.

[15] H. Jung, R. Windhaber, D. Palm, K.D. Schnackerz, *FEBS Lett.* **1995,** *358,* 133.

[16] H. Weinstein, in 'Trends in Ecological Physical Chemistry, Proceedings of the 2nd International Workshop on Ecological Physical Chemistry', Eds. L. Bonati, U. Cosentino, M. Lasagni, G. Moro, D. Pitea and A Schiraldi, Elsevier, Amsterdam, 1993, pp. 1-16.

[17] L. Oliveira, A.C.M. Pavia, C. Sander, G. Vriend, *Trends Pharmacol. Sci.* **1994,** *15,* 170.

[18] D. Timms, A.J. Wilkinson, D.R. Kelly, K.J. Broadley, R.H. Davies, *Receptors & Channels* **1994,** *2,* 107.

[19] S. Topiol, *Trends Biochem. Sci.* **1987,** *12,* 419.

[20] P.H.J. Nederkoorn, P. Vernooijs, G.M. Donné-Op den Kelder, E.J. Baerends, H. Timmerman, *J. Mol. Graphics* **1994,** *12,* 242.

[21] J.L. Spudich, *Cell* **1994,** *79,* 747.

[22] K.P. Hofmann, *Photobiochem. Photobiophys.* **1986,** *13,* 309.

[23] S. Arnis, K.P. Hofmann, *Proc. Natl. Acad. Sci. USA* **1993,** *90,* 7849.

[24] A. Schleicher, K.P. Hofmann, *Z. Naturforsch. Teil C* **1985,** *40,* 400.

[25] L.M.F. Leeb-Lundberg, S.A. Mathis, M.C.S. Herzig, *J. Biol. Chem.* **1994,** *269,* 25970.

[26] C.D. Strader, R.A.F. Dixon, A.H. Cheung, M.R. Candelore, A.D. Blake, I.S. Sigal, *J. Biol. Chem.* **1987,** *262,* 16439.

[27] W.P. Hausdorff, M. Hnatowich, B.F. O'Dowd, M.G. Caron, R.J. Lefkowitz, *J. Biol. Chem.* **1990,** *265,* 1388.

[28] R. Leurs, M.J. Smit, R. Meerder, A.M. Ter Laak, H. Timmerman, *Biochem. Biophys. Res. Comm.* **1995,** *214,* 110.

[29] O.P. Ernst, K.P. Hofmann, T.P. Sakmar, *J. Biol. Chem.* **1995,** *270,* 10580.

[30] B.K.-K. Fung, *J. Biol. Chem.* **1983,** *258,* 10495.

[31] T. Wieland, B. Nürnberg, I. Ulibarri, S. Kaldenberg-Stasch, G. Schultz, K.H. Jakobs, *J. Biol. Chem.* **1993,** *268,* 18111.

[32] S. Kaldenberg-Stasch, M. Baden, B. Fesseler, K.H. Jakobs, T. Wieland, *Eur. J. Biochem.* **1994,** *221,* 25.

[33] T. Wieland, S. Kaldenberg-Stasch, B. Fesseler, K.H. Jakobs, *Can. J. Physiol. Pharmacol.* **1994,** *72,* S5.

[34] P.H.J. Nederkoorn, H. Timmerman, G.M. Donné-Op den Kelder, *Trends Pharmacol. Sci.* **1995,** *16,* 156.

[35] M. Hohenegger, T. Mitterauer, T. Voss, C. Nanoff, M. Freissmuth, *Mol. Pharmacol.* **1996,** *49,* 73.

[36] A. Kowluru, S.E. Seavey, C.J. Rhodes, S.A. Metz, *Biochem. J.* **1996,** *313,* 97.

[37] P.H.J. Nederkoorn, H. Timmerman, G.M. Donné-Op den Kelder, in 'Signal Transduction by G Protein-Coupled Receptors. Bioenergetics and G Protein Activation: Proton Transfer and GTP Synthesis to Explain the Experimental Findings', R.G. Landes Company, Bioscience Publishers, Austin, USA, 1996.

[38] J. Heberle, J. Riesle, G. Tiedemann, D. Oesterhelt, N.A. Dencher, *Nature* **1994,** *370,* 379.

[39] R.L. Van Etten, P.P. Waymack, D.M. Rehkop, *J. Amer. Chem. Soc.* **1974,** *96,* 6782.

[40] H.M. McConnell, J.C. Owicki, J.W. Parce, D.L. Miller, G.T. Baxter, H.G. Wada, S. Pitchford, *Science* **1992,** *257,* 1906.

[41] B.R. Conklin, H.R. Bourne, *Cell* **1993,** *73,* 631.

[42] C. Kleuss, A.S. Raw, E. Lee, S.R. Sprang, A.G. Gilman, *Proc. Natl. Acad. Sci.* **1994,** *91,* 9829.

[43] E.J. Neer, *Cell* **1995,** *80,* 249.

[44] L. Birnbaumer, R. Mattera, A. Yatani, J. Codina, A.M.J. VanDongen, A.M. Brown, in 'ADP-Ribosylating Toxins and G Proteins. Insights into Signal Transduction', Eds. J. Moss, M. Vaughan, Am. Soc. Microbiol., Washington D.C., 1990, pp. 225-266.

[45] M.A. Erickson, P. Robinson, J. Lisman, *Science* **1992,** *257,* 1255.

[46] N.A. Dencher, G. Büldt, J. Heberle, H.-D. Höltje, M. Höltje, *NATO ASI SER, Ser B* **1992,** *291,* 171.

[47] S.V. Evans, *J. Mol. Graphics* **1993,** *11,* 134.

[48] P.H.J. Nederkoorn, H. Timmerman, G.M. Donné-Op den Kelder, D. Timms, A.J. Wilkinson, D.R. Kelly, K.J. Broadley, R.H. Davies, *Receptors & Channels* **1996,** *4,* 111.

[49] T.S. Kuhn, in 'The Structure of Scientific Revolutions', University of Chicago Press, Chicago, 1962.

PART IV. **3D-QSAR AND MOLECULAR MODELLING**

Chapter 34

Complex Systems: On the Simulation of Nature and the Nature of Simulation. The Chaotic Hierarchy and Interactive New Media Installations

by **Michael Klein**

INM-Institut für Neue Medien, Daimlerstrasse 32,
D-60389 Frankfurt/Main

Complex Systems:
On the Simulation of Nature and the Nature of Simulation

The Chaotic Hierarchy and Interactive New Media Installations

by **Michael Klein**

INM-Institut für Neue Medien, Daimlerstrasse 32, D-60389 Frankfurt/Main

---

With a first superficial glance Computer-Assisted Lead Finding and Optimization has nothing in common with Interactive New Media Art Installations though we would accept that Quantitative Structure-Activity Relationships are a problem of Complex Systems analysis. This article is based upon a second glance on the scientific simulation of models of complex systems and some problems arising from the need of numerical computation and visualization to handle nonlinear systems. With the lack of a formal scientific methodology to cope with the entanglement of reality and virtuality in computational theory and computer-based experiments it will only outline some ideas of the New Media Art discussion concerning interactive experiences.

On the Simulation of Nature. - Since the invention of differential calculus in the late 17th century (Newton, Leibniz) differential equations became the dominating tool to model nature and simulate dynamical behavior.The promise of the Laplace demon [1] to predict dynamical behavior of the past and the future by only knowing the differential equations of motion and the exact initial conditions has born the paradigm of classical determinism. The mechanistic clockwork universe dominating philosophical and scientific thinking over the past three hundred years was born. Many fundamental physical systems are modelled by linear and, therefore, mathematically analytically solvable ordinary differential equations (ODEs). Though the dynamics of linear ODEs is limited to fixed-point behavior and period or quasiperiodic oscillations classical mechanics was extremely successful because it worked fine for many physical problems.

In 1978 Otto E. Rössler [2] invented the science of blending as a serious part of mathematics when studying the mechanical mixing of a so called "taffy puller". He put the taffy puller on a platform, rotated in synchronity with it, to build the "lazy susan taffy puller" *(Fig.1)*. The three-dimensional mixing machine is one of the most simple examples of chaos. The sketches of some consecutive stroboscopic views of a complete turn of the rotating taffy puller *(Fig.2)* illustrate the main mechanism of chaos: stretching plus folding in a bounded region implies perfect mixing.

Beginning with the taffy puller and related systems (i.e. the water-wheel of Lorenz [3]) a totally new dynamical quality was described, deterministic chaos. It came as a surprise that not only external stochastic forces may generate irregular dynamics. Three-dimensional nonlinear ODEs may generate deterministic chaos. Low-dimensional dissipative dynamical systems far from thermodynamical equilibrium may have arbitrary complex dynamics. Deterministic chaos is the rule, linear order only a rare ideal state of dynamical systems.

The Rössler Attractor. - To find a simple mathematical model of the rotating taffy puller one may imagine a raisin or small light bulb suspended in the taffy. Its movement in time may give a chaotic path in three dimensions. The possibly simplest mathematical system [4] modelling the qualitative features of the rotating taffy puller is the three-variable time-continuous ODE:

$$\begin{aligned} x' &= -y - z \\ y' &= x \\ z' &= a - y^2 - bz \end{aligned} \tag{1}$$

with x,y,z,a,b in **R**, and appropriate initial values. *Figure 3* shows a projection of the flow of a chaotic attractor of Eq.(1) with a=0.27675 and b=0.2. An attractor is some portion of state-space such that any point which starts nearby gets closer and closer in time. In the long run an attractor is the stable settled state of dissipative dynamics.

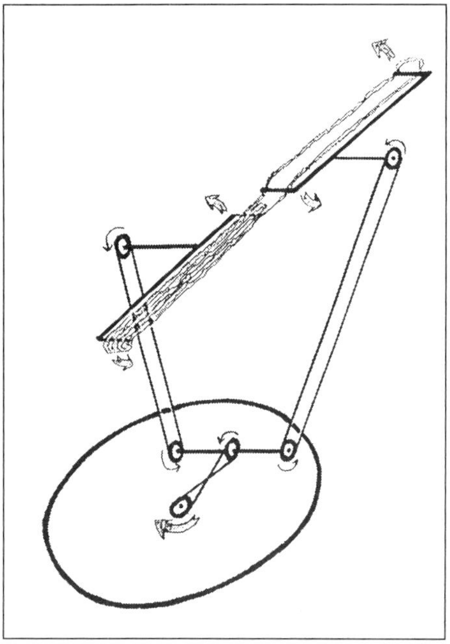

Fig.1. *„The Lazy Susan"*

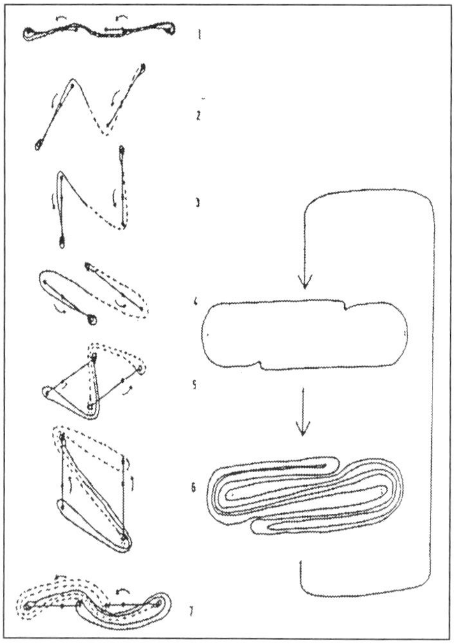

Fig.2. *Folding and stretching*

Poincare Maps. - Due to the necessity of numerical integration the analysis of nonlinear ODEs is at least time consuming. To Poincare we own the idea of the Poincare cross-sections (PCS) [5]. The PCS consists of the transversal intersection points of the N-dimensional flow of an ODE with an (N-1)-dimensional hyperplane. It is equivalent to a stroboscopic view of the continuous trajectories. The resulting Poincare maps are time-discrete maps with the same qualitative features as the corresponding flows (one may think of analog and digital representation of time). Instead of numerically integrating N-dimensional ODEs it is possible to gain the same dynamics simply by iterating explicit (N-1)-dimensional time-discrete maps.

A Time-Discrete Toy-Model. - Here a most simple yet remarkable potential toy-model of a three-dimensional time-discrete map should be introduced to just give an overview of possible low-dimensional complex dynamics and structural features. Instead of giving an extended collection of mathematical analysis of the system which can be found elsewhere [6] the presentation will concentrate on the underlying principles of chaotic dynamics and fractal patterns derived from a structural understanding of the dynamics. The following generic three-variable time-discrete map [7] exhibits spheroidal dynamics:

$$x_{i+1} = ax_i + f(z_i, b)$$
$$y_{i+1} = x_i$$
$$z_{i+1} = y_i \qquad (2)$$

whereby x,y,z,a,b in **R**; i in **N** and (a>0, b³1). The quadratic nonlinearity in variable x is governed by a control parameter a. The dissipation parameter b controls the „dissipation function"

$$f(z,b) = -bz + (b-1)z^3 \qquad (3)$$

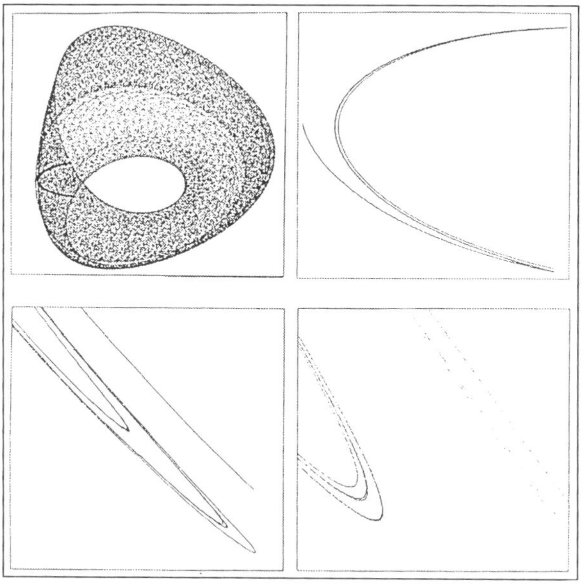

Fig.3. *Chaotic attractor and three Poincare cross-sections of Eq. (1).*

The Conservative Case. - In the case f(z,b=1)=-z, the second-order nonlinear function in variable x is combined with a linear delay mechanism to yield a 3D map with high symmetry and a volume-preserving („conservative“) dynamics under iteration. Here „conservative“ means conservation of state-space volume. The conservative map is the generating mechanism for the dynamics. The solutions of *Equation (2)* live on non-toroidal but smooth spheroidal spheres of constant energy around the origin. The linear functions in the second and third variable y and z are two discrete „delay steps“. The delayed resulting variable z is coupled back to x via a dissipation function f(z,b). Depending on the initial conditions (x_0,y_0,z_0) and the control parameter a, periodic as well as quasiperiodic and volume-preserving chaotic dynamics can be found (*Fig.4*).

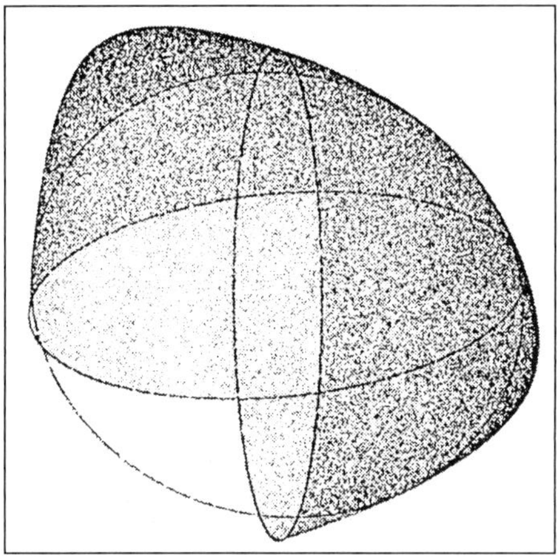

Fig.4. *A conservative chaotic solution of Eq.(2).*

The Dissipative Case. - In the dissipative case of *Equation (2)* with (b>0) the dissipation function *Equation (3)* has a destabilizing linear part and a stabilizing cubic part. The cubic dissipation function f(z,b) and the linear second-order delay give rise to stable bounded solutions around the origin.

Every type of dynamics of a system like (2) is characterized by the spectrum of its Lyapunov characteristic exponents (LCEs). The spectrum of LCEs is the ordered series of all LCEs ($\lambda 1 \geq \lambda 2 \geq \lambda 3$) of one particular dynamical state (see *Fig. 5*).

An overview of the dynamical complexity can be gained from a scan of the parameter space (a,b) (*Fig.6a*). Each parameter point (a,b) is color-coded according to the individual spectrum of LCEs of the corresponding attractor dynamics. Depending on the values of parameters (a,b) periodic, quasiperiodic, simple chaotic (one positive LCE), hyperchaotic (two positive LCEs) and hyper^2-chaotic (three positive LCEs) stable attractors can be found.

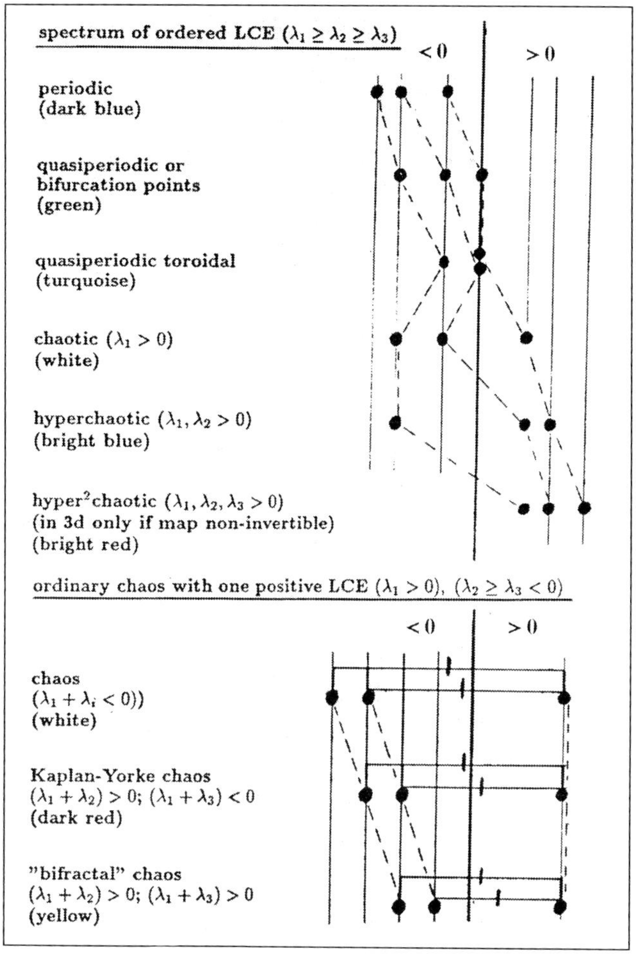

Fig.5. *Spectrum of ordered LCE ($\lambda 1 \geq \lambda 2 \geq \lambda 3$) (top) and conditions for ordinary chaos with one positive LCE ($\lambda 1 > 0$), ($\lambda 2 \geq \lambda 3 < 0$) (bottom)*

The Chaotic Hierarchy. - The idea of the chaotic hierarchy [2] orders different chaotic states according to their number of positive LCEs. The map of (2) incorporates a delay mechanism which has proved to be one way to gain arbitrarily complex chaotic dynamics, i.e. the chaotic hierarchy [4]. Thus the first steps on the ladder towards higher complexity may be found in the present 3D toy-model. Beyond the the so-called Feigenbaum point there are simple chaotic states with an LCE-spectrum (+,-,-). Hyperchaotic states with an LCE-spectrum of (+,+,-) will eventually follow. Because the map of (2) is not necessarily invertible, there are also stable chaotic solutions with three positive LCEs (+,+,+).

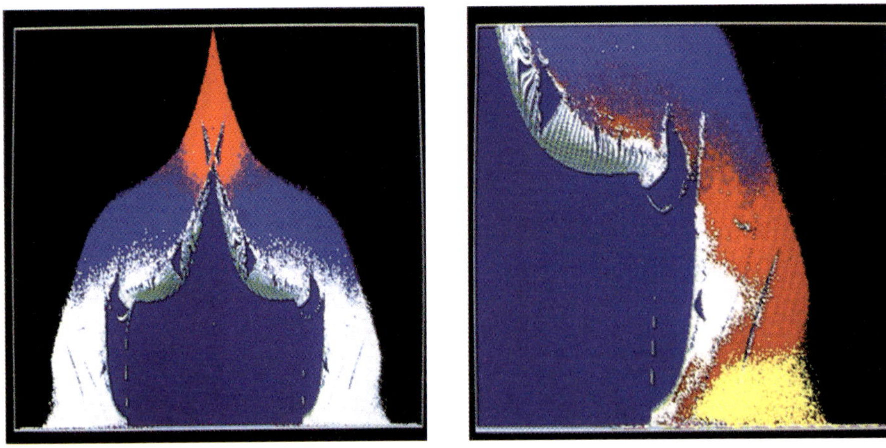

Fig.6. *Scan in parameter space (a,b), left 6a overview, right 6b zoom, color-coded according to the LCE-spectrum of the corresponding attractors. Black: no stable attractors; dark blue: periodic solutions; bright green: quasiperiodic solutions or bifurcations; white: simple chaos (left) or ordinary chaos (right); dark red: Kaplan-York chaos; yellow: bifractal chaos; bright blue: hyperchaos; bright red: hyper^2-chaos.*

Ordinary Chaos typically yields striated, infinitely-often folded line structures with fractal dimensions $1 \leq$ D_frac <2. The corresponding cross-sections (numerically calculated as slices through the state space structure) are Cantor-like patterns and possess - due to the reduction of dimension by one - fractal dimensions of less than unity $0<$ D_frac <1.

Hyper-Chaotic Attractors resemble „folded towels" which are fractals with fractal dimension $2 \leq$ D_frac < 3. These structures can be viewed as hyperplanes of sets of points which are stretched and folded infinitely-often along two directions of state space *(Fig. 7a)*. The corresponding slices show the folded line patterns of Henon-type attractors [8] with fractal dimensions D_frac <2, again according to the reduction of dimension by one.

Hyper^2-Chaotic Dynamics in the case of *Equation (2)* yield space-filling cube-like attractors *(Fig.7b)* with dimension D_frac = 3, whose cross-sectional slices show folded-towel-like patterns. One may think of these attractors as the final state of a procedure in which a state-space volume is stretched and folded infinitely often along three directions of state-space.

The building-block principle of the chaotic hierarchy, which means stretching and folding extended to higher and higher topological dimensions, is clearly visible. In the spirit of the chaotic hierarchy system (2) can be generalized to higher embedding dimensions [9].

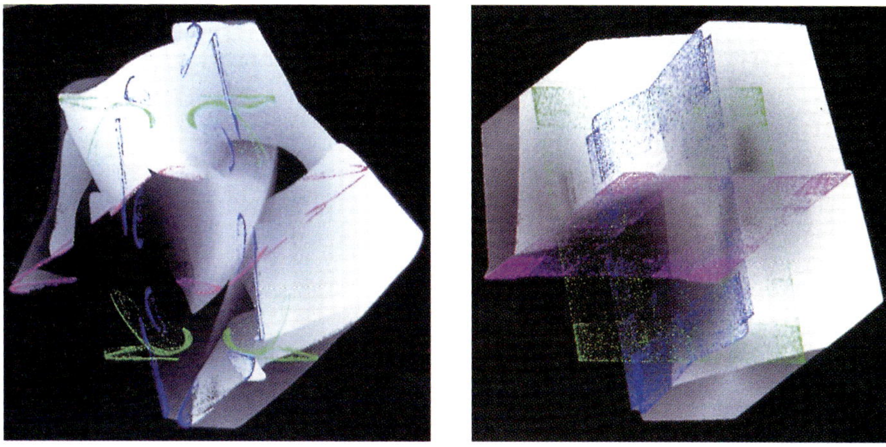

Fig.7. *A hyper-chaotic solution (left 7a) and a hyper^2-chaotic orbit (right 7b) of Eq.(2)*

Three Cases of Simple Chaotic Dynamics. - It came as a surprise that there are much more structurally different solutions to the map of (2). Three different types of simple chaotic dynamics with only one positive LCE ($\lambda 1>0$; $\lambda 2$, $\lambda 3 < 0$) can be found. In the case of the attracting dynamics, the transitions between these three different types are found in parameter space (a,b) when the sums over two exponents ($\lambda 1+\lambda 2$) or ($\lambda 1+\lambda 3$) of the ordered LCEs ($\lambda 1 \geq \lambda 2 \geq \lambda 3$) change their sign *(Fig. 6b)*. The only interpretation of the behavior for the time being is a correlation between local features of the dynamics and the global fractal structure of the overall attractor, i.e. there is a competition between local exponential divergence (which is the source of chaos in the attractive case) and the local exponential convergence (as source of the attractiveness) [10]. In a 3D map (2) one will find local convergence with two negative local exponents along directions tangential to the stable manifolds which are locally transversal to the unstable manifold(s). Three types of simple chaotic dynamics may be distinguished:

1. Ordinary Chaotic Dynamics is found when ($\lambda 1+\lambda 2<0$) (white areas in *Fig.6b)*. On average the local contractions dominate over the local divergence. State-space is thus locally contracted to lines. These are parallel to the unstable manifold, giving rise to the striated fractal structures known as Henon-type attractors [8] with fractal dimension between one and two *(Fig.8a)*.

2. Kaplan-Yorke Chaos [11] is generated if ($\lambda 1+\lambda 2>0$) but ($\lambda 1+\lambda 3<0$) (dark red areas in *Fig.6b)*. The local divergence dominates over one of the the local convergences. In the vicinity of the unstable manifold the orbits are no longer contracted toward a line before they diverge. As a consequence the attractor blows up to form a „folded curtain“ structure, which looks like a hyper-plane infinitely often folded along one topological direction of state-space. The attractor has a dimension between two and three *(Fig.8b)*.

3.„Bi-fractal“ Chaos (according to the classification of Röhricht *et al.* [12]) can apparently be found if ($\lambda 1+\lambda 2>0$) and ($\lambda 1+\lambda 3>0$) (yellow areas in *Fig. 6b)*. On the average the local convergences are both smaller than the local divergence. Therefore, the attractor appears now fractalized in two directions of state-space giving a „nowhere-differentiable“ structure along these directions. In comparison to the Kaplan-Yorke chaos one will find an even more space-filling picture with fractal dimension around 2.4 *(Fig.8c)*.

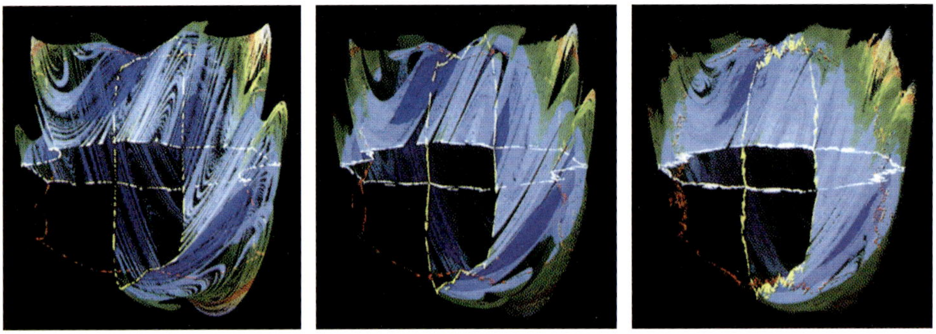

Fig.8a-c. *An ordinary chaotic solution (left), a Kaplan-Yorke chaos (middle) and a bifractal solution (right) of Eq.(2).*

From a purely mathematical point of view, the existence of different fractal structures and especially the proof of nowhere differentiable patterns existing is far from being analytically accomplished. Global nowhere-differentiability is proven only for Weierstrass functions [13], although a generalization to iterative functions using the present idea of competition of local dynamical exponents as a formal aid in explaining the resulting structures is discussed by Peinke et al. [14].

A second problem in the given explanation for the different fractal patterns of the simple chaotic states is the not yet mathematically proven correlation between local dynamical features, i.e. local exponents, of the attractive dynamics and the Lyapunov characteristic exponents of the overall attractor, only defined as an average. One should emphasize that the generating mechanism of different types of fractal structures obtained from different spectral features of the underlying dynamics may be universal in the sense that the relations of local exponents of contraction and divergence determine global geometrical properties. The fundamental property of stretching and folding acting on a contracting dynamics can be expected to always generate the outlined qualitative features [15],[16]. The corresponding theory would be of much interest for many similar problems following from a correlation between local dynamics and global geometric patterns.

Problems of a Theory of Chaos. -- There is no generally accepted definition of deterministic chaotic attractors yet. Therefore the best one can do is to give a collection of some characteristic features:

Chaos is Irregular. Attractive chaotic dynamics is a bounded steady-state behavior which is neither an equilibrium point, nor periodic nor quasiperiodic oscillations.

Chaos is Mixing. The best understanding of the structural features of chaotic attractors can be achieved from Rössler´s science of blending. Chaotic attractors are the final state of a process of an infinite stretching, squeezing and backfolding of segments of state-space. If one labels the initial piece of taffy, "black" molecules left,´"white" molecules right, one can calculate that it takes only 24 rotations of the taffy puller *(Fig.1,2)* to achieve a "grey molecular sandwich" of "black" and "white" molecules (assuming the initial hight of the taffy to be one centimeter and $(\gamma_2)^{-24}$ centimeter is the width of a molecule).

A Positive Lyapunov Exponent Identifies a Chaotic Dynamics. A simple chaotic attractor of a time-continuous equation (1) or a time-discrete map (2) will have at least one positive Lyapunov Characteristic Exponent, giving the feature of stretching, and one negative LCE, the squeezing of state-space volumes towards the attracor (time continuous ordinary differential equations always have an additional zero LCE along the linear progressing trajectory in time). Lyapunov exponents are well defined numbers which are the mean value over time of the local dynamical features along typical trajectories of the dynamical system.

Chaos is Unpredictable. Chaos has sensitive dependence on initial conditions (SDIC) [17] which means exponential divergence of neighboring states. The fundamental assumption similar reason--similar result does not hold anymore in nonlinear chaotic systems. One may think of two raisins suspended in the taffy *(Figs. 1,2)*. Due to the stretching and folding, initially neighboring raisins (points) will exponentially diverge whereas previously distant points are brought closely together.

Chaos Creates Fractals. Attractive chaotic states are "strange attractors" [18] with fractal dimensions. They can be seen as infinitely often folded "layers" of sets of points of state space (puff-paste). Fractal means non-integer valued dimension (for extended treatment of fractal dimension refer to [19],[20]).

Chaos and Images. - Besides all the incompleteness in the theory of chaos there is a good chance to proceed in the formalism of the theory . Yet some more and very fundamental problems followed from the research of chaos. Here only one consequence which is of fundamental importance to all scientific fields dealing with complex systems should be mentioned. By definition there is no chance for explicit analytical solutions of nonlinear dynamics. Nonlinear systems need to be solved numerically either by numerical integration or by numerical iteration. Numerically computed solutions need visualization besides the experience that certain features of complex systems are only accessible through images as could be shown above. In this sense nonlinear dynamics is computational imagery. But there is no formal theory for the interpretation of images. Thus images need human observers to give meaning to the visual content. Yet the human observer is neither objective nor is there a theoretical model to deal with an observer in any mathematical or physical theory. The simulation of model systems by computer is essential to either the theory or the experiment in scientific research.

As we have seen there are many more diversifications of chaotic features, not to be described by a simple number following from a statistic calculation of the LCEs but from visualizations of computational experiments which need an human interpretation. Due to the lack of a theory of the visual the solution to that problem cannot be given here. Instead the autohor would like to present another approach to computer simulated model worlds. Interactive new media art may be a good source to find some new ideas of how to deal with the problems of numerical and visual simulations, science is already facing in everyday research.

On The Nature of Simulation. - ***Interactive New Media*** art is generating a new art of the virtual. Its essence is binary informations, its media are the dataspaces of computational simulations and its realities are the interactive environments of dynamical systems. Formerly static pictures or sculptures are transformed to spatio-temporal processes. Digital objects, states and images are understood as autonomous but interactive simulated model worlds. The traditionally only inactive and external recepients will become internal actors entangled with the virtual environments via men-machine interfaces. The vision of new media art is the imagination of unexpected experiences challenging the traditional understanding of space, time and correlations. The computer is used as a space-time simulator of dynamical processes. Hypermedia machines invent new worlds of experiences which are sensually conceivable to the human observer.

New Media Evolution. The development of complex technologies shows close resemblances to biological evolution. Therefore the present new media are the electronic devices capable of handling digital informations, as there are digital video and audio, computer simulations and multimedia animations as well as the hypermedia network matrix. In the sense of an evolution of the machine supported generation of images the new media of today incorporate all the historical achievements of the technology and the resulting concepts, while adding a new feature. The historical steps were photography, film, audio, television and video. The actual multimedia machines open the universe of computer-generated interactive environments.

Interactive new media art installations should be understood as environments of individual experiences of human actors. The reality of human experiences is entangled with virtual computer generated simulations. The postulats of modern theories of physics, first of all quantum mechanics and the theories of complexity today, are again rising the question of a fundamental unity between observation and objects. Interactive new media installations give an artistic interpretation of these entanglements.

„The Thread“. The Installation „The Thread“ by Alba D'Urbano, Nicolas Reichelt and the author is the latest art-work following from the working complex „Rosa Binaria“. It expresses the transformation of a real world object, the rose, into its complete abstract digital existence.

The interactive installation is presented as a sculptural environment. It uses traditional pictures as surfaces of spacial objects, digital images as textures and interactive realtime computer animations to define a multimedia space of experiences. The observer enters the installation through a door, he steps into the space-time defined by the piece of art *(Fig.9a)*. No longer is the object of art simply a static picture freezing the experiences of an individual artist. It is a shaped world-process of its own. The observer is an integral part of the work. Without his interaction the installation is incomplete. The passive reception is extended to the active participation. The time-depending experiences of the individual observer define the art installation as a dynamical system depending on the process of interaction. On purpose there is an element of a game incorporated in the work. Another aspect of the installation is the missing external point of view for the observer. He is always in an internal position. „The Thread“ defines a observer-based perspective to the visual experiences. The actor enters a real space filled with sculpural objects, he may walk through the human-scale sculptures of letters and in the end he finds himself in front of an interactive computer animation as a virtual extension of the real space. Here again he is forced to take an internal point of view giving him no overview to the system. His only chance is to step into the time-depending process of an interactve experience. He has to solve a walk through a computer simulated model defined by letters which code for a sentence which he only can grap from an inner position.

The Letter Objects. - The real space of the exhibition room is penetrated by either three-dimensional objects or two-dimensional plates obviously coding for a meaningful sentence. Parts of the installation are the three-dimensional sculptures of letters of the word ROSE *(Fig.9b)*. Other elements are the two-dimensional fragments of the letters ROSE IS A. All surfaces are covered with ASCII code prints, which carry some information only meaningful to electronic machines but unreadable to the human observer.

The Virtual Extension. - Inside the last letter E of the sculpture there is a door, presenting a back-projection of a virtual extension to the real world *(Fig.9c)*. The computer animation can be manipulated with an interactive interface, a navigational device with three dimensions of freedom (left and right, forwards and backwards, up and down). If the observer activates the interface he may experience a virtual labyrinth of simulated objects. But his conventional state-space experinces is challenged by a strange metric topology and an unconventional point of view of the simulated world. After a while of navigating within the simulation the actor will interpret the simulated objects as inside-views of letters which are again covered with the formerly mentioned ASCII code surfaces. Obviously he is inside a model world of simulated letter-objects. His position is observer centered, no external overview of the whole scenario is possible. This idea expresses the parable of Plato´s cave, a problem well-known in scientific simulations. We are looking at one projection of the outer world without knowing much about reality *(Fig.9d)*.

Metric Topology. - Instead of defining the higher-dimensional metric topology which is used in the simulated model world it is more efficient to describe some experiences during the process of investigation. The observer can leave each letter object through either side by just walking through the hull, there are no doors or windows. Each time one leaves a virtual object one will find oneself in the next letter. There are seven objects and after a while one will have learned that one is reading a sentence from the inside. But the space-time geometry of the linked objects is complex. On the one side there is no way back, the arrow of time is always directed forwards. On the other side the geometrical involvement of the virtual letters does not follow an Euclidian geometry. Leaving one object to the left will not result in entering the next letter from the right. Topologically adjacent areas in one object are not necessarily adjacent in the following object. There is no linear correlation but sensitivity to initial conditions along the trajectory of the dynamical process. Nevertheless the simple fact of a time-dependent experince of the simulated world guaranties the understanding of the observer.

Fig. 9a-d. *Internal and external views on the interactive installation „The Thread".*

The Memory Effect. - A third feature very well known from the analysis of nonlinear complex systems is built in the vitual extension of the installation, a certain kind of history effect or memory effect. In the beginning of the journey all letters carry an individual symbol, a digital image of a rose floating somewhere in the letter objects. Touching the floating objects starts a morphing sequence of the picture of the rose transforming the ASCII code textures of the inner surface to ist underlying image. At the end of each morphing sequence the inner surface of the individual letter object is covered with that new texture. Now each letter gets its individual fingerprint. This effect is to be understood as the influence of observations on the objects under investigation. The research on complex systems uncovered the problem of the generation of states of a system which are not existing apriori but are effects of the investigation/observation itself. The problem of observation is no longer a fact only in quantuum scale problems but in classical mechanical systems, too. There is an ongoing discussion about the existence of second order processes in complex systems.

The Final Conclusion. - To conclude the possible experiences one will have while walking through the virtual world of simulated objects there finally is a solution to the dynamical process of reading a sentence from the inside. If the actor has decoded all letters of the simulation the computer will give him a short glimpse from an external point of view to the underlying sentence. For a moment in time he can read the sentence by Gertrude Stein: „rose is a rose is a rose is a rose“ just before the game starts again from scratch.

Closing Remarks. - The analysis of complex systems is an area of scientific research which is founded on the existence of numerical computations and computer-based visualizations. In general simulations are not only used to study most of todays theoretical models but also to interpret most experimental results. The intensive use of simulations on both the theoretical and the experimental side of scientific research should be accompanied by a fundamental understanding of the changes of scientific methodology. Until now science has only accepted the duality between experiment and theory, which are correlated to each other due to the fact of reciprocal verification and falsification of the theoretical predictions and the experimental results. The ubiquitous use of computer simulations challanges these established scientific procedure. Here only one aspect was highlighted, the problem of observation. Whereas classical theories are formalized with mathematical analysis and thus are accepted as objective science, complex systems for example can only be treated with numerical simulations. These simulations result in computational imagery which is not necessarily independent of the -by definition-subjective observers. Especially the lack of an acceptable formal theory for the interpretation of images rises some serious questions. One possible step ahead would be the embedding of simulations in the scientific methodology as it is already suggested by Hoßfeld [21] *(Fig.10)*:

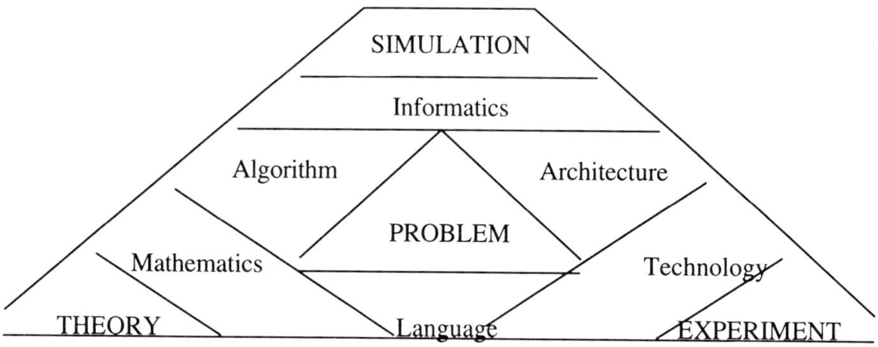

Fig.10 *The tripod of scientific methodology from [21].*

REFERENCES

[1] P.S.de Laplace, Philosophical Essays on Probabilities, 18th century.

[2] O.E. Rössler, The chaotic hierarchy, Z. *Naturforsch.*, **1983**, *38a*, 788, reprinted in: A Chaotic Hierarchy, 25, G. Baier, M. Klein (Eds.), World Scientific, Singapore, 1991.

[3] J. Gleick, Chaos, Making a New Science, Viking, New York, 1987.

[4] M.Klein, G.Baier, Hierarchies of dynamical systems, in: A Chaotic Hierarchy, 1-23, G.Baier and M.Klein (Eds.), World Scientific, Singapore, 1991.

[5] H.Poincare, Sur les courbes definies par des equations differentielles, *J.Math.Pures Appl.* 1, 167-224 (1885), and H.Poincare, Les Methodes Nouvelles de la MÈchanique Celeste, Guthier-Villas, Paris, 1892.

[6] G.Baier, M.Klein, Chaotic attractors derived from spheroidal chaotic dynamics, *Phys. Lett.*, **1993**, *A177*, 32.

[7] M.Klein, G.Baier, Fractal structures of spheroidal chaotic attractors, *Physica,* **1992**, *A191*, 564.

[8] M.Henon, A two-dimensional mapping with a strange attractor, *Comm.Math.Phys,.* **1976**, *50*, 69.

[9] M.Klein, G.Baier, Fractal features of invariant sets in a three-dimensional map, *Chaos, Solitons & Fractals*, **1994**, *4/10*, 1889.

[10] G.Baier, A.Kittel, M.Klein, Competition of local exponents and the fractal structure of chaotic attractors, *Phys.Rev.*, **1993**, *E48*, 3409.

[11] J.L.Kaplan, J.A.Yorke, Chaotic behavior of multidimensional difference equations, in: 'Lecture Notes in Mathematics'. Eds. H.-O. Peitgen, H.O. Walter, Springer-Verlag, Berlin, 1978, Vol. 730, pp. 228

[12] B.Röhricht, W.Metzler, J.Parisi, J.Peinke, W.Beau, O.E.Rössler, The Classes of Fractals, in: 'The Physics of Structure Formation', Eds. W.Güttinger, G.Dangelmayr, Springer Series in Synergetics 37, Springer-Verlag, Berlin, 1987, Vol. 275.

[13] G.H.Hardy, Weierstrass's non-differentiable function, *Trans.Am.Math.Soc.*, **1916**, *17*, 301.

[14] J.Peinke, M.Klein, A.Kittel, G.Baier, J.Parisi, R.Stoop, J.L.Hudson, O.E.Rössler, Nowhere differentiable boundaries in differentiable systems -- A proposed explanation, *Europhys. Lett.*, **1991**, *14*, 615.

[15] G.Baier, A.Kittel, M.Klein, Competition of local exponents and the fractal structure of chaotic attractors, *Phys. Rev.,* **1993**, *E48/5*, 4409.

[16] J.Peinke, M.Klein, A.Kittel, C.Kahlert, R.Stoop, J.Parisi, O.E.Rössler, How two competing characteristic exponents generate different classes of fractal boundaries, *Bifurcation & Chaos*, **1991**, I/3, 599.

[17] D.Ruelle, Sensitive dependence on initial conditions and turbulent behavior of dynamical systems, in: 'Bifurcation Theory and its Applications in Scientific Disciplines', Eds. O.Gurel, O.E.Rössler, *Ann. New York Acad. of Sci.*, Vol. 316, 1979, pp. 408.

[18] D.Ruelle, Strange attractors, *Math. Int.*, **1980**, *2*, 126.

[19] B.B.Mandelbrot, The Fractal Geometry of Nature, Freeman, San Francisco, 1982.

[20] Dimensions and Entropies in Chaotic Systems, Ed. G.Mayer-Kress, Springer-Verlag, Berlin, 1986.

[21] F.Hoßfeld, W.E. Nagel, Per Aspera ad Astra: On the Way to Parallel Processing, KFA-ZAM-IB-9507, Jülich (1995).

AUTHOR INDEX

SUBJECT INDEX

A

B

C

E

F

G

H

I

N

O

P

Q

R

S

T